Exploring Creation

with

Biology

by Dr. Jay L. Wile and Marilyn F. Durnell

Exploring Creation With Biology

Manufactured in the United States of America
Fifth Printing 2002

Published By
Apologia Educational Ministries, Inc.

Printed by
The C.J. Krehbiel Company
Cincinnati, Ohio

Cover Photos
Fish: © Gary Bell   Others: from the MasterClips collection

Need Help?

Apologia Educational Ministries Curriculum Support

If you have any questions while using Apologia curriculum, feel free to contact us in any of the following ways.

By Mail:

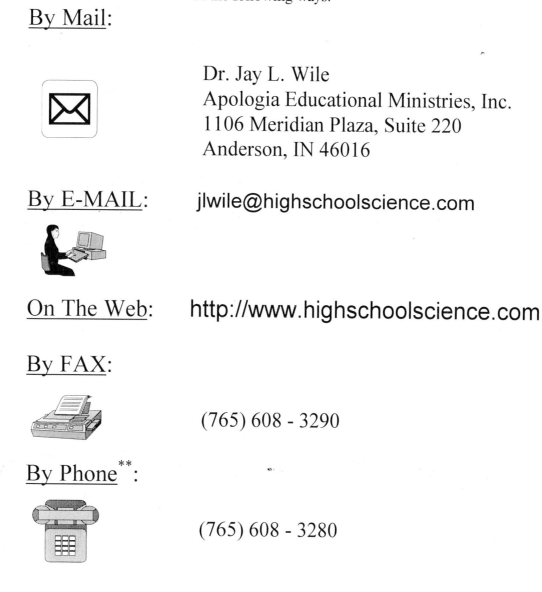

Dr. Jay L. Wile
Apologia Educational Ministries, Inc.
1106 Meridian Plaza, Suite 220
Anderson, IN 46016

By E-MAIL: jlwile@highschoolscience.com

On The Web: http://www.highschoolscience.com

By FAX:

(765) 608 - 3290

By Phone[**]**:**

(765) 608 - 3280

Exploring Creation With Biology
Dr. Jay L. Wile and Marilyn F. Durnell

Are you ready to be impressed? If not, then you'd better get ready. Why? Well, in this course, you are going to get a broad overview of God's Creation. As you begin to learn its secrets, you will become more and more impressed with its majesty and complexity. The sheer grandeur of it all should leave you in awe of God's mighty power. If you learn nothing else in this course, learn to appreciate the wonder of God's Creation!

If you are like most students, this will be the first truly rigorous science course that you have ever taken. Thus, you might find it difficult to adjust to the time and patience required by a course like this. If you find yourself getting frustrated or discouraged, remember that whether you decide to go to college, go straight into the workforce, or become a full time parent, there will be many tasks more rigorous than the experience of studying biology. Thus, you need to stick with it, because life is full of challenges!

Pedagogy of the Text

There are 16 modules in this course. In general, you should try to finish each module in two weeks' time, with the exception of Module #6. That module contains an enormous amount of information, so you should spend 3 weeks on it. If you keep on that schedule, you will complete the course in 33 weeks. Since most people's school year is longer than 33 weeks, you have some built-in flexibility.

How will you know how much to do in order to spend only 2 weeks per module? Well, start by spending one half hour per day with the course. At the end of 2 weeks, if you have not completed the module, you know that you need to spend more time each day on it. If you finish a module in less than 2 weeks, then you know that you can spend less time per day on it. In the end, then, try to find the pace that will keep you on track.

There are two types of exercises that you are expected to complete: "on your own" problems and an end-of-the module study guide.

- You should answer the "on your own" problems while you read the text. The act of answering these problems will cement in your mind the concepts you are trying to learn. Detailed answers and explanations are provided for you at the end of the module, so that you may check your own work. DO NOT look at the answer to a question until AFTER you have tried to answer it!

- You should complete the study guide in its entirety after you have finished the module. The solutions to the study guides are in a separate volume which your parent/teacher has.

All definitions presented in the text are centered. The words will appear in the study guide and their definitions need to be memorized. Words that appear in bold-face type in the text are important terms that you should know.

The study guide gives you a good feel for what you need to know for the test. Any information needed to answer the study guide questions is information that you must know for the test. Sometimes, tables and other reference material will be provided on a test so that the student need not memorize it. You will be able to tell if this is the case because the questions in the study guide which refer to this information will specifically tell you to use the reference material.

Experiments

The experiments in this course are designed to be done while you are reading the text. We recommend that you keep a notebook of these experiments. This notebook serves two purposes. First, as you write about the experiment in the notebook, you will be forced to think through all of the concepts that were explored in the experiment. This will help you cement them into your mind. Second, certain colleges might actually ask for some evidence that you did, indeed, have a laboratory component with your biology course. The notebook will not only provide such evidence but will also show the college administrator the quality of the biology program you took. We recommend that you perform your experiments in the following way:

- When you get to the experiment during the reading, read through the experiment in its entirety. This will allow you to gain a quick understanding of what you must do.

- Once you have read the experiment, start a new page in your laboratory notebook. The first page should be used to write down all of the data taken during the experiments and perform any exercises discussed in the experiment.

- When you have finished the experiment, you should write a brief report in your notebook, right after the page where the data and exercises were written. The report should be a brief discussion of what was done and what was learned. You should write this discussion so that someone who has never read the book can read your discussion and figure out what basic procedure you followed and what you learned as a result of the experiment.

- **<u>PLEASE OBSERVE COMMON SENSE SAFETY PRECAUTIONS.</u>**
  <u>The experiments are no more dangerous than most normal, household</u>
  <u>activity. Remember, however, that the vast majority of accidents do</u>
  <u>happen in the home!</u>

- Although many of the experiments in the course require a microscope, there are aspects to some of those experiments that do not. If you do not have a microscope, you should still read through the microscope labs and perform all non-microscope exercises that the lab discusses. For example, Experiment 2.1 is a trip to a pond. Although the main reason for the trip is to gather water samples for later microscope experiments, the experiment does instruct you to observe the habitat and make some drawings. Even if you do not have a microscope, you should still visit a pond, examine the habitat, and make the drawings. You need not gather the water, however.

Question/Answer Service

For all those who use my curriculum, I offer a free question/answer service. If there is anything in the modules that you do not understand - from an esoteric concept to a solution for one of the problems - just write down your question and send it to me by any of the means listed on the **NEED HELP?** page.

LABORATORY EQUIPMENT

Exploring Creation With Biology contains laboratory exercises for the student to perform. The laboratories come in three types: microscope labs, dissection labs, and household labs. The household labs use only household equipment and should be done by all students. The microscope labs, however, require expensive equipment, while the dissection labs require an additional kit. As a result, we do not REQUIRE you to perform those labs. They will be beneficial to the student (especially if the student is science-oriented) but they are not absolutely necessary. Thus, you should not feel pressured into purchasing the microscope or dissection equipment. Do so only if you can afford it!

If you have the financial means, I recommend that you order from Nature's Workshop. I have worked very hard with them to provide you with a quality set at a very low price. Indeed, this set is less than half the cost of the recommended set for other homeschool biology courses. To order, simply call or write Nature's Workshop:

> Nature's Workshop
> 1-888-393-5663 (toll free)
> P.O. Box 220
> Pittsboro, IN 46167-0220

The microscope set contains a microscope and a microscope slide set. If you wish, you can order the entire package. This gives you a price break, as the entire set is cheaper than the sum total of its individual components. If you wish to order the entire set, simply ask for:

 # 40030: Slide Set With Microscope $285.00 (post paid)

If, however, you have a quality microscope (it must have 400x magnification and a "fine focus" which is separate from the "coarse focus"), you can order just the microscope slide set. To order this, simply ask for:

#40026:   Microscope Slide Set $85.00 (post paid)

If, instead, you have some of the items in the slide set (listed below), you can just order the items that you need.

- Glass Slides and covers (10) $4.00
- Eyedroppers (4) $2.40
- Methylene Blue $4.50
- Iodine $2.00
- Lens paper $1.50
- Microscope Drawing Paper $1.00
- Microslide #17: Chick Embryo $5.75
- Prepared slide: Fish Blastula - Mitosis $7.30
- Prepared slide: Onion Root Tip - Mitosis $4.40
- Prepared slide: Amoeba Proteus $3.70

- Prepared slide: Paramecium $2.60
- Prepared slide: Euglena $2.80
- Prepared slide: Planarian $3.70
- Prepared slide: Spirogyra $3.50
- Prepared slide: Hydra Budding $3.00
- Prepared slide: *Zea Mays* stem cross-section $5.20
- Prepared slide: *Zea Mays* root cross-section $3.60
- Prepared slide: *Ranunculus* stem cross-section $4.80
- Prepared slide: *Ranunculus* root cross-section $5.20
- Prepared slide: *Grantia* spicules - w.m. $2.40
- Prepared slide: Leaf cross section with vein $4.80
- Prepared slide: Volvox $4.10
- Prepared slide: Diatoms $3.00

Finally, if you have all of the slides and simply want to order the microscope, ask for

#61012: High School Microscope $220.00 (post paid)

If you want to do the dissection experiments, then there is a separate dissection kit, which comes complete with specimens. You can order this kit through Nature's Workshop as well. Just ask for

#40031   Dissection Kit with Specimens $40.00 (post paid)

Once again, you can order the individual items in this kit if you already have some dissecting tools.

- Dissection Instruments $18.00
- Dissection Pan With Pad $10.08
- Specimens (worm, crayfish, fish, frog) $11.72

All prices are subject to change!!!!!!!

Exploring Creation With Biology
Dr. Jay L. Wile and Marilyn Durnell

Table of Contents

Module 1 : Biology, The Study Of Life

Introduction 1
What is Life? 1
DNA and Life 1
Energy Conversion and Life 2
Sensing and Responding to Change 6
All Life Forms Reproduce 7
Life's Secret Ingredient 8
The Scientific Method 9
Limitations of the Scientific Method 12
Spontaneous Generation: The Faithful Still Cling to It! 16
Biological Classification 17
Characteristics Used to Separate Organisms into Kingdoms 19
The Definition of Species 21
Biological Keys 22
Experiment 1.1: Using a Biological Key 26
Naming Organisms Based on Classification 28
The Microscope 29
Experiment 1.2: Introduction to the Microscope 30

Module 2 : Kingdom Monera

Introduction 37
Bacteria 37
Metabolism in Bacteria 41
Asexual Reproduction in Bacteria 44
Sexual Reproduction in Bacteria 47
Other Forms of Bacterial "Reproduction" 49
Bacterial Colonies 50
Experiment 2.1: Pond Life A 52
Classification in Kingdom Monera 53
Specific Bacteria 56
Preventing Bacterial Infections 58
Experiment 2.2: Pond Life B 60

Module 3: Kingdom Protista

Introduction 67
Experiment 3.1: Pond Life C 67
Classification in Kingdom Protista 68
Subkingdom Protozoa 71
Phylum Sarcodina 71
Phylum Mastigophora 74
Phylum Ciliophora 78
Phylum Sporozoa 80
Experiment 3.2: Subkingdom Protozoa 83
Subkingdom Algae 85
Phylum Chlorophyta 86
Phylum Chrysophyta 88
Phylum Pyrophyta 89
Phylum Phaeophyta 90
Phylum Rhodophyta 92
Experiment 3.3: Subkingdom Algae 93
Summing Up Kingdom Protista 93

Module 4: Kingdom Fungi

Introduction 99
General Characteristics of Fungi 99
Reproduction in Fungi 103
Classification in Kingdom Fungi 104
Phylum Amastigomycota 105
Experiment 4.1: Class Basidiomycetes 110
Experiment 4.2: Yeast and the Fermentation Process 113
Phylum Mastigomycota 117
Experiment 4.3: Molds and Mildews 117
The Imperfect Fungi 118
Optional Experiment 4.4: Imperfect Fungi 120
Phylum Myxomycota 121
Symbiosis in Kingdom Fungi 123
Summing Up Kingdom Fungi 124

Module 5: The Chemistry of Life

Introduction 129
Atoms: The Basic Building Blocks of Matter 129
Elements 132
Molecules 134
Changes in Matter 136
Physical Change 137
Experiment 5.1: Osmosis and Diffusion 138
Chemical Change 142
Photosynthesis 144
Organic Chemistry 146
Experiment 5.2: The Fragility of an Enzyme 157
DNA 159

Module 6: The Cell

Introduction 167
Cellular Functions 167
Cell Structure 170
Experiment 6.1: Cell Structure I 179
How Substances Travel In and Out of Cells 181
Experiment 6.2: Cell Structure II 186
How Cells Produce Energy 187
Protein Synthesis 192

Module 7: Cellular Reproduction

Introduction 207
Genes, Chromosomes, and DNA 208
Experiment 7.1: DNA Extraction 210
Mitosis: The Common Form of Asexual Reproduction 212
Experiment 7.2: Mitosis 216
Diploid Chromosomes 217
Meiosis: The Cellular Basis of Sexual Reproduction 220
Viruses 225

Module 8: Genetics

Introduction 235
Gregor Mendel 235
Mendel's Experiments 236
Updating the Terminology 242
Punnett Squares 245
Pedigrees 248
More Complex Genetic Crosses 251
Experiment 8.1: A Dihybrid Cross 256
Sex and Sex-Linked Genetic Traits 257
Experiment 8.2: Sex-Linked Genetic Traits 259
Genetic Disorders and Diseases 260
Experiment 8.3: The Environmental Factor 263

Module 9 : Evolution - Part Scientific Theory, Part Unconfirmed Hypothesis

Introduction 273
Charles Darwin 274
Darwin's Theory 276
Microevolution and Macroevolution 279
Inconclusive Evidence: The Geological Column 282
The Details of the Fossil Record: Evidence Against Macroevolution 286
Structural Homology: Formerly Evidence for Macroevolution 292
Molecular Biology: The Nail in Macroevolution's Coffin 294
Macroevolution Today... 299
Why Do So Many Scientists Believe in Macroevolution? 302

Module 10 : Ecosystems

Introduction 309
Energy and Ecosystems 310
Symbiosis 315
The Physical Environment 320
The Water Cycle 321
The Oxygen Cycle 324
The Carbon Cycle 326
Experiment 10.1: Carbon Dioxide and the Greenhouse Effect 328
Summing Up 333

Module 11 : The Invertebrates of Kingdom Animalia

Introduction 339
Symmetry 339
Phylum Porifera: The Sponges 341
Experiment 11.1: Observation of a Sponge From Genus Grantia 344
Phylum Cnidaria.. 345
Experiment 11.2: Observation of a Hydra 349
Phylum Annelida 352
Optional Dissection Experiment #1: The Earthworm 358
Phylum Platyhelminthes 360
Experiment 11.3: Observation of a Planarian 362
Phylum Nematoda 363
Phylum Mollusca... 364
Summing Up The Invertebrates 366

Module 12 : Phylum Arthropoda

Introduction 373
General Characteristics of Arthropods 373
Class Crustacea 376
The Crayfish 376
Optional Dissection Experiment #2: The Crayfish 384
Class Arachnida 386
Classes Chilopoda and Diplopoda 391
Class Insecta 392
A Few Orders in Class Insecta 396
Experiment 12.1: Insect Classification 400

Module #13: Phylum Chordata

Introduction 407
Subphylum Urochordata 408
Subphylum Cephalochordata 409
Subphylum Vertebrata.. 410
Class Agnatha 417
Class Chondrichthyes 420
Class Osteichthyes 424
Optional Dissection Experiment #3: The Perch 431
Class Amphibia 433
Optional Dissection Experiment #4: The Frog 437
Summing Up 437
Experiment 13.1: Field Study II 438

Module #14: Kingdom Plantae: Anatomy and Classification

Introduction 445
Basic Plant Anatomy 445
The Macroscopic Structure of a Leaf 447
Experiment 14.1: Leaf Identification 450
The Microscopic Structure of a Leaf 452
Leaf Color 454
Experiment 14.2: Anthocyanin and Leaf Color 455
Roots 458
Stems 461
Experiment 14.3: Microscopic Investigations of Leaves and Stems 465
Classification of Plants 466
Phylum Bryophyta: The Non-Vascular Plants 467
Vascular Plants 468

Module 15 : Kingdom Plantae: Physiology and Reproduction

Introduction 477
Plant Physiology 477
How a Plant Depends on Water 477
Water Absorption in Plants 479
Water Transport in Plants 480
Plant Growth 483
Insectivorous Plants 486
Reproduction in Plants 487
Vegetative Reproduction 487
Sexual Reproduction in Plants 489
Experiment 15.1: Flower Anatomy 492
The Reproductive Process in Plants 493
Seeds, Fruits, and Early Plant Development 499
Experiment 15.2: Fruits 501

Module 16 : Reptiles, Birds, and Mammals

Introduction 509
Class Reptilia 509
Classification of Reptiles 512
Order Rhynchocephalia 513
Order Squamata 514
Order Testudines 517
Order Crocodylia 518
Dinosaurs 519
Class Aves 521
Experiment 16.1: Bird Embryology 523
A Bird's Ability to Fly 524
Classification in Class Aves 529

Experiment 16.2: Bird Identification 532
Class Mammalia 533
Classification in class Mammalia 536
Summing It All Up 542

Glossary 547

Appendix 563

Index 569

Module #1: Biology, The Study Of Life

Introduction

In this course, you're going to take your first detailed look at the science of biology. Biology, the study of life itself, is a vast subject, with many subdisciplines that concentrate on specific aspects of biology. Microbiology, for example, concentrates on those life forms and biological processes that are too small for us to see with our eyes; biochemistry studies the chemical processes that make life possible, and population biology deals with the dynamics of many life forms interacting in a community. Since biology is such a vast field of inquiry, most biologists end up specializing in one of these subdisciplines. Nevertheless, before you can begin to specialize, you need a broad overview of the science itself. That's what this course is designed to give you.

What is Life?

Well, if biology is the study of life, we need to determine what life is. Now to some extent, we all have an idea of what life is. If we were to ask you whether or not a rock is alive, you would easily answer "No!" On the other hand, if we were to ask you whether or not a blade of grass is alive, you would quickly answer "Yes!" Most likely, you can intuitively distinguish between life and non-life.

Even though this is the case, scientists must be a little more deliberate in defining what it means to be alive. Thus, scientists have developed several criteria for life. If something meets all of these criteria, then we can scientifically say that it is alive. If it fails to meet even one of the criteria, it is not alive. These criteria are:

1. **All life forms contain deoxyribonucleic (dee ahk' see rye boh noo klay' ik) acid, which is called DNA.** (or RNA) (viruses <u>are</u> living)

2. **All life forms have a method by which they extract energy from the surroundings and convert it into energy that sustains them.**

3. **All life forms can sense changes in their surroundings and respond to those changes.**

4. **All life forms reproduce.**

When put together, these criteria define life, as far as science is concerned. Now if you're not sure exactly what each of these criteria mean, don't worry. We will discuss each of them in detail.

DNA and Life

Our first criterion states that all life contains **DNA**. Now we're sure you've at least heard about DNA. It is probably, however, still a big mystery to you at this point. Why is DNA so special when it comes to life? Basically, DNA provides the *information* necessary to take a bunch of lifeless chemicals and turn them into a living system. You see, if we were to analyze an organism and determine every chemical that made up the organism, and if we were then to go into a laboratory and make all of those chemicals and throw them into a big pot, we would not have made something that is alive. We would not have even made something that resembles the organism we studied. Why?

In order to make life, we must take the chemicals that make it up, and we must *organize* them in a way that will promote the other life functions mentioned in our list of criteria for life. In other words, just the chemicals themselves cannot extract and convert energy (criterion #2), sense and respond to changes (criterion #3), and reproduce (criterion #4). In order to perform those functions, the chemicals must be organized so that they work together in just the right way. Think about it this way: suppose you went to a store and bought a bicycle. The box said "some assembly required." When you got it home, you unpacked the box and piled all of the parts on the floor. At that point, did you have a bicycle? No, of course not. In order to make the bicycle, you had to assemble the pieces in just the right way, according to the instructions. When you got done, all of the parts were in just the right place and worked together with just the right parts. This made your bike.

In the same way, DNA is the set of instructions that takes the chemicals which make up life and arranges them in just the right way so as to produce a living system. Without this instruction set, the chemicals that make up a life form would be nothing more than a pile of goo. However, directed by the information in DNA, these molecules can work together in just the right way to make a living organism. Now of course, the exact way in which DNA does this is a little complicated. Nevertheless, in an upcoming module, we will spend some time studying DNA and how it works in detail.

Energy Conversion and Life

In order to live, organisms need energy. This is why our second criterion states that all life forms must be able to absorb energy from the surroundings and convert it into a form of energy that will sustain their life functions. This process is called **metabolism** (muh tab' uh liz um).

Metabolism - The process by which a living organism takes energy from its surroundings and uses it to sustain itself, develop, and grow

How is metabolism accomplished? It's a long process that actually begins with the sun.

Almost all of the energy on this planet comes from the sun, which bathes the earth with its light. When you take chemistry, you'll learn a lot more about light. For right now, however, all you need to know is that light is pure energy. Thus, the light that comes from the sun is, in fact, the main energy source for all living organisms on our planet. Green plants (and some other things you will learn about later) take this energy and, by a process called **photosynthesis** (foh' toh sin thuh' sis), convert that energy into food for themselves.

<u>Photosynthesis</u> - The process by which a plant uses the energy of sunlight and certain chemicals
        to produce its own food. Oxygen is often a by-product of photosynthesis.

Now we'll be looking at photosynthesis in great detail in a later module. Thus, if the definition is
a little confusing to you, don't worry about it. What you need to know at this point is that
photosynthesis allows plants to convert the energy of sunlight into food.

If plants absorb their energy from the sun, where do other life forms get their energy?
Well, that depends. Some organisms eat plants. By eating plants, these organisms take in the
energy that plants have stored up in their food reserves. Thus, these organisms are indirectly
absorbing energy from the sun. They are taking the energy from plants in the form of food, but
that food ultimately came from sunlight. Organisms that eat only plants are called **herbivores**
(hur bih' vors).

<u>Herbivores</u> - Organisms that eat plants exclusively

So you see that even though herbivores don't get their energy directly from sunlight, without
sunlight, there would be no plants, and therefore there would be no herbivores.

If an organism does not eat plants, it eats organisms other than plants. These organisms
are called **carnivores** (kar nih' vors).

<u>Carnivores</u> - Organisms that eat only organisms other than plants

Even though carnivores eat other organisms, their energy ultimately comes from the sun. After
all, the organisms that carnivores eat have either eaten plants or have eaten other organisms who
have eaten plants. The plants, of course, get their energy from the sun. In the end, then,
carnivores also indirectly get their energy from the sun.

Finally, there are organisms that eat both plants and other organisms. We call these
omnivores (ahm nih' vors).

<u>Omnivores</u> - Organisms that eat both plants and other organisms

Ultimately, of course, these organisms also get their energy from the sun.

Think about what we just did in the last few paragraphs in this module. We took all of
the organisms that live on this earth and placed them into one of three groups: herbivores,
carnivores, or omnivores. This kind of exercise is called **classification**. When we classify
organisms, we are taking a great deal of data and trying to organize it into a fairly simple system.
In other words, classification is a lot like filing papers. When you file papers, you place them in
folders according to similarities that they have. In this case, we have taken all of the organisms
on earth and put them into one of three folders based on what they eat. This is one of the most
important contributions biology has made in understanding God's Creation. Biology has taken
an enormous amount of data and has arranged it into many different classification systems.

These classification systems allow us to see the similarities and relationships that exist between organisms in Creation. Figure 1.1 illustrates the classification system you have just learned.

FIGURE 1.1
Herbivores, Carnivores, and Omnivores

Horses eat only plants. We therefore call them herbivores. *Photo from the MasterClips collection*	Lions eat only meat. This makes them carnivores. *Photo from the EXPERT 3000 collection*	Humans eat both plants and meats. Humans are omnivores. *Photo from the MasterClips collection*

In biology, there are hundreds and hundreds of different ways that we can classify organisms, depending on what kind of data we are trying to organize. For example, the classification system we just talked about groups organisms according to what they eat. Thus, organisms that eat similar things are grouped together. In this way, we learn something about how energy is distributed from the sun to all of the creatures on earth.

This is not, however, the only way we can classify organisms to learn how energy is distributed from the sun to all of the creatures on earth. We could, alternatively, classify organisms according to these groups: **producers, consumers**, and **decomposers**.

Producers - Organisms that produce their own food

Consumers - Organisms that eat living producers and/or other consumers for food

Decomposers - Organisms that break down the dead remains of other organisms

In this system, plants are producers because they make their own food from chemicals and the sun's light. Omnivores, herbivores, and carnivores are all consumers, because they eat producers and other consumers. Certain bacteria and fungi (the plural of "fungus"), organisms we'll learn about in detail later, take the remains of dead organisms and break them down into simple chemicals. Thus, these creatures are the decomposers. Once the decomposers have done their job, the chemicals that remain are once again used by plants to start the process all over again.

This classification scheme, illustrated in Figure 1.2, gives us a nice view of how energy comes to earth from the sun and is distributed to all creatures in God's Creation.

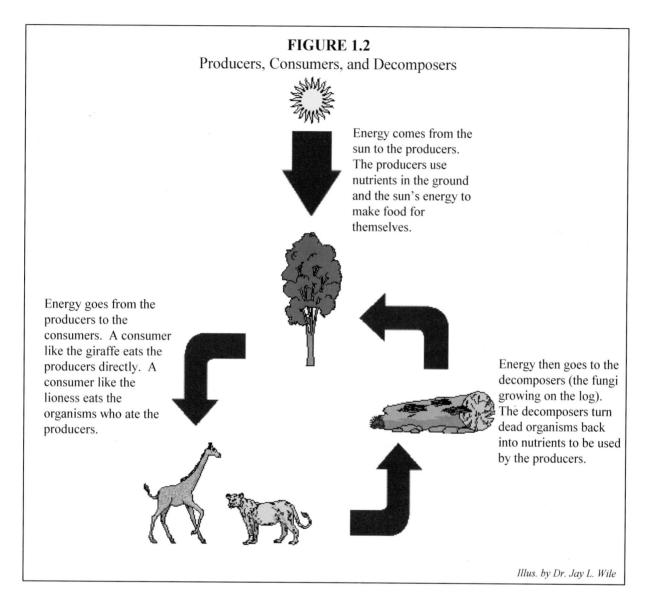

FIGURE 1.2
Producers, Consumers, and Decomposers

Energy comes from the sun to the producers. The producers use nutrients in the ground and the sun's energy to make food for themselves.

Energy goes from the producers to the consumers. A consumer like the giraffe eats the producers directly. A consumer like the lioness eats the organisms who ate the producers.

Energy then goes to the decomposers (the fungi growing on the log). The decomposers turn dead organisms back into nutrients to be used by the producers.

Illus. by Dr. Jay L. Wile

There are, of course, differences between this classification (producers, consumers, and decomposers) system and the one you learned previously (omnivores, herbivores, and carnivores). The first difference you should notice between this classification scheme and the one you just studied is that, in this case, we include plants, bacteria, and fungi in the classification. In the previous classification system, we could only classify organisms that ate plants or ate other organisms. There was no grouping in which to put the plants, the bacteria, or the fungi. Does this mean that the second classification system is better than the first? Not really. They each tell us different information. For example, if we need to look at the differences that exist between animals, then the first classification scheme is best. Some animals are herbivores (cows, for example), some animals are carnivores (tigers, for example), and some animals are omnivores (gorillas, for example). In the second classification system, all animals are consumers. So the second classification system doesn't tell us much about the differences that exist between animals. If, however, we want to study how energy flows from the sun to every creature in Creation, the second classification system is the one to use.

As a point of terminology, producers are often called **autotrophs** (aw' toh trohfs), because they make their own food. Consumers and decomposers, on the other hand, are often called **heterotrophs** (het' er uh trohfs) because they have to eat other organisms for food.

Autotrophs - Organisms that are able to make their own food

Heterotrophs - Organisms that depend on other organisms for their food

In a little while, these two terms will become very important, so you need to know them.

Before you go on to the next section, answer the "on your own" questions below. These questions will be scattered throughout the modules in this course. They allow you to reflect on the things you have just read about, cementing the concepts into your mind.

(in your head)

ON YOUR OWN

1.1 Classify the following organisms as herbivores, carnivores, or omnivores:

    a. tigers   b. cows   c. humans   d. sheep

1.2 Classify the following organisms as producers, consumers, or decomposers:

    a. flowers   b. yeast (a fungus)   c. lions   d. humans

Sensing and Responding to Change

Our third criterion for life is that it senses and responds to changes in its surroundings. It is important to realize that in order to meet this criterion, an organism's ability to sense changes is just as important as its ability to respond. After all, even a rock can respond to changes in its environment. If a boulder, for example, is perched on the very edge of a cliff, even a slight change in the wind patterns around the boulder might be enough for it to fall off of the cliff. In this case, the boulder is responding to the changes in its surroundings. The reason a boulder doesn't meet this criterion for life is that the boulder cannot sense the change.

Living organisms are all equipped with some method of receiving information about their surroundings. Typically, they accomplish this feat with **receptors**.

Receptors - Special structures or chemicals that allow living organisms to sense the conditions of their surroundings

Your skin, for example, is full of receptors. Some allow you to distinguish between hard and soft substances when you touch them. Other receptors react to hot and cold temperatures. Thus, if you have your hand under a stream of water coming from a water faucet, your receptors react to the temperature of the water. The receptors send information to your brain and you can then react to the temperature. If the water is too hot or too cold, you can remove your hand from the stream to avoid the discomfort.

A living organism's ability to sense and respond to changes in its surrounding environment is a critical part of survival, because God's Creation is always changing. Weather changes, seasons change, landscape changes, and the community of organisms in a given region changes. As a result, living organisms must be able to sense these changes and adapt, or they would not be able to survive.

All Life Forms Reproduce

Our final criterion for life says that all living organisms reproduce. Although the necessity of reproduction for the perpetuation of life is rather obvious, it is truly amazing how many different ways God has designed the organisms on earth to accomplish this feat. Some organisms, for example, can split themselves apart under the right circumstances. The two parts can then grow into wholly separate organisms. This is called **asexual reproduction**.

Asexual reproduction - Reproduction accomplished by a single organism

Other organisms, however, require a male and female together in order to reproduce. This method of reproduction (which occurs in most of the life forms with which you are familiar), is called **sexual reproduction**.

Sexual reproduction - Reproduction that requires two organisms, a male and a female

As we go along in this course, we will be studying both of these methods a bit closer, because there is a great deal of variety among the different means of sexual and asexual reproduction.

Reproduction always involves the concept of **inheritance**. Although this word has several different meanings, in biology the definition is quite specific.

Inheritance - The process by which physical and biological characteristics are transmitted from the parent (or parents) to the offspring

In asexual reproduction, the characteristics and traits inherited by the offspring are, under normal circumstances, identical to the parent. Thus, the offspring is essentially a "copy" of the parent. In sexual reproduction, under normal circumstances, the offspring's traits and characteristics are, in fact, some mixture of each parents' traits and characteristics. Of course, the organism's parents' traits and characteristics are a mixture of each of their parent's traits and characteristics, and their parents' characteristics are a mixture of each of their parent's traits and characteristics, and so on. In the end, then, the inheritance process in sexual reproduction is quite complicated, and leads to offspring that often can be noticeably different than both parents.

Notice that in describing inheritance for both modes of reproduction, we used the phrase "under normal circumstances." This is because every now and again, offspring can possess traits that are incredibly different than the offspring's ancestors. These incredibly different traits are called **mutations**.

Mutation - An abrupt and marked difference between offspring and parent

As we will see in a later module, this is actually not the best definition for mutation, but we'll use it for right now. The study of mutations is quite interesting, and we will focus on it in an upcoming module.

ON YOUR OWN

1.3 A biologist studies an organism and then two of its offspring. They are all identical in every possible way. Do these organisms reproduce sexually or asexually?

Life's Secret Ingredient

Well, now that we have a good idea of whether or not something is alive, another question should come to mind. What gives life the characteristics that we learned in the previous sections? As we said before, if we chemically analyzed an organism, gathered together all of the chemicals contained in it, and threw them in a pot, we would not have a living organism. Those chemicals would be useless without the information stored in the organism's DNA. However, even if we were able to isolate a full set of the organism's DNA and were to throw it into the pot as well, we would still not have a living organism.

You see, life is more than a collection of chemicals and information. There is something more. Scientists have tried to understand what that "something more" is, but to no avail. The secret ingredient that separates life from non-life is still a mystery to modern science. Of course, to believers, that secret ingredient is rather easy to identify. It is the creative power of God. In Genesis 1:20-27, the Bible tells us that God created all creatures, and then He created man in His own image. Think about it this way. Suppose you had a bunch of engine and metal parts and you also had instructions that led you through all of the steps necessary to take those parts and make a working motorcycle. Could you just throw the parts and the instructions into a pile and make a motorcycle? No, of course not. Even though you had all of the necessary parts as well as all of the instructions, you still need to exercise some of your own creative power to follow those instructions and make the motorcycle.

If we were talking about a living organism instead of a motorcycle, we could say that chemicals are the "parts" that make up the organism, and DNA is the instruction set that contains the information necessary to assemble the parts properly. Nevertheless, if you just threw the chemicals and the DNA into a big pot, you would not make a life. Some creative power must be exercised in order to take lifeless chemicals and use the information in DNA to make a living organism. Of course, only God has such creative power, and that is why all life comes from Him.

So you see, science will never be able to uncover the "secret ingredient" that makes life possible. At some point in the future, scientists might be able to catalog every chemical that makes up a living organism. At some farther-off point in the future, scientists might even decode the information stored in DNA and determine all of the instructions necessary to form those chemicals into a living organism. Even after those incredible feats, however, science would be no closer to creating life. Without the creative power of God, lifeless chemicals will never become a living organism.

This little discussion brings us to probably the most important thing that you will ever learn in your academic career: *science has its limitations*. We say that this is probably the most important thing that you will ever learn because we know a great many scientists whose lives have been ruined because they put too much faith in science. They think that because of all the wonderful advances we have made in recent years, science has no limitations. As a result, they live their lives looking to science as the ultimate answer to every question. This leads them down a path of spiritual destruction. Had they only placed their faith in God, who has no limitations, they would have lived fulfilling lives and spent eternity with the ultimate Life-Giver! Read the next section carefully, so that you will understand the limitations of science.

The Scientific Method

Real science must conform to a system known as the **scientific method**. This system provides a framework in which scientists can analyze situations, explain certain phenomena, and answer certain questions. The scientific method starts with **observation**. Observation allows the scientist to collect data. Once enough data has been collected, the scientist forms a **hypothesis** that attempts to explain some facet of the data or attempts to answer a question that the scientist is trying to answer.

Hypothesis - An educated guess that attempts to explain an observation or answer a question

Once he or she forms a hypothesis, the scientist (typically with help from other scientists) then collects much more data in an effort to test the hypothesis. If data are found which are inconsistent with the hypothesis, the hypothesis might be discarded, or it might just be modified a bit until it is consistent with all data that has been collected. If a large amount of data is collected and the hypothesis is consistent with all of the data, then the hypothesis becomes a theory.

Theory - A hypothesis that has been tested with a significant amount of data

Since a theory has been tested by a large amount of data, it is much more reliable than a hypothesis. As more and more data relevant to the theory gets collected, the theory can be tested over and over again. If several generations of collected data are all consistent with the theory, it eventually attains the status of a scientific law.

Scientific law - A theory that has been tested by and is consistent with generations of data

An example of the scientific method in action can be found in the life of Ignaz Semmelweis, a Viennese doctor who lived in the early-to-mid 1800's. He was put in charge of a ward in Vienna's most famous hospital, the Allegemeine Krakenhaus. He noticed that in his ward, patients were dying at a rate which far exceeded that of the other wards, even the wards with much sicker patients. Semmelweis observed the situation for several weeks, trying to figure out what was different about his ward as compared to all others in the hospital. He finally determined that the only noticeable difference was that his ward was the first one that the doctors and medical students visited after they performed autopsies on the dead.

Based on his observations, Semmelweis hypothesized that the doctors were carrying something deadly from the corpses upon which the autopsies were being performed to the patients in his ward. Thus, Dr. Semmelweis exercised the first step in the scientific method. He made some observations and then formed a hypothesis to explain those observations.

Semmelweis then developed a way to test his hypothesis. He instituted a rule that all doctors had to wash their hands after they finished their autopsies and before they entered his ward. Believe it or not, up to that point in history, doctors never thought to wash their hands before examining or even operating on a patient! Dr. Semmelweis hoped that by washing their hands, doctors would remove whatever was being carried from the corpses to the patients in his ward.

Well, the doctors did not like the new rule, but they grudgingly obeyed it, and the death rate in Dr. Semmelweis' ward decreased *to the lowest in the hospital!* This, of course, was good evidence that his hypothesis was right. You would think that the doctors would be overjoyed. They were not. In fact, they got so tired of having to wash their hands before entering Dr. Semmelweis' ward that they worked together to get him fired. His successor, anxious to win the approval of the doctors, rescinded Semmelweis' policy, and the death rate in the ward shot back up again.

Semmelweis spent the rest of his life doing more and more experiments to confirm his hypothesis that something unseen but nevertheless deadly can be carried from a dead person to a live person. Although Semmelweis' work was not appreciated until after his death, his hypothesis was eventually confirmed by enough experiments that it became a scientific theory. At that point, doctors began washing their hands before examinations and surgery.

As time went on, more and more data was gathered in support of Semmelweis' theory. With the advent of the microscope, scientists were finally able to see the deadly bacteria and germs that can be transmitted from person to person, and Semmelweis' theory became a scientific law. Nowadays, doctors do all that they can to completely sterilize their hands, clothes, and instruments before performing any medical procedure.

Before we leave this story, it might be interesting to note that the Old Testament contains meticulous instructions concerning how a priest is to cleanse himself after touching a dead body. These rituals, some of which are laid out in Numbers 19, are more effective than all but the most modern methods of sterilization. In fact, Dr. S. I. McMillen, a medical doctor and author of *None of These Diseases*, states that "In 1960, the Department [of Health in New York State] issued a book describing a method of washing the hands, and the procedures closely approximate the Scriptural method given in Numbers 19." This, of course, should not surprise you. After all, God knows all about germs and bacteria; He created them. Thus, it only makes sense that He would lay down instructions as to how His people can protect themselves from germs and bacteria. If only doctors would have had the sense to follow those rules in the past centuries. Countless lives would have been saved!

So you see, the scientific method (summarized in Figure 1.3) provides a methodical, logical way to examine a situation or answer a question. If a theory survives the scientific method and becomes a law, it can be considered reasonably trustworthy. Even a scientific theory which has not been tested enough to be a law is still pretty reliable, because it is backed up by a lot of scientific data.

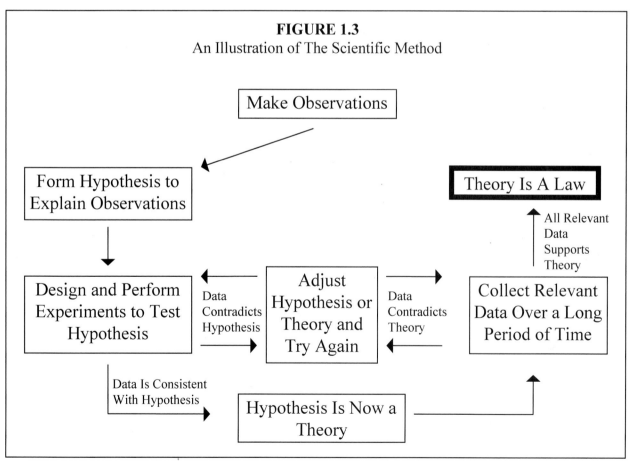

FIGURE 1.3
An Illustration of The Scientific Method

ON YOUR OWN

1.4 When trying to convince you of something, people will often insert, "Science has proven..." at the beginning of a statement. Can science actually prove something? Why or why not?

1.5 A scientist makes a few observations and develops an explanation for the observations that he or she has made. At this point, is the explanation a hypothesis, theory, or scientific fact?

Limitations of the Scientific Method

At the end of the previous section, we said that if a theory survives the scientific method and becomes a scientific law, it is "reasonably trustworthy." Why did we say "reasonably?" Aren't all scientific laws completely trustworthy? If a hypothesis survived scientific scrutiny and became a theory, and that theory went on through more significantly scientific scrutiny and became a law, isn't it 100% reliable? No, not at all. You see, in order to test hypotheses and theories, scientists must gather data. In order to gather data, they must perform experiments and observations. Since these experiments and observations are designed and performed by imperfect humans, the data collected might, in fact, be flawed. As a result, even though there might be an enormous amount of data supporting a scientific law, if the data is flawed, the law is most likely wrong! In addition, it is simply impossible, even after centuries of experimentation, to test all implications of a scientific law completely. Thus, even though years and years of experimentation exist in support of a scientific law, some clever person somewhere might devise an experiment in which the data contradict the law. So we see that scientific laws can be demonstrated false when the experiments that support them are shown to be flawed or when someone finds a new kind of experiment that contradicts the law. Both of these situations occur frequently in the pursuit of science, and they are best studied by example.

Scientific laws are constantly being overthrown due to the fact that it is impossible to completely test them. For example, prior to 1938, it was considered scientific law that the coelacanth (a bony fish) was extinct. After all, many, many fossils of the fish had been uncovered, but no live specimen had ever been found. Many biologists and paleontologists (those who study fossils) exhaustively searched for living coelacanths. Since almost 100 years of searching for this fish never turned up a live specimen, the hypothesis that it was extinct was eventually accepted as a theory and then as a scientific law. All scientists agreed: the coelacanth was extinct. Imagine their surprise when, in 1938, a fisherman who was fishing off the coast of Africa in the Indian Ocean caught a live specimen! It turns out that the coelacanth is relatively plentiful in the Indian Ocean; thus, a scientific law was overthrown due to the fact that it is impossible to test a law completely. One would think that since 100 years of careful searching for the coelacanth had never turned up a live specimen, the law stating that it was extinct should be rather reliable. However, no one had looked carefully enough in the Indian Ocean off the coast of Africa, and therefore a scientific law turned out to be quite wrong!

Other scientific laws are overthrown because the experiments that support them are flawed. For example, in about 330 BC, the famous Greek philosopher Aristotle observed that if one left meat out in the open and allowed it to decay, maggots would appear on the meat within a few days. From that observation, he formed the hypothesis that living maggots were formed from non-living meat. He called this process "spontaneous generation," and he postulated that this is where many life forms originate. He made many other observations that seemed to support his hypothesis. For example, he showed that eels have a similar smell and feel as the slimy ooze at the bottom of rivers. He considered this evidence that eels spontaneously formed from the ooze.

As time went on, many more experiments were performed that seemed to support the hypothesis of spontaneous generation. As a result, the hypothesis was quickly accepted as a theory. Of course, the experimentation did not stop there. As late as the mid-1600s, a biologist named Jean Baptist van Helmont performed an experiment in which he placed a sweaty shirt and some grains of wheat in a closed wooden box. Every time he performed the experiment, he found at least one mouse gnawing out of the box within 21 days. Think about it. A hypothesis that was formed in 330 BC was quickly accepted as a theory due to the fact that all experiments performed seemed to support it. Experiments continued for a total of 1900 years, and they all seemed to support the theory! As a result of this overwhelming amount of data in support of the theory of spontaneous generation, it became accepted as a scientific law.

About that same time, however, Francesco Redi, an Italian physician, questioned the law of spontaneous generation. Despite the fact that this law was universally accepted by the scientists of his day, and despite the fact that his fellow scientists laughed at him for not believing in the law, Redi challenged it. He argued that Helmont could not tell whether the mice that he supposedly formed from a sweaty shirt and wheat grains had gnawed *into* the box or *out of* the box. He said that in order to really test this law, you would have to completely isolate the materials from the surroundings. That way, any life forms that appeared would have to have come from the materials and not from the surroundings. He performed experiments in which he put several different types of meat in sealed jars and allowed the meat to decay. No maggots appeared on the meat. He claimed that this showed that maggots appear on meat not because they are *formed by* the meat, but instead because they *get onto* the meat.

Of course, the scientists of his day said that by sealing the jars, Redi was cutting off the air supply, which would stop the maggots from forming. Thus, Redi redesigned his experiment. Instead of sealing the jars, he covered them with a fine netting. The netting was fine enough to keep maggots out but allow air in. Still, no maggots formed on the meat, even long after it was decayed. What these experiments showed was that the previous experiments which purportedly demonstrated that maggots could form from decaying meat were simply flawed. If one were to adequately isolate the meat from the surroundings, maggots would never form.

These experiments sent shock waves throughout the scientific community. A scientific law, one which had been supported by nearly 1900 years of experiments, was wrong! Of course, many scientists were simply unwilling to accept this. Yes, they agreed, perhaps maggots did not come from decaying meat, but surely there were some types of organisms that could spontaneously generate from non-living things.

Anton van Leeuwenhoek thought he had found such organisms. In 1675, he reported that he had fashioned a homemade lens which magnified whatever was observed through it. As a result, he discovered the world of **microorganisms**.

Microorganisms - Living creatures that are too small to see with the naked eye

In the next module, we will begin studying this fascinating world in more depth. For right now, you just need to know that because these creatures cannot be seen without the aid of a magnifying lens, scientists prior to 1675 had no idea that they existed.

Leeuwenhoek and many others showed that these microorganisms did, indeed, seem to generate spontaneously. For example, in the mid-1700's John Needham did experiments very similar to Redi's. Needham made a liquid broth of nutrient rich material such as mutton gravy. He called these broths "infusions." He showed that if you boiled an infusion for several minutes, you could kill all microorganisms in it. Then, if you put the infusion in a jar and covered it with a net like Redi did in his experiments, microorganisms would appear in the infusion within a few days. Needham concluded that since he had covered the jar with a net just as Redi had, the infusion was isolated from the surroundings. These experiments were hailed as support for the beleaguered law of spontaneous generation.

Lazzaro Spallanzani, a contemporary of Needham, did not like Needham's experiments. After all, he said, since we cannot see microorganisms with our eyes, perhaps they can be transported by the air, making their way through the nets that covered Needham's infusions. Spallanzani repeated Needham's experiments, but Spallanzani covered the jars with an airtight seal. In these experiments, no microorganisms formed. Of course, those who still held on to the law of spontaneous generation argued that once again, without air, nothing could live. Thus, by making an airtight seal, Spallanzani cut off the process of spontaneous generation.

In the mid-1800's, however, the great scientist Louis Pasteur finally demonstrated that even microorganisms cannot spontaneously generate. In his experiments, illustrated in Figure 1.4, Pasteur stored the infusion in a flask that had a curved neck. The curved neck allowed air to still reach the flask, but if microorganisms were present in the air, they would be trapped at the bottom of the curve. When Pasteur repeated Needham's experiments in the curved flask, no microorganisms appeared. In a final blow, Pasteur even showed that if you tipped the flask once to allow any microorganisms that might be trapped to fall into the infusion, microorganisms would appear in the infusion. Thus, Pasteur showed that even microorganisms cannot spontaneously generate. As a sidelight, he also showed that these microorganisms can be transported through the air.

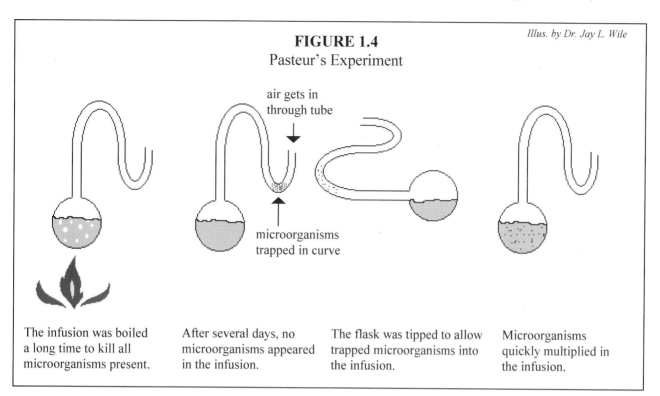

FIGURE 1.4

Pasteur's Experiment

Illus. by Dr. Jay L. Wile

air gets in
through tube

microorganisms
trapped in curve

| The infusion was boiled a long time to kill all microorganisms present. | After several days, no microorganisms appeared in the infusion. | The flask was tipped to allow trapped microorganisms into the infusion. | Microorganisms quickly multiplied in the infusion. |

The point to this rather long discussion is simple. Even though a scientific law seems to be supported by hundreds of years of experiments, it might very well still be wrong because those experiments might be flawed. All of the experiments that were used to support the law of spontaneous generation were flawed. The scientists who conducted the experiments did not adequately isolate them from the surroundings. Thus, the life forms that the scientists thought were being formed from non-living substances were, in fact, simply finding their way into the experiment.

These two discussions, then, show the limits of science and the scientific method. First, even scientific laws are not 100% reliable. We are certain that some of the things which you learn in this book will someday be proven to be wrong. That is the nature of science. Because it is impossible to fully test a scientific law, and because laws are tested by experiments that might be flawed, scientific laws are not necessarily true. They represent the best conclusions that science has to offer, but they are nevertheless not completely reliable. Of course, if you are working with something that is a theory, it is even less reliable. Thus, putting too much faith in scientific laws or theories will end up getting you in trouble, because many of the laws and theories that we treasure in science today will eventually be shown to be wrong.

Well, if scientific laws are not 100% reliable, what is? The only thing in the universe that is 100% reliable is the Word of God. The Bible contains truths that will never be shown to be wrong, because those truths come directly from the Creator of the universe. So much misery and woe has come to this earth because people put their faith in something that is not reliable, like science. In the end, they are spiritually deprived because what they believe in is, to one extent or another, wrong. Those who put their faith in the Bible, however, are not disappointed, because it is never wrong.

If science isn't 100% reliable, why study it? The answer to that question is quite simple. There are many interesting facts and much useful information that is not contained in the Bible. Thus, it is worthwhile to find out about these things. Even though we will probably make many, many mistakes along the way, finding out about these interesting and useful things will help us live better lives. Because of the advances made in science, wonderful technology like the television and the computer exist. Thus, there is nothing wrong with science. It is a good and useful endeavor. The problem occurs when certain people who are enamored with science end up putting too much faith in it. As a pursuit of flawed human beings, science will always be flawed. Because the Bible was inspired by One who is perfect, the Bible is perfect. As long as we keep this simple fact in mind, our study of science will be greatly rewarding!

Spontaneous Generation: The Faithful Still Cling to It!

After that long story, it might surprise you to learn that there are those scientists who still believe in spontaneous generation. Now of course, there is no way that they can argue with the conclusions of Pasteur's experiments, so they do not believe that microorganisms can spring from non-living substances. Nevertheless, they still do believe that life can spring from non-life! These scientists believe in a new theory known as **abiogenesis** (a' bye oh jen uh sis).

Abiogenesis -   The theory that, long ago, very simple life forms spontaneously appeared through random chemical reactions

In this theory, some scientists say that since all life is made up of chemicals, it is possible that long ago on the earth, there was no life. There were just chemicals. These chemicals began reacting and, through the random reaction of chemicals, a "simple" life form suddenly appeared.

As we go through this course, you'll see how such an idea is simply inconsistent with everything that we know about life. At this time, however, we want to make a simple point regarding abiogenesis. Back when scientists believed in spontaneous generation, they had experiments which allegedly backed up their claim. Even before Pasteur's authoritative refutation of spontaneous generation, these experiments were shown to be flawed. Rather than giving up on their law, however, those who fervently believed in spontaneous generation just said, "Well, okay, *these experiments* are wrong. However, look at *these other experiments*. Although we now know that life forms which we see with our own eyes cannot spontaneously generate, microorganisms can."

Do you see what the proponents of spontaneous generation did? Because they wanted so badly to believe in their theory, they simply pushed it into an area in which they did not have much knowledge. The whole world of microorganisms was new to scientists back then. As a result, there was a lot of ignorance regarding how microorganisms lived and reproduced. Because of the ignorance surrounding microorganisms, it was relatively easy to say that spontaneous generation occurred in that world. After about 200 years of study, however, scientists began to understand microorganisms a little better, and that paved the way for Louis Pasteur's famous experiment.

Well, nowadays, scientists have pushed the theory of spontaneous generation back to another area that we are rather ignorant about. They say that although Pasteur's experiments show that microorganisms can't arise from non-living substances, some (unknown) simple life form might have been able to spontaneously generate from some (unknown) mixture of chemicals at some (unknown) point way back in earth's history. Well, since we have very little knowledge about things that happened way back in earth's history, and since we have only partial knowledge about the chemicals that make up life, and we have no knowledge of any kind of simple life form that could spring from non-living chemicals, the proponents of spontaneous generation (now known as abiogenesis) are pretty safe. The fact that we are ignorant in these areas keeps us from showing the error in their theory.

Of course, there are a few experiments that lend some support to the theory of abiogenesis. A discussion of these experiments is beyond the scope of this module, so right now let me just say that they are not nearly as convincing as the ones that van Helmont and Needham performed. In fact, they do not even produce anything close to a living organism, as van Helmont's and Needham's experiments seemed to. They just produce some of the simplest chemicals that are found in living organisms. Nevertheless, those who cling to the idea of spontaneous generation casually disregard the flaws that can be easily pointed out in these experiments and trumpet their results as data that support their theory. However, if you look at the track record of spontaneous generation throughout the course of human history, it is safe to conclude that at some point, the version of spontaneous generation known as abiogenesis will also be shown to be quite wrong. In a later module, you will see why these authors think that such a case has already been made.

Biological Classification

Now that we've spent considerable time on the limitations of science, it's time to turn our attention to some of the strengths of science. Classification is probably one of the greatest accomplishments of science. In the study of biology, we uncover many, many facts. For example, there are many, many organisms on the earth and they have many, many properties and characteristics. Some of their characteristics they have in common with other organisms, and some of their characteristics are unique. All of these facts comprise a huge volume of data that, by itself, would be hard to understand and virtually impossible to use. Much like we have split this book into modules and have further split the modules into sections, "on your own" problems, study guides, and tests, we need to take all of the data in biology and split it up into an organized system.

Now there are many different classification systems in biology. You have already seen that all organisms can be split into three groups: producers, consumers, and decomposers. You have also seen that we can split consumers up into herbivores, carnivores, and omnivores. Those classification systems were rather simple. They took many, many different organisms and lumped them into only a few groups. Now we need to get more detailed. We need to learn a classification system that takes all organisms and splits them into several groups. The number of groups that we split the organisms into must be large enough so that we are not grouping incredibly different organisms into the same group. At the same time, however, there cannot be too many groups, because the classification system must make the data easier to understand than it was originally. With too many groups, the classification system becomes almost as complex as the data itself.

The classification system that we will use most frequently is multi-leveled. It starts by splitting all organisms up into five different groups known as **kingdoms**. The organisms within each kingdom can then be further divided into different groups called **phyla** (fie' luh), the singular of which is **phylum** (fie' lum). Each phylum can be further divided into **classes,** which can be further divided into **orders**. Within an order, organisms can be divided into **families**, which can be further divided into **genera** (jon' ruh), the singular of which is genus (jee' nus), which can finally be broken down into **species**. This multi-leveled (often called "hierarchical") classification scheme is summarized in Table 1.1.

TABLE 1.1
A Hierarchical Biological Classification Scheme

Classification Groups (In Order)
Kingdom
Phylum
Class
Order
Family
Genus
Species

To make sure that you can remember the names and orders of this classification system, you can use the following mnemonic:

King Philip Cried Out, "For Goodness Sake!"

Since the first letter of each word in this sentence can stand for a group in our classification system, you can use it to remember the order in which we place these groups. It is important to note that the classification of organisms is so complicated that we often split these groups into subgroups. Thus, do not be confused if you run across a term like "subphylum." A subphylum is simply used to split organisms in a phylum into smaller groups *before* they are split into classes. There are also subclasses, suborders, and subfamilies. Another annoying fact is that some classification schemes use "division" instead of "phylum" for kingdoms Monera, Fungi, and Plantae. Although we will not do that, you need to be aware that others might.

Now that we know the groups and their respective orders, it's time to see how we use this system to classify organisms in nature. As we mentioned before, we generally split all of the organisms in nature into five separate kingdoms. In fact, there are some applications of this classification system that use as little as three kingdoms, but most biologists use five, so that's what we'll use. The names of these kingdoms are **Monera** (muh nihr' uh)**, Protista** (pro tee' stuh)**, Fungi** (fun' jye)**, Plantae**, and **Animalia**. The proper names of all our classification groups are Latin, and when we use those names, we capitalize them to emphasize that these are proper classification names.

How do we know what organisms go into what kingdom? Well, we group organisms together based on similar characteristics. Since the first step in classification deals with placing organisms in kingdoms, the common characteristics that organisms in the same kingdom share are pretty basic.

ON YOUR OWN

1.6 Suppose you chose two organisms at random out of a list of the members of kingdom Plantae, then you chose two organisms at random out of a list of the members of family Pinaceae. In which case would you expect the two organisms to be the most similar?

1.7 You compare several organisms from different orders. You then compare organisms from different classes. In which case would you expect the differences to be greatest?

Characteristics Used to Separate Organisms into Kingdoms

The first and most basic distinction that we make between organisms is based on the number and type of cells that the organism has. Now you have probably learned a few things about cells from your earlier studies in science. You probably learned that all living creatures are made up of at least one cell, and that cells are the basic building blocks of life. We will be making a detailed study of cells throughout the next few modules, so for right now, we don't want to spend a lot of time on them. The only thing that we want to concentrate on right now is the fact that cells come in two basic types: **prokaryotic** (pro kehr ee aht' ik) and **eukaryotic** (yoo kehr ee aht' ik).

Prokaryotic cell - A cell that has no distinct, membrane-bound organelles

Eukaryotic cell - A cell with distinct, membrane-bound organelles

Now of course, these definitions mean nothing unless you know what **organelles** (or guh nelz') are and what "**membrane-bound**" means. You see, in order to live, a cell must perform certain functions. As two of our criteria for life say, living things must have an energy conversion mechanism as well as reproductive capacity. In order to carry out these functions, cells must complete many different tasks. In eukaryotic cells, the individual tasks needed to complete the functions of life are carried out by distinct structures within the cell. These structures are called organelles. In order to stay distinct, they must be surrounded by something that separates them from the rest of the cell. We call this a membrane. Thus, a "distinct, membrane-bound organelle" is simply a structure within a cell that performs a specific task. Prokaryotic cells do not contain these internal structures. Nevertheless, they still can perform all of the necessary functions of life. You might wonder how that is possible. Well, you'll learn about these fascinating organisms in the next module. For right now, just familiarize yourself with the distinction between prokaryotic and eukaryotic cells with Figure 1.5.

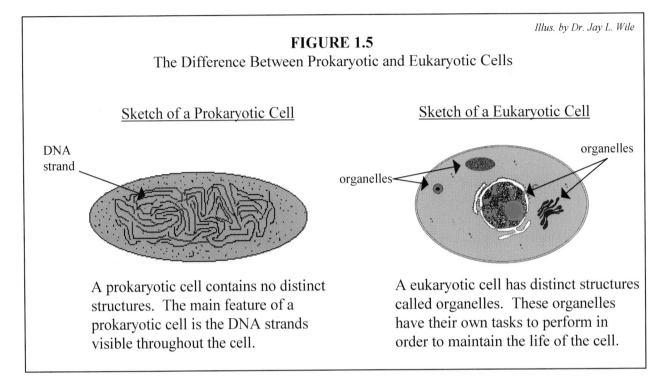

Illus. by Dr. Jay L. Wile

FIGURE 1.5
The Difference Between Prokaryotic and Eukaryotic Cells

Sketch of a Prokaryotic Cell   Sketch of a Eukaryotic Cell

DNA strand   organelles

organelles

A prokaryotic cell contains no distinct structures. The main feature of a prokaryotic cell is the DNA strands visible throughout the cell.

A eukaryotic cell has distinct structures called organelles. These organelles have their own tasks to perform in order to maintain the life of the cell.

Now that we know the distinction between these two basic cell types, we can finally discuss how to split organisms up into the five different kingdoms. Kingdom Monera contains all organisms that are composed of either one prokaryotic cell or a simple association of prokaryotic cells. What do we mean when we say "a simple association" of cells? Well, if cells work together in order to complete the tasks necessary for life, they can do so in one of two ways. They can either be highly specialized, each taking on a specific set of tasks needed for the organism to survive, or they can simply work together as a group, each performing essentially the same tasks, but doing so as a group. The cells in a person, for example, work together in the first way. The cells that make up your eyes specialize in the detection of light and the transmission of light-induced information to your brain, while red blood cells specialize in transporting oxygen to other cells. These cells perform different functions, each of which is necessary for the support of life. Blue-green algae (also known as cyanobacteria), however, simply group themselves together in chains. The cells in the chain are usually bound together by mucus, but they each do essentially the same task. They simply find strength and survivability in numbers. This is an example of a "simple association" of cells. Blue-green algae and bacteria are both members of kingdom Monera.

The next kingdom is called Protista. It contains those organisms that are composed of only one eukaryotic cell or a simple association of eukaryotic cells. Amoebae and paramecia are members of kingdom Protista. Kingdoms Monera and Protista together contain most of the microorganisms that exist on earth.

Moving out of the microscopic world (for the most part) and into the macroscopic world (the world we can see with the naked eye), we come to the kingdom Fungi. This kingdom is comprised of decomposers. If you remember from the earlier parts of the chapter, decomposers are those organisms that feed off of dead organisms, decomposing them into their constituent chemicals so that they can be used again by the producers. Most members of the kingdom Fungi have eukaryotic cells. In addition, most Fungi are multicellular, but there are a few single-celled Fungi. Mushrooms and bread molds are examples of the organisms in kingdom Fungi.

The next kingdom, Plantae, is composed of autotrophs (organisms that produce their own food). Almost all members of kingdom Plantae are multicelled organisms with eukaryotic cells. Even though we say that members of kingdom Plantae are autotrophs, there are a few exceptions. Some parasitic organisms are considered members of kingdom Plantae. As you have probably already guessed, members of kingdom Plantae are often called "plants." Thus, trees, grass, flowers, etc., are all members of kingdom Plantae.

The last kingdom, Animalia, contains multicellular organisms with eukaryotic cells. Members of kingdom Animalia are separated from kingdom Plantae by the fact that they are heterotrophs (dependent on other organisms for food) but are not decomposers (decomposers are in kingdom Fungi). Of course, members of kingdom Animalia are called "animals." Grasshoppers, birds, cats, fish, and snakes are all members of kingdom Animalia.

ON YOUR OWN

1.8 An organism is made up of one eukaryotic cell. To what kingdom does it belong?

1.9 An organism is multicellular and an autotroph. To what kingdom does it belong?

1.10 An organism is multicellular with eukaryotic cells. It is also a decomposer. To what kingdom does it belong?

The Definition of Species

After reading the last section, you should have noticed a few things about classifying organisms. It's not very easy or clear-cut. To separate organisms into five separate groups, we already ran into exceptions. Kingdom Plantae, for example, is supposed to contain autotrophs. There are, however, some parasites that belong to that kingdom as well. In addition, we use the word "mostly" quite a lot, because although the majority of the members in a kingdom might have a certain characteristic, there will be some members that do not. Thus, classification of organisms into kingdoms gets a little complicated.

As you might expect, classifying organisms in phyla, classes, orders, families, genera, and species becomes even more difficult. After all, as you move down the hierarchy in our classification scheme, you are getting more and more specific. While kingdoms have many, many members, those members are split into phyla. Thus, each phylum has fewer members than does the kingdom of which it is a part. In the same way, classes have fewer members than the phylum that they are in, orders have even fewer members, families have even fewer, and genera have still fewer. By the time you get to species, you have a very small group of organisms.

Since classification gets more and more difficult as you go down the hierarchy, splitting organisms into species becomes incredibly hard. If you thought that our definitions for what organisms go into each of the five kingdoms was bad, it is so hard to classify at the species level that *biologists can't even agree on a definition for what the classification "species" really means*! There is a lot of work going on right now in the field of biology trying to figure out a good way to define this difficult classification. For our purposes, however, we must have a definition, so we will go with the most commonly accepted one:

> Species - A unit of one or more populations of individuals that can reproduce under normal conditions, produce fertile offspring, and are reproductively isolated from other such units

Although this definition is not perfect , it is the one that we will use for now. What does it mean? Basically, if organisms can reproduce and their offspring can also reproduce (that's what "fertile" means), these organisms belong to the same species. Any other organism with which this species cannot reproduce is said to be "reproductively isolated" from this species and therefore must belong to a different species.

Notice that in the previous section, we gave you the characteristics by which you can separate all organisms on earth into the five kingdoms of our classification system. Then, in this section, we skipped over all of the other classification groups except for species. For that classification group we gave a definition. Why did we leave out the other classification groups? Mostly because we did not want to overwhelm you with information. There are (depending on whose classification system you use) at least 51 different phyla. Members of each phyla have their own characteristics, and we would have to go through each phylum individually to give you a good feel for how to classify organisms into these groups. Of course, since each phylum is split into several classes, there are even more of those. Thus, to go through and give you a view of each kingdom, phylum, class, order, family, and genus would be an incredibly long discussion! When we get to species, however, the classification is so specific that we can actually come up with a weak definition for it. That's why we skipped from kingdoms all of the way to species.

Biological Keys

Well, if a discussion of all groups in our classification system is prohibitively long, how will we ever be able to classify organisms? In order to classify organisms, biologists often refer to **biological keys**. These keys help you to classify organisms without having to memorize the characteristics of all groups within the classification scheme. A simple such biological key is given below. It also is given in the appendix at the back of the book.

FIGURE 1.6
A Simple Biological Key

1. Microscopic.. **2**
   Macroscopic (visible with the naked eye)................................. **3**
2. Eukaryotic cell.. *kingdom Protista*
   Prokaryotic cell... *kingdom Monera*
3. Autotrophic...*kingdom Plantae*..... **4**
   Heterotrophic.. **5**
4. Leaves with parallel veins ....*phylum Anthophyta*...... *class Monocotyledoneae*
   Leaves with netted veins ....*phylum Anthophyta*..... *class Dicotyledoneae*
5. Decomposer.. *kingdom Fungi*
   Consumer...*kingdom Animalia*.... **6**
6. No Backbone.. **7**
   Backbone...*phylum Chordata*.... **22**
7. Organism can be externally divided into equal halves (like a pie),
   but it has no distinguishable right and left sides......................... **8**
   Organism either can be divided into right and left sides that are
   mirror images or cannot be divided into two equal halves............ **9**
8. Soft, transparent body with tentacles .... *phylum Cnidaria*
   Firm body with internal support; covered with scales or spiny
   plates; tiny, hollow tube feet used for movement........................ *phylum Echinodermata*
9. External plates that support and protect.......*phylum Arthropoda*.. **14**
   External shell or soft, shell-less body.................................... **10**
10. External Shell...*phylum Mollusca*.... **11**
    No external shell.. **12**
11. Coiled shell.. *class Gastropoda*
    Shell made of two similar parts... *class Pelecypoda*
12. Worm-like body without tentacled receptors on head................... *phylum Annelida*
    Non-worm-like body or tentacled receptors
    on head..*phylum Mollusca*....... **13**
13. Worm-like body with tentacled receptors on head...................... *class Gastropoda*
    Non-worm-like body but 8 or more tentacles used for grasping.... *class Cephalopoda*
14. More than 3 pairs of legs... **15**
    3 pairs of walking legs..*class Insecta*...... **16**
15. 4 pairs of walking legs, body in two divisions............................ *class Arachnida*
    More than 4 pairs of walking legs...................................... *class Malacostraca*
16 Wings... **17**
   No wings... **21**
17. All wings transparent.. **18**
    Non-transparent wings.. **19**
18. Capable of stinging from back of body.................................. *order Hymenoptera*
    Cannot sting (may be able to bite).................................... *order Diptera*

19. Large, sometimes colorful wings.. *order Lepidoptera*

     Thick, hard, leathery wings... **20**

20. Pair of hard wings covering a pair of folded, transparent wings.... *order Coleoptera*

     Pair of leathery wings covering a pair of transparent wings.......... *order Orthoptera*

21. Piercing, sucking mouthparts for obtaining blood........................ *order Siphonaptera*

     Mouthparts for chewing.. *order Hymenoptera*

22. Jaws or beak... **23**

     No jaw or beak.. *class Agnatha*

23. Skin covered with scales.. **24**

     No scales on skin.. **26**

24. Fins and gills.. **25**

     No fins, breathes with lungs... *class Reptilia*

25. Mouth on lower part of body.. *class Chondrichthyes*

     Mouth on front part of body.. *class Osteichthyes*

26. No scales, no hair, no feathers; skin is slimy......*class Amphibia*... **27**

     Feathers or hair.. **28**

27. Tail.. *order Caudata*

     No tail... *order Anura*

28. Feathers on body.. *class Aves*

     Hair on body...*class Mammalia*... **29**

29. Hooves.. **30**

     No hooves.. **31**

30. Odd number of toes.. *order Perissodactyla*

     Even number of toes... *order Artiodactyla*

31. Carnivore.. **32**

     Herbivore.. **33**

32. Teeth... *order Carnivora*

     No teeth, eats insects... *order Insectivora*

33. Enlarged front teeth for gnawing.. **34**

     No enlarged front teeth for gnawing... **35**

34. Legs for crawling.. *order Rodentia*

     Hind legs for jumping.. *order Lagomorpha*

35. Enlarged trunk, used for breathing and grasping........................ *order Proboscidea*

     Tendency to stand erect on two hind limbs................................. *order Primates*

Now don't get overwhelmed by this key. It is actually quite simple to use once you are led through it. You see, this key allows you to look at the organism, seek its key features, and go through the key until you end up with a classification. For example, consider an elephant:

*Photo from the
MasterClips collection*

To classify this elephant, we would just start at the top of the key. When you find the proper characteristic, you proceed to the number that follows that characteristic. You continue to do this until you reach a classification that is not followed by a number.

So, we start at the top of the key. Key 1 asks about size. Since we don't need to magnify an elephant in order to see it with the naked eye, the elephant is macroscopic. This means that we move to key #3, because a "3" follows the term macroscopic. In key 3, we are asked whether or not the elephant is autotrophic (uses photosynthesis to make food) or heterotrophic (eats other organisms). Clearly, the elephant is heterotrophic. This means we move to key 5, where we need to determine whether it is a decomposer or a consumer. Since the elephant eats plants, it is a consumer. That tells us that our first classification is kingdom Animalia.

Now of course, this should be no surprise. An elephant is an animal. The key also tells us to move on to key 6 for a more detailed classification. Here, we determine whether or not it has a backbone. Now from the picture, you might not be able to tell, but all you have to do is think. Have you seen pictures or movies of people riding on elephants' backs or elephants carrying heavy loads on their backs? They must have a backbone to do that, so we learn that the elephant is in phylum Chordata (kor dah' tuh) and we move on to key 22.

Key 22 asks if the animal has a jaw or beak. Since the elephant's mouth opens and closes up and down, it has a jaw. Thus, we move to key 23, which asks if there are scales on the skin. There are not, so we move to key 26. This key asks about hair or feathers. The picture shows a line of hair along the back, especially on the neck. Thus, we move to key 28, which distinguishes between hair and feathers. Based on that distinction, we learn that the elephant is in class Mammalia and we move to key 29.

In key 29, we must decide whether the elephant has hooves or not. The feet have skin all the way to the bottom, so there are no hooves. This means we go to key 31, which asks whether the elephant is a herbivore or carnivore. Although not readily apparent from the picture, you should probably already know that elephants eat plants, making them herbivores. That means we move to key 33, which asks about teeth. There are certainly no enlarged teeth apparent, so we move to key 35. In this key, we are asked whether there is an enlarged trunk. Yes, there is. Thus, we know that the elephant is in order Proboscidea (pro boh seed' ee uh). This is as detailed a classification as we can make with this key. As far as this key is concerned, then, the elephant classification is:

Kingdom: Animalia
Phylum: Chordata
Class: Mammalia
Order: Proboscidea

Note that in the case of the elephant, we went all the way to the end of the key. This will rarely be the case. You continue on in the key until you run out of numbers. At that point, you have as detailed a classification as is possible with that key. Take your own shot at classification by performing Experiment 1.1.

EXPERIMENT 1.1
Using A Biological Key

Supplies:

- Photographs on the next page
- Biological Key in Figure 1.6

Object: The object of this exercise is to identify sixteen living things by using the biological key in the text. Keys vary in their style and content. This key is applicable to all five kingdoms, made especially for use in this course. A good library exercise would be to check other keys and how they are used.

The chart below gives you an example of how to identify the elephant that has been described for you in the text. Reread the section on how to identify the elephant and note how the chart has been completed for "example." Once you understand how the chart is filled in, identify each of the sixteen pictures by working through the key. As you work through the key, make a chart in your laboratory notebook and fill it in. Your chart should have all of the columns listed in the example chart below, and it should have 16 rows (one for each picture). Note how Kingdom (K.), Phylum (P.), Class (C.), and Order (O.) are written in the third column. Write them out that way in your chart as well.

Number	Specimen	Specimen Classification		Numbers From The Key
Example	Elephant	K. Animalia P. Chordata	C. Mammalia O. Proboscidea	1, 3, 5, 6, 22, 23, 26, 28, 29, 31, 33, 35
1.	Moth	K. P.	C. O.	

(continue chart for all 16 specimens in your laboratory notebook)

Once you have completed the chart in your laboratory notebook, check your work against the answers that are provided after the answers to the "on your own" problems.

Specimens For The Lab:

1. Moth

Photo by Rebecca Noelle Durnell

2. Chipmunk

Photo by Rebecca Noelle Durnell

3. Grapes

Photo by Rebecca Noelle Durnell

4. Swan

Photo from the MasterClips collection

5. Spider

Photo by Rebecca Noelle Durnell

6. Lion

Photo from the EXPERT 3000 collection

7. North American Flag

Photo from the MasterClips collection

8. Fish

Photo from the MasterClips collection

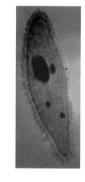

9. Paramecium
(Magnified 1,000x)

Photo by Kathleen J. Wile

10. Mushroom

Photo by Rebecca Noelle Durnell

11. Toad

Photo by Rebecca Noelle Durnell

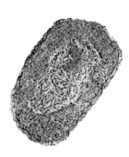

12. Bacterium
(Magnified 10,000x)

SEM image courtesy of Beth Verhostra

13. Deer
(Has hooves with an even number of toes.)

Photo from the MasterClips collection

14. Grasshopper

Photo from the MasterClips collection

15. Gibbon

Photo by Rebecca Noelle Durnell

16. Zebra

Photo by Rebecca Noelle Durnell

Naming Organisms Based on Classification

Of course, with a more complicated key, you could continue your classification of an organism right down to species. Why bother? Well, as we said before, classification is a way of ordering the diverse data in biology into some reasonably understandable system. This is such an important practice that an entire field of biology is devoted to it. We call this field **taxonomy** (tak sahn' uh mee).

<u>Taxonomy</u> - The science of classifying organisms

Taxonomy is a very important part of biology because, in order to give a scientific name to an organism, we must know both its species and its genus. In biology, we name things with **binomial** (bye no' mee ul) **nomenclature** (no' mun klay chur).

<u>Binomial nomenclature</u> - Naming an organism with its genus and species name

Humans, for example, are called *Homo sapiens*. *Homo* is the genus to which humans belong, and *sapiens* is the species. Notice that in binomial nomenclature, we italicize the genus and species name. This is to emphasize that we are using binomial nomenclature. In fact, whenever we use a genus or species name alone, we still italicize it, just to emphasize that it is a part of binomial nomenclature.

So, in order to properly name an organism, we need to know its genus and species. For example, if you were classifying oak trees, you would find that all oak trees are in the genus *Quercus*. A red oak is given the species name *rubra* while a white oak is given the species name *alba*. Notice that while we have capitalized all classification names up to this point, we do not capitalize the species name. This is a convention that makes binomial nomenclature a bit clearer. Thus, the scientific name of the red oak is *Quercus rubra* whereas the scientific name of the white oak is *Quercus alba*. As a point of notation, once we have introduced a genus name, we are allowed to abbreviate it in discussions that follow. Thus, we could say that the red oak is *Q. rubra* and the white oak is *Q. alba*.

Now why bother to do this? Why not just call a white oak a white oak and a red oak a red oak? Wouldn't that be easier? Well, yes and no. You see, English is constantly changing. What we mean by "oak" today may not mean the same thing in 100 years. That's because a spoken language continues to change. Latin, however, is a dead language. It will never change. Thus, *Q. rubra* will mean the same thing 100 years from now that it means today. Also, by using the genus name in the name of the organism, we have a start at being able to figure out other organisms that are similar to it. Any other organism that belongs to genus *Quercus* will be very similar to the red or white oaks. In addition, if we find out what family the genus *Quercus* comes from, we can find other organisms that are also similar to the white and red oaks. That's why we use this complicated naming system.

<u>The Microscope</u>

We'll be revisiting classification in nearly every module, so don't worry. It won't go away. However, this brief introduction allows us to get started exploring Creation. In the next two modules, we will be taking an in-depth look at kingdoms Monera and Protista. Since these kingdoms are composed of microorganisms, the labs we do in those two modules are heavily microscope-oriented. If you don't have a microscope, however, don't be concerned. We will have drawings or pictures of everything that you need to know, so a microscope isn't essential for taking this course. It does, however, help to make things clearer and more interesting. So for those who do have one, you need to perform Experiment 1.2. If you don't have a microscope, just skip this experiment.

EXPERIMENT 1.2
Introduction to the Microscope

Supplies:
- Microscope (preferably a National 131)
- Lens Paper
- Slides
- Coverslips
- Cotton swabs
- Dropper
- Water
- Small Pieces of Bright Thread
- Methylene Blue Stain

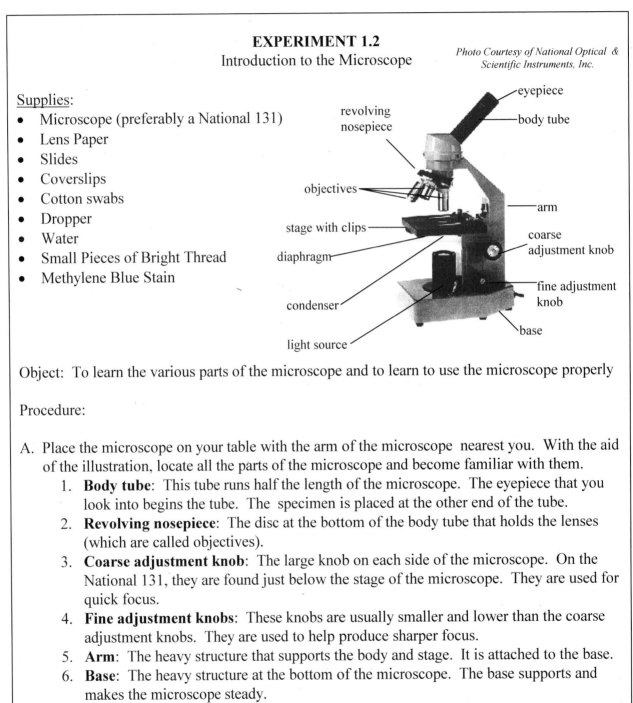

Object: To learn the various parts of the microscope and to learn to use the microscope properly

Procedure:

A. Place the microscope on your table with the arm of the microscope nearest you. With the aid of the illustration, locate all the parts of the microscope and become familiar with them.

1. **Body tube**: This tube runs half the length of the microscope. The eyepiece that you look into begins the tube. The specimen is placed at the other end of the tube.

2. **Revolving nosepiece**: The disc at the bottom of the body tube that holds the lenses (which are called objectives).

3. **Coarse adjustment knob**: The large knob on each side of the microscope. On the National 131, they are found just below the stage of the microscope. They are used for quick focus.

4. **Fine adjustment knobs**: These knobs are usually smaller and lower than the coarse adjustment knobs. They are used to help produce sharper focus.

5. **Arm**: The heavy structure that supports the body and stage. It is attached to the base.

6. **Base**: The heavy structure at the bottom of the microscope. The base supports and makes the microscope steady.

7. **Stage with clips**: A platform just below the objectives and above the light source. The clips are used to hold the slide in place.

8. **Eyepiece (called the ocular)**: Located at the top of the body tube. It usually contains a 10X lens.

9. **Objectives**: Found on the revolving nosepiece. They are metal tubes that contain lenses of varying powers.
10. **Diaphragm**: Regulates the amount of light that passes through the specimen. It is located between the stage and the light source. There are iris and disc diaphragms. The 131 has a disc diaphragm.
11. **Condenser**: Also located between the light source and stage. It is a lens system that affects resolution by bending and concentrating the light coming through the specimen.
12. **Light source**: Found at the base. It is used to provide necessary light for the examination of specimens.

The magnifications are an important feature of any microscope. In your laboratory notebook, write down the three magnifications of your microscope. You calculate the magnification by taking the power of the ocular (usually 10x) and multiplying it by the power of each objective. Thus, if your ocular is 10x and your objectives are 4x, 10x, and 40x, your three magnifications are 40x, 100x, and 400x. Label your three magnifications as Low, Medium, and High.

B. Now that you are familiar with the parts of the microscope, you are ready to use it.

1. Rotate the low-power objective so that it is in line with the eyepiece. Listen for a click to make sure it is in place.
2. Turn your light on. If you have a mirror instead of a light, look through the eyepiece and adjust the mirror until you see bright light.
3. Using the coarse adjustment, raise the stage (or lower the body tube) until it can move no more. (Never force the microscope gears.)
4. Place a drop of water on a clean slide and add several short pieces of brightly colored thread.
5. Add a cover slip (a thin piece of plastic that will cover the water and press it against the slide). This works best if you hold the cover slip close to the drops of water then drop it gently. If air bubbles form, tap the cover slip gently with the lead of your pencil.
6. Place the slide on the stage and clip it down making sure the cover slip is over the hole in the stage.
7. Looking in the eyepiece, gently move the stage down (or body tube up) with the coarse adjustment. If you do not see anything after a couple of revolutions, move your slide a little to make sure the threads are in the center of the hole in the stage. This indicates that the threads are in the field of view.
8. When you have focused as best you can with the coarse adjustment, then "fine tune" your resolution by using the fine adjustment.

   Note: Microscopes vary, carefully read the next steps.

9. Once you have the threads focused in as well as possible, then you are ready for the medium power. Place the threads in the very center of the field of view by moving the slide as you look at it through the microscope. Make sure that the threads are at the center of the field, or you will lose them when you change to a higher magnification.

10. Turn the nosepiece so that the medium power objective is in place. Until you are very familiar with any microscope, do not turn the nosepiece without checking to make sure it will not hit the slide. Always move the nosepiece slowly, making sure that it does not touch the slide in any way. A lens can easily be damaged if it hits or breaks a slide.
11. Once the medium power objective is in place, you should only have to move your fine adjustment slightly to be back in focus. Again, place the specimen in the center of the field.
12. Again, watching to make sure you don't hit the slide, turn the nosepiece so that the high magnification objective is in place. To re-focus, you should only use the fine adjustment.

Using the procedure laid out in steps 7-12, you can now view the thread under the highest magnification of the scope. Had you tried to bring the threads into focus under high magnification without first looking at them under low and then medium magnification, you almost certainly would have never found the threads. This is the procedure that we will always use to look at things under high magnification. We will start with the lowest magnification and then work our way up, centering the specimen in the field of view at each step.

C. Now that you have had some experience working with your microscope, it is time to get your first look at cells!

1. Collect some cheek cells by rubbing a cotton swab back and forth on the walls of your cheek inside your mouth. Use only one side of the swab.
2. Remove the swab carefully without getting a lot of saliva on the swab.
3. Rub the side of the swab with the cells on the slide.
4. If we were to look at the cells under the microscope right now, it would be hard to find them, because they are almost transparent. To help make them easier to see, we will add a dye to them. This dye is called a "stain," and it will help contrast the cells against the light, making them much easier to see. Place a drop of methylene blue stain on the area where you placed the cells. (This stain will not come out of most things, so use with care.)
5. Add the cover slip carefully.
6. Now place the slide on the microscope and begin the procedure outlined in steps 7-12 of section B, looking at the cells under low, then medium, and then high magnifications. In your laboratory notebook, draw what you see at each magnification level. You should see a dark blob (the nucleus) and a ring outlining the cell (the plasma membrane). Note the irregular shape of the cells.

Believe it or not, we are at the end of your first module in biology. Now you need to take a look at the study guide. On a separate sheet of paper, write out all of the definitions listed in the study guide, and answer any questions. Then, check your work with the solutions. When you are confident that you understand any mistakes you might have made, you are ready to take the test.

ANSWERS TO THE ON YOUR OWN QUESTIONS

1.1 a. <u>Carnivores</u> - Tigers eat only meat; thus, they are carnivores.
   b. <u>Herbivores</u> - Cows eat grass. This makes them herbivores.
   c. <u>Omnivores</u> - Humans eat plants and meat. This makes us omnivores.
   d. <u>Herbivores</u> - Sheep graze on grasses. This makes them herbivores.

1.2 a. <u>Producers</u> - Flowers have green stems and leaves to produce food via photosynthesis.
   b. <u>Decomposers</u> - Almost all fungi are decomposers.
   c. <u>Consumers</u> - Lions depend on other organisms for food.
   d. <u>Consumers</u> - Humans depend on other organisms for food.

1.3 <u>These organisms reproduce asexually</u>. If they reproduced sexually, then the offspring's traits would be a blend of both parents' traits. Since these offspring are identical to the organism that produced them, this must be asexual reproduction.

1.4 <u>Science cannot prove anything</u>. The best science can say is that all known data support a given statement. However, since data contradicting the statement might be uncovered, there is no way that science can prove anything.

1.5 <u>This is a hypothesis</u>. The explanation will have to be tested with a significant amount of data before it can even be considered a theory.

1.6 In a hierarchical classification scheme like ours, the further you go down the classification groups, the more similar the organisms within the groups become. This is because each group is made by splitting the previous group into smaller groups. Thus, since kingdoms are split into several phyla, we expect the organisms within the phyla to be more similar than those in the entire kingdom. Since family is several steps down from kingdom, <u>the organisms in the same family should be much more similar</u>.

1.7 Since going down the hierarchical scheme tells us that the organisms are getting more similar, going up the hierarchical should enhance the differences. Since class is one step higher than order, <u>the organisms from different classes should have more differences</u>.

1.8 <u>Protista</u> - This kingdom has the single-celled eukaryotes.

1.9 <u>Plantae</u> - Almost all autotrophs belong in this kingdom.

1.10 <u>Fungi</u> - Decomposers are in this kingdom.

ANSWERS TO EXPERIMENT 1.1

Number	Specimen	Specimen Classification		Numbers From The Key
1.	Moth	K. Animalia P. Arthropoda	C. Insecta O. Lepidoptera	1, 3, 5, 6, 7, 9, 14, 16, 17, 19
2.	Chipmunk	K. Animalia P. Chordata	C. Mammalia O. Rodentia	1, 3, 5, 6, 22, 23, 26, 28, 29, 31, 33, 34
3.	Grapes	K. Plantae P. Anthophyta	C. Dicotyledonae O.	1, 3, 4
4.	Swan	K. Animalia P. Chordata	C. Aves O.	1, 3, 5, 6, 22, 23, 26, 28
5.	Spider	K. Animalia P. Arthropoda	C. Arachnida O.	1, 3, 5, 6, 7, 9, 14, 15
6.	Lion	K. Animalia P. Chordata	C. Mammalia O. Carnivora	1, 3, 5, 6, 22, 23, 26, 28, 29, 31, 32
7.	North American Flag	K. Plantae P. Anthophyta	C. Monocotyledonae O.	1, 3, 4
8.	Fish	K. Animalia P. Chordata	C. Osteichthyes O.	1, 3, 5, 6, 22, 23, 24, 25
9.	Paramecium	K. Protista P.	C. O.	1, 2
10.	Mushroom	K. Fungi P.	C. O.	1, 3, 5
11.	Toad	K. Animalia P. Chordata	C. Amphibia O. Anura	1, 3, 5, 6, 22, 23, 26, 27
12.	Bacterium	K. Monera P.	C. O.	1, 2
13.	Deer	K. Animalia P. Chordata	C. Mammalia O. Artiodactyla	1, 3, 5, 6, 22, 23, 26, 28, 29, 30
14.	Grasshopper	K. Animalia P. Arthropoda	C. Insecta O. Orthoptera	1, 3, 5, 6, 7, 9, 14, 16, 17, 19, 20
15.	Gibbon	K. Animalia P. Chordata	C. Mammalia O. Primates	1, 3, 5, 6, 22, 23, 26, 28, 29, 31, 33, 35
16.	Zebra	K. Animalia P. Chordata	C. Mammalia O. Perissodactyla	1, 3, 5, 6, 22, 23, 26, 28, 29, 30

STUDY GUIDE FOR MODULE #1

1. On a separate sheet of paper, write down the definitions for the following terms. You will be expected to have them memorized for the test!

a. Metabolism
b. Photosynthesis
c. Herbivores
d. Carnivores
e. Omnivores
f. Producers
g. Consumers
h. Decomposers
i. Autotrophs
j. Heterotrophs
k. Receptors
l. Asexual reproduction
m. Sexual reproduction

n. Inheritance
o. Mutation
p. Hypothesis
q. Theory
r. Scientific law
s. Microorganism
t. Abiogenesis
u. Prokaryotic cell
v. Eukaryotic cell
w. Species
x. Binomial nomenclature
y. Taxonomy

2. What are the four criteria for life?

3. An organism is classified as an carnivore. Is it a heterotroph or an autotroph? Is it a producer, consumer, or decomposer?

4. An organism has receptors on tentacles that come out of its head. If those tentacles were cut off in an accident, what life function would be most hampered?

5. A parent and two offspring are studied. Although there are many similarities between the parent and the offspring, there are also some differences. Do these organisms reproduce sexually or asexually?

6. What is wrong with the following statement?

    "Science has proven that energy must always be conserved."

7. Briefly explain the scientific method.

8. Why does the story of spontaneous generation illustrate the limitations of science?

9. Where does the wise person place his or her faith: science or the Bible?

10. Why is the theory of abiogenesis just another example of the idea of spontaneous generation?

11. Name the classification groups in our hierarchical classification scheme in order.

✳ 12. An organism is a multicellular consumer made of eukaryotic cells. To what kingdom does it belong?

✳ 13. An organism is a single-celled consumer made of prokaryotic cells. To what kingdom does it belong?

14. Use the biological key in the appendix to classify the group of organisms living inside the log.

Photo from the EXPERT 3000 collection

Module #2: Kingdom Monera

[handwritten: is all bacteria is under this]
[handwritten: -All prokaryotic cells are under this]
[handwritten: (a cell w/ no distinct membrane)]

<u>Introduction</u>

With this module, you will begin to explore the incredible microscopic world that exists all around us. This world, unknown to human beings until 1675 when Anton van Leeuwenhoek crafted a crude magnifying lens, is home to an incredibly large number of microorganisms. Amazingly enough, the combined weight of all microscopic organisms far exceeds the combined weight of all other living organisms on earth! Bringing this fact a little closer to home, the number of organisms in kingdom Monera that live in your gut and on your skin is larger than the number of cells in your body! Thus, even though microorganisms are small, they are an important part of life on earth.

In this module, we will concentrate on the microorganisms that make up the kingdom Monera, which some biologists have renamed Prokaryota (pro kehr ee aht' uh) . Since, as we learned in Module #1, members of kingdom Monera are all composed of prokaryotic cells, this name does make sense. Nevertheless, in this course, we will still refer to the kingdom of prokaryotic cells by its traditional name Monera.

Organisms in kingdom Monera are interesting on many different levels. First of all, these organisms are not well-understood by biologists. There are several facets of their structure and function that we simply do not understand. Secondly, these organisms can survive in habitats that are deadly to other organisms. For example, they live on dust particles floating 6 kilometers (20,000 ft) above the surface of the earth. They thrive and multiply in temperatures that are too extreme for any other organism. In fact, microbiologists have found certain organisms from kingdom Monera living on the core of a nuclear reactor, *which can attain a temperature of up to 2000 °F*! Clearly, these interesting organisms are worth studying.

<u>Bacteria</u> *[handwritten: = all]*

The general name **bacteria** (singular is bacterium) can essentially be applied to all of the organisms in kingdom Monera. When you hear this term, you probably think about the disease and suffering caused by bacteria. Although bacteria are responsible for many illnesses that plague humanity (and other organisms), there are also some forms of bacteria which are beneficial to humans. For example, there are bacteria in your colon that help synthesize B vitamins and vitamin K which your body uses to stay healthy. To make the distinction between harmful and beneficial bacteria, we use the term **pathogen** (path' uh jen).

<u>Pathogen</u> - An organism that causes disease

Thus, a "pathogenic bacterium" is a bacterium that causes disease. It is important to realize, however, that there are many, many forms of bacteria which are not only non-pathogenic, but are, in fact, quite useful.

Bacteria from the genus *Lactobacillus* (lak' toh buh sil us) help us make cheese. When a starter bacterium is added to milk, it turns the milk into a chemical called lactic acid, and it begins to separate the milk solids (called "curds") from the liquids in the milk (called "whey"). If most of the whey is drained away, the moist curds left over are cottage cheese. Adding butter to the curds will make creamed cheese. Finally, if other bacteria are allowed to grow and breed on the curds, cheese forms. Cheddar cheese, for example, is the result of a certain bacteria type growing for a long period of time at moderate temperatures. If the bacteria type, temperature, or time of growth is changed, a different type of cheese will be produced. Cheese isn't the only thing that we use bacteria to help us make, however. Bacteria are used in the making of sauerkraut, vinegar, butter, and buttermilk.

Since bacteria belong to kingdom Monera, these organisms are prokaryotic. Their cells have no distinct, membrane-bound organelles. Nevertheless, the cell of a bacterium does have a few recognizable components. A sketch of a "typical" bacterium is given in Figure 2.1:

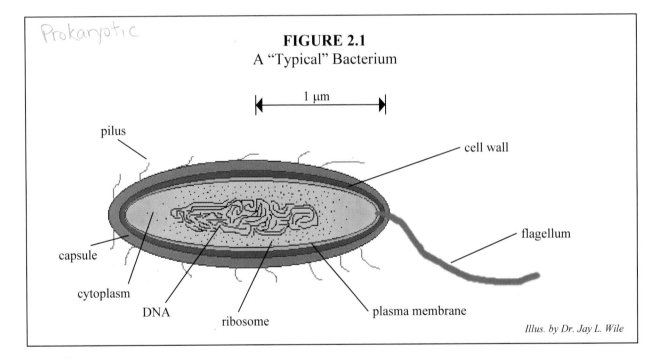

Prokaryotic

FIGURE 2.1
A "Typical" Bacterium

1 μm

pilus

cell wall

capsule

flagellum

cytoplasm

DNA

ribosome

plasma membrane

Illus. by Dr. Jay L. Wile

The word "typical" is in quotes because bacteria are so diverse that no sketch could be made which represents all bacteria. Thus, as you read through this discussion, you will find a lot of exceptions to this "typical" bacterium. Nevertheless, it is worthwhile to try and develop some general characteristics for bacteria, so we will discuss this figure in depth.

First of all, most (but not all) bacteria have a **cell wall**. This cell wall holds the contents of the bacterium together and holds the cell into one of three shapes:

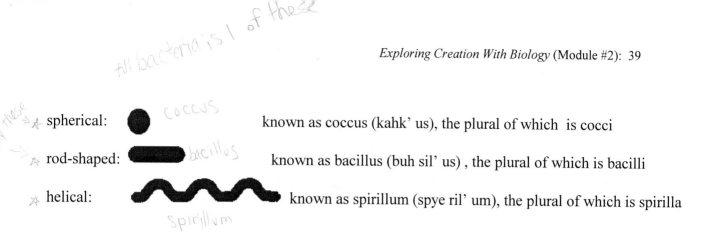

All bacteria is 1 of these (handwritten)

Mostly these (handwritten)

spherical: *coccus* (handwritten)   known as coccus (kahk' us), the plural of which is cocci

rod-shaped: *bacillus* (handwritten)   known as bacillus (buh sil' us) , the plural of which is bacilli

helical: *spirillum* (handwritten)   known as spirillum (spye ril' um), the plural of which is spirilla

Now remember, despite the fact that these are two-dimensional drawings, bacteria are three-dimensional, so there is depth to each of these shapes as well.

colonies - a group of individual bacteria (handwritten)

Most (but not all) bacteria that have a cell wall will also have a **capsule** that surrounds the cell wall. This capsule is usually made up of a sticky substance which helps the bacteria adhere to surfaces. For example, the bacteria that cause tooth decay are able to cling to teeth using their sticky capsules. The capsule usually does more than just that, however. In general, the capsule is a protective layer than tends to deter infection-fighting agents. When a pathogenic bacterium enters an organism, that organism has an infection-fighting mechanism which activates. The capsule of the pathogenic bacterium, however, can offer a significant amount of protection from the agents of this infection-fighting mechanism.

Underneath the bacterium's cell wall is the **plasma membrane**. As we will learn in Module #6, this membrane is structured much like a sandwich. Certain chemicals called phospholipids are "sandwiched" between other chemicals called proteins. These chemicals regulate what the bacterium takes in from the outside world. If the bacterium is taking in nutrients, the plasma membrane aids in the metabolism of those nutrients. In addition, it controls what substances are allowed to pass through the cell wall to the interior of the cell.

Inside the plasma membrane, we find a semi-fluid substance known as **cytoplasm** (sy' tuh plaz um). Cytoplasm exists throughout the interior of the cell, supporting the **DNA** and the **ribosomes** (rye' buh sohms). As we already learned, DNA holds all of the information required to make this mass of chemicals a living entity. Ribosomes, on the other hand, are chemical factories. They make special chemicals known as proteins, which we will take an in-depth look at when we reach Module #5. Floating throughout the cytoplasm are thousands and thousands of different types of chemicals which aid these parts of the cell in their tasks.

Outside of the cell wall and capsule, many bacteria (but not all) have **pili** (singular is pilus). Although these might look like legs or paddles, they are not used to move the bacterium about. Instead, pili are typically used for grasping. Either they grasp surfaces to help the bacteria adhere to them (as an aid to the sticky capsule), or they grasp other bacteria as a part of reproduction (for some bacteria).

Locomotion (moving the bacterium from one place to another) is accomplished with the **flagellum** (fluh jel' um), the plural of which is flagella. Not all bacteria have flagella, and some have more than one. If a bacterium does not possess a flagellum, then it is not capable of locomotion. The motion of substances in its surroundings might push it back and forth, but it is unable to move towards a goal. Now when you look at this flagellum, it looks a lot like a tail. Thus, you might think that the bacterium moves by swishing the flagellum back and forth. That is not correct, however. The flagellum is much more efficient than that.

The flagellum is truly a remarkable feat of engineering. It is composed of three parts: the **filament**, the **hook**, and the **basal body**. The filament (which is all that you see in Figure 2.1) attaches to the hook, a small tube shaped like an "L." The hook slides onto a rod that sticks through the cell wall. This rod, and all of the structures that attach it to the cell wall, make up the basal body. The rod and hook fit together so smoothly that the hook can actually spin around the rod in circles. When the bacterium wants to move, a series of amazingly complex chemical reactions make the hook spin. Since the filament is attached to the hook, it begins to spin as well. The design of this marvelous structure looks remarkably similar to the design of the turbines in a jet engine!

Now you might think at this point that the filament acts like a propeller and pushes the bacterium around. That wouldn't work well, however. The filament, in order for the bacterium to be able to move it easily, is very lightweight. As a result, it does not make a great propeller. However, when the flagellum spins in one direction, the *rest of the cell* spins in the other direction! When you take physics, you will learn that this is a consequence of the Law of Angular Momentum Conservation. The cell, then, becomes its own propeller, allowing the bacterium to move about at will!

Now before we go any further, we want you to go back and take a look at the length marked off in Figure 2.1. Basically, the figure indicates that 1 μm is about half the length of a bacterium. Of course, this little bit of information does you no good if you don't know what a "μm" is! The symbol "μm" stands for "micrometer," a metric length measurement. If you aren't familiar with the metric system, you will get a thorough introduction to it when you reach chemistry. For right now, you just need to know that a meter is about 3 feet in length, and a micrometer is one millionth of that size. Thus, bacteria are *really* small. In fact, to give you an idea of exactly how small bacteria are, 1,000 of them would easily fit side-by-side in the dot of this "i." Now *that's* small!

Let all of this information sink in for just a moment. In the last couple of pages, we have discussed some pretty complicated stuff. Bacteria have capsules that protect the cell and, along with the help of pili, allow it to attach to surfaces. They have a cell wall that keeps the shape of the bacterium. Inside the cell wall there is a plasma membrane that aids in metabolism and regulates the passage of substances in and out of the cytoplasm. They have DNA which stores all of the information needed for the cell to be alive, and they have the ability to copy this DNA in a mere 20 minutes without a single mistake! They also have ribosomes which can produce complicated chemicals called proteins. Finally, they have an incredibly advanced system of locomotion which, in fact, strongly resembles the design of a jet engine. All of these things are contained in a cell that is barely 2 μm long!

Now here's the really amazing part: *Bacteria represent the "simplest" form of life on earth!* Think about it. This tiny, complex organism is as simple as life gets. Are you beginning to see why the whole idea of spontaneous generation is so absurd? Whether we are talking about mice forming from sweaty shirts and grain or a "simple" life form being generated by random chemical reactions, life is just too complicated to form itself spontaneously. Even the simplest life form has obviously been designed by a Superintellect. Think about it this way. It took human beings nearly 3,000 years of scientific inquiry to develop the technology that could make a propeller which would move a boat. The simplest life form has a much more efficiently designed propeller as a part of its "standard equipment." Clearly, life has to be the result of highly-intelligent design. To believers, of course, this is rather obvious. Since God is all-powerful and all-knowing, it only makes sense that He could design and create some incredibly complicated life forms. What is really amazing is that scientists can study these life forms all of their lives and not recognize such a simple fact!

ON YOUR OWN

2.1 Two different species of bacteria attempt to infect an organism. One bacterium succeeds, while the other is destroyed by the organism's infection-fighting mechanisms. What is most likely the major difference between these two bacteria?

2.2 A bacterium is poisoned by a substance that is allowed into the interior of the cell. What bacterial component did not do its job?

2.3 If a bacterium cannot move, what structure is it missing?

Metabolism in Bacteria

As we already mentioned, there is really no such thing as a "typical" bacterium. From a cellular point of view, kingdom Monera is probably the most diverse kingdom in Creation. This is best illustrated by the diversity of metabolism among bacteria. Remember from Module #1 that all life must have a means of converting energy from its surroundings into energy that will sustain the processes necessary for life. We call this energy conversion "metabolism," and virtually every possible form of it exists in kingdom Monera.

The vast majority of bacteria are decomposers. Since decomposers rely on other (dead) organisms for food, we can also say that they are heterotrophs. These bacteria are referred to as **saprophytes** (sap' roh fytes).

Saprophyte - An organism that feeds on dead matter

Saprophytic bacteria are an integral part of nearly every aspect of nature. Without the aid of these microscopic decomposers, all of the materials on earth that are necessary for life would be contained in the bodies of only a few generations of dead organisms. Since bacteria decompose dead organisms, however, these materials can be recycled back into Creation so that they can be used to make new, living organisms.

Not all heterotrophic bacteria are saprophytic, however. Some bacteria (the ones that cause disease, for example) are **parasites** (pehr' uh sytes).

Parasite - An organism that feeds on a living host

Most parasitic bacteria lack the ability to digest nutrients, so they need to absorb nutrients that have already been digested. In addition, they often lack the ability to manufacture the complex chemicals necessary for life. As a result, they must also absorb these chemicals from their host.

Even though most bacteria are heterotrophic, there are many forms of autotrophic bacteria as well. In general, there are two different means by which autotrophic bacteria manufacture their own food: **photosynthesis** or **chemosynthesis**. In Module #1 you learned that photosynthesis uses the energy from sunlight along with certain chemicals to make food for the autotrophic organism. In green plants, the byproduct of photosynthesis is oxygen. The photosynthesis that occurs in some bacteria, however, does not have oxygen as a byproduct. This is because the chemicals that those bacteria use in photosynthesis are quite different than the chemicals used in the photosynthesis which is carried out by plants. Nevertheless, the byproducts of photosynthesis from these bacteria are useful to certain organisms. You'll learn more about the details of photosynthesis in Module #5.

Chemosynthetic bacteria use a different process for the manufacture of their food. The main difference between that process and photosynthesis is the source of energy. In chemosynthesis, rather than using energy that comes from sunlight, the bacteria promote chemical reactions which release energy. The bacteria then use that energy, along with another set of chemicals, to manufacture their food. Usually, the chemical reactions which provide energy to chemosynthetic bacteria also convert chemicals that living organisms can't use into chemicals that living organisms can use. Thus, even though there are only a few types of chemosynthetic bacteria, they perform an essential function for many living organisms.

Of course, once a bacterium (or any living organism) has food, there must be some process by which it can then convert that food into energy to be used in the support of life. This part of the metabolic process is called **respiration**.

Respiration - The process by which food is converted into useable energy for life functions

In human beings, we usually think of respiration as breathing; but for people, breathing is only one step in a long process that converts food into useful energy.

There are two types of respiration in nature: **aerobic** (ehr oh' bik) and **anaerobic** (an uh ro' bik).

Aerobic respiration - Respiration that requires oxygen

Anaerobic respiration - Respiration that does not require oxygen

Aerobic respiration is by far the most common form of respiration, which is why the earth must have a plentiful supply of oxygen in order to support life. There can be food aplenty but, without oxygen, there would be no aerobic respiration. Without respiration, all the food in the world would do you no good, because food is useless if it cannot be converted into useable energy.

There are certain types of bacteria (and other cells that we will learn about later), however, that do not use oxygen in their respiration. These bacteria are called anaerobic bacteria. Typically, they live in areas that are barren of oxygen such as deep in lakes or oceans or in the muck at the bottom of a swamp. These bacteria are essential to life, because they either decompose dead organisms or convert useless chemicals into chemicals that can be used by other life forms. So even though anaerobic respiration is uncommon, it is another essential part of the system that Creation uses to support life.

Before we leave this section, we want you to once again stop for a moment and think about what you have just learned. Bacteria have many different forms of metabolism, but they all, in some way, promote life on earth. Without the saprophytic bacteria, the chemicals necessary for life would not be recycled from dead organisms to live ones. As a result, in just a few generations, it would be impossible for life to exist on the planet. Without certain parasitic bacteria, many forms of life would not be able to fully digest their food. Photosynthetic (autotrophic) bacteria, on the other hand, produce certain useful chemicals that no other process on earth can produce. In addition, chemosynthetic (autotrophic) bacteria are adept at converting chemicals that are useless to living organisms into substances which are essential for life. Even pathogenic bacteria have a role in supporting life. Often, a group of organisms gets too large and could destroy the balance of nature. Typically, disease caused by pathogenic bacteria will kill off a large number of the organisms, making sure that the balance of nature remains intact.

Now if you think about it, these kinds of processes need to occur everywhere on earth. For example, when organisms die in a swamp they might sink to the bottom of the swamp and slowly be engulfed by the muck at the bottom. If there were no decomposing bacteria in that muck, the chemicals in that dead organism would never be recycled back into Creation. Because certain bacteria can be anaerobic, even though there is no oxygen down there, there are bacteria that will decompose the dead organism. Thus, the diverse metabolic procedures present in bacteria, along with the two different types of respiration, assure that there are bacteria everywhere doing their job.

Isn't that amazing? Bacteria are needed everywhere to do many different jobs in order to make sure that life can continue on earth. As a result, bacteria have many forms of metabolism and two forms of respiration so that they can, indeed, be everywhere that they are needed. Can you think of a more well-designed system for the support of life? Suppose a team of brilliant scientists spent their careers trying to devise a system that recycles the materials necessary for life, manufactures the chemicals essential for life, and assists other life forms in digestion. Do you think that they could come up with a plan this elegant and this well-designed? Of course not!

Indeed, a team of scientists tried to design and build a self-contained system for supporting life. It was called "Biosphere 2" and was supposed to be a microcosm of life on earth. It contained a variety of animals and plants and was designed to be completely self-supporting. The scientists thought they had it all worked out. They spent 7 years and 200 million dollars designing and building this air-tight, enclosed facility which spanned 3.15 acres in Arizona. Despite the best that technology and science had to offer, Biosphere 2 could not support life for even two years! After about 1 year and 4 months, oxygen levels could not be maintained. They had to start pumping oxygen in from the outside. Many of the animal species that had been put in Biosphere 2 became extinct. In the end, Biosphere 2 was a failure.

Of course, none of this should be surprising to anyone. The planet earth is an intricate web of hundreds of millions of processes that work together to support life. The best talent and technology that humans have could not possibly mimic what the earth does naturally. Why? The answer is very simple. An omniscient, omnipotent God designed the earth. He could foresee all of life's needs, even the tiny bacteria needed to support it. Limited human beings, on the other hand, just do not have the ability to design and create what God has designed and created, even on a very small scale. Although Biosphere 2 was a complete failure, it stands as a strong testament for the grandeur and elegance of God's Creation!

ON YOUR OWN

2.4 Can saprophytic bacteria be autotrophic?

2.5 Can an aerobic bacterium be chemosynthetic?

Asexual Reproduction in Bacteria

As you should recall from Module #1, another criterion for life is reproduction. For an organism to be considered alive, it must be able to reproduce so that life can continue. We also learned in Module #1 that there are two modes of reproduction: sexual and asexual. As you might expect, bacteria can engage in both. Since asexual reproduction is, by far, the most common form of reproduction among bacteria, we will start there. Figure 2.2 on the next page illustrates this asexual reproduction.

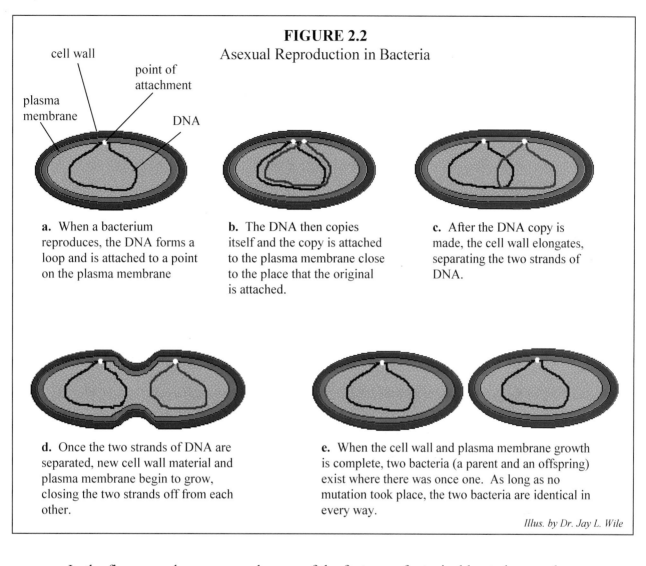

FIGURE 2.2
Asexual Reproduction in Bacteria

cell wall

point of
attachment

plasma
membrane

DNA

a. When a bacterium reproduces, the DNA forms a loop and is attached to a point on the plasma membrane

b. The DNA then copies itself and the copy is attached to the plasma membrane close to the place that the original is attached.

c. After the DNA copy is made, the cell wall elongates, separating the two strands of DNA.

d. Once the two strands of DNA are separated, new cell wall material and plasma membrane begin to grow, closing the two strands off from each other.

e. When the cell wall and plasma membrane growth is complete, two bacteria (a parent and an offspring) exist where there was once one. As long as no mutation took place, the two bacteria are identical in every way.

Illus. by Dr. Jay L. Wile

In the figure, we have removed many of the features of a typical bacterium so that we can concentrate on just those parts that are essential to reproduction. Thus, the only bacterial components shown in the figure are the cell wall, the plasma membrane, and the DNA. As you can see from the figure, reproduction begins when the DNA forms a loop attached to the plasma membrane. The DNA then copies itself. It should make sense to you that in order to reproduce, the bacterium must create new DNA. After all, DNA contains the information that helps to organize chemicals into a living organism. Since reproduction forms a new living organism, new DNA must be formed.

Once the DNA is copied, the copy attaches to the plasma membrane close to the original. The cell wall and plasma membrane then elongate, separating the copy from the original. Once they are separated, the cell wall and plasma membrane begin to grow. They grow in between the two strands of DNA, eventually closing them off from one another. At that point, there are two bacteria. As we learned in Module #1, asexual reproduction usually results in offspring that are identical to the parents. Now you should see why. The DNA in the offspring (sometimes called the "daughter") is the same as the DNA in the parent. Since DNA holds all of the information that makes up an organism, the two bacteria must be identical. If, however, a mistake was made while the DNA was being copied, then there will be a marked difference between the offspring and the parent. This is what we called a "mutation" in Module #1. Thus, mutations in asexual reproduction result when the DNA of the parent is not copied correctly.

One of the interesting aspects of bacterial reproduction is the speed at which it occurs. Under ideal conditions, a bacterium can grow to full size and divide in about 30 minutes. Of course, once it divides, the new bacterium *and the old one* can divide again in about 30 minutes. If this process continued indefinitely, one bacterium could multiply into more than a billion bacteria in about 15 hours. If this continued for a whole week, the bacteria formed from that first, single bacterium would have a combined weight that is larger than the entire planet! A situation like this can never occur, however, because of several factors.

Suppose, for example, a saprophytic bacterium found its way to a dead cow. The bacterium would feed on the cow and begin to reproduce rapidly. In just a day or so, there would be billions and billions of them. They would begin to spread out across the remains of the cow, so that they could all have access to the nutrients that they need. At some point, however, bacteria would cover all of the remains of the cow. What then? Well, some bacteria would be cut off from their food supply, and they would die, decreasing the population. Some bacteria would actually try to eat other bacteria, which would further decrease the population. Now, of course, while all this is going on, bacteria would still be reproducing, so new bacteria would be forming and other bacteria would be dying. For a while, the bacteria population would reach a point where the number of new bacteria that form equals the number of bacteria that die. We call that the **steady state** of the population.

Steady state - A state in which members of a population die as quickly as new members are born

The bacteria population would reach a steady state for a while, but then something would happen. The cow remains would start to dwindle due to all of the bacteria feeding on them. At some point, the remains of the cow would not support the population, and more bacteria would begin to starve to death. In addition, the incidents of bacteria eating other bacteria would increase. Thus, the population of bacteria would dwindle along with the remains of the cow. Eventually, when all of the cow remains were gone, all of the bacteria would be dead. This situation is illustrated in Figure 2.3.

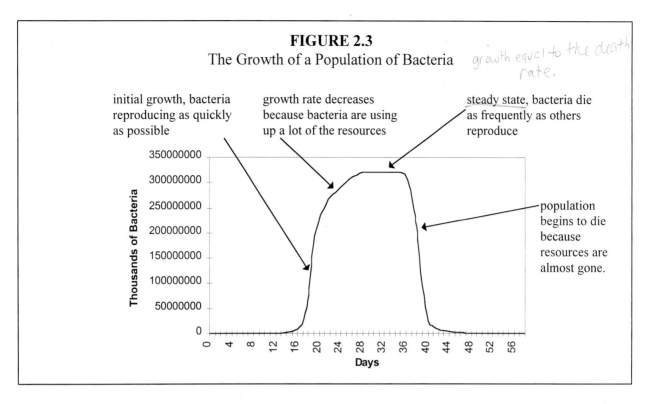

FIGURE 2.3
The Growth of a Population of Bacteria

growth equal to the death rate.

initial growth, bacteria reproducing as quickly as possible

growth rate decreases because bacteria are using up a lot of the resources

steady state, bacteria die as frequently as others reproduce

population begins to die because resources are almost gone.

So, even though bacteria can populate an area quickly, a limitation of resources will eventually slow the growth of the population until, when all resources run out, the population dies.

ON YOUR OWN

2.6 Many bacteria populations are rather fragile. If conditions change in their habitat, they often die quickly. Based on what you have just learned, develop an hypothesis for why this is the case.

2.7 A population of bacteria reaches a steady state and then, after several days, the population actually increases dramatically. What could cause such an event?

Sexual Reproduction in Bacteria

Sexual reproduction in kingdom Monera is quite different than the sexual reproduction found in other kingdoms. As a result, we give it a slightly different name: **conjugation**.

Conjugation - A temporary union of two organisms for the purpose of DNA transfer

How is conjugation different from sexual reproduction? First of all, when two bacteria sexually reproduce, no offspring is formed. Wait a minute, you might think, doesn't reproduction always result in an offspring? Well, in most cases it does. However, in kingdom Monera, it does not. Why do we call it reproduction, then? Well, even though no offspring is formed, traits are passed on from one bacterium to another. Since traits are passed on, we consider this a form of reproduction.

How does this strange process occur? Well, in a population of bacteria, there may be a few individuals with traits that are different than the others. If these traits are desirable, then it would benefit the population if they were passed on to its other members. In order to do this, sexual reproduction occurs. The desirable traits that are going to be passed on are usually contained in a small, circular strand of DNA called a **plasmid**. The plasmid is an extra component, separate from the rest of the bacterium's DNA.

Plasmid - A small, circular section of extra DNA that confers one or more traits to a bacterium

If the traits conferred to the bacterium are beneficial, then that bacterium (usually referred to as the "donor"), will sexually reproduce with a bacterium that does not possess these traits (usually referred to as the "recipient"). The mechanism of this form of reproduction is illustrated in Figure 2.4.

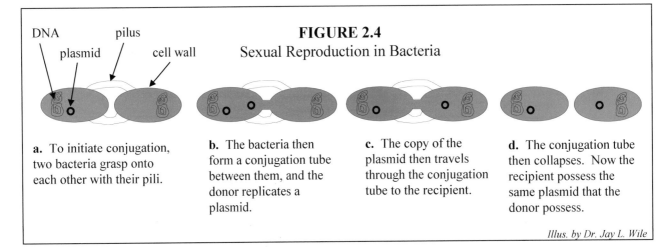

FIGURE 2.4
Sexual Reproduction in Bacteria

DNA   pilus
   plasmid   cell wall

a. To initiate conjugation, two bacteria grasp onto each other with their pili.

b. The bacteria then form a conjugation tube between them, and the donor replicates a plasmid.

c. The copy of the plasmid then travels through the conjugation tube to the recipient.

d. The conjugation tube then collapses. Now the recipient possess the same plasmid that the donor possess.

Illus. by Dr. Jay L. Wile

Once again, in order to make this figure a little more clear, we have removed most of the components of the bacteria. In this figure, we show only the cell wall, the DNA, a plasmid, and some pili.

The first step in sexual reproduction among bacteria involves the donor and recipient grasping onto one another using their pili. Typically, they use many more pili than are pictured here, but for clarity we have only shown a few. Interestingly enough, we still have no idea *how* the bacteria know when and with whom to sexually reproduce. In fact, we don't even know which bacterium initiates sexual reproduction. Does the donor somehow "sense" that the recipient lacks the traits it needs and therefore initiates reproduction? Instead, does the recipient somehow "sense" that the donor has a trait it needs? Is there some mutual initiation? This is still an area of intense research.

Once the bacteria grasp onto each other, the donor replicates the plasmid that the recipient needs. At the same time, a **conjugation tube** is formed that allows components from one bacterium to be transferred to the other. The copied plasmid then travels down the conjugation tube into the recipient bacterium. Once the recipient has the plasmid that it needs, the conjugation tube collapses, and the bacteria are separated.

Besides the fact that there is no real offspring formed by this union, there is another difference between conjugation among bacteria and sexual reproduction among the vast majority of organisms in Creation. In most cases of sexual reproduction, there is a mutual exchange of DNA. Each parent contributes DNA, and the offspring has a blend of characteristics from both of them. In bacterial conjugation, this is not the case. No DNA travels from the recipient to the donor. Thus, the traits conferred by the donor's plasmid are given to the recipient, and the recipient gives nothing back. Rather than being a mutual exchange of DNA, then, bacterial conjugation involves the traits of the donor being passed on directly to the recipient.

ON YOUR OWN

2.8 A population of bacteria are living in a lake. Due to volcanic activity nearby, the lake's temperature begins to increase. In the population, there are some bacteria that possess a resistance to low temperatures (call them type A) and another type that are resistant to high temperature (call them type B). Which type will be the donor and which the recipient as the population begins to sexually reproduce?

Other Forms of Bacterial "Reproduction"

If conjugation wasn't strange enough for you as a means of reproduction, there are two other (even stranger) ways that bacteria can "reproduce." The first is called **transformation**.

Transformation - The transfer of a "naked" DNA segment from a nonfunctional donor cell to that of a functional recipient cell

If a bacterium dies, its cell wall falls apart and the components of the cell (including the DNA) flow into the surroundings. By yet another process that we do not understand, living bacteria might "sense" that the DNA from the dead bacteria contains traits that they could use. They then absorb the DNA, like they are taking it in as food. Instead of eating it, however, the bacteria then incorporate it as a plasmid in their DNA. Of course, once a bacterium has such a plasmid, it might then engage in conjugation to pass the plasmid on to other bacteria. Transformation, then, is still considered a reproductive mechanism because traits have been passed from one bacterium (a dead one) to another bacterium (a living one). That living bacterium, then, can continue to pass on these traits.

The last "reproductive" mechanism is not really reproduction at all. It does not pass any traits on from one bacterium to another. Nevertheless, it is a means by which bacteria survive as a population. Thus, we group it under reproduction. In this process, a bacterium that is experiencing extreme conditions (such as high temperatures) can form an **endospore**.

Endospore - The DNA of a bacterium that is coated with several hard layers

This endospore, which forms in the bacterium's plasma membrane, is a means that the bacterium uses to survive conditions which it otherwise could not.

An endospore can withstand extreme situations for a lot longer than the bacterium itself can. For example, endospores can withstand several hours of boiling or freezing. In addition, they can withstand up to a few days of extreme dryness, pressure, or exposure to toxic chemicals. If the endospore can survive through the extreme situation, when favorable conditions return, the hard layers surrounding the DNA will deteriorate, and the cell will burst from the endospore, ready to grow and reproduce again. Once again, our understanding of how bacteria know when to form endospores and our knowledge of their formation are still incredibly limited. Much exciting research is being done in these areas.

Bacterial Colonies

Up to this point, our discussion has focused on bacteria as individuals. However, many bacteria exist in colonies. Remember in Module #1 when we discussed a "simple association" of single-celled creatures? A colony is an example of such a simple association. In a bacterial colony, the individual bacteria group together, but they all still exist as individuals. If the colony gets broken apart, the individual bacteria can still live and function. Thus, they do not *have* to be a part of the colony to survive. Nevertheless, since there is strength in numbers, the bacteria's ability to survive is usually enhanced when they form a colony.

As you once again might expect, bacteria form colonies in a variety of different ways. In our discussion of a "typical" bacterium, we learned that bacteria take on one of three basic shapes: coccus (spherical), bacillus (rod-shaped), and spirillum (helical). The most common bacterial colonies are made up of either cocci or bacilli bacteria. Figure 2.5 shows the most common forms of bacterial colonies that exist.

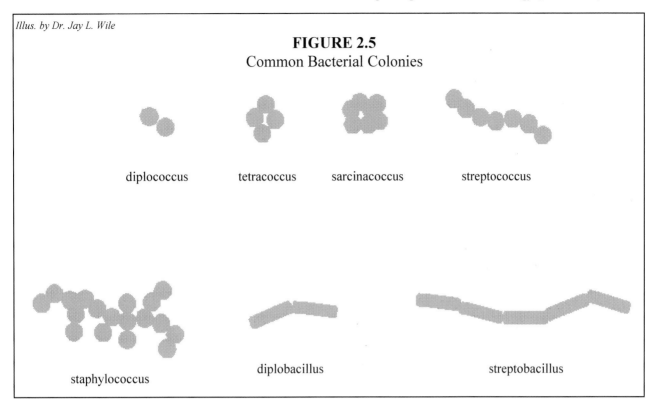

FIGURE 2.5
Common Bacterial Colonies

Illus. by Dr. Jay L. Wile

diplococcus   tetracoccus   sarcinacoccus   streptococcus

staphylococcus   diplobacillus   streptobacillus

Notice that the shape of the bacteria is contained in these names. If the bacteria are spherical, the name ends in "coccus." If they are rod-shaped, the name ends in "bacillus." This is often the case. Bacteria names often contain the bacterium shape so that you automatically know something about the organism when you see its name. Although we do not require you to memorize the names in Figure 2.5, we do expect you to be able to determine the shape of a bacterium when it is a part of the name. Thus, we require you to be able to answer questions such as the following "on your own" problem.

ON YOUR OWN

2.9 A bacterial colony is called staphylobacillus. What shape do the bacteria in the colony possess: spherical, rod-shaped, or helical?

As we mentioned before, a bacterial colony is really just a simple association of individual bacteria. Each bacterium in the colony could survive on its own if separated from the rest of the colony. Despite this fact, however, some colonies do possess strikingly collective behavior. Some colonies, for example, will secrete a capsule-like substance that surrounds and protects the entire colony. This substance gives the bacterial colony a slimy feel, and it holds the colony together. Bacterial colonies encapsulated like this often look like bluish-green mats floating on the top of water. Stagnant ponds are a common place to find such colonies. Other colonies have bacteria that actually work together to capture and eat prey. A *Myxococcus xanthus* colony, for example, moves as a unit in search of prey, typically other bacterial colonies.

When they find their prey, they engulf it and, as a group, secrete a substance that digests it. The entire colony then feeds on the nutrients.

Another interesting aspect about the bacteria that make up colonies is the fact that their DNA seems to tell them what colony they should form. In other words, if two spherical bacteria end up close to one another, they will not necessary form a diplococcus colony. Their DNA determines whether they want to exist as an independent bacterium or as the member of a colony. Furthermore, if the bacterium has the DNA of a colony-dweller, the DNA determines what kind of colony it will live in. Thus, if a streptococcus bacterium is separated from its colony, it won't join up with just any other colony. It will continue to live independently until it can find another streptococcus colony. If it never finds one, it will live out its life as an independent bacterium.

Before we move on to the next section, we need to begin an experiment that will last through the next module. Most of the experiment deals with collecting specimens that we will later view under a microscope. If you do not have a microscope, you should still visit a pond and study the ecosystem, making drawings in your lab notebook of all that you see. You just don't need to collect the samples.

Experiment 2.1
Pond Life A

Supplies:
- Four jars with lids (The brown glass or plastic jars that creamer comes in work best.)
- Small amount of chopped hay (dried grass will work, but not as well)
- White rice
- Egg yolk
- Small amount of rich soil
- Long handled ladle (A good one can be made by attaching a kitchen ladle to a broom handle with duct tape.)
- A pond or small body of water (A still creek will do in a pinch, but you won't see all of the specimens that we would like you to see.)
- Something to rest your lab notebook on while you draw in it

Object: To study the ecosystem of a pond and collect specimens for the next two experiments

Procedure:

A. After locating a proper body of water, plan a field trip of one to two hours.

B. Before leaving, prepare your four jars as follows:

1. Label one jar hay and place a tablespoon of hay in the jar.
2. Label one jar rice and place a teaspoon of rice in it.
3. Label the third jar egg yolk and place 1/4 teaspoon egg yolk in it.
4. Label the last jar soil and place 2 teaspoons of soil in it.

C. At the pond, use your dipper to collect pond water. Take your samples near the bottom of the pond. Fill each of the four jars 1/2 full with the water. Put the lids back on the jars when you are done.

D. Set the jars aside and walk slowly around the pond, sitting occasionally to observe everything that is about you. Make sure you not only look, but also listen carefully.

E. As you note those things about you, draw each of them in your notebook. Those that wish to do more detail could use colored pencils. Don't forget to look under rocks and other hiding places.

F. Take the jars home and place them in an area of subdued light. The items that you placed in the jars are food for the microorganisms that are present in the pond water. As time goes on, the microorganisms will grow and reproduce, increasing the population. When we take a few microorganisms and try to increase their population this way, biologists say that we are **culturing** the microorganisms. Thus, these jars are often called **cultures**, because microorganisms are being cultured inside them.

G. The cultures are to be used in the next two experiments. They will reach their peak of growth usually between day 3 and 5, but even after 2 weeks some organisms will still be active.

Classification in Kingdom Monera

Now that we have discussed aspects that are common to many of the organisms in kingdom Monera, it is time to start talking about the differences between individual members. To do this, we will discuss how to classify bacteria. Now one of the first things that you must learn is that since kingdom Monera is so incredibly diverse, the classification of its members is a rather difficult task. As a result, we will not delve as deeply as we might into the classification groups that make up kingdom Monera. In fact, our discussion of classification will not go much deeper than splitting bacteria into classes. Thus, we will not really talk about the different orders, families, genera, and species classifications that exist in kingdom Monera.

To learn how we split bacteria into different phyla, you need to learn a little history. In the late 1800's, a Danish physician named Hans Christian Gram was studying bacteria. In order to make them show up better under a microscope, he developed several different types of stains. In Experiment 1.2, you should have performed your own stain in order to make your cheek cells show up better under the microscope. This was the kind of staining that Gram was doing. He noticed that for a certain type of stain, later called the **Gram stain**, certain bacteria looked blue when viewed under the microscope whereas others looked red. Since Gram quickly realized that this could be used as a means of classifying bacteria, he began studying many types of bacteria and calling them **Gram-Negative** (if they looked red after the Gram stain) or **Gram-Positive** (if they looked blue following the Gram stain).

Now realize that Gram's classification was completely arbitrary. He had no idea what was causing some bacteria to be red and others to be blue. He just assumed that there was some fundamental aspect of the bacterium that caused it to react with the Gram stain in a certain way. Thus, he assumed that this would be a way of classifying bacteria. Well, it turns out that he was right. Today, we still use the terms "Gram-negative" and "Gram-positive" to classify bacteria. The only real difference is that today, we know *why* bacteria react positively or negatively to the Gram stain. The difference in reaction to Gram stain is caused by differences in the cell walls of these bacteria. Although those differences are a bit too detailed to go into in this course, we can say that bacteria which are Gram-positive have cell walls which retain the Gram stain so that it stays in the cell. Gram-negative bacteria, however, have cell walls that do not retain the Gram stain. As a result, the stain leaves those cells.

The first way we separate the organisms in kingdom Monera, then, is by their cell walls. If a bacterium has a complex cell wall (thus it is Gram-negative), it belongs to phylum Gracilicutes (gruh' sil uh kyoo' teez). If the cell wall is rather simple (thus the bacterium is Gram-positive), the bacterium belongs to phylum Firmicutes (fir' muh kyoo teez). Since the time of Hans Christian Gram, we have determined two other ways to classify bacteria. Some bacteria have no cell wall at all. We put these bacteria into phylum Tenericutes (ten' uh ruh kyoo' teez). Finally, some bacteria possess a cell wall, but the compounds which form these walls are rather different than the compounds that form the cell walls of Gram-positive and Gram-negative bacteria. Thus, we give them a separate phylum, called Mendosicutes (men' duh suh kyoo' teez).

You need to be a bit familiar with these phyla, so please complete the following "on your own" problems to make sure you understand each phyla and its characteristics.

ON YOUR OWN

2.10 A bacterium has no cell wall. To what phylum does it belong?

2.11 A bacterium is classified as Gram-positive. To what phylum does it belong?

2.12 A bacterium has a complex cell wall. To what phylum does it belong?

2.13 A bacterium has a cell wall that is different from both Gram-positive and Gram-negative bacteria. To what phylum does it belong?

As we know from our mnemonic, "King Philip cried out, 'for goodness sake!'," we take the organisms from each phylum and separate them into individual classes. Phylum Gracilicutes (Gram-negative bacteria) has three classes. We separate members of this phylum based on their metabolism. The first, class Scotobacteria (skoh' toh bak tehr'ee uh), is composed of the non-photosynthetic bacteria. The bacterium that causes Lyme disease (*Borrelia burgdorferi*) is a member of this class. In fact, many pathogenic bacteria can be found here. The second class,

Anoxyphotobacteria (an' ox ee foh' toh bak tehr'ee uh), is composed of photosynthetic bacteria that do not produce oxygen. Typically, these bacteria live in the sediments of lakes or rivers. The last class, Oxyphotobacteria (ox' ee foh' toh bak tehr'ee uh), is comprised of photosynthetic bacteria that produce oxygen. The most common example of this class is the blue-green algae (known as cyanobacteria) that you see floating on the top of stagnant ponds.

In phylum Firmicutes, there are only two classes. We do not separate the bacteria into these classes based on their metabolism; rather, we do so based on their shape. Class Firmibacteria (fir' muh bak tehr'ee uh). is comprised of cocci and bacilli, while class Thallobacteria (thal' oh bak tehr'ee uh) is made up of any other shape. One of the bacteria that cause pneumonia (*Streptococcus pnemoniae*) is an example of a Firmibacteria while many of the members of class Thallobacteria are responsible for dental and gum disease.

You'll be happy to know that the last two phyla (Tenericutes and Mendosicutes) have only one class each. This is because there are few bacteria that fall into these phyla. Phylum Tenericutes has the class Mollicutes (mol' uh kyoo teez) which contains another kind of pneumonia-causing bacteria, and phylum Mendosicutes contains the class Archaebacteria (ar chee' uh bak tehr'ee uh), which holds all of the bacteria with exotic cell walls. These exotic cell walls allow members of class Archaebacteria to live in rather odd locations, such as deep-ocean hydrothermal vents or incredibly brackish seas like the Great Salt Lake. Many places that are uninhabitable to other organisms will be populated with members of this class.

Now working through this classification scheme might have been a little tedious, but it is important for you to get an overview of the different types of bacteria that exist. The classification scheme, as we have discussed it, is summarized in Table 2.1:

TABLE 2.1
Partial Classification of Kingdom Monera
(taken from *Bergy's Manual of Systematic Bacteriology,* 9th Edition)

Phylum Gracilicutes: Gram-negative bacteria
    Class Scotobacteria: Non-photosynthetic bacteria
    Class Anoxyphotobacteria: Non-oxygen producing photosynthetic bacteria
    Class Oxyphotobacteria: Oxygen-producing photosynthetic bacteria

Phylum Firmicutes: Gram-positive bacteria
    Class Firmibacteria: Rods or cocci
    Class Thallobacteria: All others that are not rods or cocci

Phylum Tenericutes: Bacteria lacking a cell wall
    Class Mollicutes: The only class in this phylum

Phylum Mendosicutes: Bacteria with exotic cell walls
    Class: Archaebacteria: The only class in this phylum

ON YOUR OWN

2.14   Construct a biological key that separates bacteria into the different classes. You can assume that the only organisms the key will be used to analyze are bacteria. (HINT: Key one should determine whether or not the bacterium has a cell wall.)

Specific Bacteria

Now that we've gone through all of the generalities, it is time to look at some specific types of bacteria. It is important for you to realize that what follows is just a sampling of the many different kinds of bacteria that exist. Entire books have been written devoted solely to discussing the various species of bacteria and what they do. We, of course, will not go into that kind of detail. Rather, we discuss these bacteria because either we think that they are important representatives of kingdom Monera, or because we think that you might be partly familiar with them.

Class Archaebacteria

As we mentioned before, the members of this class contain exotic substances in their cell walls that allow them to live in conditions that would be uninhabitable to any other organism. For example, a species of bacteria from this class was found living on the core of a nuclear reactor. Anaerobic chemosynthetic bacteria also belong to this class. Certain species of these bacteria live in the muck at the bottom of marshes and swamps. Because they are anaerobic, they do not use oxygen in their respiration. Instead, they use hydrogen and carbon dioxide. One of the byproducts of this respiration is methane, a pungent-smelling gas. If you have ever passed a swamp or marsh and smelled a foul odor that someone called "marsh gas," that was the methane produced by these bacteria. Under the right conditions, methane is explosive. Many campers and outdoorsmen have died because they struck a match in a swamp or marsh at the wrong time!

Blue-Green Algae (al' jee)

These organisms live in freshwater lakes and ponds and are most prevalent in stagnant waters. They form the blue-green mats that float on the surface of the water. Later on, you will learn that algae are, in fact, colonies of eukaryotic cells. However, since these organisms looked like other algae, they were at first incorrectly named as such. Today, we recognize them as colonies of prokaryotic cells and thus classify them in kingdom Monera. The new (better) name for them is **cyanobacteria** (sye' an oh bak tehr'ee uh). This properly identifies them as bacteria. Nevertheless, some biologists still refer to them as "blue-green algae," so you need to be aware of both names.

Cyanobacteria are photosynthetic, and are one of the few types of bacteria that require light in order to survive. Many bacteria cannot live under intense light, and most like their habitat to be dim or completely dark. Since cyanobacteria require light energy for metabolism, however, they represent a notable exception. These bacteria are also colonial, living in long, thin strands of cells. The strands encapsulate themselves, giving cyanobacteria a slimy feel.

Clostridium (claw strid' ee um) *botulinum* (bot' yool in um)

This bacterium, which belongs to class Firmibacteria in phylum Firmicutes, is one of the principal agents of food poisoning. If food has not been properly cooked to rid it of bacteria, this is one of the most likely contaminates. Eating food with this bacteria will cause botulism. Mild cases of botulism involve severe nausea, diarrhea, and high fever. However, severe cases of botulism can cause respiratory failure and death.

Salmonella (sa muh nihl' uh) *typhimurium* (tye' fim ur ee um) **and *S. enteriditis*** (en' ter uh dye' tus)

These two species belong to the same genus (remember, if we write the genus name once we can abbreviate it afterwards). We place them in phylum Gracilicutes and class Scotobacteria. You have probably heard them both referred to by their genus name, *Salmonella*. These bacteria are common contaminates of eggs and poultry. Most people are infected by one of these two bacteria because the poultry or eggs that they have eaten have not been cooked thoroughly. Since these bacteria are present in nearly every fowl, about the only way to make sure you get rid of them is by really cooking the meat or eggs thoroughly. The extreme heat will then kill the bacteria. Once again, while mild cases of *Salmonella* poisoning cause just nausea and diarrhea, severe cases can cause death.

Escherichia (es' kur ee kee uh) *coli* (koh' lye)

This species also belongs in phylum Gracilicutes and class Scotobacteria. It is a very common bacteria, living in your gut. One of the by-products of *E. coli*'s metabolism is Vitamin K, a substance your body needs. It also secretes a chemical that helps your body digest fat. In addition, its activities in your gut actually keep food-borne pathogenic bacteria from colonizing there. Thus, the presence of this bacterium in your gut keeps out other, pathogenic bacteria!

Although the *E. coli* bacterium found in your gut is non-pathogenic, there are pathogenic forms of *E. coli*. If you are infected with pathogenic *E. coli*, it can give you severe diarrhea. Infants infected with such *E. coli* can actually die of dehydration due to the diarrhea. In some developing countries, this is the leading cause of infant death! You see, when bacteria develop in different habitats, they develop different traits. We will discuss this more in a later module. For right now, you just have to realize that the traits one organism has might be different than the traits of another organism *from the same species,* if these organisms developed in different habitats. As a result, not all *E. coli* are the same.

When two organisms from the same species have markedly different traits, we say that they are from two different **strains**.

<u>Strains</u> - Organisms from the same species that have markedly different traits

We will learn in a later module why strains develop. For right now, just realize that there can be many strains of a bacterial species. The different traits between these strains can be so extreme that one strain is beneficial while another strain is pathogenic.

Conditions For Bacterial Growth

Even though bacteria are hardy creatures and can survive in a lot of habitats, there are still some conditions that must be present in order for bacteria populations to grow. These are:

1. *Moisture* - Almost all bacteria require some moisture to grow. Some grow best submerged in water while others can only survive underwater for a few minutes. Nevertheless, almost all bacteria need some level of moisture. It is important to note, however, that the capsules which some bacteria form will protect them during periods of dryness. In addition, if a bacterium can form an endospore, it can withstand even longer dry spells.

2. *Moderate temperature* - Most bacteria prefer moderate temperatures, 27 $^{\circ}$C - 38 $^{\circ}$C (80 $^{\circ}$F - 100 $^{\circ}$F). This makes summer the most ideal time for bacterial growth. Once again, it is worth noting that many bacteria (especially those from class Archaebacteria) can survive in quite extreme temperatures.

3. *Nutrition* - Bacteria must have nutrition in order to grow. Since most are heterotrophic, removing the nutrition will destroy them. In the case of photosynthetic bacteria, you must remove light in order to kill them. Chemosynthetic bacteria can be killed by removing the chemicals that they convert to energy.

4. *Darkness* - Most bacteria grow best in darkness. There are, of course, exceptions to this rule, photosynthetic bacteria being the obvious one.

Once again, since there is great diversity among bacteria, these conditions are only the most popular conditions for bacterial growth. In general, we find that bacteria are very specialized in terms of their habitat. One bacterium might be able to withstand extreme temperatures, but it might not be able to survive in even the slightest light. Others might not need oxygen at all, but, if you introduce oxygen into the environment, it will kill them. Thus, unlike many species, bacteria often can only live in a certain habitat. Take them out of that habitat, and they will die.

Preventing Bacterial Infections

Since many forms of bacteria are pathogenic, it is worthwhile to spend a few minutes discussing how to prevent bacterial infections. A common means of becoming infected by bacteria is through the food we eat. Since bacteria are everywhere, even in the dust particles that fly through the air, all food is exposed to bacteria. One way we can eliminate the bacteria is to expose the food to extreme heat or radiation. This will kill the bacteria and then, if we seal the food in a container before exposing it to fresh air, no bacteria will be present. This is what happens in canned foods. Thus, canned foods can sit on a shelf for a long time, because all bacteria that was in the food has been destroyed, and the can does not let any air in, so no new bacteria can get to the food.

Another way to prevent bacterial infection in food is to dehydrate it. When a food is dehydrated, almost all of the water is removed. Since moisture is a condition for bacterial growth, as long as the dehydrated food stays dry, no bacteria can grow. Many campers who stay in the wilderness for a long time take dehydrated food with them. These foods are lightweight and bacteria-free, making them ideal for long trips.

If you can't dehydrate or can something, then you can freeze it. Prolonged freezing will kill many of the bacteria that exist in the food. Of course, many foods do not taste good after freezing, so it is not the solution for all kinds of food. If, instead, you put food in the refrigerator, the rate of bacterial growth will slow, but it will not stop. Thus, food placed in the refrigerator will only keep for a short time.

One of the most bacteria-prone foods that exists is milk. From the time milk leaves the cow, it is laden with bacteria. If something is not done, the bacteria count will rise to a dangerous level in just a couple of days, even if it is stored in a refrigerator. If you heat milk to a temperature sufficient for killing the bacteria, the milk curdles. This used to be a real problem. Louis Pasteur (the scientist who debunked the law of spontaneous generation), however, developed a process called **pasteurization**, which he originally used to keep wine from souring. This process is now applied to milk, hence the term "pasteurized milk."

In the pasteurization process, milk is either heated to a moderate 63 $^{\circ}$C (145 $^{\circ}$F) for thirty minutes or 72 $^{\circ}$C (161 $^{\circ}$F) for 15 seconds. This temperature is not sufficient to kill the bacteria, however, after the heating is finished, the milk is then quickly cooled to refrigerator temperature. This rapid change in temperature is sufficient for killing about 95% of all bacteria in the milk, because the bacteria do not have time to adapt to the new, cooler temperature. Thus, with most of the bacteria gone, it takes many more days for the milk to sour. Still, exposure to air will allow bacteria back into the milk, so it should be covered and refrigerated at all times.

Even though we have come to the end of the second module, we are not done with the microscopic world. In the next module, we will look into the tiny world of kingdom Protista. To give you an introduction into this world, as well as to let you see some live bacteria like those that we have been studying, perform the following experiment. If you do not have a microscope, you should still spend some time looking over the drawings in the experiment. This will give you some idea of what kinds of microorganisms exist.

Experiment 2.2
Pond Life B

Supplies:

- Microscope
- Slides
- Coverslips for the slides
- 4 jars of water collected at the pond (Remember, they are called **cultures.**)
- 4 droppers (one for each jar)
- Small amount of cotton

Object: You have studied the things that you were able to see by just using your eyes and ears while you were at the pond. This would be your macroscopic observation of the pond. In this experiment, you will begin making microscopic observations of the pond.

Procedure:

A.   Set up the microscope as described in Experiment 1.2.

B.   Prepare a slide by adding a drop or two of your first culture (it doesn't matter which one) to the slide and add a coverslip (carefully). When you open each jar, be prepared for a *mighty* stench!

C.   Begin scanning the slide on low power moving the slide slowly by creating a pattern as you move the slide. A good pattern is to use is as follows:

First, scan up and down the slide

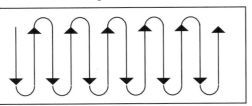

Next, scan left and right

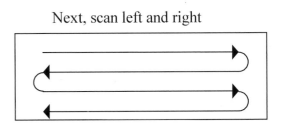

D.   As you scan, you will see many organisms. If the organisms are moving too fast for you to view easily, add a few fine strands of cotton. Add the cotton by lifting the coverslip, dropping the cotton on the slide, and replacing the coverslip. You might need to add another drop of the culture before replacing the coverslip.

E.   As you locate an organism, examine it on medium and high powers. Use the same technique that you learned in Module 1. Sketch in your laboratory notebook and try to identify each organism using the sketches on the next page. In this experiment, look especially for the presence of bacteria and algae. As shown in the sketches on the next page, bacteria will be the smallest things that you see. When you find something that is moving, try to determine how it moves.

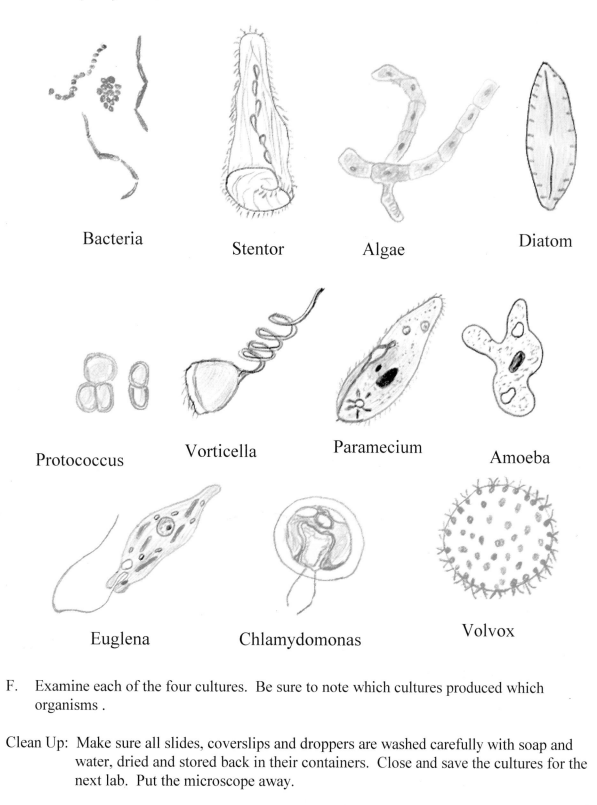

Bacteria

Stentor

Algae

Diatom

Protococcus

Vorticella

Paramecium

Amoeba

Euglena

Chlamydomonas

Volvox

F.   Examine each of the four cultures. Be sure to note which cultures produced which organisms .

Clean Up: Make sure all slides, coverslips and droppers are washed carefully with soap and water, dried and stored back in their containers. Close and save the cultures for the next lab. Put the microscope away.

ANSWERS TO THE ON YOUR OWN QUESTIONS

2.1 <u>The bacterium that succeeded most likely has a capsule while the other does not.</u> In bacteria, the capsule not only helps the organism cling to surfaces, but it also protects the organism from infection-fighting mechanisms.

2.2 <u>The plasma membrane</u> did not do its job. Since it regulates what moves in and out of the cell, it should not have allowed the toxin to enter.

2.3 <u>It is missing a flagellum.</u> The pili are not for movement; thus, if a bacterium has no flagellum, it cannot move.

2.4 <u>No.</u> Saprophytic means that it feeds on dead matter. Autotrophic organisms make their own food.

2.5 <u>Yes.</u> Aerobic and anaerobic deal with how the organism converts its food into useful energy. Chemosynthetic deals with how the organism gets the food to begin with. Thus, the bacterium can make the food chemosynthetically and then convert it to useable energy aerobically.

2.6 <u>Since asexual reproduction allows no variation in the DNA, an entire population of bacteria will have essentially the same traits. If the environment changes, an organism might need new traits to survive. Since the whole population has essentially the same traits, there is no way to get the needed new traits and the population dies.</u> The variability that exists in sexual reproduction usually makes a population much more resistant to change in the habitat.

2.7 <u>More resources (most likely food) were added to the habitat.</u> Since a steady state is usually followed by a decrease in population due to scarce resources, the only way you can get population growth *after* the steady state would be due to an influx of new resources. In our cow example, maybe another cow died and fell on top of the first cow's body!

2.8 <u>Type B will be the donors and Type A will be the recipients.</u> Since the bacteria will need to survive in high temperatures, they need resistance to high temperatures. In bacterial conjugation, the traits of the donor go to the recipient.

2.9 <u>They are rod-shaped</u>, since "bacillus" means rod-shaped.

2.10 <u>Tenericutes</u> is the phylum for bacteria without a cell wall.

2.11 <u>Firmicutes</u> contains all Gram-positive bacteria.

2.12 <u>Gracilicutes</u> contains all Gram-negative bacteria. These are the ones with complex cell walls, because they are able to keep the Gram stain out of the cell's interior.

2.13 <u>Mendosicutes</u> contains all bacteria with exotic cell walls.

2.14 To construct the biological key, we must ask a series of questions that separate out the bacteria. The hint tells us to start with whether or not the bacterium has a cell wall. If we do that, then we already have one classification. After all, if it has no cell wall, we know the phylum and the class. If it does have a cell wall, however, we will have to ask other questions. Thus, key 1 looks like this:

1. Cell wall **2**
    No cell wall ....*phylum Tenericutes*....... *class Mollicutes*

Now if we have a cell wall, it is either Gram-positive, Gram-negative, or neither. If it's either Gram-positive or negative, then we will need to ask more questions. If it is neither, however, we are done. Thus, the next one is:

2. Gram-positive or Gram-negative.. **3**
    Neither ....*phylum Mendosicutes*....... *class Archaebacteria*

Now we deal with Gram-negative or Gram-positive. We'll have to ask more questions either way, so both alternatives will have to lead to more keys. We have no idea how many more, so we'll lead to key 4 on one and leave the other with question marks that we will go back and replace later. I'll send Gram-positive to key 4 because it has fewer classes, thus it will get done quicker.

3.   Gram-Positive..*phylum Firmicutes*........... **4**
    Gram-Negative..*phylum Gracilicutes*......... ??

Key 4 needs to make the distinction between the two classes in Firmicutes, so we just have to ask about shape.

4.   Bacillus or Coccus .... *class Firmibacteria*
    Neither .... *class Thallobacteria*

Now we are ready to deal with Gram negatives. Thus, we know that the question marks on key 3 should be replaced with a "5." In key 5, we need to make the distinction between three classes. Two have photosynthetic bacteria in them, so a quick delineation would be whether or not the bacterium is photosynthetic:

5.   Non-Photosynthetic... *class Scotobacteria*
    Photosynthetic... **6**

Now we just determine whether or not the bacteria produce oxygen, and we are done.

6.   Produces oxygen...*class Oxyphotobacteria*
    Does not produce oxygen..*class Anoxyphotobacteria*

The final key, then, is as follows:

1. Cell wall **2**
   No cell wall ....*phylum Tenericutes*....... *class Mollicutes*

2. Gram-positive or Gram-negative... **3**
   Neither ....*phylum Mendosicutes*....... *class Archaebacteria*

3. Gram-Positive..*phylum Firmicutes*............ **4**
   Gram-Negative...*phylum Gracilicutes*......... **5**

4. Bacillus or Coccus... *class Firmibacteria*
   Neither .... *class Thallobacteria*

5. Non-Photosynthetic.. *class Scotobacteria*
   Photosynthetic.. **6**

6. Produces oxygen..*class Oxyphotobacteria*
   Does not produce oxygen...*class Anoxyphotobacteria*

Now your key does not have to be in the same order, but it should contain all of these questions. It might have a few more steps, if you weren't as efficient as me in setting up the key.

STUDY GUIDE FOR MODULE #2

1. On a separate sheet of paper, write down the definitions for the following terms:

a. Pathogen
b. Saprophyte
c. Parasite
d. Respiration
e. Aerobic respiration
f. Anaerobic respiration
g. Steady state
h. Conjugation
i. Plasmid
j. Transformation
k. Endospore
l. Strains

2. Label all of the important structures on the bacterium below:

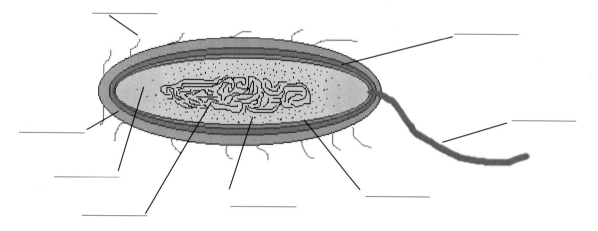

Illus. by Dr. Jay L. Wile

3. Describe the functions of each of the components you labeled in problem #2.

4. What is the most popular form of metabolism among bacteria?

5. If a bacterium is parasitic, is it heterotrophic or autotrophic?

6. List the 4 modes of reproduction in bacteria and briefly describe each.

7. Does the population of bacteria typically increase or decrease before it reaches steady state? What about after?

8. Even though sexual reproduction among bacteria does not result in offspring, it can significantly affect the population of a bacteria growth. Why?

9. What are the technical names of the three common bacterial shapes?

10. A bacterium is heterotrophic and Gram-negative. To what phylum and class does it belong?

11. A spirillum bacterium is Gram-positive. To what phylum and class does it belong?

12. A bacterium has no cell wall. To what phylum and class does it belong?

13. What conditions are necessary for most bacteria to grow and reproduce?

14. What methods exist for reducing the chance of bacterial contamination of food?

Module #3: Kingdom Protista

Introduction

In the previous module, we looked at the diverse kingdom Monera. In this module, we will move on to an equally fascinating kingdom: Protista (pro tee' stuh). Both of these kingdoms are made up of microorganisms, but they differ rather dramatically. The main reason they differ is the fact that organisms in the kingdom Monera are prokaryotic, whereas those in Protista are eukaryotic. Now as you recall from Module #1, eukaryotic cells have distinct, membrane-bound organelles whereas prokaryotic cells do not. This fact adds a level of complexity to the organisms in kingdom Protista that doesn't exist in kingdom Monera.

Why do we say this? Well, both prokaryotes and eukaryotes must perform certain basic functions (those discussed in Module #1) in order to live. In prokaryotic cells, these functions are performed by the cell as a whole. In eukaryotic cells, however, each organelle performs its own set of tasks. Only when each organelle cooperates with the other organelles will the cell be able to perform all of the functions necessary for life. Thus, in eukaryotic cells, the organelles must be able to work together. This adds a level of complexity to the system.

This added level of complexity makes these organisms even more fascinating to study, but it also causes a bit of a problem for us. You see, when studying kingdom Monera, we could discuss the "typical" bacterium. Of course, there were many exceptions, but most bacteria did resemble that which was labeled as "typical." Because of the increased complexity in kingdom Protista, however, there is just no way to describe what is typical for these organisms. Thus, in order to study this fascinating kingdom, you should start with another microscope experiment. We will then give you a short introduction to classification within Protista, which will be followed by a detailed discussion of the organisms in each phylum.

If you do not have a microscope, skip the experiment. You can, however, get a glimpse of this fascinating world with a couple of videos. These videos can be found at many libraries, or they are available from Nature's Workshop Plus (see the introduction of the book for ordering information). The first video is called *The Power of Plants*. It contains a segment called "It's a Small World" which would be very appropriate to see at this time. In addition, the video *The Clown-Faced Carpenter* has a segment entitled "Water Water Everywhere" which also works nicely at this point. These videos can be valuable even for students who perform the experiment.

EXPERIMENT 3.1
Pond Life C

Supplies:
- Microscope
- Slides
- Coverslips
- The 4 culture jars used in Experiment 2.2
- 4 droppers (one for each jar)
- Small amount of cotton

Object: To further the study of the cultures set up in Experiment 2.2. You will hopefully see different organisms this time.

Procedure:
A. Set up your microscope and make slides of each culture as you did in Experiment 2.2. Prepare the slides one at a time, as you are ready to observe them.
B. Try taking your sample from the very bottom of the jar. Do not disturb the water in the process. This might produce some good diatoms and other interesting organisms.
C. Make sure you scan the slide systematically, as you did in Experiment 2.2, to observe all organisms in the sample.
D. Examine each species as you find them (on all three powers). Remember, you can use cotton to hinder an organism's movement, as you did in Experiment 2.2. When you find an organism different than the ones you saw in Experiment 2.2, examine it with all three magnifications and draw it. Note which culture the organism came from. Remember to note how the organism moves.
E. Try to identify the organism by looking at the drawing in Experiment 2.2 and the pictures in this module.

Clean up: Unless you wish to study the cultures further on your own, take your cultures outside and dump them. Discard the containers. Clean all slides, slide covers and droppers carefully. Put the microscope away.

Classification in Kingdom Protista

Kingdom Protista is divided into two main groups: protozoa and algae. These groups are *not* phyla. Instead, each group contains several phyla; thus, they are often called subkingdom Protozoa (proh' tuh zoh' uh) and subkingdom Algae (al' jee). Most protozoa exist as individual, single-celled creatures, while most algae form colonies. Whereas most protozoa are heterotrophic, most algae are autotrophic. Finally, most protozoa have a means of locomotion, while most algae simply float on the top of a body of water.

Within the two subkingdoms, there is still an amazing amount of diversity. There are four phyla in subkingdom Protozoa: phylum Mastigophora (mas tih gah' fore uh), phylum Sarcodina (sar kuh die' nuh), phylum Ciliophora (sil ee ah' fuh ruh), and phylum Sporozoa (spoor' uh zoe uh). These phyla are distinguished from one another based on the organisms' method of locomotion. Subkingdom Algae, on the other hand, contains 5 phyla: phylum Chlorophyta (klor' uh fie tuh), phylum Chrysophyta (cry' so fie tah), phylum Pyrrophyta (pie' roh fie tuh), phylum Phaeophyta (fay' uh fie tuh), and phylum Rhodophyta (roh duh' fie tuh). Organisms are separated into these phyla based on habitat, organization, and type of cell wall.

Now you need not worry about memorizing the names of the phyla that we have just discussed. As time goes on, you will get more familiar with them, but we will not require you to remember them. When answering questions on an exercise or a test, we will provide you with a table that will list the subkingdoms and phyla in kingdom Protista. You should, instead, concentrate on trying to remember what types of organisms go into which phylum. In order to help you get an idea of the organisms that exist in kingdom Protista, study Figure 3.1. Don't worry that some of the terms are unfamiliar; we'll get to them in a while. For now, just try to get an overview of what the organisms in kingdom Protista look like and how they are separated into phyla.

Figure 3.1
Representatives of Kingdom Protista

Subkingdom Protozoa:

Phylum	Type of Locomotion	Representative
Sarcodina	Pseudopods	genus *Amoeba* (ah mee' buh) 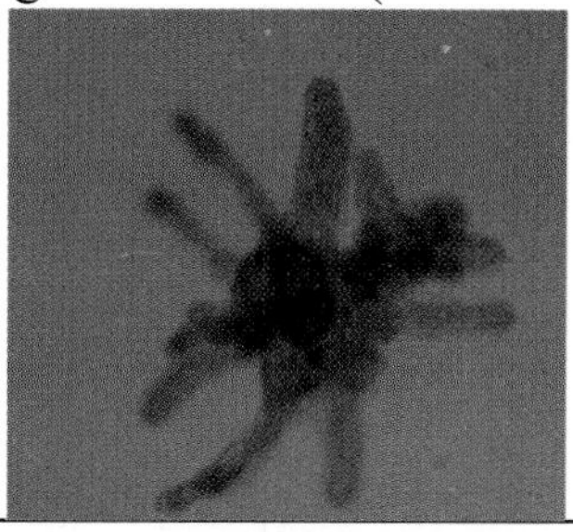 *Photo by Kathleen J. Wile*
Mastigophora	Flagellum	genus *Euglena* (yoo glee' na) 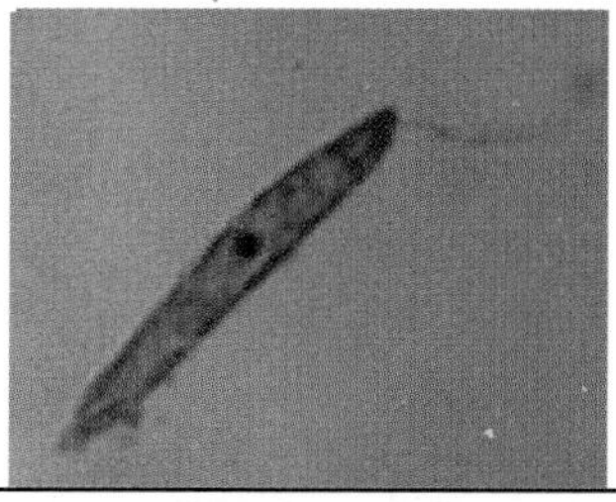 *Photo by Kathleen J. Wile*
Ciliophora	Cilia	genus *Paramecium* (pehr uh mee' see um) 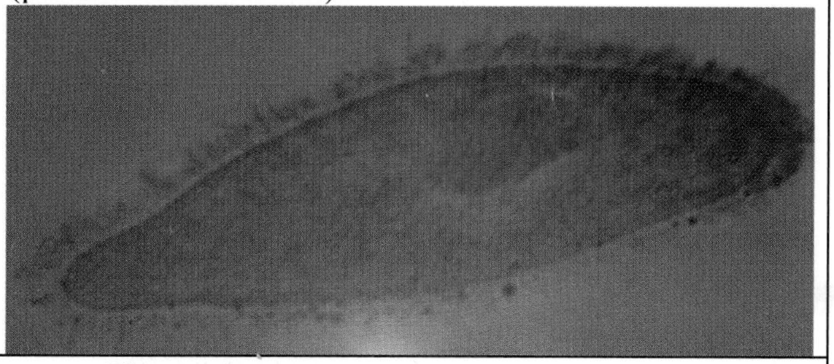
Sporozoa	None	genus *Plasmodium* (plas moe' dee um) 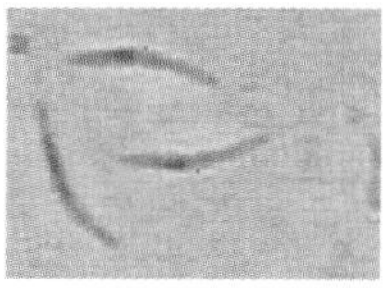 *Photo courtesy of the Pathologists Associated*

Subkingdom Algae:

Phylum	Habitat	Organization	Cell Wall	Representative
Chlorophyta	Fresh Water	Single-celled individuals as well as some colonies	Cellulose	genus *Spirogyra* (spy row' ji ruh) *Photo by Kathleen J. Wile*
Chrysophyta (contains the diatoms)	Marine and Fresh Water	Single-celled individuals as well as some colonies	Silicon dioxide	genus *Cocconeis* (cohc cohn' is) *Photo by Kathleen J. Wile*
Pyrrophyta (also called dinoflagellates)	Marine	Single-celled individuals	Cellulose or atypical	genus *Ceratium* (sir ah' tee uhm) *Photo by Dr. Jay L. Wile*
Phaeophyta (also called brown algae)	Marine: cold waters	Multi-celled organism	Cellulose and alginic acid	genus *Laminaria* (lam' uh nair ee uh) *Photo © Mike Guiry, NUI, Galway*
Rhodophyta (also called red algae)	Marine: warm waters	Multi-celled organism	Cellulose	genus *Porphyra* (poor fie' ruh) *Photo © Mike Guiry, NUI, Galway*

ON YOUR OWN

3.1 Construct a biological key for separation of organisms in kingdom Protista into phyla. You can assume that any organism for which you use the key is already known to be in kingdom Protista. You can also assume that if an organism is completely autotrophic it belongs in subkingdom Algae, but if it is at all heterotrophic it belongs in subkingdom Protozoa.

As we discuss these phyla in a little more detail, you will notice something about the way that we refer to organisms within this kingdom. Sometimes, we refer to organisms using just the name of a classification group to which they belong, typically their genus. For example, when we talk about phylum Sarcodina, we will study the *Amoeba proteus* (ah mee' buh pro tee' us). It turns out that most of the species within genus *Amoeba* are so similar that when we see them under the microscope, they look almost identical. Thus, we often refer to all members of this genus as simply, "amoeba." When we do this, we do not capitalize or italicize the word because we are not talking about the classification groups used in binomial nomenclature, we are instead using a generic term to refer to all members within the classification group.

We often use a generic term with other classification groups as well. When we were discussing the two subkingdoms Protozoa and Algae, we began referring to their members with the generic term "protozoa" and "algae." This happens quite a bit when we study kingdom Protista, so you will just have to get used to it.

Subkingdom Protozoa

In general, members of subkingdom Protozoa have a method of locomotion. This is one aspect that sets them apart from the algae, which in general have no locomotion. Another quality that separates protozoa from algae is the fact that most protozoa are heterotrophic, while most algae are autotrophic. There are, of course, exceptions to both of these rules, which adds to the confusion when it comes to classifying organisms within kingdom Protista. For example, you will soon learn that the genus *Euglena* belongs to phylum Mastigophora, which means that we consider creatures in this genus to be protozoa. Genus *Euglena* is placed in phylum Mastigophora because its members use a flagellum for locomotion. Nevertheless, members of genus *Euglena* can use photosynthesis to produce their own food. Thus, some biologists consider them algae, because they are considered autotrophic. So, even though we might lay down some general rules for classification, please realize that there will be exceptions to them!

Phylum Sarcodina

The most striking feature of the organisms in this phylum is that they have no standard body shape. They are enclosed in a flexible plasma membrane which allows them to change shape at will. When resting, they are usually spherical. However, when they wish to move, they form extensions of their bodies called **pseudopods** (soo' doe pods), or "false feet."

<u>Pseudopod</u> - A temporary, foot-like extension of a cell, used for locomotion or engulfing food

You will learn more about pseudopods in a moment. First, however, let's take a look at a typical member of the phylum Sarcodina, *Amoeba proteus*, the common amoeba (plural: amoebae).

Amoeba: A Typical Member of Phylum Sarcodina

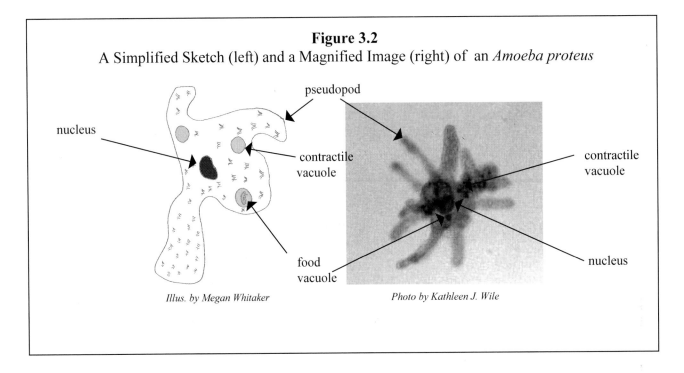

Figure 3.2
A Simplified Sketch (left) and a Magnified Image (right) of an *Amoeba proteus*

Illus. by Megan Whitaker   *Photo by Kathleen J. Wile*

As you can see from the figure, this organism is considered eukaryotic because it has a distinct, membrane-bound **nucleus** (new' clee us):

<u>Nucleus</u> - The region of a eukaryotic cell which contains the DNA

In the case of the *Amoeba proteus* (as well as other members of phylum Sarcodina) the nucleus not only holds the DNA and therefore controls reproduction, but it also controls the organism's metabolism. The other main organelle found in the amoeba is the **vacuole** (vac' you ol).

<u>Vacuole</u> - A membrane bound "sac" within a cell

The amoeba has two types of vacuoles. The **food vacuoles** hold and store food while it is being digested, and the **contractile vacuoles** regulate the amount of water in the cell. If the cell absorbs too much water, it could potentially become over-pressurized and end up exploding! To release the pressure, the contractile vacuoles collect excess water in the cell and pump it back out into the amoeba's surrounding environment.

The cytoplasm (that jelly-like substance inside the cell) of the amoeba is divided into two parts. Near the plasma membrane, we find the **ectoplasm** (ek' toh plas uhm), which is thin and watery.

Ectoplasm - The thin, watery cytoplasm near the plasma membrane of some cells

In the interior of the cell, the cytoplasm becomes much more dense. In this region, it is called the **endoplasm** (en' doh plas uhm).

Endoplasm - The dense cytoplasm found in the interior of many cells

The most striking feature of the amoeba is its method of locomotion. If you have a microscope, you most likely had the opportunity to observe this movement, which is truly fascinating. While at rest, the amoeba is usually spherical. When it needs to move, the amoeba forces endoplasm to a region near the plasma membrane. This dense liquid deforms the membrane, causing a bulge, called a pseudopod, to form. If the amoeba wants to move, all of the endoplasm can be forced into the pseudopod, carrying the cell contents with it. Thus, the amoeba moves by forming a pseudopod and pushing itself into that pseudopod. This can be done over and over, resulting in a laborious form of locomotion.

The amoeba uses its pseudopods for tasks other than motion. When it comes into contact with a food source (mostly other, smaller microorganisms like bacteria), it can use its pseudopods to surround and then engulf its prey (see Figure 3.3). Once engulfed, the food is held in a food vacuole where digestion can begin. When the food is digested, substances that can be metabolized are sent into the endoplasm while substances that cannot be used are ejected from the cell.

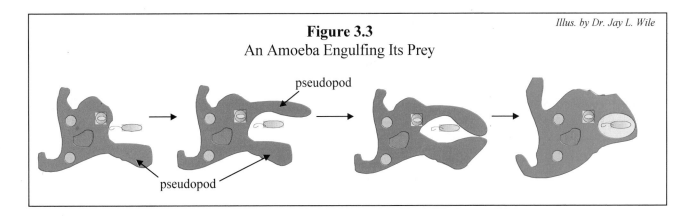

Figure 3.3
An Amoeba Engulfing Its Prey

Illus. by Dr. Jay L. Wile

pseudopod

pseudopod

Although amoebae inhabit watery environments such as ponds, lakes, and rivers with little current, amoebae do not swim. Their method of locomotion is simply too laborious for such activity. As a result, amoebae typically adhere to surfaces in the water. When you touch a rock that is in a pond, the slimy coating that you feel is often the result of amoebae which are stuck to the rock. In fact, when an amoeba is floating in water, it will try to adhere to most surfaces that it contacts.

When an amoeba reaches its maximum size (typically in about three days), it can reproduce asexually. As far as we know, amoebae do not sexually reproduce. Amoebae can form **cysts** (much like the endospores that a bacterium forms) in an attempt to survive life-threatening conditions such as dryness or a lack of food.

Other Sarcodines

Although *Amoeba proteus* is a typical sarcodine, there are many other organisms in this phylum. Genus *Entamoeba* (ent' uh mee' buh), for example, contains many organisms that live inside of human beings. *Entamoeba gingivalis* (jin' jih val us) is a harmless sarcodine that lives in your mouth, while *E. coli* is a sarcodine that lives in your gut. Notice that this *E. coli* is not the same as the *E. coli* that we discussed in the previous module. The bacterium that we discussed in the previous module is *Escherichia coli* while the sarcodine is *Entamoeba coli*. They both inhabit the human gut, but they are quite different organisms. One is prokaryotic (*Escherichia coli*) and the other is eukaryotic (*Entamoeba coli*). One is a bacterium (*Escherichia coli*) and the other is a protozoa (*Entamoeba coli*). This potential point of confusion illustrates that although you *can* abbreviate the genus name in binomial nomenclature, it is not always wise to do so because many different genera begin with the same letter.

One other organism in this same genus, *Entamoeba histolytica* (his' toh lih tih' cuh), is a gut-dwelling parasite. This particular species is pathogenic and causes severe dysentery. It is often spread through contaminated water. Countries that do not carefully sterilize their water supply often have lots of *E. histolytica* in their water. This is where the phrase, "don't drink the water" comes from. If you drink water that is contaminated with *E. histolytica*, you will experience diarrhea, nausea, exhaustion, and dizziness, which are all symptoms of dysentery. In some cases, *E. histolytica* can spread to the liver and brain, causing death!

ON YOUR OWN

3.2 Suppose you were observing an amoeba under the microscope and it suddenly exploded. What organelle was probably not working properly in the amoeba?

3.3 A member of genus Sarcodina is floating in the water when it collides with a twig. What will the sarcodine immediately try to do?

Phylum Mastigophora

As we see from Figure 3.1, one of the characteristic features of organisms in this phylum is a flagellum, which is used for locomotion. Because of this characteristic feature, members of this phylum are often called **flagellates** (flah' gel ates).

<u>Flagellate</u> - A protozoan that propels itself with a flagellum

Although most flagellates use their flagellum to swim freely in the water, some use it to attach themselves to a solid object. Interestingly enough, this phylum contains all sizes of organisms. Some flagellates are among the smallest protozoa in kingdom Protista, while others form elaborate colonies that are large enough to see with the unaided eye. Members of this phylum can be found in a variety of habitats, including salt water (which is referred to as a **marine** environment), fresh water, and moist soil.

Euglena: A Typical Member of Phylum Mastigophora

The genus *Euglena* contains interesting examples of organisms from phylum Mastigophora. Members of this genus have the ability to produce their own food via photosynthesis. Despite this fact, we do not necessarily consider them autotrophic because they also ingest and decompose the remains of dead organisms. Thus, euglena are both autotrophic and saprophytic! Even when there is plenty of light for photosynthesis, euglena still tend to absorb some food from their surroundings. When their surroundings become dim, their photosynthesis shuts down and they become completely saprophytic. In fact, if their environment stays dim for several days or more, the euglena's photosynthetic capability degenerates. After that happens, they can never produce their own food again, regardless of how bright their surroundings might later become.

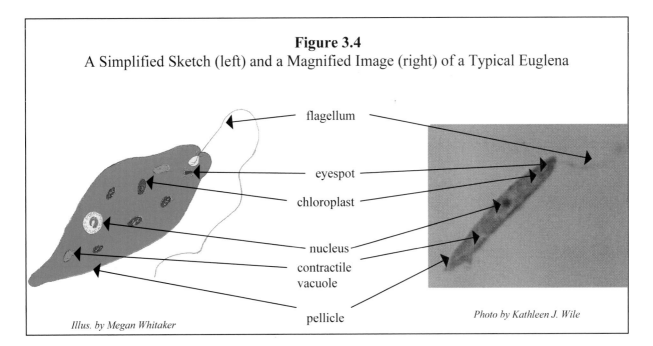

Figure 3.4
A Simplified Sketch (left) and a Magnified Image (right) of a Typical Euglena

flagellum

eyespot

chloroplast

nucleus

contractile vacuole

pellicle

Illus. by Megan Whitaker

Photo by Kathleen J. Wile

You should notice that there are many prominent organelles in this creature. Like the sarcodines, the euglena have contractile vacuoles that help expel excess water in order to reduce the pressure inside the cell. They also have a nucleus that holds the DNA and controls both reproduction and metabolism.

Unlike the sarcodines, euglena have a firm but flexible shape-sustaining **pellicle** (pel' ick ul).

Pellicle - A firm, flexible coating outside the plasma membrane

The pellicle usually keeps the euglena in its spindle-like shape. When they want to move, however, euglena can use a worm-like method of locomotion in addition to their flagella. While whirling their flagella, euglena can also draw their cytoplasm into the central region of the cell. This deforms the euglena, making it look like a "+" sign. The euglena then re-extend themselves forward. This mode of locomotion is used in addition to whirling the flagellum, as illustrated in Figure 3.5.

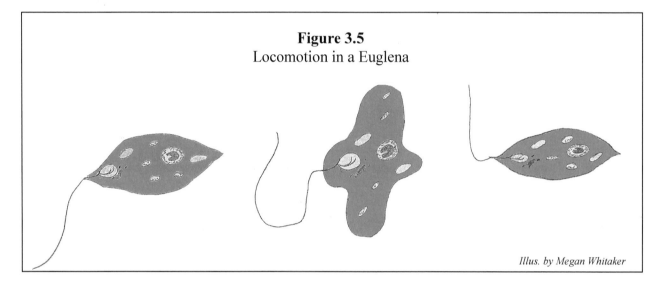

Figure 3.5
Locomotion in a Euglena

Illus. by Megan Whitaker

In order to create food by photosynthesis, euglena have **chloroplasts** (klor' oh plasts) which contain a green pigment called **chlorophyll** (klor' oh fill). You will learn more about this pigment in an upcoming module. For right now, just understand that chlorophyll is a green pigment necessary for photosynthesis.

Chloroplast - An organelle containing chlorophyll for photosynthesis

Chlorophyll - A pigment necessary for photosynthesis

Since photosynthesis requires light, the euglena also has a red, light-sensitive region known as the **eyespot**.

Eyespot - A light-sensitive region in certain protozoa

Since the eyespot is light-sensitive, euglena use it to move towards regions of bright light, increasing their photosynthetic output. Understand that although we call this the "eyespot," the euglena do not really see with it. The eyespot simply allows them to determine where there is a lot of light and where there is little light. The biological structures and processes necessary to see are far beyond the reach of a single-celled organism!

Euglena reproduce only asexually. Under ideal conditions, a single euglena can reproduce once each day.

Other Mastigophorites

Other rather interesting members of phylum Mastigophora come from the genus *Volvox*. Also photosynthetic, members of this genus are colonial and possess two flagella. They join together and create an elaborate latticework out of strands of cytoplasm. The entire colony moves by rolling the latticework. The standard mode of reproduction among *Volvox* is asexual. When this occurs, a daughter colony is formed which often stays within the original parent colony. Under certain conditions, however, individual members of a *Volvox* colony can reproduce sexually. A magnified image of a *Volvox* is shown in Figure 3.6.

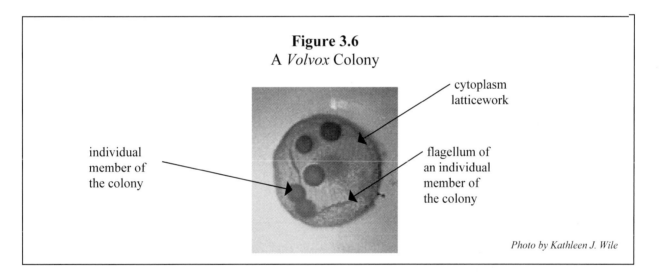

Figure 3.6
A *Volvox* Colony

cytoplasm latticework

individual member of the colony

flagellum of an individual member of the colony

Photo by Kathleen J. Wile

As is typical with most phyla in this subkingdom, some mastigophorites are harmful to human beings. A small fly known as the **tsetse fly** carries mastigophorites that belong to the genus *Trypanosoma* (try' pan oh soh muh). These protozoa are passed to humans when the tsetse fly bites a person's skin or when a person eats meat from an animal that has been bitten. The *Trypanosoma* then travel into the bloodstream and cause **African sleeping sickness**, a disease which often ends in death.

Another interesting mastigophorite comes from genus *Trichonympha* (try' coh nim fah). These protozoa live in the gut of a termite and actually allow the termite to survive. You see, termites eat wood. However, they are unable to digest a substance called cellulose that is in all types of wood. The *Trichonympha* that live in the termite's gut feed on this cellulose. If it were not for this protozoa, cellulose would continue to build up within the termite until it reached toxic levels and killed the termite. The *Trichonympha*, however, keep this from happening and, in the process, feed themselves.

The relationship between the termite and the *Trichonympha* is an example of what biologists call **symbiosis** (sim by oh' sis).

at least one

Symbiosis - Two or more organisms of different species living together so that ~~each~~ benefits ~~from the other~~ (Mutualistic is where both benefit)

Symbiosis is very common in God's Creation, and is yet another testament to His divine handiwork. You see, termites must eat wood in order to provide balance in nature. The biological processes necessary to digest wood, however, make it impossible to digest cellulose, a principle component of wood. The Creator, fully aware of this apparent inconsistency, simply created another organism that could digest cellulose and designed that organism to live in the termite's gut. That way, wood and cellulose could both be digested together. Truly, God's handiwork in Creation is astounding!

ON YOUR OWN

3.4 A euglena is sitting in dim light. There is a bright spot of light not too far away, but the euglena does not move towards it. Instead, it seems to wander aimlessly. What organelle is not functioning properly in the euglena?

3.5 According to most biologists, an organism must always perform photosynthesis or chemosynthesis to be considered autotrophic. Why do these biologists say that the euglena is not autotrophic?

3.6 A student says that members of genus *Trypanosoma* living within the bloodstream of a human is an example of symbiosis. Is the student correct? Why or why not?

Phylum Ciliophora

Phylum Ciliophora is comprised of protozoa that use **cilia** (sil' ee uh) to move. As a result, these organisms are called **ciliates**.

Cilia - Numerous short extensions of the plasma membrane used for locomotion

The cilia beat rhythmically to move the organism from place to place. Ciliates can be quite large (for a protozoan), with some species attaining lengths of three millimeters (about one-ninth of an inch). They vary in shape, ranging from cone-shaped to bell-shaped to foot-shaped. Members of this phylum mostly live in fresh water, preferring stagnant lakes and ponds.

Paramecium: A Typical Ciliate

Members of the genus *Paramecium* (shown in Figure 3.7) offer a good view of the typical ciliate.

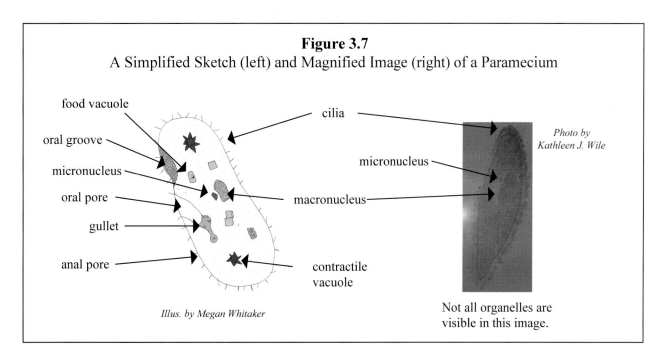

Figure 3.7
A Simplified Sketch (left) and Magnified Image (right) of a Paramecium

food vacuole
oral groove
micronucleus
oral pore
gullet
anal pore

cilia
micronucleus
macronucleus
contractile vacuole

Photo by Kathleen J. Wile

Illus. by Megan Whitaker

Not all organelles are visible in this image.

The most obvious feature of the paramecium is its kidney-shaped **macronucleus**. This large nucleus controls the metabolism. The paramecium and most ciliates require a large amount of energy to live, mostly because of the rapid movement of their cilia. As a result, the metabolic control center must be large. In fact, the metabolic demands on the macronucleus are so large that there is a smaller **micronucleus** which controls reproduction so that the macronucleus can spend all of its resources controlling metabolism.

Like the flagellates, the paramecium and other ciliates have a pellicle which sustains the cell's shape. Like the sarcodina and the flagellates, ciliates have contractile vacuoles that control water pressure within the cell. Like the sarcodina, paramecia (plural of paramecium) have food vacuoles to hold food while it is being digested. The paramecium, however, has a unique method of feeding. Paramecia have an **oral groove** that is lined with cilia. These cilia sweep food (bacteria, algae, and other ciliates) into the **gullet** through the **oral pore**. When it is full of food, the gullet then pinches off, becoming a food vacuole. The food vacuole then travels throughout the cell, floating in the cytoplasm. That way, the food is spread throughout the cell as it is digested. Any undigested portions of the food are expelled through the **anal pore**.

In addition to the asexual reproduction that is typical among the protozoa, paramecia can also engage in a form of **conjugation**. In this process, two paramecia attach to one another at the openings of their oral grooves. Then, they both exchange DNA with each other. Thus, unlike the conjugation that occurs among bacteria (which we studied in the previous module), this conjugation results in a mutual exchange of DNA. After the conjugation is complete, the two paramecia separate from each other and then immediately reproduce asexually, each making a new paramecium. Since there was exchange of DNA in the conjugation, the offspring produced in this way will not be exact duplicates of their parents. Instead, their characteristics will be a blend of the characteristics of the two original paramecia.

Other Members of Phylum Ciliophora

While looking at the pond water under the microscope, you might have seen another important ciliate, one from the genus *Stentor*. This ciliate can grow as large as one-tenth of an inch. Common in fresh-water habitats, members of this genus are shaped like a trumpet. They have cilia which surround their gullet. As the cilia beat back and forth, a micro-current in the water is created. This current sweeps food into the gullet, sometimes at the rate of a hundred protozoa each minute! Stentor eat all kinds of protozoa (including paramecia) and bacteria.

Although most ciliates are not pathogenic, one particular species, *Balantidium* (bal' an tid' ee um) *coli* is a parasite of many species including pigs, rats, and guinea pigs. These ciliates can form cysts that survive in the fecal matter of their host. If an individual eats food or drinks water contaminated with such infected fecal material, then the individual will become infected as well. In humans, infection by *B. coli* can result in severe dysentery. *B. coli* infection in humans is not common in the US, but it is common in tropical regions of the world where malnutrition occurs and where pigs and humans live in close proximity to one another.

ON YOUR OWN

3.7 A paramecium cannot conjugate. What organelle is not functioning properly?

3.8 A biologist studies a group of bacteria (all one species) and a group of paramecia (all one species). She notices that while the bacteria all seem to be almost exact duplicates of each other, there is a great deal of variation between the paramecia. Why?

Phylum Sporozoa

This phylum contains the protozoa that have no real means of locomotion. The main characteristic of the sporozoa, however, is the fact that they form **spores** at some point in their life.

Spore - A reproductive cell with a hard, protective coating

Much like a cyst, a spore can survive for quite a while, even in unfavorable conditions. Sporozoa typically form their spores as a result of a unique form of asexual reproduction, illustrated in Figure 3.8.

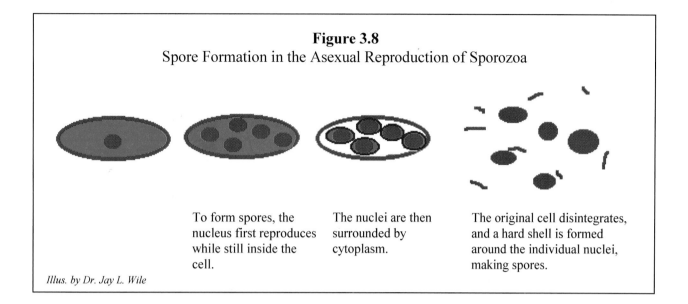

Figure 3.8
Spore Formation in the Asexual Reproduction of Sporozoa

To form spores, the nucleus first reproduces while still inside the cell.

The nuclei are then surrounded by cytoplasm.

The original cell disintegrates, and a hard shell is formed around the individual nuclei, making spores.

Illus. by Dr. Jay L. Wile

In this form of reproduction, the nucleus divides while still inside the cell. This may happen several times, resulting in several different nuclei in one cell. Cytoplasm gathers around each nucleus, and then the cell disintegrates. The nuclei that are surrounded by cytoplasm then form hard, protective layers, making spores. Spore formation is usually not the only type of reproduction in which sporozoa engage, however. The best way to illustrate this is by example.

Plasmodium: Sporozoa That Cause Malaria

The genus *Plasmodium* (plaz' moh dee um) is home to one of the most deadly parasites know to man. These parasites cause **malaria**, a disease that has claimed countless lives over the years. Spread by the mosquito, *Plasmodium* have a very interesting and complex lifestyle, illustrated in Figure 3.9.

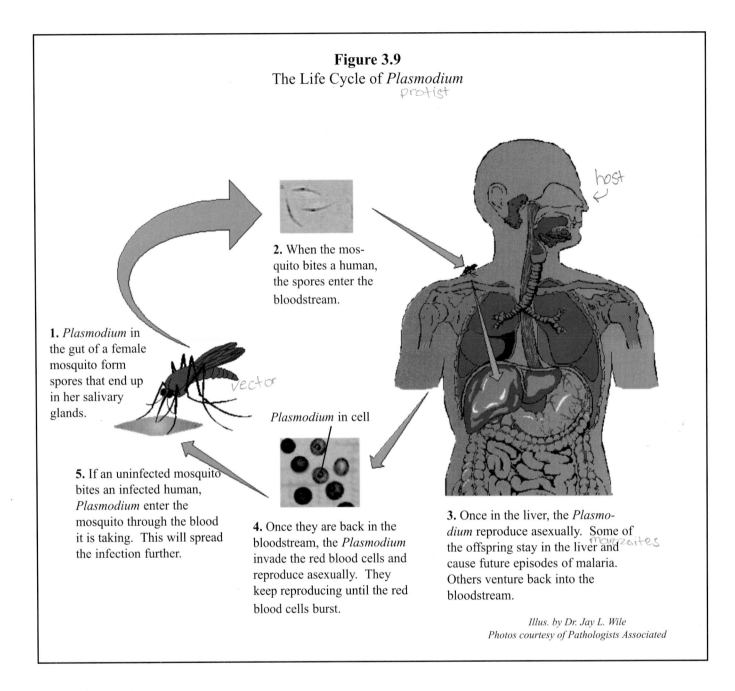

Figure 3.9
The Life Cycle of *Plasmodium*
protist

host

1. *Plasmodium* in the gut of a female mosquito form spores that end up in her salivary glands.

vector

2. When the mosquito bites a human, the spores enter the bloodstream.

Plasmodium in cell

5. If an uninfected mosquito bites an infected human, *Plasmodium* enter the mosquito through the blood it is taking. This will spread the infection further.

4. Once they are back in the bloodstream, the *Plasmodium* invade the red blood cells and reproduce asexually. They keep reproducing until the red blood cells burst.

3. Once in the liver, the *Plasmodium* reproduce asexually. Some of the offspring stay in the liver and merozoites cause future episodes of malaria. Others venture back into the bloodstream.

Illus. by Dr. Jay L. Wile
Photos courtesy of Pathologists Associated

In the gut of a female mosquito, *Plasmodium* reproduce and form spores which end up in the mosquito's salivary glands. When the mosquito bites a human, it injects saliva into the bite wound. This saliva keeps the blood from clotting, making it easy for the mosquito to ingest. The saliva also contains spores of *Plasmodium*. These spores lose their hard shells, and they reproduce in the human's liver. They then enter the bloodstream, where they invade the red blood cells (cells that carry oxygen throughout the body). The *Plasmodium* reproduce again inside the red blood cells, causing them to burst. As the infection progresses, the symptoms (fever, chills, and shaking) become more and more severe until death occurs. *Plasmodium* spreads because, once in the bloodstream, they will infect any other mosquito that bites an infected human. Thus, infected humans infect mosquitoes, which in turn infect more humans!

Plasmodium must form spores while it is in the mosquito because the salivary glands do not provide a habitat that is suitable for its survival. However, once inside the infected human, conditions are ideal for the *Plasmodium,* so future reproduction does not require the formation of spores. When a female mosquito bites an infected human and becomes infected herself, the *Plasmodium* must once again form spores in order to survive until they can be injected into another human. This is why we say that *Plasmodium* sometimes engage in reproduction that forms spores and sometimes engage in reproduction that does not.

Another typical sporozoa comes from the genus *Toxoplasma* (tox' oh plaz muh). These organisms live in the intestines of cats and reproduce there sexually. They do not harm cats, but often exit the animal with the feces (solid waste). While in the feces, *Toxoplasma* form spores so that they can survive. These spores are then spread through houseflies, cockroaches, insects, and direct contact with the cat feces. Once they infect a human, they reproduce asexually, causing the disease **toxoplasmosis** (tox' oh plaz moh sis). This disease causes severe birth defects in pregnant women, which is why doctors tell pregnant women that they should never empty litter boxes or otherwise clean up after cats.

ON YOUR OWN

3.9 What is the difference between spores and cysts?

3.10 One way that people fight the spread of malaria is to significantly reduce the population of mosquitoes in their vicinity. Why does this work?

Before we go on to discuss subkingdom Algae, perform the following experiment to review subkingdom Protozoa. If you do not have a microscope, go on to the next section.

EXPERIMENT 3.2
Subkingdom Protozoa

Supplies:
- Microscope
- Prepared slides of amoebae, paramecia, *Euglena* and *Volvox*

Object: To become familiar with Protozoa, a subkingdom under the Kingdom Protista. You will examine the structure inside the organisms which will help you understand how these organisms live, breathe, feed, move about, and rid themselves of waste.

Procedure:
As you begin looking at each of the prepared slides, try to find several individuals and try to notice differences between them. Observe them at all three magnifications, in order to get a really good feel for what they look like. Once you have gotten a good feel for the general nature of the genus that you are observing, then find an individual whose organelles are pronounced and easily visible. Concentrate on that individual when you make your drawing.

A. **Slide 1: Amoeba, Phylum Sarcodina**
1. Observe the prepared slide of amoebae. This slide has been stained to make the organelles easier to view. Scan the slide until you find an amoeba whose organelles are easy to see. Observe it on high magnification.
2. Draw one of the amoebae found on the slide.
3. Label the following structures:
   a. pseudopod
   b. food vacuole
   c. contracting vacuole
   d. ectoplasm
   e. nucleus
   f. endoplasm

B. **Slide 2: Paramecium, Phylum Ciliophora**
1. Observe the stained slide of paramecium. Find a paramecium whose organelles are easy to see. Observe it on high magnification.
2. Sketch the paramecium, labeling the following structures.
   a. macronucleus
   b. micronucleus
   c. food vacuole
   d. contractile vacuole
   e. oral groove
   f. gullet
   g. mouth pore
   h. anal pore
   i. cilia

C. **Slide 3: Euglena, Phylum Mastigophora**
1. Observe the stained slide of euglena. Find a euglena whose organelles are easy to see. Observe it on high magnification.
2. Sketch the euglena, labeling the following structures:
   a. eyespot
   b. contractile vacuole
   c. pellicle
   d. chloroplast
   e. flagellum
   f. nucleus

D. **Slide 4: Volvox, Phylum Mastigophora.**
1. Observe the stained slide of volvox.
2. Sketch the volvox.
3. Notice that the individuals within the colony have their own flagellum. This is what places them in phylum Mastigophora.

Clean up: Put away the slides and the microscope, cleaning the microscope lenses with lens paper before you put it away.

Subkingdom Algae

Often called the "grass of the water," algae are organisms that can produce their own food by means of photosynthesis. If you've ever had a fish tank or a swimming pool, you probably have had a bit of experience with algae. If algae is allowed to grow unchecked in a body of water, the water becomes cloudy, and takes on a greenish tint. In swimming pools, we get rid of algae by killing it with a chemical, usually chlorine. In fish tanks, we usually do not get rid of *all* algae, but instead impede the growth of the algae population with fish or other creatures that eat it.

Of course, natural bodies of water are the best places to find algae. Whether we are talking about salt water (marine) environments or freshwater environments, algae abound. In fact, the "fishy smell" and "slimy feel" of a body of water is usually due not to the fish under the water, but the algae floating on it. In any natural body of water, there are tiny floating organisms called **plankton** (plank' ton).

Plankton - Tiny organisms that float in the water

Biologists separate plankton into two groups: **zooplankton** (zoo plank' ton) and **phytoplankton** (fye toe plank' ton).

Zooplankton - Tiny floating organisms that are either small animals or protozoa

Phytoplankton - Tiny floating photosynthetic organisms, primarily algae

Based on these two definitions, then, we could say that up to this point, we have mostly been discussing zooplankton. At this point, however, we are ready to discuss phytoplankton in depth.

As the definition states, phytoplankton are photosynthetic organisms, using the energy of sunlight to make their own food. As we mentioned in Module #1, oxygen is often the by-product of photosynthesis. This process is, in fact, responsible for replenishing the oxygen that we need to survive. After all, virtually all of the organisms on earth need to breathe oxygen in order to survive. Well, with all of these organisms breathing oxygen, why doesn't the earth run out of it? Because photosynthetic organisms are constantly making more. We will learn a lot more about photosynthesis in Module #5, so for now, you just need to understand that photosynthetic organisms continually replenish the earth's supply of oxygen.

Now when biologists mention photosynthesis, most people think of green plants, which are classified in kingdom Plantae. As we mentioned in Module #1, green plants do indeed use photosynthesis to create their own food. What you might not know, however, is that the *vast majority* of photosynthesis on earth is not done by green plants; instead, it is done by phytoplankton. This, of course, means that most of the oxygen on earth has been produced not by green plants such as trees and grasses, but instead by phytoplankton such as algae. In fact, biologists estimate that nearly 75% of all oxygen on earth is produced by phytoplankton. This is a rather important point. Contrary to popular belief, we could probably continue to survive even if all green plants were destroyed, because most of the earth's oxygen is produced by phytoplankton. Conversely, earth would quickly become uninhabitable if the waters of earth became too polluted for phytoplankton!

Algae are important for more reasons than the oxygen they produce. Algae are also a major food source for many aquatic (water-living) organisms. In addition, humans have found quite a number of uses for algae. Japanese people cultivate it as a food crop. In New England and Hawaii, certain types of marine algae are eaten as vegetables. Other algae are used as food additives. For example, a substance extracted from a species of algae commonly called Irish moss is used to thicken such things as Jell-O, pudding, and ice cream. In addition, algae can be used as a natural kind of fertilizer. Finally, many useful products such as potash, iodine, nitrogen, vitamins and minerals can be made from processed algae.

Many algae exist as individuals cells, but most form simple colonies that are held together with slime. A few species of algae form highly complex colonies. A colony of algae is often referred to as a **thallus** (plural is thalli), but that term actually has a much broader definition.

Thallus - The body of a plant-like organism that is not divided into leaves, roots, or stems

We often refer to algae colonies as thalli because they function like a big plant, but there are no distinct parts. There is just one big mass of algae.

Algae, whether in colonies or existing as individuals, have both asexual and sexual reproduction at their disposal. This is one reason that they are so abundant in aquatic environments. In fact, when conditions are ideal, algae will reproduce so rapidly that they essentially "take over" their habitat, making the water appear the same color as the algae itself. When this happens, it is referred to as an **algal bloom**. Now that you've had a general introduction to algae, it is time to examine each individual phylum in subkingdom algae so that we can learn more about this interesting subkingdom of Protista.

Phylum Chlorophyta

Members of phylum Chlorophyta are mostly found in fresh water, although some marine species do exist. The most visible feature of these algae is that they contain the pigment chlorophyll, which is green. As a result, they are often referred to as **green algae**. Just like the *Euglena*, members of this phylum store their chlorophyll in organelles called chloroplasts. Despite their name, most members of this phylum also have a yellowish pigment called carotenoids (kuh raht' ee noyds), making them appear yellowish green.

The other distinguishing feature of this phylum is that its members have cell walls made of **cellulose** (sell' you lows). We will learn more about cellulose in Module #5. For right now, you just need to know that cellulose is a substance which is composed of certain types of sugar.

Cellulose - A substance made of sugars. It is common in the cell walls of many organisms.

The cells that make up plants usually have walls made of cellulose. Since green algae have cell walls similar to plants and chlorophyll like most plants, some biologists actually consider them plants and place them into kingdom Plantae. Most biological classification schemes tend to place them in kingdom Protista, though, because they are microscopic and tend to exist as individual cells or simple colonies of cells.

Figure 3.10 shows magnified images of algae from three different genera in this phylum.

Figure 3.10

All photos by Kathleen J. Wile

Examples of Green Algae

Chlorella   *Desmid*   *Spirogyra*

The algae to the far left of Figure 3.10 are from the genus *Chlorella*. These algae exist as individual cells. Although they might clump together, the group does not cover itself with a common slime covering, nor do the individual cells work together in any way. Many of the species in this genus actually live inside other organisms, as another example of symbiosis. These algae use photosynthesis to produce food for both themselves and the organism in which they live. In return, they are protected from predators.

A species from the genus *Desmid* is shown in the middle of Figure 3.10. Members of this genus sometimes form simple colonies, but they mostly exist as individual cells. Most species of *Desmid* have interesting shapes, usually being comprised of two halves that are pinched in the middle. These halves are usually mirror images of each other.

One of the more interesting genera of green algae is called *Spirogyra*, shown at the far right of Figure 3.10. Members of this genus get their name and unusual appearance from their spiral chloroplasts. They form colonies of slender, chain-like threads of cells. These colonies, called **filaments**, can reach up to 2 feet long.

ON YOUR OWN

3.11 If an organism is in phylum Chlorophyta, it must have a chloroplast. Why?

3.12 Of the three genera of green algae discussed above, which would you consider the most complex?

Phylum Chrysophyta

Although the green algae are responsible for a large portion of the photosynthesis that occurs on the planet, the greatest producers of oxygen are in phylum Chrysophyta. This phylum contains many different species which are collectively referred to as **diatoms**. Diatoms are a unique type of algae, mostly because their cell wall is composed of **silicon dioxide**, which is the principal component of glass. This makes their cell wall very hard, providing excellent protection. The cell wall is so hard that it remains long after the diatom dies. When the cell wall remains of many dead diatoms clump together, they form a crumbly, abrasive substance called **diatomaceous earth**. The videotape "Water, Water Everywhere" (available at Nature's Workshop Plus) contains a nice segment on diatomaceous earth.

fresh water & marine

Huge deposits of diatomaceous earth exist in most regions of the world. Many creation scientists think that these deposits were laid down in the worldwide Flood described in the Bible (Genesis chapters 6-9). This makes sense because the catastrophic nature of the flood would be responsible for killing huge numbers of organisms, including diatoms. The currents caused by the flood would then tend to sweep the remains of the diatoms together and lay them in one place. Interestingly enough, scientists who do not believe in the Bible have no real explanation for these huge deposits of diatomaceous earth.

Diatomaceous earth is actually quite useful for people. Large amounts of diatomaceous earth are used by industry as a means of filtering liquids. It is also used as an abrasive. For example, most toothpastes contain an abrasive that helps clean and polish your teeth. *polish silver & kills bugs* Diatomaceous earth is often the abrasive of choice. Figure 3.11 contains magnified images of some diatoms.

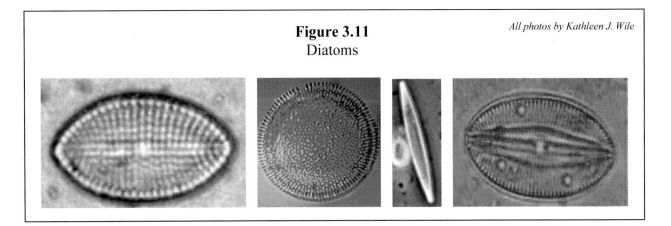

Figure 3.11
Diatoms

All photos by Kathleen J. Wile

If you look at the diatom on the far right side of the figure, you will see that there are what appear to be bubbles inside the organism. Although some people mistake these for flaws in the picture, they are, in fact another unique feature of diatoms. When diatoms have excess food, they do not store it in food vacuoles as many other protists do. Instead, they convert it into oil. The drops that you see in the figure are, in fact, oil drops that are stored food for the diatom.

Although diatoms make up a large part of phylum Chrysophyta, there are other organisms that belong to this phylum. The genus *Dynobryon* (dye noh' bree uhn), for example, is an algae that forms colonies. These colonies typically contain a few special cells called **holdfasts** which are designed to hold on to objects in the water such as rocks.

<u>Holdfast</u> - A special structure used by an organism to anchor itself

These holdfasts form long strands that attach to a surface in the water, acting like an anchor. The colony is then not at the mercy of the currents. When a colony uses holdfasts, it is usually called a **sessile colony**.

<u>Sessile Colony</u> - A colony that uses holdfasts to anchor itself to an object

If you were unfortunate enough to drink water that contained *Dynobryon*, it would taste fishy and feel quite slimy on the tongue.

ON YOUR OWN

3.13 Many biologists consider diatoms the most important form of algae in the world. Why?

3.14 If you observed a sessile colony of algae, would it tend to move about or stay in one place?

<u>Phylum Pyrophyta</u>

Phylum Pyrophyta contains a group of single-celled creatures that are often referred to as the **dinoflagellates**. They get their name because most species have two flagella of unequal length. Figure 3.12 shows a magnified image of a dinoflagellate from genus *Ceratium*. Notice the two flagella of unequal length on the bottom of the organism.

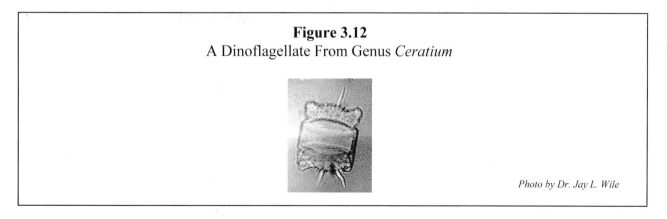

Figure 3.12
A Dinoflagellate From Genus *Ceratium*

Photo by Dr. Jay L. Wile

Some of these organisms are heterotrophic, some are photosynthetic, and they mostly inhabit marine waters. Like the green algae, their cell walls are composed of cellulose. The dinoflagellates that are photosynthetic are an important source of food for many aquatic organisms, including other forms of plankton.

The most important thing to remember about the dinoflagellates, however, is that certain species (*Gymnodinium brevis*, for example) frequently bloom in nutrient-rich waters. Because the species are reddish-brown in color, their bloom tends to turn the sea red in their immediate vicinity. As a result, these blooms are often called **red tide**.

cause

Red tides are very deadly to other marine creatures. Hundreds of thousands of fish can be killed in a single bloom of *Gymnodinium brevis*, because these organisms emit a toxin into the water. Under normal conditions, there are few enough dinoflagellates that the toxin never reaches levels which are dangerous to marine creatures. During a dinoflagellate bloom, however, the toxin reaches deadly proportions.

 nerve poison

Interestingly enough, although fish and most other marine life find red tides deadly, clams, oysters, and mollusks are immune. Unfortunately, the toxin emitted by the dinoflagellate does build up in their bodies. As a result, heterotrophs who eat clams, oysters, or mollusks that were exposed to a red tide can become poisoned. There are many human deaths that can be attributed to eating clams, oysters, or mollusks that have been in a red tide. This is why seafood restaurants refuse to serve these dishes when a red tide occurs in the area from which they get their supplies.

ON YOUR OWN

3.15 In the book of Exodus (Chapter 7), God (through Moses) caused several plagues to befall Egypt. In the first plague, all of the rivers turned to blood, the fish died, and the Egyptians could not drink from the rivers. Some have said that algae offer a natural explanation for this miracle. What algae are they referring to and why do they think this? Why is this not a good explanation?

Phylum Phaeophyta *macroscopic*

Up to this point, the members of kingdom Protista that we have studied have all been single-celled creatures that either exist as individuals or as members of simple colonies. In fact, one of the classification rules you learned in Module #1 says that kingdom Protista is made up of eukaryotic organisms made up of a single cell or a simple association of single-celled organisms. Well, hopefully by now you have seen that every classification rule has its exception. We can hardly discuss anything about a kingdom or phylum without using phrases like "usually" or "most of the organisms." Well, the last two phyla we will discuss are exceptions to the general description of the members of kingdom Protista.

Phylum Phaeophyta is comprised of multicellular organisms that inhabit the cold ocean waters. Now let's make sure that you understand the difference between a colony of single-celled organisms and a multicellular organism. In a colony of single-celled organisms, each organism can exist on its own. Although they tend to group together for mutual protection and other such benefits, if a cell is split away from its colony, it can survive. In addition, although the cells tend to live together, they mostly perform life functions independently. They may secrete mucus that covers the entire colony, and they may move together as a group, but they perform the majority of life functions without help from the other members of the colony. In a multicellular creature, the individual cells are designed to specialize in individual tasks. The cells work together, each performing the tasks that they are designed to perform. In the end, the organism survives because the cells work together. Since the cells are specialists in only one or a few of the tasks necessary for life, a single cell that is separated from a multicellular creature usually cannot exist on its own.

As shown in Figure 3.13, members of phylum Phaeophyta, also called the **brown algae**, look a lot like plants. In fact, some biologists classify them as such. Most biologists, however, still think that they have more in common with algae than with plants, so we classify them within kingdom Protista, subkingdom Algae.

If you enjoy ice cream, pudding, salad dressing, or jelly beans, you can thank the members of this phylum. You see, one of the unique characteristics of the organisms in this phylum is that their cell walls contain **alginic acid**, commonly called **algin**. This substance is *& cellulose* extracted from brown algae and used as a thickening agent in the foods described above. In addition, you can find algin in cough syrup, toothpaste, cosmetics, paper, and floor polish.

Remember

Figure 3.13
Two Genera Within Phylum Phaeophyta

Macrocytis

Photo by Cyber Sea Photography

Fucus

Photo © Mike Guiry, NUI, Galway

The most commonly known genus within phylum Phaeophyta is genus *Macrocytis* (ma' kroh sye tus). Species within this genus are called **kelp** or seaweed, although those terms also seem to be used for many species within phylum Phaeophyta. Kelp and most members of phylum Phaeophyta form holdfasts that allow them to anchor themselves to rocks which sit at the bottom of the ocean. Some kelp can reach lengths of up to 100 feet! *(2ft·a·day)* Kelp is cultivated as a vegetable in many cultures.

Species in genus *Fucus* are often called **rockweed**. These algae are thick and have a leathery feel. They live in the shallow waters along the shoreline, and are generally one to two feet long. One interesting structure on this kind of algae is the **air bladder**, which can fill with air to allow the organism to float on top of the water. You can see the air bladders on the *Fucus* in Figure 3.13. They look like brownish bulbs.

ON YOUR OWN

3.16 A biologist has a sample of what looks to be a marine plant. He thinks, however, that it might be an unknown species of brown algae. To test this, he takes part of the "plant," dries it, crushes it into a powder, and mixes it with water. The solution thickens. Is this evidence that the organism is a plant or an algae? Why?

Phylum Rhodophyta   *macroscopic*

cellulose The last phylum in subkingdom algae is Rhodophyta. Members of this phylum are often called **red algae** because of their strikingly red color. People often get these algae confused with the dinoflagellates, because they know that dinoflagellates cause red tides. The only thing that the red algae have in common with dinoflagellates, however, is the color. Like the brown algae, members of phylum Rhodophyta are multicelluar. Unlike the brown algae, however, they tend to live in warm marine waters rather than cold.

Figure 3.14 shows a typical member of phylum Rhodophyta, an organism from genus *Palmaria*.

FIGURE 3.14
Palmaria

Photo © Mike Guiry, NUI, Galway

These algae are a source of **agar**, which is often a food source. Biologists also use agar to culture bacteria.

To finish your study of subkingdom Algae, perform the following experiment. If you do not have a microscope, just skip the experiment and move on to the last section of the module.

EXPERIMENT 3.3
Subkingdom Algae

Supplies:
- Microscope
- Prepared slides of spirogyra and diatoms

Object: To observe the subkingdom Algae by observing 2 of the five phyla in this group.

Procedure:
As you did in Experiment 3.2, get a good general feel for each genus by looking at several individuals on all three settings. Then, when you find an individual with pronounced organelles, concentrate on it.

A. **Phylum Chlorophyta: The green algae. Example - genus *Spirogyra***
   1. Observe the prepared slide of spirogyra.
   2. To get a feel for the three-dimensional nature of this organism, find a single filament and observe it on medium and then high magnification. At each magnification, gently adjust the fine focus. You should see parts of the filament blur out and other parts become more defined.
   3. Sketch part of one filament, at least 4 cells.

B. **Phylum Chrysophyta: The yellow-green algae, the golden brown algae, and the diatoms. Example- Diatoms**.
   1. Observe the prepared slide of diatoms. Note the incredible variety.
   2. Sketch 3 or 4 different diatoms. In each case, look for oil spots which are stored food.

NOTE: You do not have prepared slides of phylum Phaeophyta, phylum Rhodophyta, or phylum Pyrrophyta: The first two contain the macroscopic algae, and the last one contains the dinoflagellates.

Summing Up Kingdom Protista

As you can see, kingdom Protista is quite diverse. The organisms within it range from single-celled individuals (like the amoeba), to colonies of single-celled individuals (like the *Volvox*), to multicellular creatures (like the red and brown algae). Some are heterotrophic (like the paramecium) and some are autotrophic (like the green algae). Some are pathogens (like the

Plasmodium), some are absolutely necessary for life (like the diatoms), and some produce substances which are incredibly useful to human beings (like kelp). Some of the organisms resemble animals (like the paramecium) and some resemble plants (like kelp).

Although there is a lot of diversity in this kingdom, there are a couple of things that all of these organisms have in common. First of all, they are all composed of eukaryotic cells. Secondly, they are all designed by God. As you look at the intricate features that exist in these organisms, God's incredible handiwork is readily apparent. Only He could produce the amazing diversity and complexity that exist in this kingdom. Remember, however, that the creatures in kingdoms Monera and Protista are, in fact, the **simplest** life forms that exist on the planet! As we study more of God's Creation, you will become more and more appreciative of the awesomeness of His power!

ANSWERS TO THE ON YOUR OWN QUESTIONS

3.1 Biological Key for separating members of kingdom Protista into phyla:

1. Organism is completely autotrophic............*subkingdom Algae*............. **2**
   Organism is heterotrophic ....*subkingdom Protozoa*........ **6**

2. Single-Celled.. **3**
   Multicellular **5**

3. Cell wall made of silicon dioxide .... *phylum Chrysophyta*
   Cell wall made of cellulose or is atypical **4**

4. Marine habitat .... *phylum Pyrrophyta*
   Fresh water habitat .... *phylum Chlorophyta*

5.* Cell wall made of cellulose and alginic acid .... *phylum Phaeophyta*
   Cell wall made just of cellulose .... *phylum Rhodophyta*

6. Possesses a means of locomotion **7**
   Possesses no means of locomotion .... *phylum Sporozoa*

7. Uses pseudopods for locomotion .... *phylum Sarcodina*
   Does not use pseudopods for locomotion **8**

8. Uses cilia for locomotion .... *phylum Ciliophora*
   Uses flagella for locomotion .... *phylum Mastigophora*

* This key could ask whether habitat is warm water or cold water. That would separate the two phyla as well

Now your key does not have to be in the same order, but it should contain all of these questions. It might have a few more steps, if you weren't as efficient as me in setting up the key.

3.2 Since the <u>contractile vacuoles</u> control pressure in the cell by collecting and removing excess water, those organelles must not have been working in that poor amoeba.

3.3 Members of phylum Sarcodina do not like to swim because of their laborious means of locomotion. As a result, a sarcodine will <u>try to attach itself</u> to any surface with which it collides.

3.4 Since it is wandering around, its flagellum is working fine. Since it can't find the light, however, its <u>eyespot</u> must not be working. NOTE: Even if the photosynthetic mechanism of the euglena is destroyed, it will continue to seek light as long as the eyespot is working.

3.5 *Euglena* can obtain food either autotrophically (by photosynthesis) or heterotrophically, depending on environmental conditions. Thus, it does not *always* perform photosynthesis.

3.6 The student is not correct. Symbiosis must benefit both organisms. The *Trypanosoma* hurts the human, so the relationship is not symbiotic.

3.7 Since conjugation occurs via the oral groove, that organelle must not be functioning properly in the paramecium.

3.8 Paramecia can engage in a form of conjugation that allows DNA to be mixed between organisms. As a result, the offspring formed will not be an exact duplicate of the parents, because the parents have mutually exchanged DNA. When bacteria conjugate, DNA transfer is one-way and the recipient ends up looking like the donor.

3.9 Spores are formed as a natural part of an organism's life. An organism that produces spores will always produce spores, at least once in its lifetime. Cysts, on the other hand, are formed only when life-threatening conditions occur. If no life-threatening conditions occur over the course of an organism's lifetime, it will never form a cyst.

3.10 Mosquitoes carry the Plasmodium and spread it by biting humans. Control the mosquito population, and the spread of the disease is controlled as well.

3.11 Chloroplasts hold the chlorophyll that is a part of photosynthesis. Since all members of phylum Chlorophyta have chlorophyll, they must also have chloroplasts to hold it.

3.12 Spirogyra are the most complex because they exist as colonies. Thus, to a very limited degree, the cells work together. The other two exist as individual cells. Any time you get a group of individuals to work together (even to a very small degree), you are adding complexity.

3.13 Diatoms are responsible for the majority of photosynthesis on earth. Without this photosynthesis, the earth's oxygen supply would quickly dwindle, killing off everything. Thus, diatoms are pretty important!

3.14 It would stay in one place, because a sessile colony, by definition, anchors itself to an object.

3.15 They refer to the *Gymnodinium brevis*, because it causes red tide. Since the water in a red tide turns red, it might appear to be blood. Also, red tides are toxic to humans, so people cannot drink the water in a red tide. This is not a good explanation, however, because the Bible does not say that the waters looked like blood, it says they *turned into* blood. It is very dangerous to look for naturalistic explanations for clearly supernatural events!

3.16 This is evidence that the organism is an algae. Since the water thickens, the plant most likely contains alginic acid, a substance that is in the cell walls of brown algae.

STUDY GUIDE FOR MODULE #3

1. Give definitions for the following terms:

a. Pseudopod
b. Nucleus
c. Vacuole
d. Ectoplasm
e. Endoplasm
f. Flagellate
g. Pellicle
h. Chloroplast
i. Chlorophyll
j. Eyespot

k. Symbiosis
l. Cilia
m. Spore
n. Plankton
o. Zooplankton
p. Phytoplankton
q. Thallus
r. Cellulose
s. Holdfast
t. Sessile colony

2. Study the images of the organisms found in Figure 3.1. You will be expected to be able to place each of these organisms into the correct subkingdom and phylum by just looking at its picture. On the test, you will have a list of the subkingdoms and phyla, you will simply have to match them to the picture.

3. Which of the following genera contain organisms with chloroplasts?

Amoeba, Euglena, Paramecium, Spirogyra

4. What is the function of a contractile vacuole? What is the difference between this and a food vacuole?

5. What is the difference between endoplasm and ectoplasm?

6. The amoeba and euglena each have different means of locomotion. How are they different? How are they similar?

7. Name at least three pathogenic organisms from kingdom Protista.

8. For each of the phyla below, list the means of locomotion employed by the organisms in that phyla:

Sarcodina, Mastigophora, Ciliophora

9. What are the main features that separate organisms into phylum Sporozoa?

10. A tapeworm is a parasite that feeds on the nutrients which the host eats, depriving the host of that nutrition. *Trichonympha* is a mastigophorite that lives in the gut of a termite, helping break down chemicals that the termite cannot break down on its own. Which is an example of mutualistic symbiosis? Why is the other not an example of symbiosis?

mutualistic

11. Why do the ciliates have two nuclei (plural of nucleus)? What is the purpose of each?

✳ 12. What is the difference between the conjugation that occurs between paramecia and the conjugation that occurs between bacteria?

13. Two microorganism groups are studied. In the first group, the organisms form hard shells around themselves when exposed to life-threatening conditions. If not exposed to those conditions, however, these organisms never form hard shells. The second group form hard shells around themselves as a natural part of their life cycle. Which group would be classified as coming from phylum Sporozoa?

14. What is unique about the way a euglena obtains food?

✳ 15. Which phylum (see list in question #2) contains the organisms responsible for most of the photosynthesis that occurs on earth? What generic term is used to refer to these organisms?

16. Give the main function of each of the organelles listed on the left below. Also, choose from list on the right at least one phylum that has organisms which possess the organelle.

Organelle
Food vacuole
Contractile vacuole
Flagellum
Pellicle
Chloroplast
Eyespot
Cilia
Nucleus
Oral Groove

Phylum
Sarcodina
Mastigophora
Ciliophora
Sporozoa
Chlorophyta
Chrysophyta
Pyrrophyta
Phaeophyta
Rhodophyta

✳17. What are large deposits of diatom remains called? List two uses of these deposits.

✳ 18. What is a red tide?

✳19. What two phyla principally contain macroscopic algae?

✳ 20. What substance produced by members of phylum Phaeophyta is useful for thickening ice cream, pudding, salad dressing, and jelly beans?

Module #4: Kingdom Fungi

Introduction

Have you ever gone mushroom hunting? We mean, other than in a store. Where did you search for the mushrooms? Most likely, you looked for them in decaying mats of leaves, piles of dead tree limbs, or other places where the remains of dead plants and animals are found. Why? Well, mushrooms are a part of kingdom Fungi, and most of the organisms in this kingdom are saprophytic. In other words, they are the decomposers that promote the decay of once-living matter. As we mentioned back in Module #1, the role of the decomposers in nature is very important. In a single autumn, the average elm tree will drop as much as 400 pounds of leaves on the ground. If it were not for the fungi and other decomposers, those leaves would continue to pile up, until the tree choked on its own dead leaves in just a few seasons! Because of the decomposers, however, the leaves will be broken down into chemicals that can then be re-used by the tree.

Although mushrooms are the most well-known fungi, there are many other organisms that make up this important kingdom. For example, the yeast that is used to make bread is a member of kingdom Fungi. The mold that later grows on that bread is also a fungus. There are some rather exotic creatures such as slime molds that belong to this important kingdom as well. Some fungi are beneficial to us beyond their roles as decomposers. We eat some fungi, other fungi are used in the production of cheese, others are used in baking, and some even produce important medicine. Still other members of kingdom Fungi are pathogenic. Diseases such as St. Anthony's fire, histoplasmosis, potato blight, and Dutch elm disease are all caused by fungi. A study of kingdom Fungi, then, should prove to be quite interesting.

NOTE: In order to perform Experiment 4.3, you will need to grow some mold on bread, jelly, and/or fruit. You need to start that process now. To do this, take a slice of bread (sprinkle some water on it), a sample of jelly, and a piece of sliced fruit and set them out in the open. Putting them outside works best, but you will need to find a place where the birds and other animals cannot get to them. If you can't find such a place outside, then put them somewhere like a garage or shed. They will attract fruit flies, so do not put them in the house! If you live in a very arid climate (like Arizona), then put the samples in a plastic bag with some water. This will keep them moist, which is necessary for the growth of molds. As you read through the first part of this module, the mold will start to grow. By the time you reach Experiment 4.3, at least one of the samples should have mold on it. This experiment has non-microscope components, so you should prepare for it whether or not you have a microscope.

General Characteristics of Fungi

3 charachteristics

Although very diverse, the organisms in kingdom Fungi are not nearly as widely-varying as those in kingdom Protista. As a result, we can develop a few general characteristics for all fungi. For example, all fungi are heterotrophs. Most are saprophytic (feeding on the remains of dead organisms), although there are a few species that are parasitic (feeding on living hosts). Whether saprophytic or parasitic, however, all fungi digest their food outside of their bodies.

Phylum-Amastigomycota

3 classes:
1.) basidiomycetes
  (mushrooms)
2.) ascomycetes
  (yeasts) - reproduce primarily by budding (asexual)
3.) Zygomycetes
  (breads)

The fungus grows on and in its food, secreting a chemical onto the food that digests it *before* it is ingested. The digested food is then absorbed into the cells of the fungus, an instant source of nutrition and energy. This **extracellular digestion** can be beneficial to other organisms which often absorb some of the nutrients before the fungus has a chance to absorb them.

budding process: 1.creates a new nucleus 2.creates a pouch (nucleus + some cytoplasm) move in

• Extracellular digestion - Digestion that takes place outside of the cell

3. New cell wall formed

If a fungus is harmful to other creatures, it is often the substances which the fungus excretes for extracellular digestion that are responsible.

Fermentation- yeast uses sugar + produces alchohal + carbon dioxide

Another trait common to all fungi is an aspect of reproduction. All fungi reproduce by making spores. Most fungi have other means of reproduction at their disposal, but they all have this mode of reproduction in common. If the fungus is multicellular (as most are), then the fungus will form a specialized structure (we will dwell on these structures later) to produce the spores. Once the spores are produced, they are dispersed from the fungus and begin to develop on their own. In the next section, we will look at the reproduction of fungi in detail. First, however, we need to learn a few more general characteristics of fungi.

The vast majority of fungi are multicellular creatures. The largest part of their body is the part that performs extracellular digestion and absorbs the digested food. This part of the fungus, called the **mycelium** (my sell' ee uhm) is composed of many interwoven filaments called **hypha** (hi' fuh).

• Mycelium (plural is mycelia) - The part of the fungus responsible for extracellular digestion and absorption of the digested food *(reproduction)*

• Hypha (plural is hyphae) - Filament of fungal cells

Figure 4.1 is a sketch of a typical mushroom. Notice the mycelium that exists below the mushroom's stalk. It is not uncommon for a mushroom's mycelium to be ten or twenty times as large as its stalk.

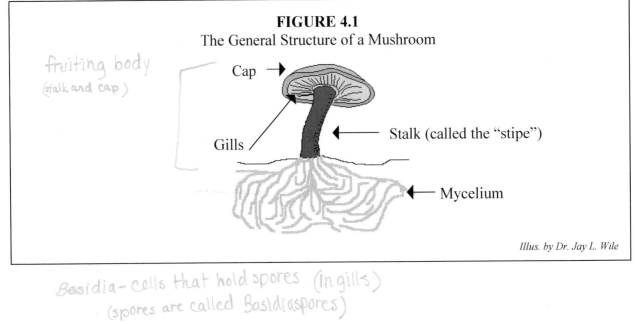

FIGURE 4.1
The General Structure of a Mushroom

fruiting body
(stalk and cap)

Cap →

Gills

← Stalk (called the "stipe")

← Mycelium

Illus. by Dr. Jay L. Wile

Basidia- cells that hold spores (in gills)
(spores are called Basidiaspores)

Now don't be fooled by this drawing. The mycelium is *not* a root system for the mushroom. Although the hyphae might look a lot like roots, there are *many* differences between a root system and the mycelium of a mushroom. A root system, for example, has one purpose: to pull nutrients and water from the soil so that they can be transported to the rest of the plant. The mycelium of the mushroom, on the other hand, is the *main part* of the mushroom. We will learn later that the stalk, cap, and gills of a mushroom exist only at a certain stage of the mushroom's life. The mycelium, however, exists throughout the entire life of the mushroom. Thus, whereas the root system is really just an extension of the tree, a mushroom's stalk, cap, and gills are really just an extension of the main body - the mycelium.

In some fungi, the hyphae are composed of individual cells separated from one another by cell walls. Such hyphae are called **septate hypha**. Even though the individual cells are separated from one another by a cell wall, there is usually a hole or pore through which cytoplasm can be passed between the cells. Other fungi have hyphae that look like one big cell. There are no walls, and the nuclei are spread throughout the hypha. These hyphae are called **nonseptate hyphae**. The difference between these two type of hyphae is illustrated in Figure 4.2

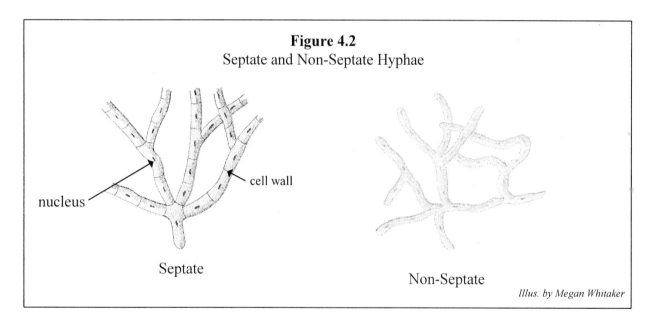

Figure 4.2
Septate and Non-Septate Hyphae

nucleus

cell wall

Septate

Non-Septate

Illus. by Megan Whitaker

Many fungi have hyphae that are designed to perform specific tasks. If a hypha is part of the mycelium, for example, it is called a **rhizoid** (ry' zoyd) **hypha**.

Rhizoid hypha - A hypha that is imbedded in the material on which the fungus grows

Rhizoid hyphae are responsible for supporting the fungus and digesting the food. As stated before, these hyphae are considered the main body of the fungus.

Other types of hyphae can exist in a fungus, depending on the species and the circumstances. For example, an **aerial hypha** is not imbedded in the material on which the fungus grows.

<u>Aerial hypha</u> - A hypha that is not imbedded in the material upon which the fungus grows

As its name implies, an aerial hypha sticks straight up in the air. It can do one of three things: absorb oxygen from the air, produce spores, or asexually reproduce to form new filaments. If the hypha performs one of the latter two jobs, it is further specified as either a **sporophore** (spor' uh for) or a **stolon** (sto' lun), respectively.

<u>Sporophore</u> - Specialized aerial hypha that produces spores

<u>Stolon</u> - An aerial hypha that asexually reproduces to make more filaments

Often, a sporophore will form within an enclosure, making one or more balls on the hypha. When this happens, the sporophore is called a **sporangiophore** (spuh ran' jee uh for'). If no enclosure is made, the sporophore is called a **conidiophore** (ko nid' ee uh for'). Sporophores are not the only way that fungi produce their spores. Likewise, stolons are not the only way a fungus asexually reproduces. In fact, many fungi do not form sporophores or stolons at all. Nevertheless, these two specialized hyphae are means by which some fungi reproduce.

In the rare case of a fungus that feeds on a living organism, a hypha can actually enter the cells of the living organism and draw nutrients directly from the cytoplasm of its cells. This kind of hypha is an extension of the mycelium and is called a **haustorium** (haw stor' ee uhm).

<u>Haustorium</u> - A hypha of a parasitic fungus which enters the host's cells, absorbing nutrition directly from the cytoplasm

Examples of these specialized hyphae are shown in Figure 4.3. As the figure implies, not all fungi have all of these structures. The only specialized hyphae that all fungi have is the rhizoid hyphae, because they make up the mycelium.

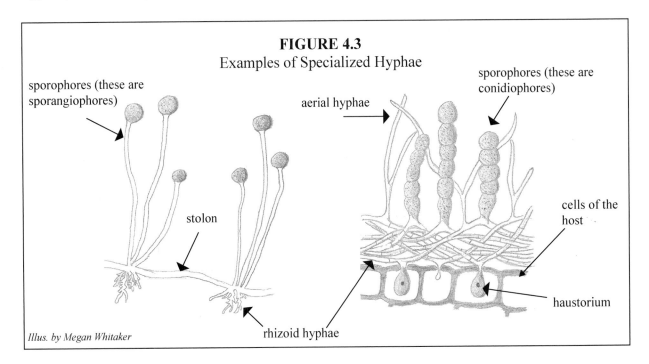

FIGURE 4.3
Examples of Specialized Hyphae

sporophores (these are sporangiophores)

aerial hyphae

sporophores (these are conidiophores)

stolon

cells of the host

haustorium

Illus. by Megan Whitaker

rhizoid hyphae

The last common characteristic of fungi refers to the cell wall. Most fungi have cell walls that contain **chitin** (ky' tin).

<u>Chitin</u> - A chemical that provides both toughness and flexibility

This chemical makes the fungi more hardy than most plants, because it provides protection, much like armor provided protection to the knights of old. Interestingly enough, this same chemical is found on the shells (called exoskeletons) of many insects, such as spiders and ants. It is also found in crustaceans like lobsters.

ON YOUR OWN

4.1 An organism eats food and then digests it. Does this organism belong to kingdom Fungi? Why or why not?

4.2 A farmer tries to remove a patch of mushrooms from his field by pulling all of the stalks and caps in the patch out of the ground. Why has the farmer really not gotten rid of the fungus?

4.3 A fungus produces haustoria (plural of haustorium). Is it saprophytic or parasitic?

<u>Reproduction in Fungi</u>

One reason that fungi are so hardy and plentiful is that they have many means of reproduction at their disposal. As we mentioned previously, all fungi reproduce by making spores. Sometimes spores are formed asexually, but all fungi are assumed to have some sexual mode of spore formation as well. Asexual spore formation is accomplished by a hypha that becomes either a sporangiophore or a conidiophore. The spores are formed and then dispersed. Some fungi produce spores that have flagella for locomotion, but most spores are just carried by the wind until they land. If a spore finds itself in a suitable environment, it can grow into a hypha and form a new mycelium.

The sexual reproduction that usually occurs in fungi involves forming specialized structures called **fruiting bodies**. These fruiting bodies can be formed out of hyphae in the mycelium or, in some fungi, as the result of sexual reproduction between two different mycelia. Once the fruiting body is formed, it rises out of the mycelium and releases its spores. The cap and stalk that we normally call a mushroom are, in fact, just parts of the fruiting body of the mycelium of a fungus.

Although spore formation is the reproductive mode common to all fungi, most can also reproduce by other asexual means. The hyphae cells in the mycelium can reproduce asexually, increasing the size of the mycelium. Also, when a stolon is formed, the cells within it will reproduce asexually, lengthening the stolon. After the stolon reaches a certain length, it will begin to reproduce into hyphae that will form the mycelium of a new fungus. Because the stolon is still attached to both the parent fungus and the new fungus, these fungi will always be attached by the stolon that gave rise to the new fungus. The new fungus can then form another stolon, which will give rise to yet another fungus. This process can be repeated over and over again, so it is not uncommon to find long chains of fungi, all linked together by stolons.

Finally, there is one more mode of sexual reproduction available to fungi. Hyphae of two different mycelia can also reproduce sexually, forming new hyphae.

ON YOUR OWN

4.4 What job does the fruiting body of a fungus perform?

4.5 Spores of a fungus give rise to offspring that are identical in every way to the parent. Were the spores formed asexually or sexually?

Classification in Kingdom Fungi

Kingdom Fungi is divided into four phyla: Mastigomycota (mass' tuh go my' ko tuh), Amastigomycota (ah mass' tuh go my' ko tuh), Imperfect Fungi, and Myxomycota (mike so' my ko tuh). Organisms are placed within each phyla based mostly on the type of spores they produce in sexual reproduction. Phylum Mastigomycota contains those fungi whose spores have flagella and can thus move from place to place on their own. We often called these spores **motile**. Phylum Amastigomycota, on the other hand, contains fungi whose spores cannot move on their own. We often refer to such spores as **non-motile**.

The phylum called Imperfect Fungi is, in fact, a phylum where we put fungi whose complete reproductive methods are not well-known. As we have said before, all fungi reproduce by forming spores. Although some fungi have a means of forming spores asexually, all fungi in phyla Mastigomycota and Amastigomycota have at least one mode of sexual reproduction which forms spores. We assume, therefore, that all fungi have at least one mode of sexual spore formation. There are some species of fungi, however, for which sexual reproduction is unknown. We place these fungi into the phylum Imperfect Fungi, expecting that more research will eventually lead us to the discovery of a sexual, spore-forming mode of reproduction. If such a mode is found for a fungus in this phylum, it is reclassified into one of the first two phyla based on whether or not the spores are motile. If someone can ever demonstrate that there are some fungi without any mode of sexual spore formation whatsoever, then another phylum will have to be formed. Some biologists have taken the first step in this direction, giving the imperfect fungi a classification name, Deuteromycetes (doo' ter oh my see' teez).

The last phylum in kingdom Fungi, Myxomycota, is rather controversial. This phylum contains the strange organisms known as slime molds. These interesting creatures grow in moist habitats on the decaying remains of living creatures (particularly trees) or in some cases exist as parasites that feed on living plants. They are typically brightly-colored and slimy to the touch. In many ways, they resemble protozoa. Most of the time, they behave like a colony of single-celled creatures. However, at some point in their life cycle, they produce sporangiophores for reproduction. Since a sporangiophore is generally associated with a fungus, they tend to resemble fungi during this stage of their life. Since the slime molds resemble protozoa during a part of their life cycle and fungi during another part of their lifecycle, there is much controversy as to whether to place these creatures in kingdom Protista or kingdom Fungi. We choose to place them in this kingdom for two reasons. First, since they reproduce by forming spores from specialized structures, they seem to fit in with the other fungi. Also, they are found in the same type of habitats as fungi, so it only seems natural to consider them a part of this kingdom. Be warned, however, that the next biology book you read might classify the slime molds as belonging to kingdom Protista! Table 4.1 sums up the classification of fungi into phyla.

TABLE 4.1
Phyla in Kingdom Fungi

Phylum	Characteristics
Mastigomycota	Reproduce with motile spores
Amastigomycota	Reproduce with non-motile spores
Imperfect Fungi	No known sexual mode of spore formation
Myxomycota	Resemble both protozoa and fungi

Do not try to memorize the names of the phyla. They will always be given to you on tests. Instead, learn the characteristics that separate organisms into these phyla.

ON YOUR OWN

4.6 Construct a biological key that separates organisms into the phyla of kingdom Fungi. You can assume that the organisms all come from kingdom Fungi.

Phylum Amastigomycota

We will devote a large amount of attention to this phylum, as it contains the fungi with which we are most familiar. We separate fungi in this phylum into three different classes: Zygomycetes (zie go my see' teez), Ascomycetes (as ko my see' teez), and Basidiomycetes (buh sid' ee oh my see' teez). We will discuss each class in detail, starting with the one which contains the most well-known of all fungi: the mushrooms.

Class Basidiomycetes: The Mushrooms

Often referred to as the "club fungi," members of class Basidiomycetes form four spores (called **basidiospores**) on club-shaped cells known as **basidia** (singular is basidium). These spores are the result of sexual reproduction between mycelia. As we pointed out in the previous sections, the cap and stipe (stalk) of the mushroom is actually only the fruiting body of a vast network of mycelia that exist below the surface upon which the mushroom grows. Examples of species from class Basidiomycetes are shown in Figure 4.4.

FIGURE 4.4
Class Basidiomycetes

Mushrooms	Puffballs	Shelf Fungi
Photo from the MasterClips collection	*Photo from the EXPERT 3000 collection*	*Photo by Kathleen J. Wile*

As you can see from the figure, most of the fungi with which you are familiar fall into this class. Most of these fungi are saprophytic, but a few (which we will discus later) are parasitic.

The lifecycle of the club fungi is rather interesting, and it is illustrative of the complex nature of most fungi. Therefore, we will study it in some detail. Begin by examining Figure 4.5 on the next page.

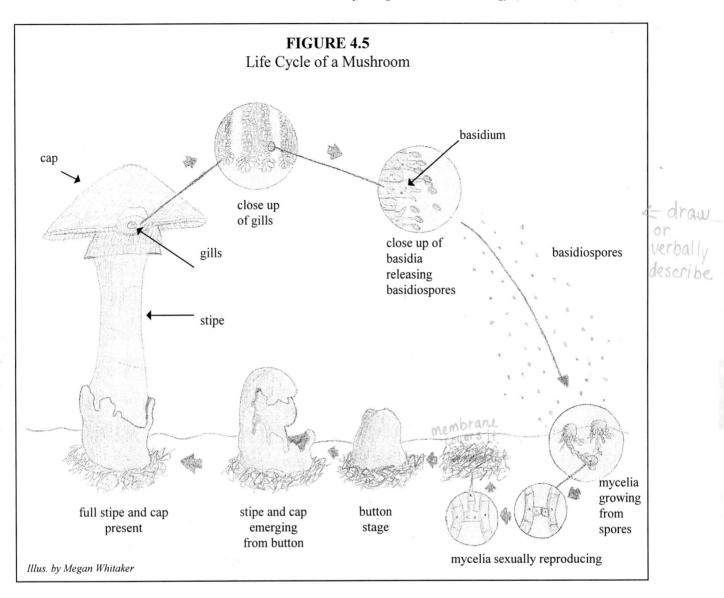

FIGURE 4.5
Life Cycle of a Mushroom

Illus. by Megan Whitaker

A mushroom begins life as a small mycelium that grows from spores which have come from another mushroom. As the mycelium begins to grow, it might encounter the mycelium of another mushroom nearby. As the two mycelia begin to intertwine, their hyphae will sexually reproduce. They accomplish this by aligning themselves parallel to each other and forming a small junction. At this point, we say that the hyphae are **fused**. Once fused, the hyphae exchange DNA and form more mycelia. Eventually, through some process that we do not understand, a group of the hyphae will form a complex web and enclose themselves in a **membrane**.

Membrane - A thin covering of tissue

This membrane-enclosed web of hyphae (often called a "button") is the beginning of the mushroom's fruiting body.

When the hyphae are formed in the membrane, we say that the mushroom has reached the **button stage** of its existence. At that point, the hyphae begin filling with water quickly, and eventually the stipe and cap of the mushroom break through the membrane. The hyphae fill with water so quickly during the button stage, that the stipe and cap of a mushroom can literally "pop up" out of the membrane overnight. Since many mushrooms are hard to see in their button stage, it gives the illusion that a mushroom formed itself overnight, when in fact, the mushroom that you see is the result of many days' growth.

Since the stipe and cap of a mushroom form the fruiting body of a fungus, their main function is that of reproduction. The cap is full of **gills**, small plates that are lined with basidia. The basidiospores contained on the basidia are flung from the mushroom, where wind and water carry them to a new location, starting the process all over again. Once the fruiting body of the mushroom releases all of its spores, it withers and dies. Of course, the fungus is still very much alive, because the mycelium is still digesting and absorbing nutrients below the surface. Eventually, more mycelia will sexually reproduce, clump together, and form another fruiting body.

The fruiting body of the mushroom, of course, is the part that we eat. Although most mushrooms are tasty and nutritious, there are some that are quite toxic. The genus *Amanita* (ah mah nee' tuh), for example, contains mushrooms which are commonly called "destroying angel" mushrooms. These pure white mushrooms carry a poison that is deadly to humans. If you eat one of these mushrooms, it tastes quite normal. However, once you eat one you will die in about 16 hours. There is virtually nothing that can be done to treat a person who has eaten these deadly mushrooms. Because there are no truly distinguishing marks that can separate poisonous mushrooms from non-poisonous ones, the only place that you should hunt them is in the grocery store. Many people who try to hunt wild mushrooms end up in a hospital or a coffin because the mushrooms that look tasty are, instead, toxic.

Sometimes, you can find mushrooms that grow in an almost perfect circle. Inside or outside of this ring of mushrooms, no other mushrooms grow. These rings, often called **fairy rings**, are believed by some to have magical properties because of their unique appearance. Of course, there is no magic associated with a fairy ring. Instead, it is just a result of the saprophytic nature of the fungus. You see, when a fungal mycelium begins to grow in an area, it eats the remains of dead organisms. As it eats, it grows and reproduces. Eventually, the mycelium will spread out in all directions, making a relatively circular patch of hyphae. Once the hyphae in the center of the circle eat up all of the remains of dead organisms, there is no more food for them, and they die. The hyphae at the edge of the mycelium, however, still have food, because they haven't existed for as long and therefore have not used up the food in their area. As a result, they continue to live, and the mycelium becomes ring-shaped. When it is time to reproduce, then, the ring produces fruiting bodies, forming a fairy ring.

As time goes on, the mycelium continues to grow outwards, and the inner hyphae continue to die because they use up their food. As a result, the mycelium retains its ring shape, but the ring gets larger in diameter. When the reproduction cycle comes again, then, a new ring of stipes and caps are formed, and this ring is larger than the old one. This happens year after year. Some fairy rings have been found in which the ring of mushrooms is only a few inches wide, but has a diameter of nearly 20 feet!

Other Members of Class Basidiomycetes:

Class Basidiomycetes is also home to the **puffball** fungi (middle of Figure 4.4). Puffballs, which are also saprophytic, produce their spores on basidia inside a membrane, rather than in the gills of a cap. When disturbed by a passing animal or a heavy wind, the membrane breaks and the spores are released. The spores, which are as fine as dust, are often carried on the wind for several miles before they hit the ground. As a result, you are unlikely to find dense patches of puffballs. They tend to be spread out over vast distances.

The **shelf fungi** (right hand side of Figure 4.4) are generally found either on dead wood or on living trees. If you find them on dead wood, they are obviously one of the saprophytic species of shelf fungi, while those found on living trees are the parasitic species. Although parasitic fungi are uncommon, they do exist, and class Basidiomycetes is home to some of them.

The spores of the shelf fungi are formed in the pores of the shelves. These spores are also very fine, so that when they are released, they can travel great distances to find another tree on which to grow. Once again, the shelves are just the fruiting bodies of these fungi. The mycelia are inside the wood of the tree. In fact, some parasitic shelf fungi actually add a new layer to the fruiting body each year, resulting in huge shelves growing out of the tree trunk.

Rusts are another form of fungi that are parasitic. They typically grow on living plants, reducing the plant's ability to grow and mature. If the rust happens to be living on a commercial crop, it can make the crop virtually useless as a source of food. One particularly bothersome form of rust is the **wheat rust**. This fungi is well-known for destroying tons of wheat crop over the course of history. Its life cycle is actually rather complex, because it requires two hosts, a **main host** and an **alternate host**. Both of them are necessary for the rust to complete its life cycle.

When wheat rust infects a wheat plant, it produces a red spore called a **uredospore** (yoo ree' duh spor). This red spore is what gives the fungus its name. These spores can travel on the wind to other wheat plants and grow on them, destroying entire fields of wheat. As the wheat season ends, however, the wheat plants turn yellow, and the rusts form a different type of spore, called a **teliospore** (tee' lee uh spor). These spores survive the winter and then grow into basidia in the spring. The basidia produce basidiospores, but these spores cannot grow into fungi on wheat. Instead, they find their way to a barberry bush and grow on the underside of the leaves of this bush. They form tiny cups in which **aeciospores** (ee' see oh sporz) are produced. These spores can then find their way to wheat plants and grow into rust there. The rust, then, lives most of its life cycle on wheat, its main host, but in the spring it must spend a certain part of its life cycle on its alternate host, the barberry bush. The need for an alternate host is rather common among parasitic fungi.

Smuts are another group of parasitic fungi that belong in class Basidiomycetes. These fungi also feed on crops such as wheat, barley, rye, and corn, resulting in millions of dollars worth of crop loss each year. Typically, farmers cannot control smuts or rusts chemically, because anything that kills the fungi also kills the plants upon which they live. Instead, agricultural scientists try to develop strains of wheat, barley, rye, and corn that are resistant to these fungi. Many advances have been made in such crop-related research and, as a result, the crops planted today are less likely (but not immune) to be ruined by fungi.

To finish our study of class Basidiomycetes, perform the following experiment. Although this lab calls for a microscope, you can really perform all of the steps with just a magnifying glass. You won't see quite as much, but you will still learn a lot!

Experiment 4.1
Class Basidiomycetes

Supplies:
- Microscope
- Magnifying glass
- Slides
- Coverslips
- Water
- Needle
- Mushrooms
- Puffballs
- Shelf fungi

The Fall, when leaves are dropping to the ground and there is abundant moisture, is a perfect time to find mushrooms, shelf fungi and/or puffballs in yards and woods nearby. Their job is to decompose all of the dying organisms about them. If you can not locate any specimens, then purchase a few mushrooms from the store. **You only need one of the above specimens to do the lab**. If you can find more, however, you will learn more.

Object: To observe the fungi that are readily found in most areas and to understand how members of class Basidiomycetes grow and reproduce.

Procedure:
A. Go for a walk and look for mushrooms, puffballs, and shelf fungi in your area. A wooded area with fallen leaves or dead logs should have some fungi. Collect samples to take back home. Handle them carefully. Do not breathe in spores, especially puffball spores. They can be an irritant to your respiratory system. Always wash your hands after handling any fungi.

B. Observe each specimen collected. Study using a magnifying glass. Sketch and label the stipe, cap and gills.

C. With a hand lens or magnifying glass, look at the gills. Try to see the basidia and the basidiospores. As shown in Figure 4.5, the basidia are on the gills, and the basidiospores are in the powdery substance attached to the basidia.

D. Cut the stipe off at the base of the cap. Tap the mushroom lightly on a microscope slide. A powder (the basidiospores) should fall out of the gills. This may not work if the mushroom isn't dry. Place a drop of water and coverslip over the powder. Observe on all magnifications of your microscope. If you do not have a microscope, just tap the mushroom against a small plate and add water. Then observe the spores with a magnifying glass. You should see oval-shaped basidiospores. Sketch a few.

E. Cut the cap vertically. Now take your needle and pull a gill off of the cap. Place on a slide. Add a drop of water and coverslip. Observe on all magnifications. Once again, if you don't have a microscope, use a small plate and a magnifying glass. Note the clear basidium and the darker basidiospores. Sketch and label.

F. Cleanup: Wash and dry all slides and droppers. Dispose of all fungi. They do not keep; do not try to save them. WASH YOUR HANDS! Clean your microscope and return all supplies to the proper places.

ON YOUR OWN

4.7 A mushroom is in its button stage. Has it released its spores yet?

4.8 One major characteristic that separates the members of class Basidiomycetes is the structure in which the fungi form their basidia. Where are the basidia formed in mushrooms? Puffballs? Shelf fungi?

Class Ascomycetes: The Sac Fungi

The members of class Ascomycetes are both single-celled creatures and multicellular organisms. They are generically referred to as **sac fungi**, because they form their spores in protective membranes (sacs) shaped like globes, flasks, or dishes. These sacs are called **asci** (as' ki) and the spores inside are called **ascospores** (as' kuh sporz). The single-celled members of this class are generically called **yeasts**. The yeast that we use in cooking is an example of a single-celled member of class Ascomycetes.

FIGURE 4.6
Members of Class Ascomycetes

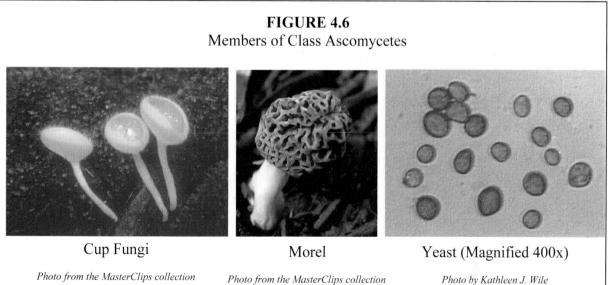

Cup Fungi	Morel	Yeast (Magnified 400x)
Photo from the MasterClips collection	*Photo from the MasterClips collection*	*Photo by Kathleen J. Wile*

Yeasts

Since yeasts are probably the most popular members of class Ascomycetes, we will begin there. Most yeasts are saprophytic, although there are examples of parasitic yeasts as well. When forming spores, they reproduce sexually, producing 4 or 8 ascospores each. Most yeasts, however, have a form of asexual reproduction, called **budding**, at their disposal as well. When a yeast buds, the nucleus of the cell reproduces inside a single cell. A section of the cell wall and plasma membrane then swell to form a pouch, into which the nucleus and some cytoplasm flow. This pouch with its nucleus is called a bud. The bud continues to grow until it is about the same size as the parent cell, and then the two cells separate. Budding is distinct from the asexual reproduction that you are used to (such as that of the bacteria studied in Module #2) because the daughter cell remains attached to the parent cell as it grows.

Yeasts are typically egg-shaped cells that are only somewhat larger than bacteria. Besides a nucleus, about the only organelle in a yeast cell is a vacuole that stores food substances and certain chemicals which the yeast needs. Certain species of yeast store substances useful to humans in these vacuoles. For example, there are many species of yeasts that tend to store vitamins in their vacuoles. Some people eat these yeasts in a ground-up, powder form to obtain the vitamins.

The yeast with which you are most familiar, however, is used in baking. Active dry baker's yeast which you can buy at the grocery store contains *Saccharomyces* (sak air' oh my seez) *cerevisiae* (sehr uh vuh say' ee) spores. When mixed with water, these spores mature into yeast cells which can carry on a process called **fermentation** (fur men tay' shun) .

Fermentation - The anaerobic (without oxygen) breakdown of sugars into alcohol, carbon dioxide, and lactic acid.

We will learn later that this is not the best definition of fermentation, but it will suit our purposes for right now. Fermentation is responsible for making bread dough rise. The yeast mixed in with the bread dough begin to feed on the sugars in the bread, breaking them down into alcohol and carbon dioxide. Since carbon dioxide is a gas, it pushes its way out of the dough, causing the dough to rise. When the dough is put into the oven, the yeast are killed and the alcohol evaporates.

In fact, the nice smell that you always associate with baking bread is a mixture of two things: alcohol and another substance called ozone. In large quantities, ozone is poisonous to humans. In small quantities, however, it simply has a distinct odor. When bread is in the oven, the heat causes the alcohol made by the yeast to evaporate. The alcohol, as it evaporates, can chemically react with other substances in the air to make ozone. The mixture of ozone and alcohol in the air makes the unique smell that we associate with bread baking.

Yeasts are also used in the manufacture of alcoholic beverages such as beer and wine. Since one of the products of fermentation is alcohol, yeasts are used to put the alcohol into the alcoholic beverages. Yeasts feed on the sugars in the hops (a plant) and barley to perform the fermentation process in the manufacture of beer. In the manufacture of wine, they feed off of sugars in the grapes that are used to make the wine. Interestingly enough, yeast cannot survive high concentrations of alcohol. Thus, as they continue the fermentation process, the increasing amount of alcohol actually ends up killing them. Wild forms of yeast can stand only a 4% level of alcohol before they begin to die off. Wineries and breweries, however, have bred strains of yeast that can survive levels of up to 12%. A mixture of wild yeast and specially-bred yeast are used in the making of most alcoholic beverages. Alcoholic drinks (such as whiskey) that have levels of alcohol much greater than 12% are made by taking a drink with 12% alcohol and boiling it. When a mixture of alcohol and water is boiled, the alcohol tends to boil off faster than the water. If you collect the vapors produced by boiling and re-liquefy them, the result is a solution with a higher concentration of alcohol. This process is called distillation.

Learn more about yeast by performing the following experiment. Despite the fact that the supplies call for a microscope, the first two steps in the procedure (A and B) should be done even if you do not have a microscope and the related equipment. Performing these two steps will allow you to observe the fermentation process.

Experiment 4.2
Yeast and the Fermentation Process

Supplies:
- Packet of active dry yeast (can be purchased at a grocery store)
- Warm water
- Tablespoon
- Measuring cup
- Glass that holds at least 2 cups of water
- Sugar
- Microscope
- Eyedropper
- Slides and coverslips
- Methylene blue

Object: To observe the fermentation process and how yeast reproduce through budding.

Procedure:

A. Mix one packet of yeast with two cups of warm water in the glass. Add a tablespoon of sugar. Stir gently. Let the mixture stand at least five minutes. Active dry yeast contains the living spores of baker's yeast for use in home baking. Adding them to water and sugar causes them to begin growing.

B. As the mixture stands, you should observe bubbles beginning to form. Most likely, the bubbles formed will be very small. The best way to observe them is to watch the top of the mixture through the side of the glass. A layer of foam caused by the bubbles will appear and grow thicker as time goes on. There should be a familiar odor for those that bake rolls or

---

bread. The bubbles are carbon dioxide produced in the fermentation process. The odor is partially caused by the alcohol produced in the fermentation process.

C. Place a drop of the yeast solution on a slide and add a coverslip. Observe the slide under low, medium and high powers. Look for oval-shaped cells that have another cell attached to them as shown in this picture. The attached cell is a bud from the budding process. Sketch an example of budding that you see in your specimen.

Photo by Kathleen Wile

D. On another slide place a drop of yeast solution and then a drop of methylene blue. The methylene blue is a stain. It will stain the yeast cells, making their features stand out. Observe on all three powers. Sketch both an individual cell and a cell that is budding .

E. OPTIONAL: Let the yeast sit for thirty minutes or more and observe under the microscope (both stained and unstained) again. You might see some budding that is forming chains, as shown in this picture.

Photo by Kathleen Wile

F. Clean up: wash and dry all slides, coverslips and droppers. Wipe lenses clean with lens paper. Put all supplies away in the proper place.

Other Members of Class Ascomycetes

Many of the tasty, edible fungi also belong to class Ascomycetes. Morels, whose fruiting bodies look like sponges, are one of the most sought-after forms of edible fungus. The ascospores of these fungi are formed within the holes that make up the sponge-like fruiting body. Wind and rain release the spores, allowing them to travel. Once again, however, just because a fungus looks like a sponge, it is not necessarily edible. Many amateur fungus-hunters have been tricked into eating toxic fungi because the fruiting body happened to look like a sponge! Cup fungi, whose fruiting bodies look like small cups, form their ascospores on the inside of the cup. When raindrops hit the cup, the force of impact releases the ascospores.

Many of the fungi that cause disease are in this class as well. *Claviceps* (kla vie' seps) *purpurea* (purr purr ee' uh), better known as **ergot of rye**, can be deadly to humans. As its popular name implies, it feeds on rye grain. If rye bread made with rye that has *Claviceps purpurea* in it is eaten, it is often deadly. History tells us that Peter the Great was thwarted in his efforts to conquer the known world because his troops were fed rye bread that contained this fungus. In addition, many historians believe that the calamities which plagued the early settlers in New England were caused by rye that contained *Claviceps purpurea*. Unfortunately, the settlers at that time had no idea about this fungus, so they blamed it on witches and started the famous Salem witch trials in colonial Massachusetts.

Have you heard about **Dutch elm disease**? What about **chestnut blight**? Both of these diseases, caused by fungi in class Ascomycetes, affect trees. The American chestnut was once one of our most important sources of hardwood lumber. The fungus that causes chestnut blight,

Cryphonectria (cry' fohn ek tree uh) *parasitica* (pare uh sit' ik uh) spread so quickly across America, however, that the American chestnut was completely wiped out! Likewise, many regions have lost their elm trees due to the fungus that causes Dutch elm disease, *Ophiostoma* (oh fee oh stoh' muh) *ulmi* (uhl me').

ON YOUR OWN

4.9 A single-celled organism asexually reproduces by duplicating its nucleus, causing a bulge to form in its plasma membrane, transferring the copied nucleus and some cytoplasm to the bulge, and then separating the bulge into a small cell. The small cell grows to the size of the parent in a day or so. How does this compare to the budding that takes place in yeasts?

4.10 Bread rises because of the fermentation process. Since this process produces both alcohol and carbon dioxide, why don't you get drunk when you eat bread?

Class Zygomycetes: The Bread Molds

The last class in phylum Amastigomycota contains those fungi which form **zygospores** (zie go' sporz).

Zygospore - A zygote surrounded by a hard, protective covering

Of course, this definition does little good if you do not know what a **zygote** (zie' goht) is. We will study this in detail later. However, for right now think of it this way: a zygote forms as a result of sexual reproduction when each parent contributes only half of the DNA necessary to form the offspring. When those two halves join together, a full set of DNA is formed and the offspring can begin development.

Zygote - The result of sexual reproduction when each parent contributes half of the DNA necessary for the offspring

If this definition confuses you, don't worry about it. When we cover reproduction in detail, you will understand it much better. For right now, just think of a zygote as a certain product of sexual reproduction. We will learn much more about what this product is and how it is different from other products of sexual reproduction later.

Because the members of this class have a variety of reproduction methods at their disposal, it is useful to study their lifecycle in detail. Begin by examining Figure 4.7, an illustration of the common bread mold's lifecycle.

Reproduction
1) Stolon (myeelium that stretches out +produces a new organism)

2.) Spores (asexual)

3) sexual hypha grow together

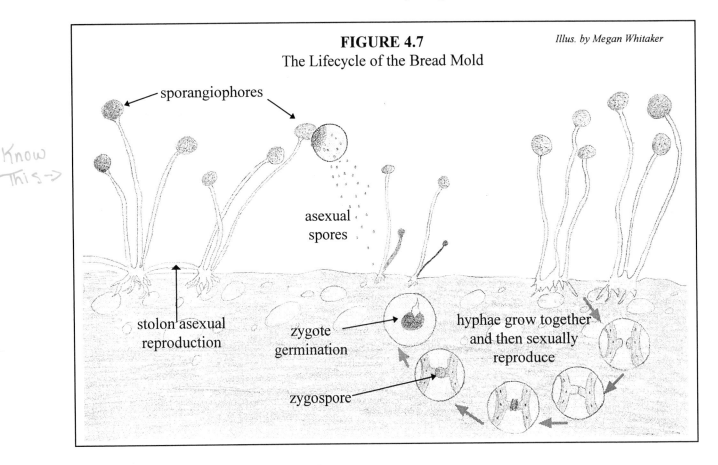

FIGURE 4.7
The Lifecycle of the Bread Mold

Illus. by Megan Whitaker

As you can see from the figure, there are three ways that these molds can reproduce. They can asexually reproduce when a stolon lengthens and forms new filaments. The new filaments become a new mycelium and thus a new fungus. Another form of asexual reproduction involves the production of sporangia (from aerial hyphae) that release spores. Finally, hyphae can fuse together and sexually reproduce to form a zygospore that can then mature into another fungus. Although the latter form of reproduction is what separates these fungi from the ones in the other classes, all three means of reproduction are used.

The most well-known members of this class come from genus *Rhizopus*, which contains most of the common bread molds. Because these fungi have so many reproductive modes at their disposal, their spores are in the air virtually everywhere. If you leave bread out in the open, bread mold spores will eventually land on it and, within a matter of days, the growing mold will be noticeable. Most molds which grow on bread and other baked goods are harmless if consumed in small quantities.

ON YOUR OWN

4.11 A bread mold forms a stolon for reproduction. Is it reproducing sexually or asexually?

4.12 A fungus forms a fruiting body. Is it likely to be a bread mold?

Phylum Mastigomycota

As we mentioned earlier, fungi in this phylum produce spores that are motile. They generally have flagella with which to move from the area in which they are released to another suitable place in which to grow. There are only two classes in this phylum: Oomycetes (oh uh my see' teez) and Chytridiomycetes (ki' truh dee oh my see' teez).

Class Chytridiomycetes contains the single-celled fungi called **chytrids** (ki' trids). Chytrids inhabit muddy or aquatic areas. They are typically saprophytic, feeding on decaying water plants. Some species of chytrids are parasitic. Even the parasitic ones, however, feed on plants, not animals.

Class Oomycetes contains the water molds and mildews. If you do not clean the tile around your bathtub, the brown or black mold that grows is a member of this class. If you let fruits (such as grapes) sit for too long, the downy, white "fuzz" that grows on them is also a member of this class. Finally, if you have ever seen white fuzz growing on a sick fish (terribly common among goldfish), that fuzz is the mycelium of a fungus from this class.

Probably the most notorious member of class Oomycetes is *Phytophthora* (fie' toe puh thor uh) *infestans* (in fest' uhns). This water mold causes a disease called **late blight of potato**. This disease can also infect tomato plants, but the potato version is the best remembered. In the early-to-mid 1800's potatoes were the *main* food crop for the peasants that lived in Ireland. Beginning in 1845, however, the weather during the growing season became ideal for the growth and spread of *Phytophthora infestans*. This fungus rotted the potatoes, ruining entire crops. For 15 years, the fungus spread its spores virtually unchecked, destroying countless potato crops and causing massive starvation. A third of Ireland's population died during that time. Another third of the population (2 million Irish) were able to immigrate to America, causing the largest influx ever of Irish into this country. After 15 years, weather patterns changed and the fungus essentially died out. For many Irish, however, it was simply too late. The years 1845-1860 are still remembered in Ireland as the "great potato famine."

To learn more about what molds and mildews really look like, perform the following experiment. Realize that when you use bread mold as a specimen, you are really looking at a member of class Zygomycetes from phylum Amastigomycota. If you use the mold from jelly or fruit, you are observing a member of class Oomycetes from phylum Mastigomycota. Even those without microscopes should perform steps A and B in the procedure.

Experiment 4.3
Molds and Mildews

Supplies:
- Mold bread, jelly, and/or fruit grown earlier. Only one specimen is necessary, but if you observe more than one specimen, you will learn more!
- Magnifying glass

- Knife
- Needle
- Microscope
- Slides and coverslips
- Water
- Eyedropper

Object: To observe various molds and mildews and the differences in how they look both macroscopically and microscopically.

Procedure:

A. Observe mold found on bread, jelly and fruit. For each specimen, note the source and describe the mold by color, shape, size and texture (is it fuzzy, smooth, flat, raised, etc.).

B. Observe with a magnifying glass. Sketch the magnified image, noting the source of each specimen.

C. Using a knife or a needle, scrape off a section of the mold and place a tiny amount on a slide. Observe dry on all magnifications. Sketch an example of your observations, noting the source of the specimen.

D. Add a drop of water to the slide and a coverslip. Observe again on all magnifications and sketch an example, noting the source.

E. If you were able to observe both a bread mold and a mold from fruit, try to note the differences between the two.

F. Clean up: wash slides, coverslips and droppers; dry and put away. Wipe microscope lenses with lens paper. Return microscope to safe place. Dispose of all molds.

The Imperfect Fungi

If a fungus is studied and scientists cannot determine a sexual reproductive phase in its lifecycle, it is placed in this "phylum" until it can be better classified. The reason we put phylum in quotation marks here is that most biologists don't really consider this classification group a true phylum. Instead, they consider it a "holding area," until more can be learned about the fungus. Most biologists are convinced that every fungus has a phase of sexual reproduction. Since some fungi are rather hard to study in detail, however, there are some whose mode of sexual reproduction eludes us. As a result, we place them in this classification until the sexual reproduction method can be found and the spores that it produces can be analyzed. At that point, the fungus will be re-classified in phylum Mastigomycota or Amastigomycota.

Since many biologists consider this classification group a temporary holding classification, they often say that this phylum "has no taxonomic status." Remember, taxonomy is the science of classification. This statement, then, is equivalent to saying that the phylum Imperfect Fungi really does not exist. It is simply a place for us to stick fungi which are not yet fully understood. We can't overemphasize, however, that the reason this classification has no taxonomic status is because biologists *assume* that all fungi have a sexual mode of spore formation. This, of course, could very well be an incorrect assumption, and perhaps a true phylum should be named for these fungi. The fact that some of the fungi in this group have been extensively studied for more than 50 years without finding a sexual mode of reproduction might be considered strong evidence that the assumption is, indeed, wrong!

One of the most useful imperfect fungi is the genus *Penicillium*, from which we get the drug penicillin. In 1929, Alexander Fleming, an English physician, discovered this wonderful drug quite by accident. He had taken a short vacation, leaving a bacteria culture open to the air. When he returned, he saw that the culture was overrun by bacteria, except in a certain place where a blue mold (from the genus *Penicillium*) was forming. The blue mold seemed to be producing a substance that killed the bacteria. Fleming isolated that chemical and called it **penicillin**. With the help of two other scientists, he demonstrated that this substance can kill the bacteria associated with many human sicknesses when ingested by a sick person. Thus, the first **antibiotic** (an tie bye ah' tik) was discovered.

Antibiotic - A chemical secreted by a living organism that kills or reduces the reproduction rates of other organisms

Because penicillin and other antibiotics have been so successful in treating many forms of sickness, these three scientists shared the Nobel Prize (the most prestigious prize in the world) in Medicine in 1945.

One interesting fact about antibiotics is that they can be very effective at killing many bacteria which cause disease. After a while, however, they can lose their effectiveness. You see, God created his creatures with the ability to adapt and change to their environment. It turns out (we will dwell on this in a later module) that God has made it possible for bacteria (and other organisms) to change significantly over the course of several generations. A colony of bacteria that is being destroyed by an antibiotic can, under certain conditions, produce offspring that are not at all affected by the antibiotic. We say that these bacteria are **immune** to its effect. As an antibiotic is used more and more, strains of bacteria develop that are immune to the antibiotic. To counter this, medical scientists must find new antibiotics to which these bacteria are not immune. This is a constant struggle. Bacteria adapt to become immune to an antibiotic, and then humans adapt and find another antibiotic. Paradoxically, the more antibiotics are used to cure disease, the more ineffective they become, because the very act of using an antibiotic can give rise to a strain of bacteria immune to it!

Other members of genus *Penicillium* are useful to us because they flavor certain cheese. Camembert and Roquefort cheese are both flavored by the growth of *Penicillium* species. Although it might at first sound strange that fungi are grown on cheese to give them flavor, it is no different than putting mushrooms in a sauce to change its flavor. Although the term "fungus" often carries a bad connotation, many fungi are completely edible and, to some, quite tasty! If you want, perform the following experiment to observe some of these cheese molds. Note that these molds do not produce penicillin, because they are not the same species as the one analyzed by Dr. Fleming. They are just within the genus *Penicillium*, so they are similar to Dr. Fleming's fungus, but they will not produce penicillin.

Optional Experiment 4.4
Imperfect Fungi

(Perform this experiment only if the student is interested and if you can find at least one sample of Camembert or Roquefort cheese.)

Supplies:
- Camembert cheese (available at large supermarkets)
- Roquefort cheese (available at large supermarkets)
- Microscope
- Slides and coverslips
- Eyedropper
- Knife
- Water

Object: To observe various molds and mildews and the differences in how they look both macroscopically and microscopically.

Procedure:
A. Refrigerate the cheese or cheeses until about six hours before performing the experiment. Then take them out of the refrigerator and let them sit for about 6 hours.

B. The mold on the Camembert cheese (*Penicillium camemberti*) generally looks like white masses which form the mycelium of the fungus. Scratch a small amount of mold loose with a knife and place on a slide. Observe on all magnifications. Sketch and label the white mycelium.

C. The blue areas on the Roquefort cheese are samples of the mold *Penicillium roqueforti*. Because of the color of the fungus, this is often called "blue cheese." Scratch some of this blue off with the knife and place on a slide. Add a drop of water and a coverslip. Try to find chains of the tiny, blue-green fungal spores. You might only see them as individuals, but they are usually found in chains. Sketch and label this mold.

D. Clean up: wash slides, coverslips and droppers; dry and put away. Wipe microscope lenses with lens paper. Return microscope to safe place. Dispose of all molds.

ON YOUR OWN

4.13 A famous biologist once said, "The only thing imperfect about the Imperfect Fungi is our knowledge of them." What does the biologist mean?

4.14 In medical journals these days, there is a lot of concern about the overuse of antibiotics. Doctors think that since antibiotics are so effective, they are prescribed far too often for patients. Why are doctors worried about overuse of antibiotics?

Phylum Myxomycota

As we mentioned previously, placing this phylum in kingdom Fungi is rather controversial. Of course, placing it in Protista, the other kingdom in which you might find it, is just as controversial. There really seems to be no clear consensus as to where the species in this phylum belong. The reason is quite simple: the members of this phylum, typically called **slime molds**, behave like fungi when they reproduce, and they behave like colonial protists when they feed. As a result, there are many arguments for placing them in either kingdom. Some biologists have even considered making a sixth kingdom in which to place them; however, there are really not enough species of slime mold to justify such a drastic step.

Although modern textbooks do tend to place the slime molds in kingdom Protista, we choose to leave it in its traditional place, as a part of kingdom Fungi. This is mainly because we consider reproduction to be a more fundamental aspect of life than feeding. After all, many species have several feeding options open to them. If you look at a population of wild cats, for example, you will find that they are all carnivores, but their choice of diet and means of finding food will be quite different depending on their environment. Their mode of reproduction, however, will always be the same. Also, the definition of species that you learned in Module #1 references reproduction, so it is obviously a very fundamental means of classification. As a result, we will place the slime molds in the kingdom which contains other organisms that predominately reproduce in a similar fashion. Another reason to classify slime molds with fungi is because you find them in similar habitats.

Realize, however, that the reasons presented simply reflect the preconceived notions of these authors. Despite what you might think, scientists are not all that objective. We all approach science with an inherent bias. As a result, the conclusions that we come to are often affected by this bias. This is clear in the means of classification that we employ. It is also clear in the general outlook that we have. For example, if you were to pick up a biology textbook written by a scientist that does not believe in God, the entire book would be quite different than this one. In order to back up such an unscientific belief, atheistic scientists will try to teach biology within the framework of the well-debunked theory of evolution. Since this theory attempts (quite unsuccessfully) to explain the formation of life without reference to a Creator, these scientists must look at life through its blinders. As a result, their entire view of biology is quite different than that of this book. In a later module, we will study evolution in great detail, showing you why it has really lost its status as a theory and is, at best, an hypothesis. The point we are trying to make now is that you should never be fooled into thinking that scientists are objective. We cannot be. Our preconceptions and our world view will always color the science that we perform!

Now that we have that out of the way, we can start learning about these interesting creatures known as slime molds. Most of the time, you will see slime molds in their feeding stage, when they resemble colonial protists. They usually can be found on the bark of decaying logs, or between the layers of leaves on the floor of a forest. Despite the fact that slime molds seem rather exotic, they are quite common. If you ever lay tree bark or dead leaves down as a mulch for a garden, you can almost always turn the mulch over during the middle of the growing season and find a slime mold, providing that the mulch has been kept moist. Figure 4.8 shows two slime molds.

Under Kingdom Protista
feeding stage
resembles protozoan
its reproductive
stage it resembles
fungus

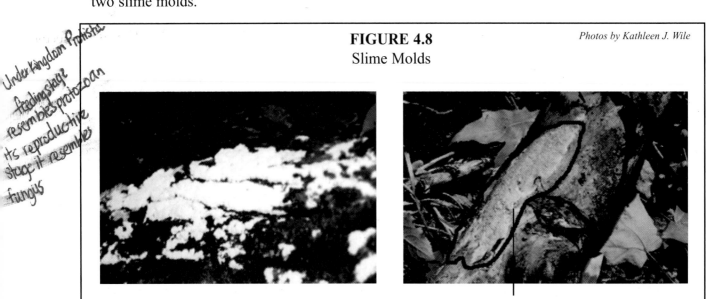

FIGURE 4.8
Slime Molds

Photos by Kathleen J. Wile

White slime mold on a rotting tree trunk   Yellow slime mold (outlined for emphasis)

Slime molds are usually white, red, orange, or yellow. They exist in their feeding stage as a mass of living matter called a **plasmodium**. This is rather confusing, however, since there is a genus in kingdom Protista that is given that same name. The term plasmodium when applied to slime molds, however, has no connection to the genus *Plasmodium* found in kingdom Protista! As their name implies, most slime molds are slimy to the touch. Despite the fact that their name and appearance are rather disagreeable, most slime molds are not harmful to plants or animals, as they are almost all saprophytic. The non-saprophytic slime molds tend to feed on bacteria, although there are some parasitic slime molds.

Slime molds can move about as a unit in search of food. When the food supply is exhausted, however, or when unfavorable conditions occur, the slime mold will produce fruiting bodies that are best described as sporangia. The sporangia contain spores that are released to find new sources of food. If they find a new source of food, they will grow and reproduce to form a new slime mold. Some slime molds produce motile spores, while others produce non-motile spores.

Probably the main feature of a slime mold's habitat is water. Wherever dead trees, bark, or leaves are kept moist, slime molds are almost certain to grow. To get rid of a slime mold, you simply need to dry the area. Gardeners often dislike slime molds because of their appearance. To rid themselves of the molds, they need only to rake the surface of their gardens daily. This will bring the moist mulch to the surface allowing it to dry.

ON YOUR OWN

4.15 A biologist observes a slime mold only during its feeding stage. In what kingdom will the biologist most likely classify the slime mold?

Symbiosis in Kingdom Fungi

Before we leave our discussion of kingdom Fungi, we must mention the different forms of symbiosis in which its members participate. The most well-known form of symbiosis in which you will find a fungus is the **lichen** (lie' kun). Lichens are produced by a symbiotic relationship between a fungus (usually of class Ascomycetes) and an algae (usually of class Chlorophyta). The algae in the relationship produces food for itself and the fungus by means of photosynthesis, while the fungus supports and protects the algae. As a result of this symbiotic arrangement, you can find lichens where other organisms just cannot survive. They are commonly found growing on dry rocks, brick walls, fences, and trees. In certain cold, snowy regions, lichens grow so large that they cover a few square miles!

Despite the fact that the algae which make up the known species of lichens can live independently, the fungi that make them up cannot. This should make sense to you, because the algae produce the food. Thus, they can live with or without the fungus. The fungus just makes their survival easier. The fungus, on the other hand, cannot live without the food that the algae produces.

Since the fungus of a lichen cannot live without the algae, you might wonder how lichens reproduce. Actually, it is rather fascinating. Most lichens reproduce by releasing a dustlike substance called a **soredium** (so ree' dee uhm). The soredium contains spores of *both* the algae and the fungus in a protective case. Thus, the soredium is like a spore that contains two different spores. Wherever the soredium lands, then, both the fungus and the algae can grow. Isn't that amazing? The two separate species work together not only to survive, but also to reproduce! Scientists still know very little about the details of this fascinating process.

A more prevalent (but less well-known) symbiotic relationship in which fungi participate is called a **mycorrhizae**, or a "fungus root." Nearly 80% of all plants with root systems participate in this symbiotic relationship with a fungus. In a mycorrhizae, the fungus forms haustoria that penetrate the cell walls of the root system's cells. The fungus absorbs food from the roots as it is transported to the plant. In return, the plant takes in certain needed chemicals, called minerals, that it cannot absorb efficiently from the soil.

You see, in order to absorb minerals from the soil, an organism must be wide and thin. This is how the mycelium of a fungus grows. This is not, however, the way that a root system grows. Roots usually grow thick and long, trying to go deep within the soil. Since the mycelium of the fungus has the ideal structure for absorbing minerals, it does so in exchange for food. In laboratories, it has even been demonstrated that fungi absorb the minerals when they are plentiful and store the excess. These excess minerals are then released slowly into the roots of the plant when the minerals are scarce in the soil! Once again, scientists have no idea how the fungus knows to do this.

ON YOUR OWN

4.16 Suppose a biologist were to isolate the fungus spores from the algae spores in a soredium. Would the biologist be able to grow the fungus by itself? What about the algae? Would the biologist be able to culture free-living algae from the spores?

4.17 There are some scientists who have studied the effect of air pollution on fungus. They conclude that air pollution destroys fungi at a much higher rate than it destroys other organisms. These same scientists say that if air pollution kills too much fungi, trees and other plants will begin to die as well. Why?

Summing up Kingdom Fungi

As you can see, we still have a lot to learn about this fascinating kingdom. Are imperfect fungi simply not completely understood, or should there be another phylum dedicated only to fungi that have no means of sexually producing spores? How do we classify slime molds? How does a mushroom know when to form a button? How in the world do lichens form soredia? How do the fungi in mycorrhizae know to store up excess minerals and release them when the tree needs them most?

Of course, despite the fact that there are many things we don't know about kingdom Fungi, there are many things that we do know. We know that the saprophytic nature of most fungi is critical for the balance that exists in nature. We also know that many fungi are useful to humans as food, flavoring, and medicine and others are deadly or toxic. Finally, we know that fungi are another wonderfully interesting part of God's Creation, bringing glory and honor to Him.

ANSWERS TO THE ON YOUR OWN PROBLEMS

4.1 <u>No, the organism is not a part of kingdom Fungi</u>. All fungi digest their food before eating it. This process, called "extracellular digestion," is common to all fungi.

4.2 <u>The stalks and the caps are not the main parts of the mushroom. The mycelium, which is underneath the ground, is the main body of the fungus.</u> Even though the farmer removed the stalks and caps, the mycelium is still there. It will produce more stalks and caps later on.

4.3 <u>Only parasitic fungi produce haustoria, so the fungus must be parasitic.</u> Since a haustorium's job is to invade a living cell and draw nutrients from it, there is no reason a saprophytic fungus would produce a haustorium.

4.4 <u>The fruiting body holds and releases the spores for reproduction.</u> Once all spores are released, the fruiting body withers and dies.

4.5 <u>The spores result from asexual reproduction,</u> because only asexual reproduction results in offspring that are identical to parents.

4.6 Your key can be ordered differently and can even have more steps in it. All of the following questions must be asked, however:

1. Organism in feeding stage resembles protozoa ....*phylum Myxomycota*
   Organism never resembles protozoa **2**

2. Fungus has an identifiable means of producing sexual spores...**3**
   Fungus has no identifiable means of producing sexual spores............*phylum Imperfect Fungi*

3. Spores are motile ....*phylum Mastigomycota*
   Spores are non-motile ....*phylum Amastigomycota*

4.7 <u>No.</u> The button stage comes before the stipe and cap are formed. Since the spores are released from the gills in the cap, a mushroom cannot release its spores until the stipe and cap are formed.

4.8 <u>In mushrooms, basidia form on the gills of the cap. In puffballs, they form inside the membrane of the fruiting body. In shelf fungi, they form in the pores of the fruiting body.</u>

4.9 <u>It is very similar, but not identical.</u> For yeasts, the bud typically does not detach itself until after it is fully grown.

4.10 <u>Baked bread does not have alcohol in it because the heat of the baking process evaporates the alcohol.</u> Thus, even though alcohol is formed during the making of bread, it gets removed by the heat of the baking process.

4.11 Stolons are an asexual means of reproduction, so the mold is reproducing <u>asexually</u>.

4.12 <u>No, bread molds do not really form a fruiting body</u>. In sexual reproduction, they form an underground zygospore that grows into a new mold. They do have sporangiophores that form asexual spores, but those are not fruiting bodies; they are just specialized aerial hyphae.

4.13 <u>The very fact that a fungus is called imperfect means that we simply *do not know about its sexual reproduction*</u>. Since we do not know about an aspect of its life, our knowledge of it is imperfect.

4.14 <u>The more an antibiotic is used, the more likely the chance of a bacteria (or other pathogen) strain developing that is immune to the antibiotic. If that strain reproduces and is spread, then a new antibiotic must be made to destroy that strain of bacteria (or other pathogen)</u>. Thus, the more you use an antibiotic, the more likely you are to get a strain of bacteria (or other pathogen) that is immune to it!

4.15 Since slime molds in their feeding stage resemble protozoa, the biologist will most likely classify it in <u>kingdom Protista</u>.

4.16 <u>The spores from the fungus could never grow into a free-living fungus</u>, because the fungus in a lichen (that's where a soredium comes from) has no food supply without the algae. <u>The algae would be able to live on its own</u>, because the fungus simply gives it support and protection. The algae can live without that. It will not be as prolific as it could be WITH the fungus, but it can live on its own.

4.17 <u>Because nearly 80% of plants have a symbiotic relationship with the mycelia of fungi, if fungi die off, the trees will no longer be able to participate in the symbiotic relationship</u>. The fungi help trees absorb vital minerals from the soil. Without the aid of the fungi, the trees will not be able to absorb enough minerals, and they will begin to die.

STUDY GUIDE FOR MODULE #4

1. Define the following terms:

a. Extracellular digestion
b. Mycelium
c. Hypha
d. Rhizoid hypha
e. Aerial hypha
f. Sporophore
g. Stolon

h. Haustorium
i. Chitin
j. Membrane
k. Fermentation
l. Zygospore
m. Zygote
n. Antibiotic

2. Which of the following characteristics or structures exist for the vast majority of fungi? Which are present in only a few species?

extracellular digestion	sporangiophores	motile spores
stolons	mycelia	septate hyphae
chitin	hyphae	cells
caps and stalks	haustoria	rhizoid hyphae

3. Many biologists say that a mushroom is much like an iceberg, because only about 10% of an iceberg is visible from the surface of the ocean. What do they mean?

4. What is the difference between septate and non-septate hyphae?

5. What is the function of the following specialized hyphae?

rhizoid hyphae   stolon   sporophore   haustorium

6. Of the hyphae listed in question 5, which are aerial?

7. What is the difference between a sporangiophore and a conidiophore?

8. Give the main characteristic associated with each of the four phyla of kingdom Fungi: Mastigomycota, Amastigomycota, Imperfect Fungi, and Myxomycota.

9. Describe each of the stages (in chronological order) associated with the life cycle of a mushroom, starting with the formation of a mycelium.

10. What is the main difference between shelf fungi, puffballs, and mushrooms?

11. What is an alternate host? List a type of fungus that uses one.

12. What type of fungus is best known for fermentation? To which class does it belong?

13. How is budding different than the asexual reproduction in bacteria?

14. Name at least 2 pathogenic fungi and the maladies that they cause.

15. List the major characteristic associated with each of the three classes of phylum Amastigomycota: Basidiomycetes, Zygomycetes, and Ascomycetes.

16. Describe the three ways a bread mold can reproduce. In each case, say whether the reproduction is sexual or asexual.

17. What puts a fungus into phylum Imperfect Fungi?

18. What can happen when an antibiotic is used too much?

19. Name the genus of the fungus that produces penicillin.

20. When a slime mold is a plasmodium, it resembles organisms from what kingdom?

21. What is the easiest way to get rid of slime molds?

22. What are the two major forms of symbiosis in which fungi participate? Describe each relationship and the job of each participant in that relationship.

23. What is a soredium?

Module #5: The Chemistry of Life

Introduction

In the past three modules, we have introduced three of the five kingdoms in God's Creation. You've learned a lot about the organisms within these kingdoms, and hopefully you've begun to develop a keen appreciation for the grandeur and complexity of the Creator's work. Now it's time to step back and look at biology from a different angle. Rather than studying life itself, in this module we will be studying the chemistry that helps make life possible.

In the future, you should spend an entire year studying chemistry. After that year, you will have only received a basic introduction into this vast field. Thus, what we cover in this module will barely scratch the surface of the science called chemistry. Nevertheless, without this brief introduction, you will probably be lost in later modules.

Atoms: The Basic Building Blocks of Matter

In chemistry, we study **matter**.

Matter - Anything that has mass and takes up space

Of course, this definition does us no good unless we know what **mass** is. Unfortunately, mass is a difficult term to define; thus, we will not try to define it in this course. Instead, we will say this: if something has mass, it will also have weight. As a result, we could say that matter is anything that has weight and takes up space. Now if you think about it, almost everything has weight and takes up space. In chemistry, then, we really study just about everything!

As you learned in Module #1, cells are the basic building blocks of life. Indeed, the type of cell or cells that make up an organism is the first thing used to determine classification. Organisms made of prokaryotic cells belong in kingdom Monera, whereas those made of eukaryotic cells belong in one of the other four kingdoms. Well, just as cells are the basic building blocks of life, **atoms** are the basic building blocks of matter. Everything from the tiniest speck of dust to the biggest mountain in the world is made up of atoms.

Although atoms make up all matter in the universe, we cannot see them because they are quite small. Indeed, in the letter "I" that begins this sentence, there are approximately 1,000,000,000,000,000,000,000 atoms. Even though we cannot see them, scientists have performed detailed experiments that provide ample evidence for their existence. As a result, the existence of atoms has become a foundational principle that guides our understanding of chemistry.

Believe it or not, atoms are made up of even smaller things called protons, neutrons, and electrons. Now you will learn a lot more about protons, neutrons, and electrons when you study chemistry. For right now, you can think of them as small, spherical particles which are arranged in a very specific way to form an atom. For example, a simplistic schematic of an atom is shown in Figure 5.1:

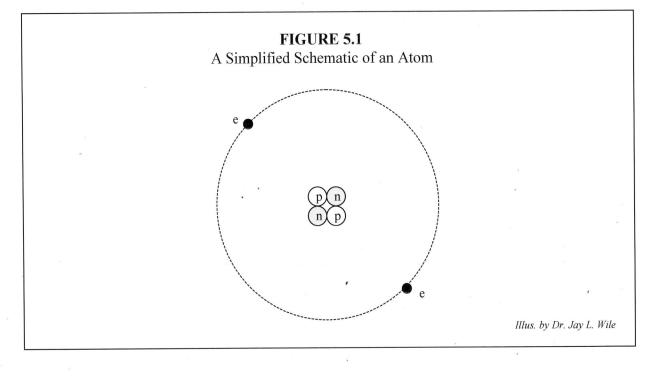

FIGURE 5.1
A Simplified Schematic of an Atom

Illus. by Dr. Jay L. Wile

In this figure, the electrons are labeled with "e," the protons with "p," and the neutrons with "n." Notice that the neutrons and protons clump together in the center of the atom, which is called the **nucleus**. Now don't get confused here. The term nucleus is also used to identify the organelle that holds the DNA in a eukaryotic cell. When discussing atoms, however, the nucleus refers to the center of the atom where the protons and neutrons clump together. Whirling around in an orbit outside of the nucleus, we find the electrons.

You might wonder how in the world we know what atoms look like if we cannot see them. The fact is, we don't really know what they look like. Scientists have, however, done many experiments designed to help us understand the structure of the atom and, as far as we can tell, the drawing in Figure 5.1 is consistent with most of those experiments. As a result, we say that this is a good **model** of the atom.

<u>Model</u> - An explanation or representation of something that cannot be seen

The fact that a model is consistent with experiments does not mean it is an accurate representation of the object being studied. Later on, someone might come up with experiments that contradict the model, or someone else might come up with a different model that is also consistent with all of the experiments done. Thus, the drawing in Figure 5.1 is instructive in that it gives us a picture of what an atom might look like, but please realize that it is not an actual drawing of an atom. It is merely a representation of what an atom *might* look like.

In fact, we already know that the model presented in Figure 5.1 is not really accurate. There are certain experiments that clearly show this model (called the "Bohr model") to be in error. As a result, a new model of the atom, called the "quantum mechanical model," is currently the model believed by most scientists. However, that model is far too complex to be shown here. Indeed, to truly understand the quantum mechanical model, you need to have several years of mathematics beyond calculus! Thus, although we know that the Bohr model is not exactly correct, we will use it. Even though you might not like the idea of using an incorrect model to learn about the atom, don't worry. Most of the concepts that you learn using the Bohr model will help you to understand the quantum mechanical model when you get to it. If you take chemistry next year, you will get a brief introduction into the quantum mechanical model of the atom.

With all of this in mind, let's take a look at the Bohr model of the atom. Protons and neutrons are packed into the nucleus which resides at the center of the atom. Electrons whirl around the nucleus in circular orbits. The number of protons, neutrons, and electrons in an atom determines all of the properties of that atom. For example, the atom pictured in Figure 5.1 is a helium atom. You probably already know that helium is a gas, and that the gas is lighter than air. Thus, if you fill a balloon with helium, it will float. Well, the fact that helium is a gas and the fact that it is lighter than air are both a result of the number of protons, electrons, and neutrons in the atom. If you change the number of protons, neutrons, and electrons in an atom, you completely change its properties. For example, an atom that contains 6 protons, 6 neutrons, and 6 electrons is called a carbon atom. Unlike helium, carbon is solid. It is black in color, and it is brittle. These properties, once again, are the result of the number of protons, electrons, and neutrons in the atom.

No matter how many electrons, protons, and neutrons make up an atom, there is one general principle that applies to them all. All atoms have equal numbers of protons and electrons. Thus, if an atom has 3 protons, you know that it will have 3 electrons. If it has 17 electrons, you know that it has 17 protons. There is, of course, a reason for this. Protons and electrons each have electrical charge. Now we won't even try to explain what electrical charge is (indeed, scientists don't really have a good idea of what it is), we will just tell you that it comes in two types: positive and negative. Electrons have a negative charge while protons have a positive charge. Well, it turns out that an atom cannot have any overall electrical charge, so the number of positive charges (protons) in the atom must always balance out the number of negative charges (electrons) in the atom. That way, the total charge is zero. Neutrons, by the way, have no electrical charge.

Here's the really interesting part: The vast majority of an atom's properties are determined by the number of electrons that it has. Although this might sound a little weird, it actually makes a great deal of sense. You see, based on the Bohr model of the atom, the protons and neutrons are tucked away in the center of the atom (the nucleus). Since the electrons orbit around the nucleus, they are, in effect, what makes up the "outer layer" of the atom. Thus, if two atoms were to come close to one another, their electrons would be the first things to begin to interact. As a result, electrons determine the vast majority of an atom's properties.

To confuse this issue a little bit, remember that all atoms have the same number of protons and electrons. Thus, even though it is in reality the *number of electrons* that determine an atom's properties, since the number of protons will always be equivalent to the number of electrons, we could also say that the number of protons can be used to determine an atom's properties. Now don't let this confuse you! The electrons are what really determine an atom's properties; however, because of some complicated history, we track the properties of atoms by their number of protons. Although this is a bit confusing, there is nothing wrong with it, since the number of protons and electrons in an atom are always the same.

So, if two atoms have the same number of protons, they will have the same number of electrons. Thus, even if the number of neutrons in the first atom is different than the number of neutrons in the second, the vast majority of the two atom's properties will be the same, because they have the same number of electrons.

ON YOUR OWN

5.1 The Bohr model of the atom is sometimes called the "planetary model" of the atom. Why?

because its plain

5.2 What determines an atom's properties? What determines the vast majority of an atom's properties?   Electrons

5.3 An atom has 13 electrons. How many protons does it have?

13

Elements

Now since the number of electrons (and therefore the number of protons) is the most important factor in determining an atom's properties, we tend to separate atoms by their number of protons. Atoms that have the same of number of protons (regardless of their number of neutrons) are said to belong to the same **element**.

Element - All atoms that contain the same number of protons

Helium, for example, is an element. One of the atoms that makes up this element is pictured in Figure 5.1. It has 2 protons, 2 electrons, and 2 neutrons. However, there are two other atoms in the element helium. One has 2 protons, 2 electrons, and 3 neutrons and the other has 2 protons, 2 electrons, and 1 neutron. Since all three of these atoms have the same number of protons and electrons, they have the same basic properties. Thus, when you fill a balloon with helium, you are actually filling it with three kinds of atoms: one that has 2 protons, 2 neutrons, and 2 electrons; one that has 2 protons, 1 neutron, and 2 electrons; and one that has 2 protons, 3 neutrons, and 2 electrons.

How do we identify these atoms? Well, first and foremost, we call them by the element to which they belong. Thus, all three of these types of atoms are called helium atoms. To distinguish between the three different atoms that make up helium, we add their protons and neutrons together. Thus, the atom with 2 protons, 2 electrons, and 1 neutron is called "helium-3," because two protons and 1 neutron add to three. The atom picture in Figure 5.1 is called "helium-4," and the atom that has 2 protons, 2 electrons, and 3 neutrons is called "helium-5."

It is often convenient for us to abbreviate the name of elements so as to reduce the amount of time it takes to write them. Usually, an element is abbreviated by the first one or two letters of its name. Helium, for example, is abbreviated "He," while carbon is abbreviated "C." Unfortunately, some elements are also abbreviated by the first one or two letters of their Latin names. Sodium, for example, is abbreviated as "Na" because its Latin name is "natrium." There is no real way to learn these abbreviations short of memorization, so you will need to learn the abbreviations of the following biologically-important elements:

TABLE 5.1
Biologically Important Elements

Element Name	Abbreviation
carbon 6	C
hydrogen 1	H
oxygen 8	O
nitrogen 7	N
phosphorus	P
sulfur 16	S

Although you are not required to remember this, you might be interested in the makeup of these elements. Carbon contains all atoms that have 6 protons and 6 electrons. Hydrogen is made up of all atoms that have 1 proton and 1 electron. Oxygen contains all atoms that have 8 protons and 8 electrons, while nitrogen is made up of the atoms that have 7 protons and 7 electrons. Finally, sulfur is made up of the atoms that have 16 protons and 16 electrons.

Now, of course, there are many, many more elements that you will not really deal with until you take chemistry. In fact, there are currently 109 known elements in God's Creation. Most likely, more will be discovered. For right now, however, we will concentrate mostly on the elements listed in Table 5.1.

Before we move on to the next section, step back and review what you have learned. Atoms are made up of protons, neutrons, and electrons. No matter how many protons an atom has, it will have the same number of electrons. When different atoms have the same number of protons (and therefore the same number of electrons), they are said to belong to the same element, because they have the same basic properties. To name an atom, we call it by the element to which it belongs, followed by the sum of its protons and neutrons. Thus, when you see a name like sulfur-32 (or S-32), you know that we are talking about a *particular* atom, the one that has 16 protons (that's what makes it sulfur) and 16 neutrons (that's how to get a 32 at the end). On the other hand, if you see just the name sulfur (or S), then you know that we are talking about an element, because there are several different atoms that all have 16 protons and thus belong to the element sulfur.

Now it turns out that for the vast majority of situations in chemistry, the number of neutrons in an atom is completely irrelevant. This is (once again) because the major properties of an atom are determined by its number of electrons (and therefore its number of protons as well). As a result, the element to which an atom belongs is, by far, the most important aspect of identifying an atom. Thus, you will usually see just the element (sulfur) instead of seeing a particular atom (sulfur-32) when you study chemistry. You always need to remember, however, that an element (such as sulfur) is composed of many individual atoms (sulfur-32, sulfur-33, sulfur-34, and sulfur-36).

ON YOUR OWN

5.4 Two atoms have slightly different properties, but they belong to the same element. What is different about them: their number of protons, neutrons, or electrons? *nuetrons*

5.5 The element carbon is composed of all atoms that have 6 protons. One of the atoms that makes up carbon is carbon-13. How many protons, neutrons, and electrons are in a carbon-13 atom? *6 protons 7 nuetrons. & 6 electrons*

Molecules

If there really are only 109 elements in God's Creation, you might think that there are only 109 different types of matter. After all, if matter is made up of atoms, and each element contains atoms that have the same basic properties, then 109 different elements make 109 different types of matter, right? Wrong! You see, atoms are only the basic building blocks of matter. In order to provide the chemical diversity necessary for life, God has designed atoms to link together much like the pieces of a jigsaw puzzle. When atoms link together, they form **molecules**.

Molecules - Chemicals that result from atoms linking together

For example, carbon dioxide is a gas that humans (and most organisms) produce as a part of respiration. This gas is formed when a carbon atom and 2 oxygen atoms link together.

Just as elements have abbreviations, molecules do, too. We call these abbreviations **chemical formulas**. Carbon dioxide, for example, is abbreviated as "CO_2." Where does this abbreviation come from? Well, carbon dioxide is made up of carbon and oxygen. The "C" stands for carbon and the "O" stands for oxygen. Numbers that appear as subscripts indicate how many of those atoms are in the molecule. The "2" that appears as a subscript after the "O" tells us that there are two oxygen atoms in a carbon dioxide molecule. If no subscript appears after an element, then we know that there is only one atom of that type in the molecule. Thus, since there is no subscript after the "C," we know that there is only 1 carbon in a molecule of carbon dioxide.

Just to make sure you really understand this, we want to give you a couple more examples. Another important molecule in the chemistry of life is methane. This molecule is abbreviated as CH_4, so you should be able to tell that there is one carbon atom linked to 4 hydrogen atoms in a methane molecule. Glucose, which is the basic food substance that most autotrophic organisms produce via photosynthesis, is abbreviated as $C_6H_{12}O_6$. This should tell you that a molecule of glucose has 6 carbon atoms, 12 hydrogen atoms, and 6 oxygen atoms all linked together.

The important thing to realize about molecules is that the properties of a molecule are determined by both the type and the number of atoms that link together. For example, as we already mentioned, carbon dioxide is formed when one carbon atom links to two oxygen atoms. However, this is not the only molecule that can be formed by carbon and oxygen atoms. If one carbon atom links to one oxygen atom, the result is carbon monoxide (CO). Carbon monoxide and carbon dioxide are two completely different molecules! Carbon dioxide, for example, is a gas that is harmless to humans. In fact, humans produce it as a result of respiration. Carbon monoxide, however, is poisonous to humans. If we breathe too much carbon monoxide, we will die of suffocation! Thus, not only the type, but also the number of atoms that link together determine the properties of a molecule. Two molecules can be made up of the same exact atoms, but if the number of even one type of atom is different between the two molecules, the molecules will have completely different properties.

To make things just a little more confusing, atoms of the same type can join together as a molecule just like atoms of different types can. For example, oxygen is an element. However, when we breath oxygen from the air around us, we are not breathing in the element oxygen. Instead, we are breathing in the molecule O_2. This molecule is formed when two oxygen atoms link together. Even though we are lazy and tend to refer to this as "oxygen," it is not the element oxygen. Instead, it is a molecule formed by two oxygens linking up. Other examples of molecules formed by the same types of atoms linking up are: ozone (O_3), nitrogen gas (N_2), and hydrogen gas (H_2). As you can see from their chemical formulas, ozone is a molecule comprised of three oxygen atoms linked together, nitrogen gas is a molecule formed by two nitrogen atoms linking up, and hydrogen gas results when two hydrogen atoms link together.

We see, then, that molecules are formed when atoms link together. Different atoms can link together (as is the case with methane, CH_4) or atoms of the same type can link together (as is the case with oxygen gas, O_2). Both the type of atoms that link together and the number of each type of atom will determine the properties of the molecule.

It is important for you to understand the distinction between atoms, elements, and molecules. Atoms are the basic building blocks of matter, and their properties are determined by the number of protons, neutrons, and electrons which make them up. A group of atoms that have the same number of protons all belong to the same element. If atoms are not linked together, they form the element for which they are named. Thus, when you fill a balloon with helium, you are filling it with a bunch of helium atoms, some of which will be helium-3, some will be helium-4, and some will be helium-5. If, instead, atoms link together, they form molecules. The type of and number of atoms that link together determine the properties of the molecule. When you blow into a balloon to fill it up, for example, you are filling it with a lot of oxygen molecules (O_2) and carbon dioxide molecules (CO_2).

ON YOUR OWN

5.6 Identify the following as either an atom, element, or molecule.

       M   E   A   E   M

     a. NH_3 b. O   c. carbon-14   d. S   e. H_2

5.7 A student is told to study the chemicals nitrogen monoxide (NO) and nitrogen dioxide (NO_2) and determine the differences. The student reports back that there are no differences between the molecules because they are comprised of the same elements. Is the student right or wrong? Why? *wrong   in NO_2 the oxygen is linked together*

5.8 Name each element and the number of atoms of that element in one molecule of acetic acid ($C_2H_4O_2$), which is the active ingredient of vinegar. *2 carbons 4 hydrogens 2 oxygens*

Changes in Matter

As we all know, God's Creation is constantly changing. Seasons change; weather changes; organisms grow and mature, bringing on many changes. On a smaller level, however, there are countless changes occurring in the elements and molecules that make up life. These changes fall into one of two broad categories: **physical changes** or **chemical changes**.

     Physical change - A change that affects the appearance but not the chemical makeup of a substance

     Chemical change - A change that alters the makeup of the elements or molecules of a substance

These definitions probably seem a little cryptic right now, but in the next couple of paragraphs, their meanings should become clear.

Let's start with physical change. Suppose you were to go to a barber shop and get your hair cut. The shape, style, and length of your hair would change (hopefully for the better). Nevertheless, once you are all done, your hair will still be hair, right? Even though the appearance of the hair is quite different, the actual substance (hair) did not change. This is because the process of cutting hair did not change the chemical makeup of the molecules of your hair. The same elements are linked together in the same way for each molecule in your hair. Some of those molecules are still on your head, arranged in a certain way, and some of those molecules are now scattered on the barbershop floor. Nevertheless, the molecules themselves did not change, just the arrangement of them. That's a physical change. We'll study physical change in more detail in the next section, so if you're still a little confused, don't worry.

Chemical change is something entirely different than physical change. Suppose you were to take a piece of paper and light it with a match. What would happen? The paper would begin to burn, making flames and smoke appear. In the end, you would be left with a small pile of ashes where once there was a piece of paper. This is an example of a chemical change. The molecules that made up the paper have been completely changed. The molecules were changed into gases (principally carbon dioxide and water vapor) and ash. One substance (paper) changed into other substances (carbon dioxide, water vapor, and ash). That's chemical change. As with physical change, don't worry if you are a little confused on this point; we will be studying chemical change in great detail soon!

One idea that is often helpful to students in determining whether a change in matter is physical or chemical is as follows: physical changes are generally reversible; chemical changes are not. For example, it would be a little silly, but you could reverse a haircut. You could take every strand of hair that was cut and glue it back together. The process would be tedious, but it could, in principle, be done. On the other hand, you cannot reverse the burning of paper. Once it has burned, there is no way to reverse the process and get the paper back. It has been changed forever. Thus, cutting hair is a physical change because it can be reversed, and burning paper is a chemical change, because it cannot be reversed.

ON YOUR OWN

5.9 Identify the following changes as chemical or physical.

    a. putting milk on cereal   b. baking bread   c. breaking a vase

---

Physical Change

Now that you have at least some understanding of the definition of physical change, it is time to study a few specific types of physical change in detail. The first important type of physical change involves the **phases of matter**.

Phase - One of three forms - solid, liquid, or gas - which every substance is capable of attaining

When we change a substance from one phase to another, it is a physical change. For example, when you put water in the freezer, it turns from its liquid phase (water) into its solid phase (ice). Despite the appearance change, the molecules in the ice are still water molecules; they are simply in a different phase. Thus, phase changes are physical changes.

In general, we all know that to change something from its liquid phase to its solid phase, we must freeze it. This involves cooling it down. When you cool a substance down, you actually remove energy from the molecules of the substance. Thus, freezing a substance is really just a matter of removing energy from its molecules. We also know that to change a substance from its solid phase to its liquid phase, we simply have to melt it. This involves heating it up.

When you heat a substance up, you are, in fact, adding energy to its molecules. Thus, changing the phase of a substance is really just a matter of modifying the energy that the molecules of the substance have. As you take energy away from the molecules of a substance, its phase changes from gas to liquid to solid. If you add energy to a substance, its phase will change from solid to liquid to gas. This relationship can be illustrated schematically as follows:

$$\text{SOLID} \underset{\text{REMOVE ENERGY}}{\overset{\text{ADD ENERGY}}{\rightleftarrows}} \text{LIQUID} \underset{\text{REMOVE ENERGY}}{\overset{\text{ADD ENERGY}}{\rightleftarrows}} \text{GAS}$$

Once again, it is important to realize that since a phase change only involves removing or adding energy to the molecules of a substance, the chemical makeup of those molecules does not change. As a result, phase changes are physical changes.

One type of physical change that is very important in biology occurs when one substance is dissolved in another. For example, if you take salt and mix it with water, what happens? The salt seems to disappear, and the resulting mixture tastes salty. What you have done, in fact, is dissolved the salt in the water. The salt is still there and it is still salt. It has just been distributed throughout the water molecule by molecule. As a result, you can no longer see it. Nevertheless it is still there, as evidenced by the salty taste. When one substance is dissolved in another, it is a physical change, and the result is called a **solution**. The substance being dissolved in the liquid is called the **solute** and the liquid is called the **solvent**. In the case of salt being dissolved in water, the salt is the solute, the water is the solvent, and the resulting saltwater is a solution.

A lot of the chemistry of life involves solutions and the changes that occur in them. It is therefore important to study one aspect of how molecules behave in solution with the following experiment. Note that once you set it up, this experiment has to sit for two hours, so plan your school day accordingly!

EXPERIMENT 5.1
Osmosis and Diffusion

<u>Supplies</u>
- A reasonably fresh potato
- Salt
- Sugar
- Water
- Three small glasses
- Plastic wrap
- A napkin
- Tape
- Tablespoon

- Cutting knife

Object: To observe and learn about the processes of diffusion and osmosis

Procedure:

A. Measure out a tablespoon of sugar and dump it into the center of an unfolded napkin.

B. Fold the napkin around the lump of sugar so that it forms a nice "package" completely surrounding the sugar. Tape the napkin so that it does not unfold. DO NOT cover the entire napkin with tape! Tape it with only one or two strips of tape just so it does not unfold. In the end, you should be able to pick up the napkin without any sugar spilling out of it.

C. Put the napkin/sugar "package" into one of the small glasses and fill the glass with water. The napkin will probably float up to the top. That's okay.

D. Cover the glass with plastic wrap and let it sit for at least 2 hours.

E. Cut the potato along its width (not along its length) so that you have two nearly circular slices of potato that are about half an inch thick. Do not use the ends of the potato, because the slices need to be flat on each side. To make sure that your potato is fresh enough, grasp the edges of one of the potato slices with both hands and try to bend it. The potato slice should not bend easily.

F. Take the next glass and fill it 3/4 of the way full of water. Place the first potato slice in the glass and cover the glass with plastic wrap. Let it sit for at least 2 hours.

G. Take the remaining glass and fill it 3/4 of the way full with water. Add enough salt so that no matter how much you stir the water, not all of the salt will dissolve.

H. Take the remaining potato slice and place it in the glass, under the saltwater. Cover the glass with plastic wrap and let it sit for at least 2 hours.

I. Once at least 2 hours are up (if you wait longer your results will be better), remove the plastic wrap and then the napkin from the first glass. Swirl the water in the glass and then take a small sip of the water. What does it taste like?

J. Remove the plastic wrap from the second glass and remove the potato slice. Hold the slice at the edges with both hands and try to bend it. Is it harder, easier, or about the same as bending the slice when you first cut it (step E)?

K. Do the same thing with the potato in the last glass. How does it bend compared to the first slice?

If everything in your experiment went well, you should have noticed that the water in the first glass tasted sweet, the potato slice in the second glass was about the same or perhaps harder to bend than when you first cut it, and the potato slice in the last glass should have been very easy to bend, almost as if it was made of rubber.

In the first glass, sugar obviously found its way out of the napkin and into the water. It did so by a process known as **diffusion**.

Diffusion - The random motion of molecules from an area of high concentration to an area of low concentration

Concentration - A measurement of how much substance exists within a certain volume

You see, when you placed the napkin in the water, there was at first no water in the napkin or in the sugar. Thus, the concentration of water inside the napkin was zero. Since there are many holes in a napkin, water molecules were able to travel through those holes and start filling the napkin up. Thus, water molecules began moving from an area of high concentration (the water in the glass) to an area of low concentration (the napkin). You probably noticed this right away because the napkin got wet.

What you probably did not notice was that the sugar was doing the same thing. Inside the napkin, there was a LOT of sugar. Outside the napkin, there was none. When in its solid form, the sugar was too large to get through the tiny holes in the napkin, so the sugar stayed where it was. However, as soon as the water got inside the napkin, some of the sugar dissolved. When dissolved, the sugar molecules could individually get through the tiny holes in the napkin. Thus, the sugar moved from inside the napkin (where the concentration of sugar was high) into the water outside of the napkin (where the concentration was low).

Now it's important for you to realize that the motion of the water molecules into the napkin and the motion of the sugar molecules outside the napkin was not directed by some mysterious force. Instead, molecules that are in their liquid phase and molecules that are dissolved in liquid tend to move about randomly as a matter of course. That's why the definition of the term diffusion includes the word "random." The water molecules simply were moving randomly. Some of them happened to find themselves in the napkin as a result of their random motion. Likewise, once the water began to dissolve the sugar molecules, the dissolved sugar molecules began randomly moving. Since, when dissolved, they could fit through the holes of the napkin, some of the sugar molecules found themselves in the water outside of the napkin. Thus, the sugar molecules did not suddenly "decide" that they had to get to an area of low sugar concentration. Instead, they just randomly ended up there. Given enough time, this random motion would have evenly mixed the sugar and water throughout the glass.

Compare this to what happened with the potato slices. A potato gets its rigidity from water that is in the potato. If you allow a potato to sit out for a long time, you will notice that it gets soft, almost rubbery. That's because as time goes on, some of the water in the potato evaporates and thus the potato loses its rigidity, becoming rubbery. When you placed the first potato slice in the water-filled glass, diffusion allowed water to move into the potato. This helped the potato replenish any water it might have lost, making it more rigid. Of course, if the potato you used wasn't too old, then it probably hadn't lost much water, so the effect was probably not noticeable. Thus, the first potato slice either was about the same as when you first cut it or more rigid.

The second potato slice, however, should have been very easy to bend, as if it were made out of rubber. This is because it lost a great deal of water. Wait a minute, though. How could the potato have lost water while it was soaking in a saltwater solution? Well, the cells in a potato are surrounded by a **semipermeable membrane**.

Semipermeable membrane - A membrane that allows some molecules to pass through but does not allow other molecules to pass through

The semipermeable membrane in a potato allows water to pass through it but does not allow salt to pass through. As a result, normal diffusion could not occur, because water was free to move in and out of the potato, but salt was not.

The salt, however, is attracted to the water (that's why it dissolves in water to begin with). Since the salt is not allowed to pass through the semipermeable membrane and get to the water that is in the potato, the attraction between the salt and the water molecules is enough to pull the water out of the potato and into the saltwater solution. In effect, since the salt could not travel to the water inside the potato, the water in the potato traveled to the salt instead. This process is called **osmosis**.

Osmosis - The tendency of a solvent to travel across a semipermeable membrane into areas of higher solute concentration

Do you see the difference between osmosis and diffusion? In diffusion, molecules move randomly so that solute and solvent spread out evenly throughout the solution. That's what happened in the first glass. In osmosis, the travel of the solute is restricted by a semipermeable membrane. As a result, only the solvent can move. Since the solvent is attracted to the solute molecules, the solvent moves across the semipermeable membrane, into the area with the highest concentration of solute. Figure 5.2 illustrates the difference between osmosis and diffusion.

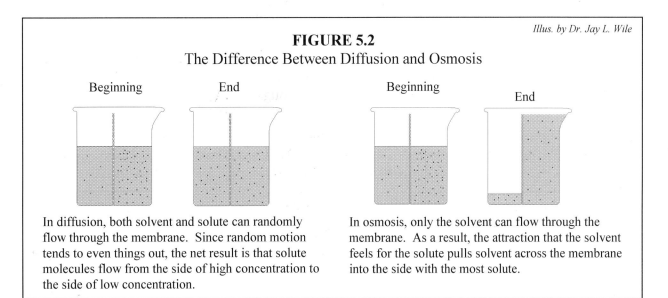

Illus. by Dr. Jay L. Wile

FIGURE 5.2
The Difference Between Diffusion and Osmosis

Beginning   End   Beginning   End

In diffusion, both solvent and solute can randomly flow through the membrane. Since random motion tends to even things out, the net result is that solute molecules flow from the side of high concentration to the side of low concentration.

In osmosis, only the solvent can flow through the membrane. As a result, the attraction that the solvent feels for the solute pulls solvent across the membrane into the side with the most solute.

In "Rhyme of the Ancient Mariner," a famous poem about a ship lost at sea, the mariner laments:

"Water, water everywhere and all the boards did shrink.
Water, water everywhere nor any drop to drink."

This is a reference to a common problem in early sea travel. Many crews died of thirst because their voyage took longer than expected and not enough fresh water had been taken along. Despite the fact that the ship was surrounded by water, none of it could be drunk. You see, ocean water has a high concentration of salt in it. If you drink this salt water, it will come into contact with the cells of your body, which are full of water that has a *much lower* concentration of salt in it. Since the cells in your body are covered with a semipermeable membrane, osmosis causes the water in your cells to leave the cell (where the solute concentration is low) and enter into the saltwater that you drank (where the solute concentration is high). This causes the cell to lose its water, which in turn causes it to shrivel up and die. Thus, if you drink a lot of saltwater, you will kill yourself one cell at a time! That's what was frustrating to the mariner in the poem. His crew was dying of thirst, despite the fact that there was ocean water all around!

It turns out that osmosis and diffusion play rather critical roles in many biological processes, so it is important for you to understand each process as well as the difference between the two.

ON YOUR OWN

5.10 A semipermeable membrane was placed in a beaker. Equal amounts of saltwater solution were placed on each side of the membrane, but the solution on one side was twice as concentrated with salt as was the solution on the other side. After one hour, the water level of the solution on the right had increased and the water level of the solution on the left had decreased. Which solution (the one on the left or the one on the right) started out with the higher salt concentration? *right*

Chemical Change

Now that we've spent some time on physical change, we need to study chemical change. Remember, chemical change occurs when the molecules in a substance change their chemical makeup. For example, most people have seen a natural gas stove or furnace flame. It is a pleasant blue flame that can be used to cook food or heat a home. Well, in order to make that flame, natural gas is burned. The proper name of natural gas is methane and it can be abbreviated as CH_4. Thus, a molecule of methane is a carbon atom linked to four hydrogen atoms. When methane burns, the oxygen in the air interacts with the methane molecules, changing them into water molecules and carbon dioxide molecules. The resulting chemical change causes energy to be released, which we detect as heat and light from the flame.

If we wanted to write out the chemical change that occurs when methane burns, we could do so as follows:

methane and oxygen interact to make water and carbon dioxide

This is a bulky way of informing someone what happened, so we will abbreviate it. We will use the chemical formulas of the substances in the reaction; we will replace "and" with a plus sign, and we will replace "interact to make" with an arrow. The result is:

$$CH_4 + O_2 \rightarrow H_2O + CO_2$$

Now it turns out that if we were able to watch this happen on a molecular level, we would see that it takes TWO oxygen molecules to interact with ONE methane molecule and the result is TWO water molecules and ONE carbon dioxide molecule. To really abbreviate what happened, then, we need to add this information:

$$CH_4 + 2O_2 \rightarrow 2H_2O + CO_2$$

What we did was put numbers to the left of the chemical formulas. These numbers indicate how many of that molecule it took to make the chemical change. If there was only 1 molecule, no number was written. This is called a **balanced chemical equation**. The chemical change that occurs is often called a **chemical reaction**. Thus, we could say that the balanced chemical equation above describes the chemical reaction that takes place between methane and oxygen.

It is important for you to understand that when a chemical reaction occurs, the molecules on the left side of the arrow are destroyed, and the molecules on the right side of the arrow are produced. Thus, in the burning of methane, we start out with 1 methane (CH_4) molecule and 2 oxygen (O_2) molecules. These molecules interact, exchanging atoms so that the methane and oxygen molecules are destroyed, making two water (H_2O) molecules and one carbon dioxide (CO_2) molecule. In chemical terminology, we call the molecules on the left side of the arrow **reactants**, because they react with one another and are destroyed. Those molecules on the right side of the arrow are called **products**, because they are produced as a result of the chemical reaction.

When looking at a chemical equation, students often get confused between the numbers that are subscripts and the numbers that exist to the left of the chemical formulas. You need to remember that the subscripted numbers refer to how many atoms of a certain type are in the molecule. If a subscripted number does not exist after an element's symbol, that means only one such element is in the molecule. Thus, the subscript "2" in H_2O tells us that there are 2 hydrogen atoms in the water molecule. The fact that there is no subscript after the "O" tells us that there is one oxygen atom in the water molecule. On the other hand, the "2" that is written to the left of the water molecule's abbreviation tells us that 2 water molecules are produced in the chemical reaction. Make sure you understand this distinction!

ON YOUR OWN

5.11 One of the main chemical reactions that makes a car run is the combustion of octane (C_8H_{18}):

$$2C_8H_{18} + 25O_2 \rightarrow 16CO_2 + 18H_2O$$

a. Write the chemical formulas of the reactants in this equation.

b. Write the chemical formulas of the products in this equation.

c. How many molecules of C_8H_{18} are used in the reaction?

d. How many molecules of H_2O are made in the reaction?

Photosynthesis

Now that you understand how to interpret chemical equations, let's talk about photosynthesis, a subject we have already discussed, but not in detail. The most popular means by which autotrophic organisms create their own food is by photosynthesis. As you already learned, this process converts carbon dioxide and water into a simple sugar called glucose and oxygen. The balanced chemical equation is:

$$6CO_2 + 6H_2O \rightarrow C_6H_{12}O_6 + 6O_2$$

In photosynthesis, then, six carbon dioxide (CO_2) molecules interact with six water (H_2O) molecules. Once they interact, a glucose molecule ($C_6H_{12}O_6$) and six oxygen (O_2) molecules are formed.

Think for a moment about what you've already learned regarding photosynthesis, and you will probably realize that something is missing from this chemical equation. What is it? Well, you've already learned that photosynthesis requires sunlight. Where is the sunlight in this equation? Sunlight isn't in the equation, but it is still necessary in photosynthesis. You see, if a bunch of carbon dioxide and water were put into a container, they would not interact to form glucose and oxygen. In order for them to interact, the carbon dioxide and water molecules must be pushed together. This takes energy, which the plant gets from sunlight.

Okay, then, in order for plants to make glucose and oxygen, they just need carbon dioxide, water, and energy, right? Well, no. Even after supplying a lot of energy to a container full of carbon dioxide and water, you would not produce much glucose. You see, even with a lot of supplied energy, the interaction between carbon dioxide and water is very slow. Thus, the reaction would have to continue for a long, long time before a reasonable amount of glucose is produced. This is unacceptable to an autotrophic organism, because it must produce a lot of food for itself. Thus, the organism must speed up the reaction, making it produce glucose much faster than it normally would.

How does an organism speed up the photosynthesis reaction? It turns out that there are two ways to speed up a chemical reaction. First, you could perform the reaction at a higher temperature. Higher temperatures increase the speed of most chemical reactions. This isn't a viable option, however, because most autotrophic organisms cannot survive unusually high temperatures. Thus, autotrophic organisms use a second method to speed up the chemical reaction. They employ a **catalyst** (kat' uh list).

<u>Catalyst</u> - A substance that alters the speed of a chemical reaction but does not get used up in the process

Most catalysts speed up chemical reactions, but there are a few which are known to slow chemical reactions down.

Now look at that definition for a moment. Catalysts alter the speed of a chemical reaction, but they do not get used up in the process. If you think about it, this is really an amazing feat. Somehow, the catalyst *affects* the reaction, but it doesn't get *used up by* the reaction. Pretty nifty, huh? How does a catalyst accomplish this feat? Unfortunately, you will have to wait until chemistry class to learn the answer to that question.

So, autotrophic organisms speed up the photosynthesis reaction by using a catalyst. What is the catalyst? For most autotrophic organisms, it is chlorophyll. Back in Module #3 we studied green algae in kingdom Protista, and we briefly discussed chlorophyll, telling you that it was necessary for photosynthesis in these organisms. Now you know why. Chlorophyll speeds up the photosynthesis reaction so that an autotrophic organism can produce a reasonable amount of glucose for its food supply. Green plants use chlorophyll as a catalyst in photosynthesis as well; however, chlorophyll is not the only catalyst that can be used in photosynthesis. Blue-green algae, for example, use a substance called blue phycobilin in addition to chlorophyll. Members of phylum Rhodophyta (the red algae) use red phycobilin as their catalyst in photosynthesis.

At this point, your head might be swimming a bit because you've taken in a lot of information. Thus, let's do a little review. Most autotrophic organisms use photosynthesis to create their own food. This process is a chemical change which takes 6 carbon dioxide molecules and 6 water molecules to make one glucose molecule and 6 oxygen molecules. The glucose is food for the organism, and the oxygen goes back into the atmosphere for other organisms to breathe. In order for this chemical reaction to occur, however, the organism needs more than just carbon dioxide and water. It needs energy, which it gets from sunlight, and it needs a catalyst, which is usually (but not always) chlorophyll. The energy from the sunlight forces the carbon dioxide and oxygen to interact, while the catalyst speeds up the interaction.

Now if you think about this, it is an incredibly complex system! It's amazing enough that an organism can actually produce its own food. Once you learn the details of how this is accomplished, however, you should be overcome with a deep sense of awe. God has designed His Creation so intricately that even a single-celled organism like the euglena knows how to feed itself by collecting carbon dioxide and water, producing a catalyst that will speed up their reaction, and trapping energy from the sun to make the reaction occur. If it hasn't already become clear to you, this should drive a fundamental point home: **there is no such thing as a simple life form**. All of God's Creation is intricate and complex, and as you learn more about it, your appreciation of His power should increase dramatically!

ON YOUR OWN

5.12 A plant loses all of its chlorophyll. Will it be able to produce any glucose at all? *yes but only a little*

5.13 A chemist is trying to speed up a chemical reaction. If the chemist does not have a catalyst, what other means can be used? *more heat*

Organic Chemistry

The most important chemicals in the study of life fall into a broad category called **organic molecules**.

Organic molecule - A molecule that contains only carbon and any of the following: hydrogen, oxygen, nitrogen, sulfur, and/or phosphorous

This is actually a simplistic definition of organic molecules, but it fits more than 99% of them, so for right now it will do. By this definition, molecules like CH_4, CO_2, and C_2H_4O are organic molecules (CH_4 contains only carbon and hydrogen, CO_2 contains only carbon and oxygen, and C_2H_4O contains only carbon, hydrogen, and oxygen) but $CaCO_3$ and H_2O are not (the element Ca is not on the list, and H_2O contains no carbon).

Most organic molecules are actually produced by living organisms. When a living organism makes molecules, we call it **biosynthesis** (bye oh sin' the sis).

Biosynthesis - The process by which living organisms produce molecules

Photosynthesis is an example of biosynthesis. Living organisms (plants, for example) make glucose and oxygen from carbon dioxide and water. That's biosynthesis. The products of biosynthesis are many and varied, but they fall into a few broad categories, most of which we will discuss in this section.

Carbohydrates

Carbohydrates are organic molecules that contain only carbon, hydrogen, and oxygen. In addition, they have the same ratio of hydrogen atoms to oxygen atoms as does water. For example, one of the most simple carbohydrates is glucose, $C_6H_{12}O_6$. Notice that it has 12 hydrogen atoms and 6 oxygen atoms. In other words, there are twice as many hydrogen atoms as there are oxygen atoms. This is the same as water (H_2O). In fact, that's where the term "carbohydrate" comes from. "Carbo" stands for the carbon in the molecule, and "hydrate," which means "to add water," stands for the fact that there are twice as many hydrogen atoms as oxygen atoms, just like water.

Now although giving you the chemical formula of glucose is instructive, there is an even more instructive means of describing glucose to you. The **structural formula** of a molecule gives you the type and number of atoms in the molecule, but it also tells you something else. It tells you which atoms are linked to which. For example, the structural formula of glucose is:

$$
\begin{array}{ccccccc}
\text{H} & \text{H} & \text{H} & \text{OH} & \text{H} & \text{H} \\
| & | & | & | & | & | \\
\text{H--C} & \text{--C} & \text{--C} & \text{--C} & \text{--C} & \text{--C} = \text{O} \\
| & | & | & | & | \\
\text{OH} & \text{OH} & \text{OH} & \text{H} & \text{OH}
\end{array}
$$

If you simply count the atoms in the molecule, you will see that there are 6 carbon atoms pictured, 12 hydrogen atoms, and 6 oxygen atoms. Thus, as we knew already, the chemical formula of glucose is $C_6H_{12}O_6$. By looking at the structural formula, however, we can see a lot more. The lines in the structural formula represent **chemical bonds**, which link atoms together in a molecule. Thus, we can see that the six carbon atoms are all linked to one another in a straight line, which we call a **carbon chain**. The first carbon in the chain (the one on the far left-hand side) has two H's attached to it, an O attached to it, and another C attached to it. This is exactly the way we would see the atoms arranged in the molecule, if we were able to see a molecule of glucose.

Looking at the structural formula of glucose a little more carefully, you will see two strange things. First, you will notice that there are two lines which link the carbon on the far right of the carbon chain to the oxygen at the end of the structural formula. This is called a **double bond**, and it is approximately twice as strong as the bonds represented by the single lines in the structural formula. For biology, it is not critical that you understand what a double bond really is. You will learn that next year in chemistry. For right now, just realize that it is a chemical bond that is roughly twice as strong as most chemical bonds. There are, in fact, **triple bonds** that exist in some molecules. In structural formulas, triple bonds are represented by three lines and are roughly three times as strong as most chemical bonds.

The second strange thing you will see in the structural formula of glucose is the fact that there are 5 times in which an oxygen atom (O) and a hydrogen atom (H) are drawn next to each other but there is no line linking them together. Does that mean they are not linked to one another? No, not really. It turns out that when an oxygen atom and a hydrogen atom are linked together in a molecule, biologists rarely draw a line to represent the bond that exists between them. This is because the combination of an O and an H linked together is fundamentally important in biology, so to represent this, we purposely do not draw the bond. This makes that part of the molecule stand out. Thus, even though it is not drawn, you need to realize that there is a bond linking the O to the H in an OH group.

Now that you have some understanding of structural formulas, it is time to throw a curveball at you. **Many molecules have more than one structural formula!** In fact, this is the case for glucose. Although some glucose molecules look like the one drawn above, other glucose molecules look like this:

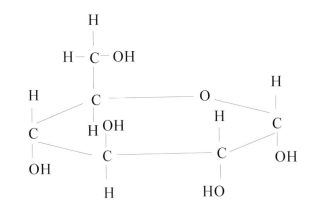

What's the difference? Well, the chemical formula is still the same. If you count the atoms, you will find 6 C's, 12 H's, and 6 O's. The main difference is that the double bond is gone between the oxygen and the carbon and, instead, the atoms have arranged themselves in a ring. Not surprisingly, the first structural formula of glucose is called the **chain structure** while this one is called the **ring structure**. In biology, the ring structure is the most prevalent form of glucose.

Are you ready for another curveball? We hope so! You see, not only can the same molecule have different structural formulas, but different molecules can also have the same chemical formulas! When two different molecules have the same chemical formula, they are called **isomers**:

<u>Isomers</u> - Two different molecules that have the same chemical formula

Now before we move on to *why* two different molecules can have the same chemical formulas, it is imperative that you know what we mean when we say "different." Have you ever noticed that both lettuce and fruit taste sweet, but they taste sweet in different ways? Well, the reason for this is that there are two different types of sugar molecules in them. Lettuce has glucose, the molecule that we have been discussing. This molecule gives lettuce its sweet taste. On the other hand, fruit has a sugar called fructose. It tastes different than glucose, so the sweetness of fruit is different than the sweetness of lettuce. These two sugars taste different because they are different molecules. They have different chemical characteristics, one of which is taste. That's what we mean when we say "different" molecules. Different molecules have different chemical properties.

What's the chemical formula of fructose? It's $C_6H_{12}O_6$, the same as glucose! If these two molecules have the same chemical formula, how can they be different molecules? Well, they have the same *chemical formula*, but they have different *structural formulas*. The ring and chain structural formulas of both glucose and fructose are shown in Figure 5.3.

FIGURE 5.3
The Difference Between Glucose and Fructose

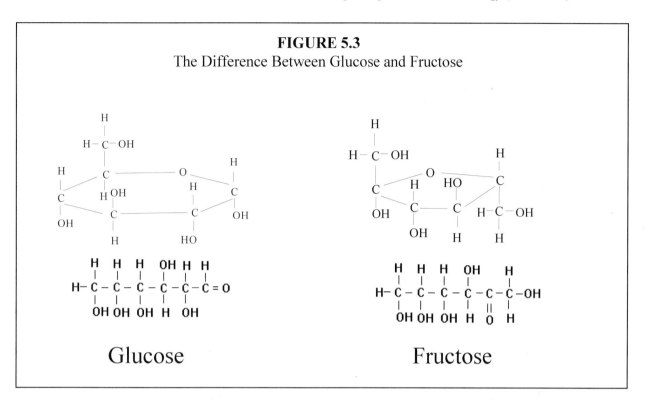

Glucose

Fructose

These different structural formulas are the reason behind the taste differences and other chemical differences between glucose and fructose.

Glucose and fructose belong to a class of compounds known as **monosaccharides** (mahn uh sak' uh rides), which are also called **simple sugars**.

Monosaccharides - Simple carbohydrates that contain three to ten carbon atoms

The reason that these are called monosaccharides is that they form the basic building blocks of more complex carbohydrates called **disaccharides** (dye sak' uh rides) and **polysaccharides** (pahl ee sak' uh rides).

Disaccharides - Carbohydrates that are made up of two monosaccharides

Polysaccharides - Carbohydrates that are made up of more than two monosaccharides

These more complex carbohydrates form the basis of much of the food that we eat.

For example, table sugar is a disaccharide called sucrose. It is formed when glucose and fructose chemically react in a process known as a **dehydration** (dee hye dray' shun) **reaction**.

Dehydration reaction - A chemical reaction in which molecules combine by ejecting water

The dehydration reaction that makes sucrose is shown in Figure 5.4:

FIGURE 5.4
Dehydration Reaction That Produces Table Sugar, a Disaccharide

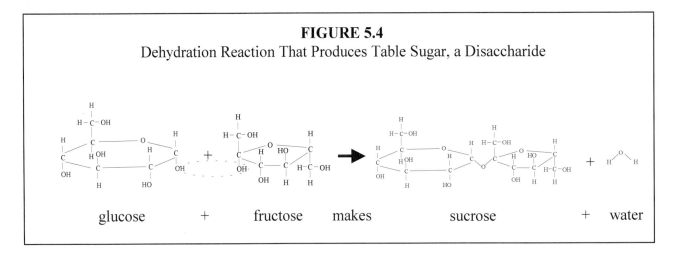

glucose + fructose makes sucrose + water

Do you see now why we call this a dehydration reaction? When you dehydrate something, you remove water. Well, in this reaction, a hydrogen from the glucose and an OH from the fructose combine to make water, "removing water" from the two molecules. The remaining oxygen atom then links the two molecules together, making a disaccharide. Many other disaccharides make up the sugars that sweeten our foods. Glucose and galactose, for example, combine in a dehydration reaction to make lactose, the sugar that gives milk its sweetness.

When several monosaccharides link together, the result is a polysaccharide. Typically, polysaccharides are not sweet, because the sweetness of the monosaccharides are lost when several combine. Nevertheless, polysaccharides are an important part of our diet. The polysaccharides known as starches, for example, are found in most plants. When a plant has extra monosaccharides, it will store them as polysaccharides by running many dehydration reactions that link the monosaccharides together. Potato starch, corn starch, and starch from wheat, rice, and other grains are a major source of food for humans. Humans and other animals make their own starch, glycogen (gly' ko jen), when they have excess carbohydrates to store. The non-digestible part of our diet (often called "roughage") is made up of the polysaccharide known as cellulose. Although most organisms cannot digest cellulose, God has created a few protozoa, bacteria, and fungi that use it as their major source of food.

As you might have already learned by studying nutrition, carbohydrates are one of the principal sources of food energy for most animals. Interestingly enough, however, most animals can only use monosaccharides for energy. Thus, when an organism eats disaccharides or polysaccharides, it must first break them down into their individual monosaccharide components. How is this done? Well, monosaccharides combine to form disaccharides and polysaccharides by dehydration. To break these complex molecules back down into their monosaccharide components, all you have to do is add water to them. This process is called **hydrolysis**.

Hydrolysis - Breaking down complex molecules by the chemical addition of water

Hydrolysis is essentially the reverse of dehydration. We will discuss it in more detail when we discuss proteins and enzymes.

ON YOUR OWN

5.14 What is the chemical formula of the molecule with the following structural formula?

$H_4 C_2$

$$H - C = C - H$$
$$\quad\;\; | \quad\; |$$
$$\quad\;\; H \quad\, H$$

5.15 A chemist takes a polysaccharide and turns it into many disaccharides. Has the chemist used dehydration or hydrolysis? hydrolysis

---

Organic Acids and Bases

Two classes of molecules that you will spend a lot more time learning about in chemistry are **acids** and **bases**. In general, acids are substances that taste sour and bases taste bitter. Many of the drinks that we enjoy (soda pop and fruit juice, for example) contain acids, and many of the substances we use for cleaning (soap and spray-cleaner, for example) contain bases. When acids and bases react together, they typically form water and another class of molecule called a "salt." Table salt is just one example of the general chemical class known as salts.

Organic acids, in general, contain the following pattern of atoms bonded together:

$$\begin{array}{c} O \\ \| \\ C - OH \end{array}$$

When a molecule has a section that looks like this, it is generally considered an organic acid. Not surprisingly, then, this is often called an **acid group**. Figure 5.5 shows the structural formula of three organic acids. Notice that even though they look quite different, they each have one thing in common. They each have at least one acid group.

FIGURE 5.5
Organic Acids

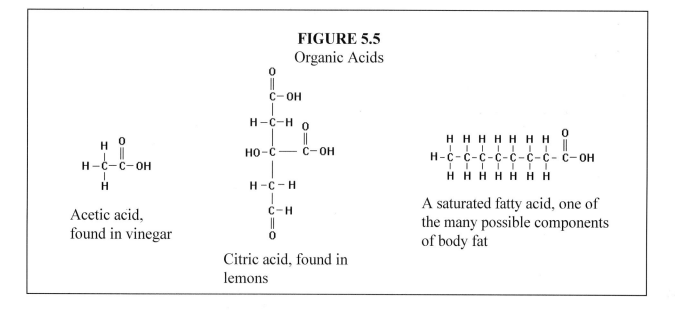

Acetic acid,
found in vinegar

Citric acid, found in
lemons

A saturated fatty acid, one of
the many possible components
of body fat

Organic bases also have a group of atoms in common. Often called the **amine group**, molecules that contain a nitrogen atom and a hydrogen atom bonded together can be an organic base.

In most of the chemical reactions that make life possible, the amount of acid or base present has a profound effect on the speed and effectiveness of the reaction. As a result, tracking the level of acid or base is quite important. Chemists have a scale called the "pH" scale to do just this. You will learn more about the pH scale in chemistry, so for right now you just need to know some basics. The pH scale runs from 0 to 14. When a solution has a pH of 7, it is considered **neutral**, having no net acid or base characteristics. Solutions with pH from 0 to just under 7 are **acidic**. The lower the pH, the stronger the acid. Solutions with pH from just above 7 to 14 are called **alkaline** and have the characteristics of a base. The higher the pH, the more like a base the solution becomes. Figure 5.6 illustrates this scale, along with some popular substances and where they fall within it.

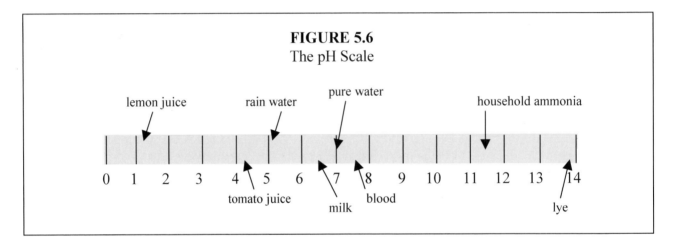

FIGURE 5.6
The pH Scale

ON YOUR OWN

A chemist measures the pH of several solutions. The results are: Solution A: 8.1, Solution B: 1.1, Solution C: 5.5, Solution D: 13.2.

5.16 Which solution is the most alkaline? *Solution D*

5.17 Which solution is the most acidic? *Solution B*

Lipids

Lipids, also known as fats, are complex molecules formed when three fatty acids, like the one shown in Figure 5.5, link to a substance known as glycerol in a dehydration reaction. A sample dehydration reaction which forms a lipid is shown in Figure 5.7.

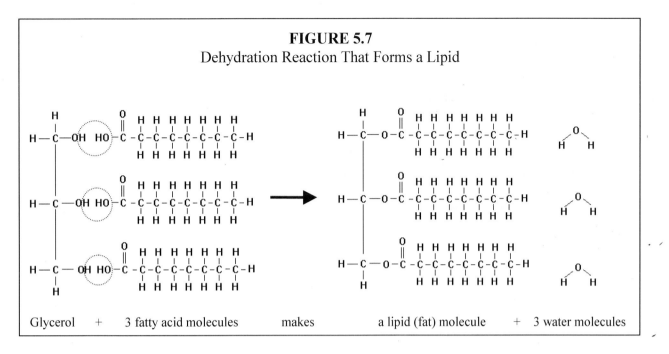

FIGURE 5.7
Dehydration Reaction That Forms a Lipid

Glycerol   +   3 fatty acid molecules   makes   a lipid (fat) molecule   +   3 water molecules

In general, lipids cannot be dissolved in water. Cooking oil, for example, is made of lipids. What happens when you try to mix water and cooking oil? They don't mix well, do they? This is because the structure and components of lipid molecules make them **hydrophobic**.

Hydrophobic - Lacking any affinity to water

Since lipids are not, in any way, attracted to water, they simply cannot be dissolved in it.

Lipids are important biologically because most animals store their excess food as lipids. Even though animals can take their excess carbohydrates and turn them into glycogen (as we learned just a few pages ago), lipids are the preferred method of storing excess food, because more than twice as much energy can be stored in an ounce of fat than in an ounce of carbohydrates. As a result, animals try to make fats whenever they have excess food. Later, when food is scarce, their bodies can digest the fat for energy. For most humans in the United States, of course, there isn't a lot of food scarcity. Thus, the fat just builds up in our bodies, making us overweight.

Typically, our bodies tend to digest any carbohydrates eaten before they digest the fats. That way, fats will only be digested if they are needed. If not, they will simply be stored with any other fats already in storage, eliminating the body's need to make fat molecules in order to store excess energy. This is why many nutritionists say that the easiest way to lose weight is to limit your intake of fat. That way, if your body does get excess food, it must make fats in which to store the extra energy. The body won't do this unless it has a significant amount of excess food, so often, no fats will be made. Also, since the body gets twice as much energy from fats as it does from carbohydrates, you can actually eat a much larger amount of carbohydrates before gaining weight than you can fats.

Nowadays, nutritionists also tell us that **saturated fats** tend to lead to heart troubles while **unsaturated fats** are less likely to. As a result, they suggest that we eat unsaturated fats. What's the difference? When a fat is made from fatty acids that have no double bonds between the carbons, it forms a saturated fat. We call the fat "saturated" because it has all of the hydrogens it can take. As a result, it is "saturated" with hydrogen. If a fatty acid has one or more double bonds between the carbons, however, it has fewer hydrogens than it possibly could. As a result, we call it "unsaturated."

Saturated fat - A lipid made from fatty acids which have no double bonds between carbon atoms

Unsaturated fat - A lipid made from fatty acids that have at least one double bond between carbon atoms

Fats made from unsaturated fatty acid molecules are much more compact than are those made from saturated fatty acid molecules. As a result, unsaturated fats tend to move around in our bloodstream rather easily. The more bulky saturated fats, however, can get stuck in the twists and turns of our bloodstream. As a result, they clog up the bloodstream, and the heart must work harder. That can lead to heart problems.

ON YOUR OWN

5.18 How many fat molecules can be made from 15 fatty acid molecules, providing you have plenty of glycerol? 5

Proteins and Enzymes

Proteins are one of the most important kinds of molecules in the chemistry of life. They are involved in virtually every chemical reaction that supports life. As you might expect, they come in many different shapes and sizes and are very complex. They are mostly made from carbon, nitrogen, hydrogen, and oxygen, but some proteins have phosphorous and sulfur in them as well.

Like polysaccharides , proteins have some basic building blocks. The basic building blocks of a protein are called amino (uh mee' no) acids. There are about 20 different amino acids in the proteins that make up life, and the type of amino acid along with the order in which the amino acids are linked up determine the shape and function of a protein. As you might expect, amino acids link up using dehydration reactions, just like monosaccharides do when they link up to form polysaccharides. As the amino acids link together, ejecting water, a bond called a **peptide bond** forms.

Peptide bond - A bond that links amino acids together in a protein

Now unfortunately, it is impossible to give you a structural formula for a protein. You see, the average amino acid has about 20-40 atoms, and even the simplest protein of life contains 124 amino acids linked together. Obviously, then, drawing the structural formula of a protein would be quite a job. The situation would be even worse for the "average" protein, which contains several thousand amino acids!

Even though it is impossible to give you a structural formula for a protein, it is possible to give you some idea of what one looks like. Figure 5.8 is a schematic representation of a protein.

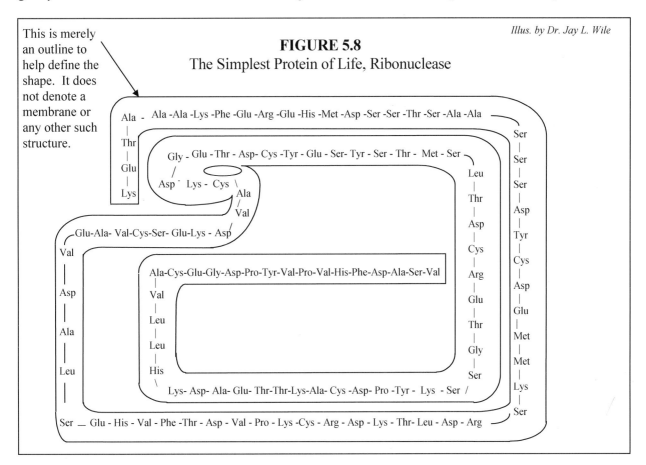

FIGURE 5.8
The Simplest Protein of Life, Ribonuclease

Illus. by Dr. Jay L. Wile

This is merely an outline to help define the shape. It does not denote a membrane or any other such structure.

What does this schematic mean? Well, each three-letter abbreviation stands for an amino acid. The letters "Lys," for example, stand for the amino acid called "lysine," while the letters "Glu" stand for glutamic acid. This schematic, then, tells you the type of amino acid and the order in which it appears in the protein ribonuclease. In addition, the shape that the amino acids are lined up in is roughly equivalent to the shape of ribonuclease. Of course, since the paper is only two-dimensional, it is impossible to draw the entire three-dimensional shape, but what is drawn here is a reasonable approximation.

Remember back in Module #1 when we promised to tell you why these authors think that abiogenesis (the formation of life from non-living chemicals) is impossible? Well, proteins are a part of the reason. In order for abiogenesis to work, proteins would have to be able to form from random chemical reactions. Without this happening, life could never appear because proteins are such a fundamental component of the chemical reactions that make life possible. It is our contention that the formation of proteins from random chemical reactions is impossible. You see, when assembled as shown in Figure 5.8, these 124 amino acids form a protein, ribonuclease, that performs a vital task in the chemistry of life. If the amino acids were to link up in the wrong order, or if even one of the amino acids were the wrong type, then ribonuclease would be *completely unable* to perform its job. Thus, if the first amino acid were "Glu" instead of "Lys," the protein would no longer be ribonuclease and would be completely useless. Think, for a moment, about the probability of such a protein forming by chance. Is it possible for such a molecule to form by chance from a mixture of lots of amino acids?

Let's make it easy on ourselves and assume that the only amino acids in the mixture are the 17 types needed to make this particular protein. In fact, there are many types of amino acids that exist naturally on earth, but adding more amino acid types would significantly reduce our chance of forming ribonuclease. So, in order to make the outcome more likely, we will restrict ourselves to using only the 17 different types of amino acids that make up this molecule. Making this assumption, we can say that the possibility of forming a protein that has "Lys" as its first amino acid is 1 in 17. Those aren't bad odds at all. However, the chance of forming a protein with "Lys" as its first amino acid and "Glu" as its second amino acid is 1 in 17 *times* 1 in 17, or 1 in 289. Suddenly the odds are looking less and less favorable.

The probability of forming a protein whose first three amino acids are "Lys," "Glu," and "Thr" in that order are 1 in 17 *times* 1 in 17 *times* 1 in 17, or 1 in 4,913. If you were to complete this calculation, you would find that the odds for making this protein by chance from a mixture of the proper amino acids is approximately 1 in 10^{152} (a 1 followed by 152 zeros). In order to illustrate just how ridiculously low these odds are, the probability for forming ribonuclease by chance is roughly equivalent to the probability of a poker player drawing a royal flush *19 times in a row!* Remember, ribonuclease is a "simple" protein. There are proteins in our bodies that contain more than 10,000 amino acids! Clearly the idea that these proteins could form by chance is absurd, which lays to waste the entire theory of abiogenesis!

Enzymes are a special class of proteins that act as catalysts for many of the chemical reactions that support life. For example, remember from our discussion of carbohydrates that in order to digest disaccharides and polysaccharides, animals must break them down into monosaccharides using hydrolysis. Well, it turns out that hydrolysis reactions like these are incredibly slow, and animals would die long before they could digest polysaccharides if it weren't for enzymes. These proteins catalyze the hydrolysis reactions, making them fast enough to allow the proper digestion of food.

It turns out that most enzymes do their job based on the shape that the enzyme molecule has. The shape of a hydrolysis enzyme, for example, complements the disaccharide or polysaccharide that it is trying to break down. As a result, the enzyme can attach itself to the molecule it is breaking down and actually pull on it a bit. This will stretch the bond that holds monosaccharides together, making hydrolysis much quicker.

Since most enzyme function is based on shape, each disaccharide or polysaccharide that we digest must have its own enzyme in order to help digest it. If a person does not have a particular enzyme, he or she might be able to digest most carbohydrates fine, but there will be one carbohydrate that he or she simply cannot digest. This is the case with people who are lactose intolerant. A person like this cannot digest milk or milk-based products. As a result, they can get very sick by eating or drinking dairy products. The whole reason for this lactose intolerance is that certain cells in their bodies are defective and cannot manufacture the enzyme necessary to catalyze the hydrolysis of lactose.

Now it turns out that enzymes, because they are so complex and because their function is shape dependent, are quite fragile. Enzymes are very easy to destroy, as can be illustrated in the following experiment.

EXPERIMENT 5.2
The Fragility of an Enzyme

Supplies:

- Part of a *fresh* pineapple (It cannot be canned. It must be fresh.)
- A blender or fine cheese grater
- Three small bowls
- Jell-O gelatin mix - any flavor
- Pot
- Stove
- Refrigerator
- Two tablespoons

Object: To see how easily enzyme function can be destroyed

Procedure

A. Cut the pineapple to remove any skin. Mix it in a blender or grate it with the cheese grater and then mash the gratings so that in the end you get a thick, pulpy mixture of fresh pineapple. You need about a cup of this thick, pulpy mixture.

B. Prepare the Jell-O as described in the directions on the box.

C. As you are boiling the water for the Jell-O, take a tablespoon of the thick, pulpy pineapple mixture and pour it into one of the three small bowls. Save that tablespoon for use with that bowl *only*. Mark the bowl as "room-temperature pineapple juice." Take the rest of the thick, pulpy mixture and pour it into a pot. You will eventually heat it, but DO NOT do that now.

D. When you have finished preparing the Jell-O up to the point where you stick it in the refrigerator, pour one-third of it into each of the three bowls. Stir the bowl that has Jell-0 and the thick, pulpy pineapple mixture with the tablespoon you saved in step (C).

E.   Take the pot of thick, pulpy pineapple mixture and heat it on high for three minutes. Keep stirring it constantly, in order to distribute the heat evenly. The thick, pulpy mixture may boil. That's fine; just keep stirring.

F.   After three minutes of heat, take one tablespoon of the hot pulpy mixture and pour it into one of the two bowls that have only Jell-O in them. Use a different tablespoon than the one you used in step (C). Stir vigorously. Label that bowl as "heated pineapple juice."

G.   Put all three bowls in the refrigerator and wait for the amount of time described on the Jell-O box.

H.   Examine the three bowls of Jell-O. What happened?

You should have seen that the Jell-O in the bowl with nothing added gelled as you would expect, as did the Jell-O in the bowl which had heated pineapple juice added to it. However, the Jell-O in the bowl that had room-temperature pineapple juice in it should not have gelled. If it did, you did not use fresh pineapple. Why did this happen?

Pineapple contains an enzyme that stops the reaction which causes Jell-O to gel. Remember, catalysts can either speed up or slow down a reaction. In this case, the enzyme catalyst slows the reaction down to essentially a halt. As a result, when pineapple juice is added to Jell-O, the Jell-O cannot gel. As is the case with all enzymes, however, this enzyme is very fragile. The heat that you added to the thick, pulpy pineapple mixture in the pot was enough to destroy the enzyme, and that's why the Jell-O in the bowl with the heated pineapple juice added was still able to gel. This should give you an idea of how fragile enzymes are. In fact, enzymes are so fragile that most food processing destroys them. That's why the experiment called for a fresh pineapple. In processed pineapple juice (canned or frozen), the enzyme used in this experiment has been destroyed by the processing. Interestingly enough, pineapple also contains enzymes which can break down your gums. That's why your gums bleed if you eat a fresh pineapple core.

Because enzymes are so fragile, they break down soon after they are formed. Even in the fresh pineapple that you used in the experiment, most of the relevant enzyme had already broken down before you used it. However, even a tiny amount of enzyme is enough to see the effect that you saw in the experiment, so even though most of the enzyme in the pineapple was gone, enough was left to make the experiment work. Since enzymes break down soon after they are formed, your body must continually produce more and more enzymes, just to replace the ones that are breaking down. That's why it's important to eat protein in your diet. The protein that you eat gets broken down into its constituent amino acids, and those amino acids are shipped to the cells in your body so that they can produce more enzymes.

ON YOUR OWN

5.19 Since dehydration reactions link amino acids in order to form proteins, why don't proteins break down into their amino acids when they are mixed with water? *There are no enzymes*

DNA

Our discussion of chemistry would not be complete, of course, without a brief description of DNA, the molecule which forms the basis of life. If you thought proteins were complex, you haven't seen anything until you have studied DNA! To begin your study of DNA, examine the schematic representations given in Figure 5.9.

FIGURE 5.9
Two Schematic Representations of DNA

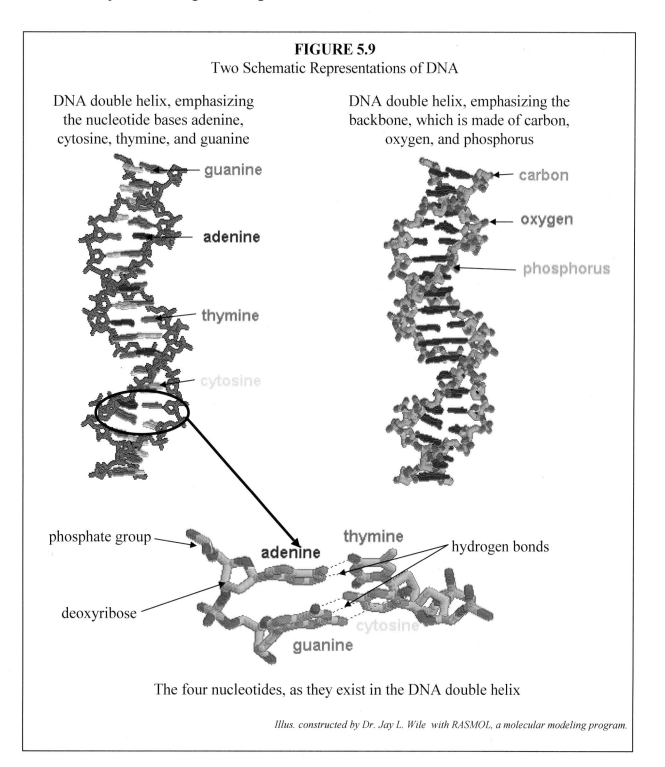

DNA double helix, emphasizing the nucleotide bases adenine, cytosine, thymine, and guanine

DNA double helix, emphasizing the backbone, which is made of carbon, oxygen, and phosphorus

The four nucleotides, as they exist in the DNA double helix

Illus. constructed by Dr. Jay L. Wile with RASMOL, a molecular modeling program.

Deoxyribonucleic (dee' ox ee rye boh noo klay' ick) acid, or DNA, is a double chain of chemical units known as nucleotides (noo' klee uh tides). These two chains twist around one another in the double-helix that is so familiar to most who have studied any amount of biology. The nucleotides that make up these two chains are comprised of three basic constituents: **deoxyribose** (a simple sugar that contains 5 carbons), a **phosphate group** (an arrangement of phosphorous, hydrogen, and oxygen atoms), and a **base**. A nucleotide's base can be one of four different types: **adenine** (add' uh neen)**, thymine** (thye' meen)**, guanine** (gwa' neen)**,** or **cytosine** (sie' toh seen).

In the lower part of the figure which shows the nucleotides, you can see that the phosphate groups link one nucleotide to another. The sugar part of the nucleotide supports the base. The two nucleotide chains are held together because the bases link together in a process known as **hydrogen bonding**.

<u>Hydrogen bond</u> - A strong attraction between hydrogen atoms and certain other atoms (usually oxygen or nitrogen) in specific molecules

Hydrogen bonding is actually a very complex process that you will learn much more about in chemistry. For right now, you just need to realize that it is very dependent on the types of molecules involved and the attraction between the atoms in hydrogen bonding is about 15% as strong as the attraction between two atoms that have a true chemical bond linking them. Thus, the hydrogen bonds in DNA are strong enough to keep the two chains together in a double helix, but are significantly weaker than a true chemical bond. Since they are weaker than a true chemical bond, it is rather easy for the two helixes in DNA to unravel. That, as we will learn in the next chapter, is VERY fortunate.

Since hydrogen bonding is highly dependent on the molecules involved, it turns out that only certain nucleotide bases can link together using hydrogen bonds. The nucleotide base adenine can only hydrogen bond to thymine. It cannot hydrogen bond to cytosine or guanine. In the same way, thymine can only hydrogen bond to adenine. Likewise, cytosine can only hydrogen bond to guanine and guanine to cytosine. As a result, DNA is often pictured as drawn in Figure 5.10.

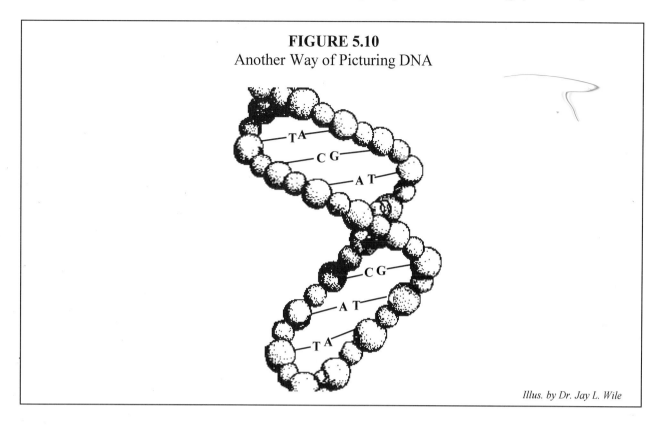

FIGURE 5.10
Another Way of Picturing DNA

Illus. by Dr. Jay L. Wile

In this figure, "T" represents thymine, "A" adenine, "G" guanine, and "C" cytosine. The balls in the figure represent the backbone of the DNA. Notice that T's are only linked to A's and C's are only linked to G's and vice-versa. This represents the fact that only adenines and thymines or cytosines and guanines link together.

Now you can finally learn how DNA stores all of the marvelous information that it uses as the instructions for making life. Just as the entire English language can be reduced to sequences of dots and dashes in Morse code, all of the information necessary for life can be reduced to sequences of nucleotide bases in a DNA molecule. Cells, as we will learn in the next chapter, have chemical machinery which decodes the sequences of nucleotide bases into instructions for what structures need to be built and where to build them. Isn't that marvelous?

ON YOUR OWN

5.20 One way that chemists can "unwind" DNA so that the individual strands can be studied is to heat a sample containing DNA. This causes the two strands of nucleotides to unravel. Based on that information, what can we say about the effect of heat on hydrogen bonds?

heat distabalizes hydroden bonds

ANSWERS TO THE ON YOUR OWN PROBLEMS

5.1 <u>Since the electrons orbit the nucleus, they look like planets orbiting a sun.</u>

5.2 <u>An atom's properties are determined by the number of electrons, protons, and neutrons that it contains. The vast majority of an atom's properties, however, are determined by the number of electrons (or protons) that an atom has.</u>

5.3 Since atoms have the same number of protons and electrons, this atom must have <u>13 protons</u>.

5.4 Atoms that belong to the same element have the same number of protons and therefore the same number of electrons as well. Thus, <u>their number of neutrons are different</u>.

5.5 Since carbon-13 is a member of the element carbon, it has <u>6 protons and 6 electrons</u>. The "13" in carbon-13 represents the sum of all protons and neutrons. Thus, there must be <u>7 neutrons</u>.

5.6 a. <u>This is a molecule</u>, because it has several atoms (1 nitrogen and 3 hydrogens) linked together.

b. <u>This is an element</u>, because it is a single abbreviation. With no other atoms linked to it, it is not a molecule. Also, since there is no number following it, we do not know its sum of neutrons and protons, so it is not a specific atom. It represents a group of atoms that all have the same number of protons.

c. <u>This is an atom</u>, because it has the element name as well as the number of neutrons and protons.

d. Like (b), <u>this is an element</u>.

e. Even though this has only one element abbreviation, the subscript indicates that there are two hydrogen atoms linked together. Thus, this is a <u>molecule</u>.

5.7 <u>The student is wrong. Even though these molecules are comprised of the same elements, there are different numbers of oxygen atoms. Different numbers of atoms result in molecules with different properties.</u>

5.8 The subscript after the "C" tells us that there are <u>2 carbons</u>. The subscript after the "H" tells us that there are <u>4 hydrogens</u>. The subscript after the "O" tells us that there are <u>2 oxygens</u>.

5.9 a. <u>Physical change</u>. This is reversible. You could pick the cereal out of the milk.

b. <u>Chemical change</u>. There is no way to "unbake" bread, so this is not reversible.

c. <u>Physical change</u>. You could glue the vase back together, so this is a reversible change.

5.10 The water level of the solution on the right increased while the water level of the solution on the left decreased. Since we are dealing with a semipermeable membrane, this is osmosis. In osmosis, the solvent travels from an area of low solute concentration to one of high solute concentration. Since the water levels indicate that the solvent traveled from left to right, the right side had a higher salt concentration.

5.11 a. Reactants appear on the left hand side of the arrow. Also, the numbers to the left of the formulas are not part of the formulas. They simply tell us how many of each molecule is in the reaction. Thus, we do not include them in our answer. The reactants, then, are C_8H_{18} and O_2.

b. Products are on the right side of the arrow and, once again, the numbers to the left of the formulas are not a part of the formulas. The products, then, are CO_2 and H_2O.

c. The number to the left of the chemical formula for C_8H_{18} tells us how many molecules are used in the reaction. Thus, the answer is 2.

d. The number to the left of the chemical formula for H_2O tells us how many molecules are made in the reaction. Thus, the answer is 18.

5.12 It can produce *some* glucose, but without the catalyst the rate will be far too slow for the plant to sustain itself.

5.13 The chemist can increase temperature. This speeds up most reactions.

5.14 In the structural formula, 2 C's are pictured along with 4 H's. The chemical formula, then, is C_2H_4.

5.15 To turn a polysaccharide into a disaccharide, you have to break it down. Hydrolysis reactions break saccharides down.

5.16 The pH scale says below 7 is acidic and above 7 is alkaline. In addition, the higher the pH, the more alkaline the solution. Based on that, then, Solution D is the most alkaline.

5.17 The pH scale says below 7 is acidic and above 7 is alkaline. In addition, the lower the pH, the more acidic the solution. Based on that, then, Solution B is the most acidic.

5.18 Each fat molecule takes 3 fatty acids along with a glycerol. As long as we have plenty of glycerol, then, we can make 5 fat molecules.

5.19 As we discussed with regards to enzymes, in order for hydrolysis reactions to work with any worthwhile rate, they need a catalyst. If proteins are mixed with just water, there are no enzymes to catalyze the hydrolysis reaction, so they will not break down into amino acids.

5.20 Since hydrogen bonding holds the DNA strands together, and since heat causes the strands to unravel, we can safely say that heat destabilizes hydrogen bonds.

STUDY GUIDE FOR MODULE #5

1. Define the following terms:

⚹ a. Matter
   b. Model
   c. Element
   d. Molecules
⚹ e. Physical change
⚹ f. Chemical change
   g. Phase
⚹ h. Diffusion
   i. Concentration
   j. Semipermeable membrane
⚹ k. Osmosis
   l. Catalyst
⚹ m. Organic molecule

   n. Biosynthesis
   o. Isomers
   p. Monosaccharides
   q. Disaccharides
   r. Polysaccharides
⚹ s. Dehydration reaction
⚹ t. Hydrolysis
   u. Hydrophobic
   v. Saturated fat
   w. Unsaturated fat
   x. Peptide bond
   y. Hydrogen bond

2. Describe where the protons, neutrons, and electrons are in an atom. *nucleus (middle)*

⚹ 3. What determines the vast majority of characteristics in an atom? *electrons*

4. What does the number after an atom's name signify? *element*

5. What is the difference between an element and an atom? *element is a group, atom is one.*

6. How many electrons are in an atom that has 32 protons? *32*

⚹ 7. How many atoms (total) are in a molecule of C_3H_8O? What atoms are present and how many of each atom? *12*

⚹ 8. Identify the following as an atom, element, or molecule:

   molecule   *atom*   *element*

   a. H_2CO_3 b. nitrogen-14   c. P

⚹ 9. If you add energy to the molecules of a liquid, will it turn into a gas or a solid? *gas*

10. A chemist wants to study diffusion. Should a semipermeable membrane be used? *no*

⚹ 11. Two solutions of different solute concentration are separated by a membrane. After a while, the water levels of the two solutions change. Has osmosis or diffusion taken place? What kind of membrane is being used? *osmosis - semipermeable*

✳12. Consider the following chemical reaction:

$$N_2 + 3H_2 \rightarrow 2NH_3$$

   a. What are the reactants? $N_2 + 3H_2$
   b. What are the products? $2NH_3$
   c. How many molecules of H_2 are used in the reaction? 3

✳ 13. What is the chemical equation for photosynthesis? What 4 things are necessary for a plant to carry out photosynthesis? $CO_2 + H_2O \rightarrow C_6H_{12}O_6 + O_2$ energy, food source, heat water.

14. Other than using a catalyst, how can a reaction be sped up? chlorophyll

15. Which of the following is a carbohydrate?

      a. NH_3 b. CO_2 c. C_2H_4O d. $C_5H_{10}O_5$ e. $C_3H_8O_3$

16. What kind of reaction is used for building disaccharides, polysaccharides, fats, and proteins? What kind of reaction can break these substances down? dehydration reaction

17. Which of the following is an acid?

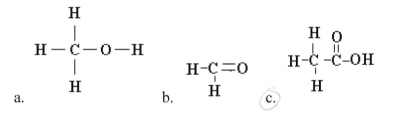

    a.   b.   c.

✳ 18. Describe the pH scale and what it measures.

19. What are the basic building blocks of proteins, lipids, and polysaccharides?

20. If two proteins contain the same type and number of amino acids, but the order in which they link up are different, are the properties of the two proteins the same?

✳21. What are enzymes and what are they usually used for?

22. What are the basic parts of a nucleotide?

✳23. How does DNA store information?

✳ 24. What holds the two helixes in a DNA molecule together?

Module #6: The Cell

Introduction

In Module #1 we started talking about the cell, and we really haven't stopped talking about it since. Indeed, in Modules #2 and #3, you learned quite a bit about the cell structures that exist in kingdoms *Protista* and *Monera*. Well believe it or not, we have barely scratched the surface of all that can be learned about the cell. That's why we are now going to devote an entire module to learning about the structure and function of cells. Since we have already studied kingdom *Monera*, however, we will not spend any time on prokaryotic cells. This module will concentrate entirely on eukaryotic cells. Take heed that there is a lot of information in this module. You might want to plan on spending an extra week or so studying it, in order to be able to absorb all of the information that it contains!

Since cells are the basic building blocks of life, it's not surprising that we need to study them in detail. What might surprise you, however, is the fact that there is really very little diversity among cells. In fact, you have already studied the most diverse eukaryotic cells that exist - those in kingdom *Protista,* and, to a lesser extent, those in kingdom *Fungi.* In the last two kingdoms (*Animalia* and *Plantae*), there is really very little difference from one cell to another. All of the incredibly diverse organisms in these two kingdoms are built from cells that are all very similar!

Cellular Functions

In the process of life, cells must carry on several different functions. First of all, they must maintain enough energy to live and perform the duties for which God has designed them. To maintain energy, cells must perform three basic functions:

Absorption - The transport of dissolved substances into cells

Digestion - The breakdown of absorbed substances

Respiration - The breakdown of food molecules with a release of energy

In **absorption**, dissolved substances must enter the cell from outside. This is actually more complicated than it sounds, because cells cannot let just anything inside them. After all, some substances are poisons which would immediately kill the cell if they were allowed to enter. As a result, absorption is a complicated process in which the cell "recognizes" the substances which are trying to enter. Useful substances are allowed in, while harmful substances are not.

Although their definitions seem rather similar, there is a big difference between **digestion** and **respiration**. In digestion, large molecules are broken down into smaller ones. For example, when polysaccharides are broken down into monosaccharides, we call this digestion. As we mentioned in Module #5, this has to happen, because most organisms can only use energy from the breakdown of monosaccharides. When a cell actually breaks down the monosaccharides and produces energy, we call that respiration. There are two reasons that we cannot lump digestion and respiration together. First, they usually occur in different places within an organism or a cell. Second, digestion can be used for other processes besides just respiration. For example, when proteins are digested, they are broken down into their constituent amino acids. Instead of being used for respiration, these amino acids can be used by the cells to produce more proteins. In this case, digestion is used to provide the building blocks of **biosynthesis**, another function of the cell.

Since a cell absorbs substances from the outside environment, it makes sense that, at some point, it must eliminate excess substances as well. There are three methods by which cells eliminate substances:

Excretion - The removal of soluble waste materials

Egestion - The removal of non-soluble waste materials

Secretion - The release of biosynthesized substances for use by other cells

Once again, even though these definitions look similar, they describe completely different processes. If a substance can be dissolved in a fluid, it is called "soluble." Thus, **excretion** involves the removal of substances that can be dissolved in the fluids of a cell. On the other hand, **egestion** involves the removal of substances that cannot be dissolved in the fluids of a cell. Egestion is a much more difficult process than excretion. Why? Well, think of Experiment 5.1. When you put solid (undissolved) sugar in a napkin, the sugar could not leave the napkin. Once water seeped into the napkin and dissolved the sugar, however, the sugar molecules could diffuse right through the napkin. Thus, transporting soluble substances is simpler than transporting non-soluble ones. Finally, **secretion** does not involve removal of wastes at all. Instead, it involves the removal of substances that the cell has manufactured for use by other cells.

Cells also must perform functions of **movement** and **irritability**. When we say "movement," we might mean the actual locomotion of a cell from one point to another, or we might mean moving things *within* a cell. Also, when we say "irritability," we do not mean the cell must get cranky from time to time. Instead, biologists use this term to mean sensing and responding to changes in the surroundings.

In order for a cell to continue its existence, there are two other functions that it must perform:

Homeostasis - Maintaining the status quo in a cell

Reproduction - Producing more cells

In order to survive, the cell must make sure that all of its organelles are functioning properly, that all organelles are supplied with the substances which they need, and that everything within the cell is running according to God's design. This is called **homeostasis** (ho mee oh stay' sis). In addition, all cells die. Thus, in order to maintain life, cells must produce other cells in the process of **reproduction**.

In fact, that last sentence in the previous paragraph is another reason why these authors think that abiogenesis could never happen. You see, scientists have been studying cells in one way or another since the 1600's. In the last century, especially, an enormous amount of scientific resources have been devoted to studying cells. In fact, the science of studying cells has become so fundamental that we have a name for it: **cytology** (sigh tahl' uh jee).

<u>Cytology</u> - The study of cells

Well, in the entire history of cytology, scientists have only seen cells produced in one way: from other cells. Never have chemicals or any other non-living substances produced cells. In fact, scientists cannot even produce cells in the lab unless they have a living cell to start with. Even the process of cloning starts with a living cell. Without that living cell, cloning would not work. If, in the history of cytology, scientists have only seen cells produced by other cells, why in the world would some scientists believe that cells could ever be produced in a different way?

Summing up, then, cells must perform at least eleven main functions in order to support and maintain life. They are absorption, digestion, respiration, biosynthesis, excretion, egestion, secretion, movement, irritability, homeostasis, and reproduction. Interestingly enough, in the single-celled life forms that you studied in kingdom *Protista*, all of these functions were performed by a single cell. In the multicellular life forms that you studied in kingdom *Fungi* and that you will study in the next semester of this course, these functions are performed by different groups of cells. Some cells have been designed to specialize in certain functions, while others have been designed to specialize in other functions. As a result, in multicellular organisms, most of the cells do not perform all of the processes listed. Instead, the functions have been assigned to certain groups of cells, and the individual groups of cells work together to make sure all of the above-listed functions are performed.

This brings up an interesting point that we have already emphasized but cannot emphasize enough. **There is no such thing as a simple life form**. Biologists are fond of calling those organisms in kingdoms *Protista* and *Monera* "simple," because they mostly consist of only a single cell. However, in some ways, these organisms are more complex than those in the other kingdoms! Certainly, having several cells working together and specializing in different life functions adds a level of complexity to the organism. However, since the cells need not perform all functions associated with life, they need not be as sophisticated as a single cell that must perform all of life's functions. Thus, multicellular organisms are certainly more complex when you look at them as a whole, but when you focus in on one cell, single-celled organisms are definitely more complex. In the end, then, there is nothing "simple" about a single-celled life form!

ON YOUR OWN

6.1 A cell makes proteins in an organelle that is near the center of the cell. It then transports the proteins to the edge of the cell and sends them into the surroundings to be used by other cells. What three of the basic life functions were employed? biosynthesis movement secretion

6.2 A cell takes in a polysaccharide, sends it to several organelles, and ends up producing energy. The soluble waste products are eliminated. What 5 of the basic life functions were performed? absorption digestion respiration movement excretion

Cell Structure

Once we get through the cells in kingdoms *Protista* and *Monera*, we generally lump the remaining cells into two distinct categories: **plant cells** and **animal cells**. Schematic, idealized examples of a plant cell and an animal cell are shown in Figure 6.1.

FIGURE 6.1
Schematic Representations of an Animal Cell (left) and a Plant Cell (right)

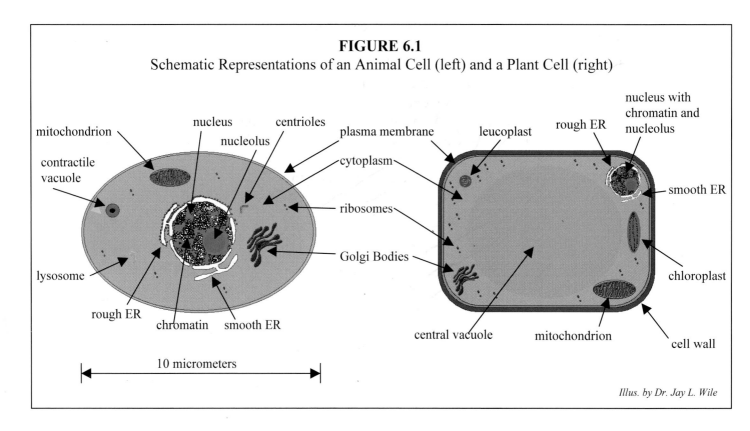

Illus. by Dr. Jay L. Wile

Before we discuss the individual components of a cell, we need to discuss size. As the legend in Figure 6.1 indicates, most cells are just a little less than 10 micrometers across. As we mentioned in Module #2, a micrometer is one millionth of a meter. Now if you aren't familiar with the metric system, don't worry about it. When you study chemistry, you will learn it in detail. For right now, just realize that about 2,000 cells can fit across your fingernail. That should give you an idea of how small a cell is.

It turns out that cells are small for a reason. You see, the volume of materials in a cell increases with the cube of the radius of the cell. Now if you haven't had geometry yet, that phrase might be a little mystifying to you. Don't worry. What it means is that when the distance across the cell (twice the radius) doubles, the cell's volume (the amount it holds) goes up by a factor of 8! Well, when a cell's volume increases, it must absorb more nutrients to survive. The amount of nutrients it must absorb actually is dependent on its volume. Thus, when a cell's radius doubles, its absorption must increase by a factor of 8. Since the amount of absorption that a cell needs is so dramatically dependent on the cell's size, there is a fundamental size limit that cells can reach. After that point, they cannot grow any larger. That's why cells are so small.

The Cell Wall

One of the major distinctions between plant cells and animal cells can be found on the outside of the cell. Plant cells usually have a **cell wall**, and animal cells do not.

Cell wall - A rigid substance on the outside of certain cells, usually plant and bacteria cells

The cell wall is typically made of cellulose and **pectin**, a substance that hardens cellulose. These materials are secreted by organelles in the cell. Although the main function of the cell wall is to protect the cell from its surroundings, the cell wall is full of small holes called **pores**. These pores allow substances (like nutrients) in its surroundings to diffuse through the cell wall and into the cell. They also allow waste products from the cell to diffuse into the surroundings.

As a plant cell grows, the cell wall must grow along with it. Thus, as a plant cell is maturing, the cell wall needs to be rather flexible so as to allow for the growth. Once a plant cell has matured and stopped growing, however, there is no longer any need for the cell wall to be flexible. Therefore, at that time, the cell starts producing **secondary cell walls**. These walls are formed on the inside of the original cell wall and are much more rigid, providing better protection for the contents of the cell. Once the secondary cell walls are formed, the original cell wall is usually referred to as the **primary cell wall**.

A single plant, of course, has many, many cells. These cells are usually arranged adjacent to one another, separated by a thin film called the **middle lamella** (luh mel' uh).

Middle lamella - The thin film between the cell walls of adjacent plant cells

The middle lamella is made primarily of pectin.

Besides plant cells, many bacteria and algae have cell walls. In addition, some cells have structures that substitute or add to the cell wall. For example, in Module #2 we learned that in addition to a cell wall, some bacteria form a capsule which surrounds the cell wall and adds protection. A few bacteria have a capsule and no cell wall. Also, some bacteria (those in genus

Mycoplasma, for example) have neither cell wall nor capsule. Finally, in Module #3 we learned that a euglena has a pellicle, which functions much like a cell wall but is thinner.

The Plasma Membrane

The **plasma membrane** is inside the cell wall if the cell has one. In the case of animal cells (which do not have a cell wall), the plasma membrane is all that separates the cell from its surroundings.

Plasma membrane - The semipermeable membrane between the cell contents and either the cell wall or the cell's surroundings

As the definition states, the plasma membrane is semipermeable. It allows certain substances (nutrients, water, and oxygen, for example) to pass through and enter the cell, but it does not allow other substances (certain toxins, for example) in. Likewise, the plasma membrane allows water and waste products to leave the cell, but it does not allow the contents of the cell itself to leave. How does it accomplish this feat? To understand this, we need to understand a bit more about how the plasma membrane is built, so we will leave that to another section. For right now, just realize that the plasma membrane is semipermeable, allowing certain substances, and not others, to pass through it.

Even though the plasma membrane is semipermeable and only allows certain molecules to pass, by itself it simply isn't enough protection from all of the potential hazards of the cell's surroundings. How, then, can animal cells exist without a cell wall? Well, the answer is twofold. Firstly, the plasma membrane is more complex in an animal cell than in a plant cell. As a result, the plasma membrane does assume some of the responsibilities of a cell wall in the case of animal cells. However, the main reason comes from the fact that cells within animals are protected by other cells. In animals, there are special cells (white blood cells, for example) that protect the other cells of the body, keeping their surroundings free of major contaminants. This allows the other cells to exist without cell walls.

The Cytoplasm

Cytoplasm (sy' tuh plaz uhm) is a jelly-like fluid in which all of the cell organelles are suspended.

Cytoplasm - A jelly-like fluid inside the cell in which the organelles are suspended

The cytoplasm is comprised of a mixture of several different compounds, including proteins, fats, and carbohydrates. In addition, there are substances called **ions** in the cytoplasm.

Ions - Substances in which at least one atom has an imbalance of protons and electrons

As you already learned, atoms must have the same number of protons and electrons in order to remain electrically balanced. Well, if an atom loses or gains an electron, it suddenly is thrown out of electrical balance. If it gains electrons (which are negatively charged), it ends up with an overall negative charge, and if it loses electrons, it ends up with an overall positive charge. Once this happens, it ceases to be an atom and is called an ion. Since ions have electrical charge, they respond to electrical stimuli.

The ions in the cytoplasm are responsible for a very important process called **cytoplasmic streaming**.

Cytoplasmic streaming - The motion of the cytoplasm which results in a coordinated movement of the cell's organelles

You see, it is important for a cell to be able to move its organelles about in order to respond to change. In addition, the cell must be able to move substances from one place to another in the cell. To accomplish these tasks, the ions in the cytoplasm cause certain parts of the cytoplasm to become more watery and less jelly-like. This causes other portions of the cytoplasm which remain jelly-like to flow into the watery areas, in essence moving that portion of the cytoplasm along with any organelles contained within it. The ions in the cytoplasm are also responsible for other chemical aspects of cell function that are simply too difficult to explain here.

The Mitochondrion (plural: mitochondria)

Suspended in the cytoplasm are all of the cell's organelles. Perhaps the most important of the organelles are the **mitochondria** (my tuh kahn' dree uh).

Mitochondria - The organelles in which nutrients are converted to energy

These bean-shaped organelles are often referred to as the "powerhouses of the cell," because they are responsible for the respiration of monosaccharides which releases energy that can be used by the cell. We will spend a great deal of time on the functions of the mitochondria in an upcoming section of this module.

The Lysosome

Another difference between plant and animal cells is the presence of an organelle called the **lysosome** (lye' soh soam).

Lysosome - The organelle in animal cells responsible for hydrolysis reactions which break down proteins, polysaccharides, disaccharides, and some lipids

Remember, in order to use polysaccharides or disaccharides, a cell must first break them down into monosaccharides using hydrolysis reactions. In addition, cells must break certain proteins and lipids down into their constituent parts (amino acids and fatty acids, respectively) in order to use them for biosynthesis. All of the necessary hydrolysis reactions take place in the lysosome. Now remember, hydrolysis reactions do not occur on their own. They need enzyme catalysts in order to make them happen. Thus, the lysosome is full of enzyme catalysts for the hydrolysis reactions which it needs to perform.

Interestingly enough, when the lysosome was first discovered, it was called the "suicide sac" because prior to the death of certain cells, the lysosomes would release chemicals that destroyed the cell. Later on, scientists realized that as the cell dies, the membrane which encloses the lysosome is generally the first thing to deteriorate. When that membrane deteriorates, the contents of the lysosome spill into the cytoplasm of the cell. Why does this kill the cell? Think about it. The cytoplasm is full of proteins, lipids, and carbohydrates. What do the contents of the lysosomes do? They break down these molecules. Thus, when the contents of the lysosome are released into the cell, they begin breaking down the cytoplasm. That spells the death of the cell.

Ribosomes

Because they control respiration and digestion, the mitochondria and the lysosomes are primarily responsible for breaking down and destroying molecules. In order to live, however, cells must also synthesize molecules. There are several organelles in the cell that work together to get that done as well. The **ribosomes** (rye' buh sohms) are non-membrane bound organelles that are found in both eukaryotic and prokaryotic cells. These organelles are responsible for production of proteins in the cell.

Ribosomes - Non-membrane-bound organelles responsible for protein synthesis

Since protein synthesis is an integral part of a cell's function, we will discuss this process heavily in an upcoming section.

The Endoplasmic Reticulum

The **endoplasmic** (en do plaz' mik) **reticulum** (rih tik' yuh lum), commonly abbreviated as ER, is composed of an extensive network of folded membranes.

Endoplasmic reticulum - An organelle composed of an extensive network of folded membranes which perform several tasks within a cell

The ER typically runs throughout the cytoplasm. In animal cells, it helps to maintain the shape of the cell, since there is no cell wall to perform this task. In addition, the ER aids in the transport of complex molecules through both animal and plant cells.

There are two types of ER, **rough ER** and **smooth ER**.

Rough ER - ER that is dotted with ribosomes

Smooth ER - ER that has no ribosomes

Since it has ribosomes, rough ER is a part of protein synthesis. Typically, specialized proteins that are secreted by certain cells are produced here. Thus, cells which are designed to specialize in the secretion of proteins are filled with rough ER. Although no protein synthesis occurs in smooth ER, many cells produce lipids in this organelle. These lipids are generally the means by which excess energy is stored. In addition, smooth ER inactivates certain harmful by-products of digestion and respiration and then sends them to the plasma membrane to be ejected.

The Golgi Body

Golgi (gole' jee) **bodies** look like a stack of pancakes. In fact, they are comprised of flattened, interconnected membrane sacs that store proteins and lipids.

Golgi bodies - The organelles where proteins and lipids are stored and then modified to suit the needs of the cell

When the cell produces proteins and lipids, they are transferred to the Golgi bodies where they are sorted and stored until needed. When a protein or lipid is needed by the cell, the Golgi bodies package the molecule so that it can be transported to the place in which it is needed. The "packaging" that goes on in the Golgi bodies is typically some process of adding a small molecule to the protein or lipid that is being processed. This chemical then can be used by the cell as a marker, telling the cell where to transport the molecule. Thus, the Golgi bodies function much like a mailing service. They take in molecules, package and address them, and then send them to where they are needed.

The Plastids

Another organelle involved in biosynthesis is found only in plant cells, algae cells, and some protozoa. These organelles, called **plastids**, are generally grouped into two categories: **leucoplasts** (loo' kuh plasts) and **chromoplasts** (kroh' muh plasts).

Leucoplasts - Organelles that store starches or oils

Chromoplasts - Organelles that contain pigments used in photosynthesis

If you recall from our discussion of starch, when plants have excess monosaccharides from photosynthesis, they are typically linked together in a long polysaccharide called starch. This allows the plant to store them for future use. The leucoplasts are where these starches are stored. Potatoes, for example, are full of starch because that's where a potato plant has its leucoplast-containing cells.

The chromoplast with which you are most familiar is the **chloroplast**. This structure contains the pigment chlorophyll, which you have already learned is a catalyst for the photosynthesis process. Photosynthesis actually takes place in the fluid which fills the chloroplast. This fluid is called the **stroma**.

Vacuoles and Vesicles

In general, a **vacuole** is a membrane-bound "sac." In Module #3, we learned about **food vacuoles** and **contractile vacuoles**. In plant and animal cells, however, there are other types of vacuoles to consider. For example, most plant cells have a **central vacuole**.

Central vacuole - A large vacuole that rests at the center of most plant cells and is filled with a solution which contains a high concentration of solutes

Because of the high concentration of solutes in the solution that fills the central vacuole, water tends to enter the central vacuole by osmosis. This makes the central vacuole bigger and bigger, causing it to push the cytoplasm and all of the organelles against the cell wall. This causes the cell to be pressurized, much like a balloon. This pressure, called **turgor pressure**, helps keep a plant rigid. When a plant begins to wilt, it is because a lack of water has resulted in a lack of turgor pressure inside the plant's cells. Fresh lettuce, for example, is crisp because its cells are highly pressurized. As the lettuce gets old, however, it loses water, decreasing the turgor pressure and resulting in wilted leaves.

Non-digestible, non-toxic waste products are contained in **waste vacuoles**.

Waste vacuoles - Vacuoles that contain the waste products of digestion

Remember, toxic waste products are typically handled by the smooth ER. Non-toxic waste products, however, are typically enclosed in waste vacuoles. These vacuoles then move by cytoplasmic streaming to the plasma membrane where the waste products can be eliminated.

Another type of vacuole is the **phagocytic** (fag uh' sih tik) **vacuole**. When some cells come across a food substance that is too large to pull through the plasma membrane, they simply surround the food substance and engulf it. Once engulfed, a vacuole is formed around the food substance so that digestion can begin. You should recall from Module #3 that this is how an amoeba feeds. This process is often called **phagocytosis** (fag uh' sigh toh' sis), and the vacuole formed is called a phagocytic vacuole.

Phagocytosis - The process by which a cell engulfs foreign substances or other cells

Phagocytic vacuole - A vacuole that holds the matter which a cell engulfs

In animals, the white blood cells that try to protect other cells feed in this way. When they recognize a pathogenic cell (such as a bacterium), they destroy it by phagocytosis.

When a vacuole is small, it is typically called a **vesicle**. Two important vesicles are the **pinocytic** (pin uh sih tik') **vesicle** and the **secretion vesicle**.

       Pinocytic vesicle - Vesicle formed at the plasma membrane to allow the absorption of large molecules

       Secretion vesicle - Vesicle that holds secretion products so that they can be transported to the plasma membrane and released

Pinocytic and secretion vesicles are, in some ways, polar opposites. When a cell needs to absorb a molecule that is large (but not so large that it needs to be engulfed), a tiny pocket is formed in the plasma membrane. That pocket is then pinched off, forming a vacuole. This process, which is similar in some ways to phagocytosis, is called **pinocytosis** (pin uh sigh toh' sis). If, instead of absorbing a molecule, the cell wants to secrete a molecule, it forms a secretion vesicle around the molecule and allows cytoplasmic streaming to carry the vesicle to the plasma membrane. When the vesicle and membrane unite, a tiny pocket is formed out of which the molecule is secreted.

Centrioles

       Centrioles are interesting organelles that have two rather different functions. In cells that possess flagella (euglena, for example) or cilia (paramecia, for example), the centrioles actually form the base of the movement organelle. This is because centrioles are comprised of structures called **microtubules**.

       Microtubules - Spiral strands of protein molecules that form a rope-like structure

When centrioles contact the plasma membrane of the cell, the microtubules are lengthened by the addition of more secreted proteins. The result is either a cilium or a flagellum, depending on the cell involved.

       Amazingly enough, however, centrioles have a completely different function as well. In many eukaryotic cells, centrioles appear in pairs. Each individual centriole is oriented at a 90 degree angle to its partner, like the corner of a square. These centrioles are involved in a process of reproduction called **cellular division**. We are going to discuss cellular division extensively in the next module, so we will put off any further discussion of centrioles until that time.

The Nucleus

       If the mitochondria are not the most important organelle in the cell, then the **nucleus** must be. Often called the "control center of the cell," the nucleus holds the DNA, which tells the cell everything it needs to know about its structure and functions. This DNA is just a copy of the DNA that was in its parent cell. Although the nucleus can make additional DNA, the new DNA will just be a copy of the old DNA. Thus, there is no way for the nucleus to make DNA that has different information than the DNA which it already has. Therefore, a cell cannot generate any new information about structure and function. It can only copy the information which is currently available.

Most likely, you still have no idea how the information that is stored in DNA can be used by the cell. In the previous module, you learned that the information was stored as a sequence of nucleotide bases, much like Morse code. However, you probably still have no idea how that information is *used*. Well, in a later section of this module, you will finally learn this fascinating process.

The nucleus of the cell has its own structure, starting with the **nuclear membrane**. This membrane has large pores in it, allowing the diffusion of materials between the nucleus and the cytoplasm.

Nuclear membrane - A highly-porous membrane that separates the nucleus from the cytoplasm

Inside the nuclear membrane, you will find clusters of DNA surrounded by a rich concentration of proteins. These clusters are called **chromatin** (kroh' muh tun).

Chromatin - Clusters of DNA and proteins in the nucleus

Also inside the nuclear membrane, you will find the **nucleolus** (noo klee' uh lus), which contains proteins and a substance called **ribonucleic** (rye buh noo klay' ik) **acid**, often referred to as RNA. We will learn a lot about RNA in an upcoming section of this module.

Before we leave this section, we need to make sure you understand something about Figure 6.1. In order to make it easy to see the organelles that we wanted to discuss, we *radically simplified* the structure of plant and animal cells in that figure. For example, we drew the cells as flat, even though they are not. Cells (and each organelle within a cell) are three-dimensional, having length, width *and* height. Also, whereas we drew only one of each organelle, there are, in reality, many examples of most organelles (except the nucleus) present in the typical cell. Thus, just to give you an idea of the complexity of a cell, Figure 6.2 shows you a less simplified drawing of a plant cell. This figure is still simplified, just not as much as Figure 6.1 was.

FIGURE 6.2
A Less-Simplified View of a Plant Cell

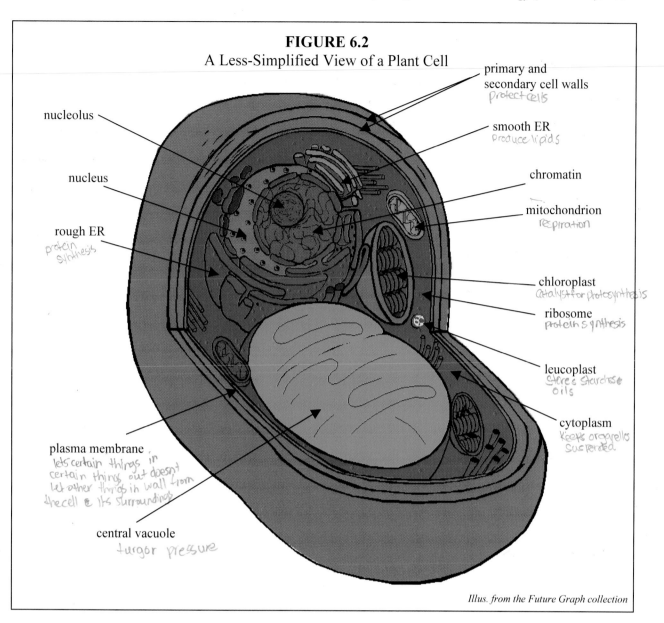

primary and
secondary cell walls
protect cells

smooth ER
produce lipids

nucleolus

chromatin

nucleus

mitochondrion
respiration

rough ER
protein synthesis

chloroplast
catalyst for photosynthesis

ribosome
protein synthesis

leucoplast
stores starches & oils

cytoplasm
keeps organelles suspended

plasma membrane
*lets certain things in
certain things out doesn't
let other things in wall from
the cell & its surroundings*

central vacuole
turgor pressure

Illus. from the Future Graph collection

To further give you an idea of how complex cell structure can be, perform the following experiment if you have a microscope.

EXPERIMENT 6.1
Cell Structure I

Supplies:
- Microscope
- Lens paper
- Slides
- Coverslips
- Droppers
- Water
- Iodine
- Onion

- Cork - any cork item can be used
- Knife or scalpel
- Prepared slide: Hydra
- Prepared slide: *Ranunculus* root
- Prepared slide: *Zea Mays* root

Object: To study the internal structures of various cells.

Procedure:

A. Prepare a wet mount of cork cells:
1. Secure the cork so that it will not slip. With a very sharp knife or a scalpel, cut a very thin slice of cork. Be careful. Do not cut yourself!
2. Place the cork on a clean slide. Add a drop of water and then a coverslip.
3. Place the slide on the microscope and observe on all three powers.
4. In your notebook, sketch what you see. Note that even dead cork cells hold their shape.

B. Prepare a wet mount of onion epidermis cells:
1. Quarter an onion and peel the outside layer off of one quarter.
2. With the point of your knife, carefully go under the skin (epidermis) on the inside of the layer that you just peeled off. Remove a paper-thin sample of the epidermis.
3. Place the epidermis sample on a clean slide and add a drop of iodine and then a coverslip. The iodine will stain the cells to make the nuclei easier to see.
4. Observe the onion epidermis on all three powers.
5. Sketch in your notebook what you have observed. Note the cell walls and the brick-like configuration of the cells. Also, note in the cytoplasm, the nucleus and the nucleolus.

C. Observe each of the following prepared slides on all three magnifications: Hydra, *Ranunculus* root, and *Zea mays* root. Draw a sample of a cell from each, and note the differences between the cells for each slide. Can you tell just by looking at the cells whether you have a plant or an animal?

D. Clean up: Wash and dry all slides and coverslips. Return all supplies to the storage area. Clean the microscope and put it away.

ON YOUR OWN

6.3 In the first section of this module, we discussed the eleven main functions of life that a cell has to perform (absorption, digestion, respiration, biosynthesis, excretion, egestion, secretion, movement, irritability, homeostasis, and reproduction). Take the organelles listed in Figure 6.1 (with the exception of the nucleolus and the chromatin) and indicate which of the eleven functions the organelle helps the cell perform. For example, for *mitochondrion*, you would list respiration. Some organelles might participate in more than one of the eleven basic life functions, so be sure to list multiple functions when appropriate.

6.4 As we mentioned in Module #5, lactose intolerant people cannot digest the disaccharide lactose due to the lack of an enzyme. Which organelle in the cells of a lactose-intolerant person does not have what it needs to get its job done? *lysosome*

6.5 Two cells eat food substances. The first takes the food in through its oral groove, while the second engulfs it. Which cell engaged in phagocytosis? *the second one*

6.6 Even though ribosomes are considered organelles, they exist in both prokaryotic and eukaryotic cells. A student claims that this contradicts the definition of a prokaryotic cell, because he says that a prokaryotic cell cannot have organelles. Why is the student wrong?

they can't have membrane bound organelles

How Substances Travel In and Out of Cells

As we learned in the previous section, the plasma membrane is responsible for substances moving in and out of a cell. If a substance is to leave the cell or enter it, it must first pass through the plasma membrane. To see how the plasma membrane works, however, we first need to see how it is made. A plasma membrane is usually constructed of proteins, cholesterol, and **phospholipids**.

Phospholipid - A lipid in which one of the fatty acid molecules has been replaced by a molecule which contains a phosphate group

Remember from Module #5 that lipids are molecules which contain three fatty acids linked together on a glycerol molecule. Well, a phospholipid has, instead of the third fatty acid molecule, a small molecule that contains a phosphate group, which is composed of phosphorus, hydrogen and oxygen. Figure 6.3 shows a lipid and a phospholipid as well as symbols which biologists often use to represent them in biological drawings.

FIGURE 6.3
Lipids and Phospholipids

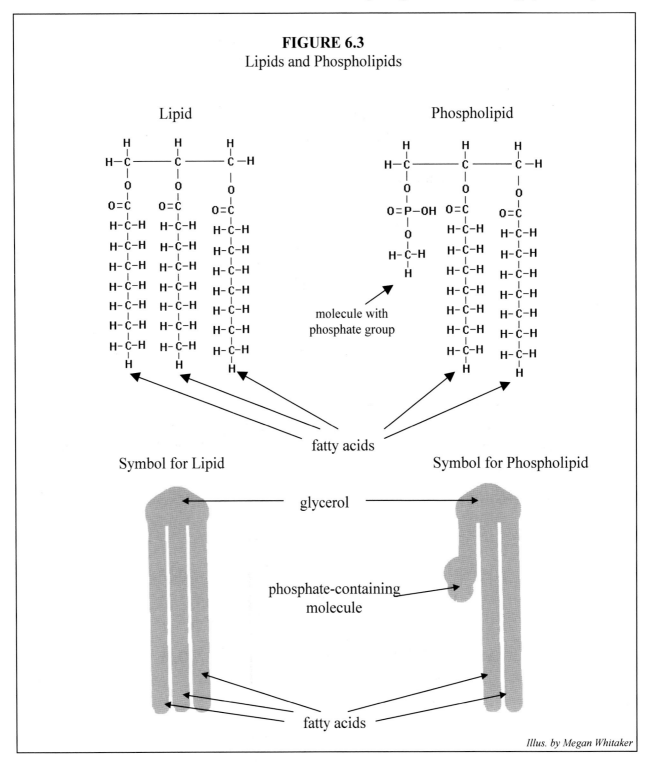

The interesting thing about a phospholipid is that the phosphate group gives the molecule a slight affinity to water. It turns out, however, that since the two fatty acid components are so long, and since the molecule containing the phosphate group is so short, the affinity to water exists only on one end of the molecule - the end that contains the glycerol. The other end of the molecule, at the ends of the fatty acid molecules, still has no affinity for water. Thus, we say that a phospholipid has a **hydrophilic end** ("hydrophilic" means attracted to water) and a **hydrophobic end** (which is repelled away from water). This is shown in Figure 6.4.

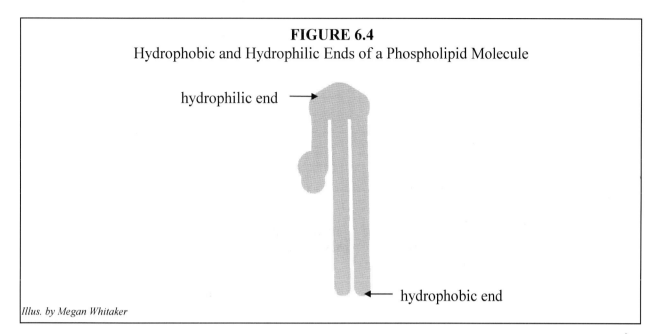

FIGURE 6.4
Hydrophobic and Hydrophilic Ends of a Phospholipid Molecule

hydrophilic end →

hydrophobic end

Illus. by Megan Whitaker

Now that you understand what a phospholipid looks like, you can learn how a plasma membrane is constructed. A plasma membrane contains proteins and phospholipids, arranged according to Figure 6.5.

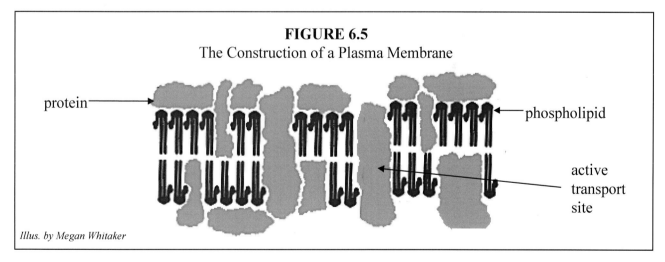

FIGURE 6.5
The Construction of a Plasma Membrane

protein

phospholipid

active transport site

Illus. by Megan Whitaker

In a plasma membrane, phospholipids are arranged in a double layer with the hydrophobic ends pointing towards each other and the hydrophilic ends pointing out towards the surroundings and in towards the cell. This should make sense, because water is both outside in the surroundings and inside in the cytoplasm. The proteins group around the phospholipids, staying near the hydrophobic ends for the most part. Every once in a while, a protein spans the entire width of the plasma membrane, leaving no room for phospholipids. This protein "bridge" is called an **active transport site**, which we will discuss in a moment.

One neat feature of the plasma membrane is that it can self-reassemble when it is broken. When a secretion vesicle, for example, reaches the plasma membrane, the membrane actually breaks to allow for the release of the secretion vesicle's contents. Afterwards, the plasma membrane reassembles by itself. This self-reassembly is due primarily to the fact that the hydrophobic ends of the phospholipids are attracted to one another and the hydrophilic ends are attracted to the water on the inside and outside of the cell. Thus, the phospholipids always end up "pointing" in the right direction, once they have time to regroup.

There are two ways that substances enter or exit a cell: **passive transport** or **active transport**.

Passive transport - Movement of molecules through the plasma membrane according to the dictates of osmosis or diffusion

Active transport - Movement of molecules through the plasma membrane (typically opposite the dictates of osmosis or diffusion) aided by a chemical process

Since we already know a bit about osmosis and diffusion, we will start with passive transport.

In passive transport, no cellular energy is expended as a result of the substance passing through the plasma membrane. Water, for example, is a very small molecule and thus passes easily through the spaces between the proteins and phospholipids in the plasma membrane. As a result, it will travel according to the dictates of osmosis. Thus, if the concentration of solutes in the solution surrounding the cell is essentially equal to the concentration of solutes inside the cell, then water will aimlessly wander back and forth across the plasma membrane, resulting in no net change in the amount of water inside or outside the cell. When cells are in a solution whose solutes are essentially equal in concentration to those of the inside of the cell, the cell is said to be in an **isotonic** (eye suh tahn' ik) **solution**.

Isotonic solution - A solution in which the concentration of solutes is essentially equal to that of the cell which resides in the solution

The red blood cells in your bloodstream, for example, are in an isotonic solution. Your kidneys get rid of excess solutes in the blood to make sure that the bloodstream stays isotonic with your red blood cells.

Why do your kidneys need to make sure that your bloodstream stays isotonic with your red blood cells? Well, remember the story about *The Ryme of the Ancient Mariner* from Module #5? If cells are placed in a solution which has a significantly higher concentration of solutes than the cell, then osmosis will drive water out of the cell and into the surroundings. This will cause the cell to implode, killing it! Solutions that have a higher concentration of solute than the inside of a cell are called **hypertonic** (hi pur tahn' ik) **solutions**. The implosion that results from placing cells in hypertonic solutions is called **plasmolysis** (plaz mahl' uh sis).

<u>Hypertonic solution</u> - A solution in which the concentration of solutes is greater than that of the cell which resides in the solution .

<u>Plasmolysis</u> - A collapse of the cell's cytoplasm due to lack of water

This is why fertilizers emphasize that you should not *over-fertilize* a plant. If you put too much fertilizer in the ground near a plant, when the fertilizer mixes with the water in the soil, a hypertonic solution could be formed. This will result in plasmolysis of the plant's cells, killing the very plant that you were trying to fertilize!

Osmosis can also cause **cytolysis** (sigh tahl' us sis), when a cell is placed in a **hypotonic** (hi puh tahn' ik) **solution**.

<u>Cytolysis</u> - The rupturing of a cell due to excess internal pressure

<u>Hypotonic solution</u> - A solution in which the concentration of solutes is less than that of the cell which resides in the solution

In a hypotonic solution, there is greater solute concentration on the inside of the cell than on the outside. As a result, water flows by osmosis into the cell, until the cell bursts from too much water pressure.

Since many organisms live in fresh water habitats, a lot of cells must deal with living in hypotonic solutions. After all, fresh water has few solutes dissolved in it; thus, it is a hypotonic solution for cells. Plant cells typically deal with living in a hypotonic solution by using their central vacuole and their cell wall. As discussed in the previous section, as water flows into a plant cell, it collects in the central vacuole. This expands the vacuole, pushing the cytoplasm against the rigid cell wall. This results in an increase in pressure, which we called turgor pressure. Once the pressure gets large enough, it counteracts the effect of osmosis, and no more water can get into the cell. Even though the water wants to get into the cell due to osmosis, the turgor pressure pushes the water away. This is what we mean when we say that turgor pressure counteracts the effects of osmosis. This keeps the cell from rupturing.

Plant cells can do this, of course, because they have a rigid cell wall against which the cytoplasm can push, causing turgor pressure. What about cells that must live in freshwater environments but do not have cell walls? Well, if you think back to Module #3, you will remember that certain protozoa have contractile vacuoles which constantly pump excess water out of the cell. Animal cells that live in hypotonic solutions also have these vacuoles. Thus, they cannot stop water from entering the cell as plant cells can, but they do get rid of it right away, so excess pressure does not build up. Actually, when scientists first observed contractile vacuoles, they thought that they were looking at some sort of heart, because the contractile vacuoles pumped constantly, just like hearts.

When the contractile vacuoles pump water out of the cell as discussed above, they are engaging in **active transport**, because the cell is expending energy to push a substance (water) across the plasma membrane against the dictates of osmosis. Another example of active transport involves the "protein bridges" discussed earlier. When a protein spans the width of the plasma membrane, it typically is an enzyme that has been fashioned to recognize a certain type of molecule. When it recognizes that molecule, it uses a series of chemical reactions to pull the molecule across the plasma membrane and into (or out of) the cell. That's why the area pointed out in Figure 6.5 is called an active transport site, because the enzyme actively works to transport molecules, with the help of energy provided by the cell.

ON YOUR OWN

6.7 If a plasma membrane were made out of regular lipids (not phospholipids) it could never self-reassemble. Why? *because the reason it can reassemble is because opposites attract and the different ends are opposites*

6.8 A scientist observes a cell and watches as it ruptures. Was the cell in a isotonic, hypertonic, or hypotonic solution? *hyptonic*

6.9 A cell's mitochondria cease to function, and the cell has no more energy. Will all transport across the plasma membrane stop? Why or why not? *no passive will continue because it doesn't need energy*

If you have a microscope, perform the following experiment before you go on to the next section. It will give you the opportunity to see some of the processes that we have been discussing.

EXPERIMENT 6.2
Cell Structure II

Supplies:

- Microscope
- Lens paper
- *Anacharis* leaves (These are aquatic plants that can be purchased at stores where gold fish are sold. You could also use thin leaves from some other plant like Impatients. The main point is that the leaves need to be alive and very thin.)
- Banana
- Methylene blue stain
- Water
- Dropper
- Slides and coverslips
- Iodine
- Cotton swab
- Salt water (one tablespoon of salt in about 1/8 cup of water)

Object: To see how cells vary in structure by examining the cells of two different living objects.

Procedure:

    A. Study of living leaf:

        1. Select a young or new leaf for your observation.

        2. Place the leaf on a slide, add a drop of water and place a coverslip on top. Be careful not to crush the leaf.

        3. Observe under all three powers.

            a. Some kinds of leaves will have distinct veins. These cells appear to be long and thin.

            b. Look for cells of different shapes as you look at various areas of the leaf.

        4. At the edge of the leaf, you should be able to see some clear cells. If you look closely (lower the brightness of your light source), you should see the nuclei. In your notebook, sketch some of the different kinds of cells.

        5. Note the green (usually somewhat oval) objects in the cells. These are chloroplasts. In some of your leaves you might see a movement of these chloroplasts in what looks like a "stream". This is the **cytoplasmic streaming** you learned about earlier. Sometimes putting the leaf close to a warm bright light makes the streaming begin.

        6. Place a drop of salt water on the slide next to the coverslip. Take a tissue and place it against the other side of the coverslip. As the tissue absorbs the water originally on the leaf, the salt water will replace it. You should be watching through the eyepiece as you do this.

        7. If your leaf is thin enough, you should see some cells begin to break and the cytoplasm, including the chloroplasts, flow out into the salt water. What is this an example of?

    B. Study banana cells and find out what the largest part of the cells contain:

        1. With a cotton swab or the end of a knife, smear a small amount of banana on a slide.

        2. Observe the slide on all three powers and sketch some cells in your notebook.

        3. Add a drop of iodine and a coverslip over the smear of banana.

        4. Observe again and sketch in your notebook.

    Note: Banana cells are high in starch content. When the iodine comes in contact with the starch in the cells, it turns the starch a blue-black color.

    C. Clean up and put away all equipment. Make sure your microscope is clean.

How Cells Produce Energy

Cells produce useable energy in a chemical reaction that, at first glance, seems rather simple:

$$C_6H_{12}O_6 + 6O_2 \rightarrow 6CO_2 + 6H_2O + energy$$

This chemical equation represents what happens when a molecule of glucose is burned. Six molecules of oxygen from the air interact with a glucose molecule. The atoms rearrange, forming six carbon dioxide molecules and six water molecules. As is the case when anything burns, energy is also released. This is, in essence, how a cell produces the energy that it uses to sustain all of the processes it must run. Of course, we say "in essence" because the actual process of burning monosaccharides in cells is actually quite a bit more complicated.

Although you've probably never tried it (and you really shouldn't - it makes a mess), sugar does burn. If you strike a match and light some sugar, the sugar will burn, producing water, carbon dioxide, heat, and smoke. The heat is a form of energy. Although you expended some energy striking the match, once you get enough sugar burning, the heat that it generates will be much, much more than the energy you spent using the match to light the sugar to begin with. You see, the chemical reaction described by the equation above needs a bit of a "push" to get it started. This push, called **activation energy**, is supplied by the match.

Activation energy - Energy necessary to get a chemical reaction going

Once the chemical reaction starts, however, it produces lots of energy. Some will be used as activation energy to start more reactions and thus keep the fire going, while the rest will be used to produce heat and light in the form of flames.

Now it turns out that all chemical reactions require some amount of activation energy in order to start. Reactions such as the burning of glucose, however, end up producing a lot more energy than that which is required to start them. The net result, then, is lots of energy that can be used for other things.

If a mitochondrion were simply to burn sugar in the way that we just described, the results would be disastrous! You see, the reaction of glucose and oxygen is very fast once it gets going. It produces a large amount of energy in a short amount of time. This large concentration of energy in a short time interval would be so violent that it would destroy the mitochondrion! So, the cell simply slows the reaction down, right? Wrong! You see, the cell needs a lot of energy. If the cell were to slow the burning of glucose down to a point where the mitochondria could withstand the heat, there would not be enough energy produced to allow the cell to run all of the processes that it needs to run. Thus, something else must happen.

Well, what happens is really quite complicated. We will give you a brief introduction to it here, but please realize that the actual processes involved are quite a bit more complicated than anything that we will do here. This is simply a short, sketchy introduction into one of the most complicated topics in biology: cellular respiration.

Cellular respiration is accomplished through a three-staged process which actually begins in the cytoplasm. When the cell wants to "burn" a monosaccharide for energy, the monosaccharide first goes into the cytoplasm. There, enzymes catalyze a reaction which causes the monosaccharide molecule to lose some of its hydrogen atoms and form two molecules of a substance called **pyruvic acid** ($C_3H_4O_3$). The chemical equation that describes this process is:

$$C_6H_{12}O_6 \rightarrow 2C_3H_4O_3 + 4H$$

The cell actually has to expend energy to get this to happen, but that's okay. Much like striking a match to start something burning, the amount of energy needed to begin this reaction is half of that which is released when the reaction finishes. This is the first stage of cellular respiration and is called **glycolysis** (glye kahl' uh sis).

Once two molecules of pyruvic acid are made, they travel to a mitochondrion, where the last two stages of cellular respiration take place. The next stage, called the **Krebs cycle**, takes the two molecules of pyruvic acid and reacts them with oxygen to make hydrogen and carbon dioxide.

$$2C_3H_4O_3 + 3O_2 \rightarrow 8H + 6CO_2$$

This is actually a long and complicated procedure that involves many reactions which are carefully controlled by enzymes.

The last stage of cellular respiration is called the **electron transport system**, and it produces the most energy of all three stages. Just like the Krebs cycle, it takes place in the mitochondria. In this stage, the hydrogen that was produced in glycolysis and the Krebs cycle is carefully reacted with oxygen to make water. Once again, like the Krebs cycle, this is actually done through a complicated series of enzyme-controlled reactions.

Now, to make things even a little more complicated, the energy produced in the three stages of cellular respiration must be stored so that it can be transported to different parts of the cell and used when the cell needs it. Energy can be stored in nearly any molecule, and when that molecule is broken down, the energy will be released. In a cell, however, the energy must be stored in such a way as to allow for a *gentle* release of energy. If the energy is not released gently enough, it will destroy a part of the cell rather than help it perform its tasks. Thus, the energy produced in cellular respiration must be stored in a molecule that, when broken down, releases only a small amount of energy. That molecule is **adenosine** (uh den' uh seen) **triphosphate** (try fahs' fate), which is usually abbreviated ATP.

ATP is composed of a molecule of substance called adenosine linked with three phosphate groups. When one of the phosphate groups is broken away from the rest of the molecule, a gentle release of energy occurs. This energy is sufficient to accomplish most tasks in a cell, but gentle enough so that it will not harm the cell. When the phosphate group breaks off from the rest of the molecule, there is an adenosine molecule attached to two phosphate groups. This molecule is called **adenosine diphosphate** (ADP). This reaction is usually represented as follows:

$$ATP \rightarrow ADP + P + energy$$

The products of this reaction (ADP and the lone phosphate group) find their way back into either the cytoplasm or a mitochondrion so that they can be re-assembled into ATP to store more energy.

Now its really important that you sit back and understand what's going on here. In each stage of cellular respiration, energy is released. This energy cannot be transported to the parts of the cell that need it unless it is "packaged" in some way. It must be packaged so that it can be gently released directly in the place in which it is needed. That's the job of ATP. Thus, the energy produced in each stage of cellular respiration is used to make ATP from ADP and a phosphate group. The ATP is then transported to the place where energy is needed and, at that point, enzymes break the ATP down into ADP and a phosphate group. The resulting gentle release of energy is used by the cell, and the ADP and phosphate are returned so that more ATP can be produced.

Since each stage of cellular respiration releases different amounts of energy, you might be interested to know how much energy is released in each stage. The first stage, glycolysis, actually requires two ATPs in order to get going. Thus, when a monosaccharide needs to be broken down, two ATPs are broken down and the resulting energy released supplies the activation energy required to start glycolysis. The process of glycolysis, however releases enough energy to make four ATP molecules from the ADP and phosphates that are just "lying around" in the cytoplasm. Thus, two ATPs are used in glycolysis but four are produced, for a net gain of two ATPs. The Krebs cycle releases enough energy to produce two more ATPs from the ADP and phosphate in the mitochondrion. Finally, the electron transport system produces enough energy to make 32 ATPs from the ADP and phosphates in the mitochondrion. Adding it all up, a single glucose molecule produces enough energy in cellular respiration to make 38 ATP molecules. The process requires 2 ATPs to get going, however, so the net gain is 36 ATPs. These 36 ATP molecules can then be transported to where the cell needs energy. Once there, enzymes will break a phosphate group off of the ATP, providing a gentle release of energy for the cell.

To sum up, cellular respiration can be illustrated by the schematic in Figure 6.6 on the next page.

FIGURE 6.6
A Simplified Schematic of Cellular Respiration

STAGE 1: GLYCOLYSIS
Occurs in Cytoplasm - Needs no oxygen - Requires 2 ATPs to start, but produces 4 ATPs in the end, for a net gain of 2 ATPs. It also produces 4 hydrogen atoms, and 2 pyruvic acid molecules

$C_6H_{12}O_6 \rightarrow 4H$ (to electron transport system)

$+ 2C_3H_4O_3$ (to Krebs cycle)

(The released energy can make 4 ATPs from ADP and phosphate in cytoplasm. Since the process needs 2 ATPs to start, the net gain is 2 ATPs.)

STAGE 2: KREBS CYCLE
Occurs in Mitochondria - Requires 3 oxygen molecules and 2 pyruvic acid molecules from stage 1- Produces 2 ATPs, 8 hydrogen atoms, and 6 carbon dioxide molecules

$2C_3H_4O_3 + 3O_2 \rightarrow 8H$ (to electron transport system)

$+ 6CO_2$

(released energy can make 2 ATPs from ADP and phosphate in mitochondria)

STAGE 3: ELECTRON TRANSPORT SYSTEM
Occurs in Mitochondria - Requires 3 oxygen molecules and 12 hydrogen atoms (4 from stage 1 and 8 from stage 2) - Produces 32 ATPs and 6 water molecules

$12H$ (4 from stage 1 and 8 from stage 2) $+ 3O_2 \rightarrow 6H_2O$

(released energy can make 32 ATPs from ADP and phosphate in mitochondria)

Notice that, in the end, the complex process of cellular respiration has the same result as does the simple chemical equation with which we started out this section. A single glucose molecule is reacted with 6 oxygen molecules (3 in the Krebs cycle and 3 in the electron transport system). The result is 6 carbon dioxide molecules (produced in the Krebs cycle), six water molecules (produced in the electron transport system), and energy (produced in all three stages). That energy is stored in the form of ATP, which can then be transported to the cells.

Now as we mentioned, when a single molecule goes through all three stages of cellular respiration, the cell gains 36 molecules of ATP. Now although that sounds like a great trade (one glucose for 36 ATPs), it actually isn't as good as it first sounds. Remember, each ATP holds only a small amount of energy. This is necessary so that energy is released to the cell gently. Well, 36 ATPs actually store only about 55% of the energy contained in a glucose molecule. Where does the other 45% go? It is lost in the process of respiration. Although this suddenly doesn't sound very good, it is still better than anything humans have come up with! For example, a top-of-the-line automobile is only about 20% efficient in converting the energy of its fuel into energy of motion! Thus, although a 45% loss of energy sounds bad, it is significantly better than anything 3000 years of human science has produced. This, of course, should not surprise you. After all, the process of cellular respiration was designed and implemented by God. Anything that He produces will always be superior to the best that humankind has to offer!

One final note regarding cellular respiration. Notice that the first stage of the process (glycolysis) uses no oxygen. Thus, glycolysis is often called the anaerobic step in cellular respiration. Sometimes, cells exist in an environment where there is little to no oxygen. Many bacteria, for example, live in areas with no oxygen. In addition, during heavy exercise, you cannot supply your muscle cells with enough oxygen, no matter how hard you breathe. In both cases, the cells can still produce energy, even though they have no oxygen. They simply stop after glycolysis, due to the lack of oxygen. Of course, since the cells gain only 2 ATPs in this step, they aren't getting nearly as much energy as they could from the glucose that they are using. Nevertheless, it is still better than producing no energy at all. When this happens, we say that the cell is performing **anaerobic respiration**.

When anaerobic respiration takes place, the pyruvic acid molecules produced in glycolysis cannot continue into the Krebs cycle, because there is no oxygen. As a result, the cell eventually converts the pyruvic acid into alcohol or lactic acid. Since the production of alcohol is often called fermentation, anaerobic respiration in cells is sometimes referred to as **cellular fermentation**.

ON YOUR OWN

6.10 Despite the fact that a cell has all of the enzymes necessary for cellular respiration, plenty of glucose, functioning mitochondria, and a plentiful supply of oxygen, it cannot produce energy which is useable by the cell. What is it missing? *missing ADP and/or phosphate*

6.11 How many glucose molecules would have to undergo anaerobic respiration in order to produce the same amount of energy that one glucose molecule produces in aerobic respiration? *18*

6.12 A cell has plenty of oxygen but can only perform glycolysis. If it has all of the necessary enzymes, what isn't working? *mitochondria*

Protein Synthesis

The last thing that we need to cover in this module is perhaps the most important. In this section, we discuss how a cell produces protein. Why is this the most important thing to learn? Well, as you should have already surmised, proteins are a major part of the chemistry of life. Most of the cell's structures have proteins as an integral part of their makeup. Proteins known as enzymes control virtually every chemical reaction that occurs in the working of the cell. From active transport across the plasma membrane to controlling the reactions of cellular respiration, enzymes are a fundamental part of life. In addition, cells produce proteins to be used by other cells. Obviously, then, proteins are important.

Perhaps, however, you still do not understand how important proteins are. Remember, in multicellular organisms, specific cells have specific tasks which they must accomplish so that the organism will survive. In your body, for example, you have skin cells that shield the inside of your body from contaminants and provide insulation from the cold. You also have cells in the retina of your eye which detect light and send electronic messages to your brain based on the light that it detects. How in the world do these cells perform such radically different tasks? The answer is simple: they produce different types of proteins. By and large, the tasks that a cell can complete are dependent on the proteins that it produces. If it produces one set of proteins, a cell functions as a skin cell. If it produces another set of proteins, it acts as an eye cell. So, how does a cell know what proteins to produce? It uses the DNA in the nucleus as a guide! In this section of the module, you will learn the basics of how that happens.

Before we can show you how a cell uses DNA to produce proteins, we first need to give you a little more background on an important chemical of life, ribonucleic acid (RNA). We briefly mentioned this chemical when we were discussing the nucleus, but now we need to spend a little more time on it. As you might have already guessed by virtue of its name, RNA must have some relationship to DNA. Well, it does. In fact, there are a lot of similarities between them. RNA is also made up of nucleotides, but the individual structure of the nucleotides is a bit different. First of all, the sugar that makes up the foundation of the nucleotides is ribose, not deoxyribose (as is the case with DNA). This is a minor difference.

Like DNA, the nucleotides of RNA join together in long strands. Unlike DNA, however, they do not form a double helix. Instead, RNA is a single strand of joined nucleotides. Like DNA, RNA has 4 nucleotides. While three of them are the same as DNA's (adenine, cytosine, and guanine), one of them is different (uracil). Uracil performs the same tasks in RNA that thymine does in DNA. Thus, RNA has uracil in place of thymine. This can be confusing to students. Don't let it stump you. When you are thinking about RNA, just remember that uracil replaces thymine.

So, RNA has a lot of similarities to DNA but also some differences. Why is it important in the process of how the cell makes proteins? Well, RNA acts a lot like a camera. You see, the DNA of a cell is in its nucleus. Proteins, however, are made in the ribosomes, which are located outside of the nucleus. In order to get information from the DNA in the nucleus to the ribosomes, RNA takes a "snapshot" of the DNA. It then takes that information out of the nucleus and to the ribosome.

How in the world does RNA do this? Examine Figure 6.7 on the next page.

FIGURE 6.7
How RNA Takes a Snapshot of a DNA Section

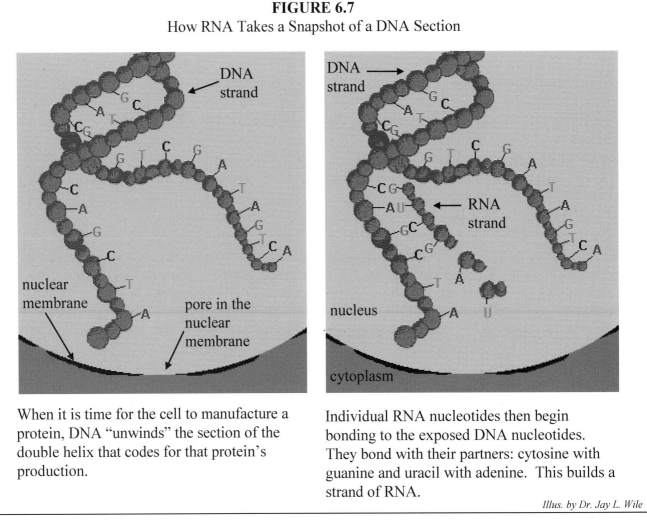

Illus. by Dr. Jay L. Wile

When it is time for the cell to manufacture a protein, DNA "unwinds" the section of the double helix that codes for that protein's production.

Individual RNA nucleotides then begin bonding to the exposed DNA nucleotides. They bond with their partners: cytosine with guanine and uracil with adenine. This builds a strand of RNA.

So, because RNA has nucleotides that bond only with certain other nucleotides (guanine with cytosine and adenine with uracil), RNA produces a "negative" of the DNA section that it is copying. Everywhere the DNA has a cytosine, the RNA will have a guanine. Everywhere the DNA has a guanine, RNA will have a cytosine. Everywhere that DNA has a thymine, RNA will have an adenine, and everywhere DNA has an adenine, RNA will have a uracil. This is much like the way a camera takes a picture. When the film is developed, a negative first appears. This negative is dark everywhere that the picture is supposed to be light, and it is light everywhere that the picture is dark. Based on this negative, then, a picture can be made. Based on the "negative" produced by RNA, a protein can be made. Before we go on to see how this is done, perform the following "on your own" problems to make sure that you understand this concept.

ON YOUR OWN

6.13 An RNA strand has the following sequence of nucleotides:

uracil, adenine, adenine, guanine, cytosine, cytosine

What was the nucleotide sequence in the DNA that it copied?

6.14 A DNA strand has the following sequence of nucleotides:

thymine, thymine, thymine, adenine, guanine, cytosine

What will the RNA sequence be when this DNA section is copied?

The process by which RNA produces a negative of the DNA nucleotide sequence is called **transcription**. After transcription, the RNA leaves the nucleus and moves to the ribosome, giving the DNA's information to the organelle that produces the protein. Because the RNA that performs transcription is essentially a messenger (sending instructions from the DNA of the nucleus to the ribosome), we usually call it **messenger RNA (mRNA)**.

Messenger RNA - The RNA that performs transcription

Transcription - The process in which mRNA produces a negative of a strand of DNA

Now if you're getting a bit confused at this point, don't worry; we'll sum it all up in a nice figure at the end. For right now, just understand that in order to get the DNA's instructions from the nucleus to the ribosome, RNA produces a negative of the DNA's nucleotide sequence and takes that negative to the ribosome.

The ribosome is surrounded by amino acids, enzymes, and a different kind of RNA called **transfer RNA (tRNA)**. Transfer RNA is a big molecule that contains a special sequence of three nucleotides called an **anticodon**.

Anticodon - A three-nucleotide sequence on tRNA

In addition to this anticodon, tRNA has a molecular "hook" which carries an amino acid. The amino acid carried by this "hook" is determined by the three nucleotides in the anticodon. For example, when the anticodon is made up of uracil, uracil, and cytosine (in that order), the tRNA carries the amino acid lysine on its "hook." If, instead, the anticodon is made up of uracil, cytosine, and cytosine, the tRNA carries the amino acid arginine on its "hook." Everywhere the tRNA goes, the amino acid goes as well.

When a mRNA molecule comes to the ribosome, the tRNA strands are attracted to three nucleotides that will bond with the 3 nucleotides in their anticodon. Thus, if a tRNA strand has guanine, cytosine, and adenine, it will be attracted to a section of the mRNA that has cytosine, guanine, and uracil in succession. If such a sequence exists, the tRNA will go and link up with it, carrying the amino acid on its "hook" the whole way. The three-nucleotide sequence which attracts the tRNA anticodon is called, reasonably enough, a **codon**. Since a codon attracts a tRNA with a particular amino acid on it, we say that the codon refers to a specific amino acid.

Codon - A sequence of three nucleotides on mRNA that refer to a specific type of amino acid

Now don't get lost in all of this terminology. A strand of mRNA can be thought of as a bunch of three-nucleotide sequences. Each three-nucleotide sequence is called a codon. A strand of tRNA contains a three-nucleotide sequence called an anticodon. A certain anticodon on tRNA results in a certain amino acid carried by the tRNA. Since the tRNA anticodons are attracted by the mRNA codons, the net result is that a codon on mRNA attracts a specific amino acid. Figure 6.8 illustrates this process.

FIGURE 6.8
Transfer RNA Strands Linking to a Messenger RNA Strand On a Ribosome

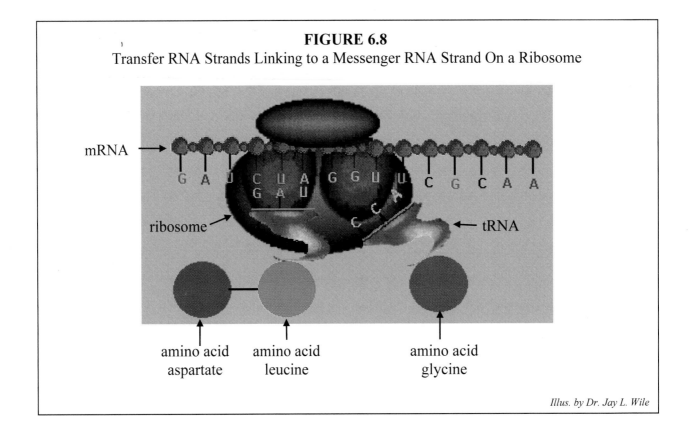

Illus. by Dr. Jay L. Wile

Now look at what happens. When the tRNA strands have linked up to the mRNA, there are two amino acids lined right up next to each other. What, by the way, is a protein? It is a bunch of amino acids linked together in a particular sequence. Well, at this point, the amino acids sitting right next to each other link up. This happens again and again and again, so that many, many amino acids link up together. When all the amino acids called for by the codons on mRNA are linked together, the result is a protein! The process of mRNA coming to the ribosome and attracting the proper tRNA molecules followed by the amino acids that come along with the tRNA linking together is called **translation**.

ANSWERS TO THE ON YOUR OWN PROBLEMS

6.1 In making the proteins, the cell engaged in <u>biosynthesis</u>. In transporting the proteins to the edges of the cell, it engaged in <u>movement</u> (movement includes moving things *inside* the cell-not just moving the cell itself). Finally, in sending the materials out into the surroundings, it engaged in <u>secretion</u>. Secretion is different from excretion and egestion in the sense that secretion sends *useful* things outside the cell, whereas the others send out waste products.

6.2 In taking in the polysaccharide, the cell engaged in <u>absorption</u>. A polysaccharide cannot be used directly in respiration, however, because it first must be digested. Thus, the cell also performed <u>digestion</u>. Once the cell had broken the polysaccharide down into useful monosaccharides, it performed <u>respiration</u> on the monosaccharides. In order to move all of these things to the proper organelles, as well as to move the waste products to the edge of the cell, it also engaged in <u>movement</u>. Finally, since the waste materials are soluble, the cell performed <u>excretion</u> to get rid of them.

6.3 Before you look at the answers, do not be too concerned if you did not list all of the life functions with each organelle. It's tough to think of them all. In addition, <u>homeostasis</u> could probably be added to every organelle's function, because, after all, these organelles all help the cell maintain the status quo.

Organelle	Function(s)
Cell wall	Absorption, secretion, excretion, egestion
Plasma membrane	Absorption, secretion, excretion, egestion
Cytoplasm	Really, the cytoplasm allows <u>all functions</u>, but <u>movement</u> is probably the most important.
Mitochondrion	Respiration
Lysosome	Digestion
Ribosome	Biosynthesis
Smooth endoplasmic reticulum	Biosynthesis, movement, excretion, egestion
Rough endoplasmic reticulum	Biosynthesis, movement
Golgi body	Movement, biosynthesis
Chloroplast	Biosynthesis. You could also include <u>absorption</u> here, since the chloroplasts absorb energy from sunlight.
Leucoplast	Homeostasis. By storing food for later use, this organelle helps to maintain the status quo. You could also include <u>biosynthesis</u> here, because without the leucoplasts, there would be nowhere to put the products of biosynthesis.
Central vacuole	Homeostasis. By maintaining turgor pressure, this helps maintain the status quo.
Contractile vacuole	Homeostasis, excretion
Centrioles	Movement, reproduction
Nucleus	Since the nucleus really holds all of the information regarding cell structure and function, it allows <u>all functions</u> of life.

6.4 The <u>lysosome</u> is where digestion of disaccharides and polysaccharides occurs. If a person lacks the enzyme to digest lactose (a disaccharide), it must be the lysosomes which lack it.

6.5 Phagocytosis is defined as one cell engulfing a foreign substance or another cell. The first cell did not engulf its food, but the second did. Thus, the <u>second cell</u> engaged in phagocytosis.

6.6 <u>A prokaryotic cell cannot have *membrane-bound* organelles. Ribosomes, however, are not membrane-bound. Thus, they can exist in prokaryotic cells.</u>

6.7 <u>The reason that a plasma membrane can self-reassemble is that the phospholipids in the plasma membrane have both a hydrophobic end and a hydrophilic end. Since regular lipids have no hydrophilic end, they do not have the ability to self-reassemble.</u>

6.8 <u>The cell was in a hypotonic solution</u>. In hypotonic solutions, the solute concentration is lower in the solution than in the cell. Thus, osmosis forces the water in the solution to flow into the cell. This builds up cell pressure, causing the cell to explode.

6.9 <u>Not all transport will cease. Active transport will cease, because it requires energy. However, passive transport requires no energy from the cell. Thus, passive transport will continue.</u>

6.10 <u>The cell is missing ADP and/or phosphate</u>. In order to make energy that is useable by the cell, respiration must make ATP from ADP and phosphate. If all of the components necessary to make energy are there, respiration must be making energy, but without ADP and phosphate to make ATP, the cell cannot use the energy.

6.11 One molecule of glucose in aerobic respiration allows the cell to gain 36 molecules of ATP. In anaerobic respiration, one molecule of glucose allows the cell to gain only 2 molecules of ATP. Thus, <u>18 molecules of glucose</u> must undergo anaerobic respiration to make as much energy as one molecule of glucose does in aerobic respiration.

6.12 <u>The mitochondria must be malfunctioning</u>. Since glycolysis occurs in the cytoplasm, it is independent of the mitochondria. Also, since glycolysis is occurring, we know that the cell has everything it needs (enzymes, ADP, phosphate, etc.) for respiration. Now the last two stages of respiration do require oxygen, unlike glycolysis. However, the problem tells us that oxygen is present. Thus, it must be the mitochondria.

6.13 Guanine can bond to cytosine and vice-versa. Thus, when the RNA has guanine, the DNA must have had cytosine. When the RNA has cytosine, the DNA must have had guanine. Adenine bonds with uracil and vice-versa. Thus, when you see a uracil in RNA, there must have been an adenine in the DNA. Now, when you see an adenine in RNA, you might think that there was a uracil in the DNA. However, DNA has no uracil. It has thymine instead. Thus, when you see an adenine in RNA, there was a thymine in DNA. Thus, the DNA sequence must have been:

<u>adenine, thymine, thymine, cytosine, guanine, guanine</u>

6.14 The same rules apply here, realizing that RNA has uracil. Thus, when DNA has cytosine, RNA will have guanine. When DNA has guanine, RNA will have cytosine. When DNA has thymine, RNA will have adenine. When DNA has adenine, RNA will have uracil.

<u>adenine, adenine, adenine, uracil, cytosine, guanine</u>

6.15 The scientist must be studying <u>RNA</u>, because only RNA has uracil.

6.16 a. Now remember, if this codon calls for alanine, then it must be on the mRNA. Thus, the sequence given is on the mRNA. The DNA, then, must have the complement of this sequence. We can therefore use the same reasoning that we did in 6.13. Thus, the DNA must have had:

<u>cytosine, guanine, thymine</u>

b. The tRNA must also have the complement to this sequence. Unlike DNA, however, RNA uses uracil rather than thymine. Thus, the tRNA must have:

<u>cytosine, guanine, uracil</u>

STUDY GUIDE FOR MODULE #6

1. Define the following terms:

a. Absorption

b. Digestion

c. Respiration

d. Excretion

e. Egestion

f. Secretion

g. Homeostasis

h. Reproduction

I. Cytology

j. Cell wall

k. Middle lamella

l. Plasma membrane

m. Cytoplasm

n. Ions

o. Cytoplasmic streaming

p. Mitochondria

q. Lysosome

r. Ribosomes

s. Endoplasmic reticulum

t. Rough ER

u. Smooth ER

v. Golgi bodies

w. Leucoplasts

x. Chromoplasts

y. Central vacuole

z. Waste vacuoles

aa. Phagocytosis

bb. Phagocytic vacuole

cc. Pinocytic vesicle

dd. Secretion vesicle

ee. Microtubules

ff. Nuclear membrane

gg. Chromatin

hh. Phospholipid

ii. Passive transport

jj. Active transport

kk. Isotonic solution

ll. Hypertonic solution

mm. Plasmolysis

nn. Cytolysis

oo. Hypotonic solution

pp. Activation energy

qq. Messenger RNA

rr. Transcription

ss. Translation

tt. Codon

uu. Anticodon

2. What is the plasma membrane made of?

3. What is the difference between a phospholipid and a regular lipid?

4. What makes it possible for the plasma membrane to self-assemble?

5. A cell begins running low on food and its energy output decreases by 20%. What kind of plasma membrane transport (active or passive) is affected?

6. If a cell dies by implosion, what kind of solution (isotonic, hypertonic, or hypotonic) was it in?

7. What are the three stages of cellular respiration? Which one produces the most energy?

8. What is ATP's purpose in the cell?

9. If a cell has no oxygen, what stage(s) of cellular respiration can still run?

10. A scientist determines a means to extract all ATP from a cell. Why will this kill the cell?

11. A DNA strand has the following sequence of nucleotides:

   guanine, cytosine, adenine, adenine, thymine, guanine

   a. What will the mRNA sequence be?

   b. How many amino acids will the mRNA code for?

12. Fill in the blanks:

                  a. _____   b. _____

        DNA ⟶ RNA ⟶ Protein

13. An RNA strand has an anticodon. Is it tRNA or mRNA?

14. What is the difference between rough ER and smooth ER?

15. Is chlorophyll stored in a leucoplast or a chromoplast?

16. How many ATPs does a cell gain in anaerobic respiration? How many does it gain in aerobic respiration?

17. What organelle is responsible for breaking polysaccharides into monosaccharides?

Module #7: Cellular Reproduction

Introduction

As we learned way back in Module #1, all living organisms reproduce. Indeed, the ability to reproduce is one of the major criteria for life. It makes sense, then, that we should study this process in great detail. As is the case with many of the concepts in biology, however, reproduction is a difficult and detailed process to study. Thus, we will spend both this and the next module exploring some of the facets of reproduction.

Now the process of reproduction has always fascinated scientists. Throughout the history of biological inquiry, scientists have tried to understand how an organism's offspring receive the traits and characteristics that make them what they are. Sometimes, an organism's offspring might look identical or nearly identical to its parent or parents. In other cases, the offspring and parents look so different that you can hardly believe the offspring came from those parents.

When we look at a child, for example, we usually try to find the characteristics that the parents and child have in common. When we see that a boy and his father both have thick, brown hair, we say that the boy has his father's hair. When we see that he and his mother share vividly green eyes, we say that the boy has his mother's eyes. Sometimes, however, we have a hard time finding such similar traits. It is not unusual for two brown-haired, brown-eyed parents to have a child with blonde hair and blue eyes. How does that happen?

Well, the explanation for these facts eluded scientists for centuries. To be sure, there have always been theories which attempted to explain how characteristics passed from parent to offspring, but many were quite ridiculous. For example, there were those that thought characteristics like height, strength, and weight were determined by the activities of the parents. These scientists, called Lamarkian biologists, thought that if parents spent a lot of time in strenuous activity, their children would be born stronger. Somehow, they thought that the parents' activities would send "messages" to the infant, telling it how to develop.

The beginning of a real explanation for how traits are passed on from generation to generation was developed by a monk name Gregor Mendel in the mid 1800's. He did careful experiments with peas, determining how pea plants reproduced and passed on characteristics to their offspring. Unfortunately, this magnificent work was tucked away in a monastery for more than 50 years! When it was finally discovered, it laid the groundwork for all of our modern studies of **genetics**.

Genetics - The science that studies how characteristics get passed from parent to offspring

Because Mendel's work was so fundamental in our understanding of genetics, we often refer it as "Mendelian genetics," in an attempt to honor him.

Genes, Chromosomes, and DNA

Building on the work of Gregor Mendel, scientists today have determined that each organism contains a storehouse of information that determines all of its traits and characteristics. As we have already mentioned, this storehouse is called DNA. Now remember, DNA's main function is to tell the cell what proteins to make. Although it sounds hard to believe, that is all it really takes to determine the traits and characteristics of a person. Your eye color, for example, is completely dependent upon what proteins are produced in some of your eye cells. Thus, by coding for the production of certain proteins, the DNA in your eye cells determines your eye color.

Now, of course, not all of your traits and characteristics are completely determined by your genetic makeup. For example, if you tend to lift weights and work out, you will develop strong muscles. You might think, then, that your muscle strength is not determined by your DNA. Well, this is only partially correct. Even though you can increase your strength by working out, there is a fundamental limit to how strong you can become. For example, a person might lift weights and work out every day but never become as strong as another individual who really does not exercise seriously at all. In the case of muscle strength, your DNA determines a general range of how strong your muscles can become. If you do not exercise your muscles, you will stay on the weak side of the muscle strength range determined by your DNA. If, on the other hand, you work out seriously, you can increase your muscle strength to the strong side of what your DNA has determined for you. You might never become as strong as you want to become, however, because there is some inherent limit set up by your DNA.

It turns out that the majority of characteristics in your body are determined in this way. The DNA sets a range of possible characteristics (sometimes called the "genetic tendency"), and your activities determine what portion of that range actually manifests itself in your body. Some people have a genetic tendency to be overweight. If they eat just a little more than they need, they will immediately gain weight. Others, however, have a genetic tendency to stay slim. They can eat almost as much as they like and never gain a pound. Although such tendencies exist, they can nevertheless be controlled. Even if you have a genetic tendency to be overweight, by strict diet and exercise, you can keep your weight down. Thus, your DNA has set a general tendency, but you can work to fight that tendency, if you wish.

You might have read or heard that scientists have discovered that alcoholism is genetic. This is, in fact, partially true. This makes some people believe that alcoholics are not at fault for their behavior. This notion, however, is completely false. Just as is the case with a tendency for being overweight, your DNA can give you a general tendency to be an alcoholic. People with this genetic tendency find it very easy to get drunk, and they have fewer side effects the next morning. They also tend to enjoy the experience of drunkenness more. These traits are determined by their DNA. However, their DNA does not make them drink. With the proper self-control, even those with the strongest tendency for alcoholism can completely abstain from alcohol, fighting the tendency that their DNA has given them. This is evidenced by the fact that more than 30% of all non-alcoholics have a strong genetic tendency to alcoholism. In addition, only 65% of alcoholics have this genetic tendency. These numbers indicate that some people have a genetic tendency towards alcoholism but manage to control it, while others do not even have the genetic tendency to begin with and are still alcoholics.

The main point to this discussion, then, is that your DNA alone does not determine who you are or what you will become. There are some traits (eye color, for example) that are completely determined by your DNA. For most of your characteristics, however, your DNA just sets up a general framework. Within that general framework, what you do with yourself will ultimately decide what you become. Since DNA is not the only factor in determining what kind of person you will be, we generally say that there are three factors in determining the characteristics of a person: **genetic factors, environmental factors,** and **spiritual factors**.

Genetic factors - The general guideline of traits determined by a person's DNA

Environmental factors - Those "non-biological" factors that are involved in a person's surroundings such as the nature of the person's parents, the person's friends, and the person's behavioral choices

Spiritual factors - The quality of a person's relationship with God

These three factors each work together to determine what kind of person you are. There is no real consensus in the scientific community as to which of these three factors are the most important, if any. Needless to say, however, from the Christian point of view, spiritual factors are of the utmost importance. Since spiritual factors do not really apply to any organism other than humans, we generally say that the characteristics of animals and plants are determined just by genetic and environmental factors.

Even though we do not know much about the relative importance of genetic factors and environmental factors, we can make one statement for sure. The genetic factors are laid down first. From a scientific point of view, these are the factors most easily studied, so that is what we will concentrate on. The information in an organism's DNA is split up into little groups known as **genes** (jeens). Although a firm definition for a gene is rather hard to pin down, the following will work for the purposes of this course:

Gene - A section of DNA that codes for the production of a protein or a portion of protein, thereby causing a trait

As we mentioned before, your genetic traits or tendencies are determined by what proteins are produced in your cells. Thus, by coding for the production of a certain protein, a gene is actually coding for a particular genetic trait or tendency. In fact, some proteins are so complex that just a portion of the protein is enough to determine a genetic tendency or trait.

Remember from Module #6 that certain cells in your body produce a certain set of proteins while other cells in your body produce other proteins. Indeed, this is the major characteristic that distinguishes a skin cell from an eye cell. The proteins that each of these cells produce are different, so the cells function differently. Interestingly enough, all of the cells in your body have the exact same set of genes. In other words, each cell has the instructions to produce every protein that your body needs. Nevertheless, they do not. For some reason that scientists cannot explain at this time, your skin cells know that they should disregard all genes that do not code for the proteins that they need to produce. Likewise, eye cells disregard all information related to proteins that they do not need to produce. This is truly an amazing ability, and no one knows how the cells accomplish it!

As we already learned, DNA is stored in the nucleus of the cell. It is stored there in long, thin strands. If all of the DNA from one of your cells were strung together end to end, the string would dangle from your head to your toes. How does such a large amount of DNA get packed into a tiny cell? The DNA of a eukaryotic cell is tightly bound together with a network of proteins. Certain proteins, called **histones**, act as spools, which wind up small stretches of DNA. Other proteins (including other histones) stabilize and support these spools, making a complex network of DNA coils and proteins. This network is called a **chromosome** (krohm' uh zohm).

Chromosome - A strand of DNA coiled around and supported by proteins, found in the nucleus
              of the cell

Most of the time, these chromosomes appear as long, fuzzy strands in the nucleus. The chromatin material that we discussed in Module #6 is, in fact, a collection of all chromosomes in the nucleus.

Now as you might expect, it takes several strands of DNA to carry all of the genes necessary to code for the proteins of a living organism. Thus, there are several different chromosomes in the nucleus of the cell. What you might not expect, however, is that the number of chromosomes is dependent on the organism studied. For example, humans have 46 chromosomes in the nuclei of their cells, while a cat has only 38. Don't get the idea that the more complex an organism is the more chromosomes it will have, however. A horse has 64 chromosomes in the nuclei of its cells, and a crayfish (which you might call a crawdad) has 200!

In order to get an idea of the physical characteristics of DNA, perform the following experiment.

EXPERIMENT 7.1
DNA Extraction

Supplies:
- Blender
- Plastic bowl
- Toothpick
- Clear dish soap
- Salt
- Water
- Strainer

- Coffee filter
- Small glass
- Meat tenderizer (must be fresh!)
- Rubbing alcohol
- Medium-sized onion
- Measuring cups and spoons
- Flashlight

Object: To extract DNA from onions so that you can see what it looks like

Procedure:

1. Coarsely chop the medium onion, including the skin.
2. Dissolve 1/2 teaspoon salt in 1/4 cup of warm water.
3. Place the chopped onion and the saltwater into the blender and blend it for 15-25 seconds. The result should have the consistency of pea soup and should still contain a lot of onion pieces.
4. Pour the mixture into the plastic bowl and add 1/4 cup of clear dishwashing soap. Mix everything with a spoon for at least 5 minutes. Try not to make a lot of bubbles. The soap will dissolve the lipids in the plasma membrane of the cells. This will destroy the plasma membrane, opening the cells. The contents of the cells will then flow into the solution.
5. Once you have mixed the soap, onions, and saltwater for at least 5 minutes, place a coffee filter into the strainer and hold it over the small glass.
6. Pour the onion/soap/saltwater mixture into the coffee filter. Allow the liquid to drip into the small glass until the top of the liquid level is about 1/4 inch above the bottom of the glass.
7. Once you have this thin layer of liquid in the glass, discard the filter and its contents. Take 1/4 teaspoon of meat tenderizer and mix it thoroughly with the small amount of liquid in the glass. The meat tenderizer is full of enzymes that destroy proteins. This is what, in fact, tenderizes meat. In this experiment, the meat tenderizer will destroy the proteins that coat the DNA in the chromosomes, exposing the strands of DNA.
8. Add twice as much rubbing alcohol as you have liquid. Wait for a few minutes, and you should see clear/white strands appear near the middle/top of the solution. To aid you in seeing the strands, shine the flashlight down onto the surface of the solution and look at the solution from the side of the glass. These strands are the onion's DNA! If you do not see any strands, mix the solution gently and wait again.
9. Depending on your experimental technique, the strands might be long enough for you to twist them onto a toothpick and remove them from the solution.

Now you know what the strands of DNA look like. If all of the DNA from a single human cell were extracted in this way, you would have about 2 yards of strands.

ON YOUR OWN

7.1 Two identical twins have exactly the same set of genes. They are separated at birth and grow up in different households. If a scientist were to study the twins as adults, would he find them to be identical in every way, since they have the same genes? Why or why not?

enviromental factors & spiritual factors

7.2 If a chromosome was missing the proteins that coat and wind the DNA, what would happen to the chromosome? *it woul unravel & fall apart*

Mitosis: The Common Mode of Asexual Reproduction

As we have already learned, there are two types of reproduction in God's Creation: sexual and asexual. The common mode of asexual reproduction is called **mitosis** (mye toh' sis)

Mitosis - The duplication of a cell's chromosomes to allow daughter cells to receive the exact genetic makeup of the parent cell

Now remember, asexual reproduction results in offspring that have exactly the same genetic code as that of the parent. Since this is the case, we often call mitosis "cellular division," because, in the end, the process makes two duplicate cells from a single cell.

When a cell is busy with its normal life functions, it is said to be in a state of **interphase**.

Interphase - The time interval between cellular reproduction

During this time, the cell is simply maintaining the status quo. The chromosomes are in the nucleus of the cell. From time to time, sections of the DNA on the chromosomes unwind to begin the process of protein synthesis. Otherwise, the DNA remains tightly coiled.

When the cell decides that it is time to reproduce (we do not know how it decides this), two things happen. First, the chromosomes in the cell coil tightly and duplicate. Although we have learned a lot in recent years about how this happens, many of the details regarding how chromosomes duplicate are still a complete mystery. Thus, we will not try to explain that process here. Once the DNA strands duplicate, they attach themselves to their copy at a point called the **centromere** (sen troh' mer).

Centromere - Constricted region of a chromosome and the point at which duplicate DNA strands attach themselves

As you can see in Figure 7.1, when the duplicates attach themselves at one point to the original strand of DNA, they form an "X"-shape. This is the common representation of a chromosome. You have to understand, however, that chromosomes only look like this during modes of reproduction. During interphase, the chromosomes are simply single strands, as illustrated on the left side of Figure 7.1.

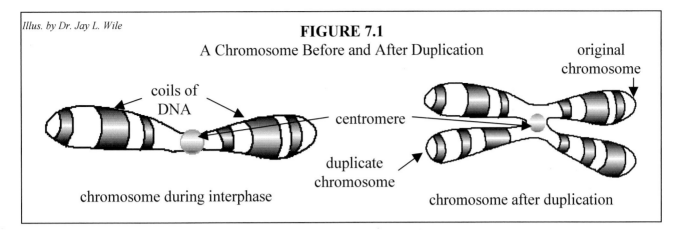

FIGURE 7.1
A Chromosome Before and After Duplication

Illus. by Dr. Jay L. Wile

coils of DNA

centromere

duplicate chromosome

original chromosome

chromosome during interphase

chromosome after duplication

   The second thing that happens to initiate mitosis is that the centriole, which is outside the nucleus, duplicates itself as well. Now that the DNA and the centriole are duplicated, the cell is ready to begin mitosis. At this point, it is called a **mother cell**.

   Mother cell - A cell ready to begin reproduction, containing duplicate DNA and centriole

The mother cell then enters the first stage of mitosis, called **prophase**.

   In prophase, the centrioles migrate towards opposite ends of the cell. As they migrate, they begin to assemble long strands of proteins into microtubules. These microtubules, sometimes called the **aster**, radiate out in all directions from the centrioles. As the aster is forming and the centrioles are moving, the nuclear membrane breaks apart, exposing the chromosomes to the rest of the cell. Microtubules from each centriole attach themselves to the chromosomes and move them to the center of the nucleus. They line the chromosomes up along an imaginary line called the **equatorial plane**, which runs down the center of the cell, equidistant from the two centrioles. At this point, the microtubules that are attached to the chromosomes make up the structure known as the **spindle**.

   The brief time in which the chromosomes are lined up along the equatorial plane is called **metaphase**. During metaphase, the duplicate chromosome acts as if it is repelled by the original chromosome to which it is attached. This gives the two chromosomes the distinct "X"-shape that many students incorrectly think is the standard shape of a chromosome. As we mentioned before, chromosomes only look like this during reproduction.

   At the conclusion of metaphase, the microtubules on the spindle pull the chromosomes towards the centrioles. Since each chromosome and its duplicate have a microtubule pulling on them from opposite centrioles, the duplicates separate from the originals. The original chromosome gets pulled one way and the copy gets pulled the other. This phase of mitosis is called **anaphase**.

   When the chromosomes are pulled towards the centrioles as far as they can go, the last phase of mitosis, **telophase**, begins. During this time, the spindle disintegrates, leaving only a small aster around the centrioles. The plasma membrane begins to constrict along the equatorial plane, eventually forming two cells where there was once only one. A nuclear membrane begins to form around each group of chromosomes, and the chromosomes uncoil, forming the chromatin material of the nucleus. At this point, there are two **daughter cells** where there was just one mother cell before. Please note that this phase is different in plants, as we will see in a moment.

The process of mitosis is rather long and complicated, so it certainly deserves some review. Examine Figure 7.2 to get a different view of mitosis.

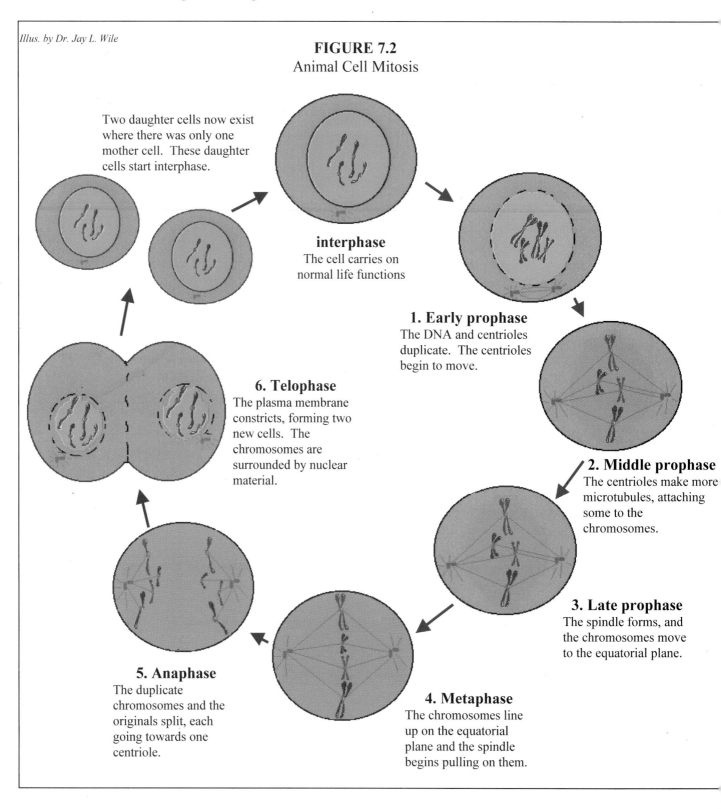

Illus. by Dr. Jay L. Wile

FIGURE 7.2
Animal Cell Mitosis

Two daughter cells now exist where there was only one mother cell. These daughter cells start interphase.

interphase
The cell carries on normal life functions

1. Early prophase
The DNA and centrioles duplicate. The centrioles begin to move.

2. Middle prophase
The centrioles make more microtubules, attaching some to the chromosomes.

3. Late prophase
The spindle forms, and the chromosomes move to the equatorial plane.

4. Metaphase
The chromosomes line up on the equatorial plane and the spindle begins pulling on them.

5. Anaphase
The duplicate chromosomes and the originals split, each going towards one centriole.

6. Telophase
The plasma membrane constricts, forming two new cells. The chromosomes are surrounded by nuclear material.

Now you might wonder what happened to all of the other organelles in the cell during mitosis. Well, they are all still present while mitosis is occurring, we simply chose to ignore them in our discussion, because we wanted to concentrate on just those organelles which are important to mitosis. So, when the plasma membrane constricts itself, whatever position each organelle in the cell finds itself in determines which of the two daughter cells that the organelle ends up in.

But wait, you might ask, what if all of the mitochondria in the cell are on one side of the equatorial plane during telophase? Does this mean that one cell will get all of the mitochondria and the other will not have any? Well, yes, at least for a short time. Remember, however, that the DNA in each of the cells has all of the information necessary to construct any organelles that the cells need. Thus, if one of the two cells does not have enough mitochondria, it will just make more. That's the beauty of mitosis! As long as the DNA is duplicated properly, it doesn't matter what happens to the rest of the organelles in the process. The new cells will make any organelle that they need!

Figure 7.2 might have been easy to follow since it was an idealized, schematic drawing. In real life, of course, things are a little more difficult to see. Examine Figure 7.3 to see what mitosis looks like in animal cells when it is viewed under a microscope.

Photos by Kathleen J. Wile

FIGURE 7.3
Animal Cell Mitosis as Observed with a Microscope (1000x magnification)

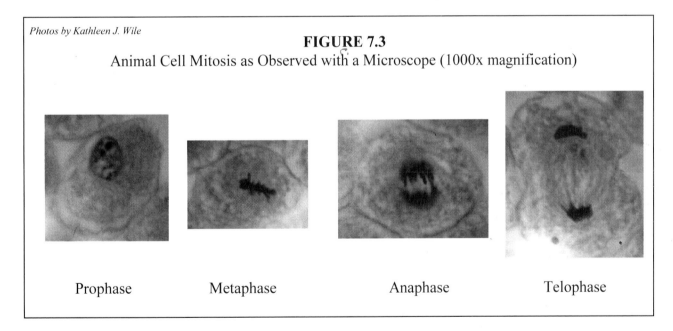

| Prophase | Metaphase | Anaphase | Telophase |

When you think of asexual reproduction, you probably think of organisms like bacteria and protozoa, and perhaps fungi. Most of the organisms with which you are familiar (dogs, cats, humans, etc.), however, reproduce sexually. This means mitosis is irrelevant in the study of these organisms, right? Wrong! It turns out that nearly every life form uses mitosis. After all, when you are born, you are tiny. As time goes on, you grow. How does this happen? Well, your body produces more cells. How are those cells produced? By mitosis! Also, even after you are fully grown, your cells must continually go through mitosis in order to replace cells that have died. For example, when you scratch an itch, you are actually destroying several hundred to several thousand skin cells. Your body must replace those cells, and it does so by mitosis.

Some cells can carry on mitosis relatively quickly. Skin cells, for example, can complete mitosis in about 10 minutes. This is important, because your skin cells die at such a high rate, that they must be replaced quickly. Other cells take quite a long time. Bone cells can go through mitosis, but they do so very slowly. This is, in fact, why it takes a broken bone so long to heal. Cuts and scrapes heal rather quickly because skin cells can reproduce so quickly. Broken bones take a long time to heal, however, because mitosis is a very slow process for them. Some cells in your body cannot perform mitosis at all. Brains cells are an excellent example. Brain cells grow in size over time to increase your thinking capacity. However, they cannot perform mitosis. Thus, if a brain cell dies, it cannot be replaced. This is one of the extreme dangers of drugs such as alcohol, marijuana, and cocaine. These drugs destroy brain cells. Since brain cells cannot be replaced, the damage is permanent!

So you see that mitosis is relevant to nearly every creature in God's Creation. Now the mitosis you just learned about occurs in animals. A slightly different kind of mitosis occurs in the telophase of plant cells. Because of the rigid cell wall, the plasma membrane cannot just close down as we learned in animal cell mitosis. Instead, once the chromosomes have been pulled to the centrioles, vesicles containing the building blocks for the plasma membrane line up along the equatorial plane. These vesicles begin fusing together, forming a new plasma membrane. As the new plasma membrane forms, cellulose is formed in the middle, forming a new cell wall that splits the plasma membrane in two. Eventually, even a middle lamella is formed, resulting in two cell walls, two plasma membranes, and thus, two cells.

If you have a microscope, you can get more experience identifying the stages of mitosis by performing the following experiment. If you do not have a microscope, you can skip the experiment.

EXPERIMENT 7.2
Mitosis

Supplies:
- Microscope
- Prepared slide of *Allium* (onion) root tip
- Prepared slide of Whitefish blastula

Object: To observe how a cell divides during mitosis.
　　　　To observe the difference between plant and animal cell division.

Procedure:

1. Observe the Whitefish blastula prepared slide.
　　a. Look up the word blastula and write the definition.

b. Observe the Whitefish blastula under low power. Look for cells that have stained (blue/purple) objects in the cell. Those stained objects are the chromosomes. As you find a cell in one of the stages of mitosis (use the pictures in Figure 7.3 as a guide), switch to high power to observe. Draw examples of each stage that you find.

2. Observe the onion root tip prepared slide. Once again, begin under low power. As you find cells in the stages of mitosis, switch to high power so that you can observe them and draw them in your notebook. Note that telophase is different in the onion root tip than that shown in Figure 7.3. In plant cells, the plasma membrane does not constrict. Instead, vesicles build up the cell wall. Thus, you will distinguish telophase in this slide by looking for two nuclei that have split apart but there is little or no cell wall between them.

3. Clean up and return all equipment to the proper place.

ON YOUR OWN

7.3 The phases of a cell's life are listed below. Which one is not a part of mitosis? Take the remaining phases and order them according to when they occur in the mitosis process.

anaphase, prophase, interphase, telophase, metaphase

7.4 A cell uses vesicles during the telophase of mitosis. Is it a plant cell or an animal cell?
Plant cell

7.5 In which phase of mitosis are the chromosomes separated from their duplicates?
anaphase

Diploid Chromosomes

As we mentioned before, humans have 46 chromosomes. What we failed to mentioned, however, is that these chromosomes come in pairs. In a human cell, there are only 23 *different* chromosomes, but each of these chromosomes has a partner. When chromosomes pair up like this, we call them **homologous** (ho mahl' uh gus) **chromosome pairs**. Each member of a homologous chromosome pair is called, reasonably enough, a **homologue** (hahm' uh lawg).

Now this can get a little confusing, so we will try to illustrate it with a figure. Suppose I were to take the chromosomes from a human cell during the metaphase of mitosis. Remember, at that point, the chromosomes are attached to their copies at the centromere. This gives them an "X"-shape. If I were to look at the chromosomes at that point, I would get something that looks like Figure 7.4.

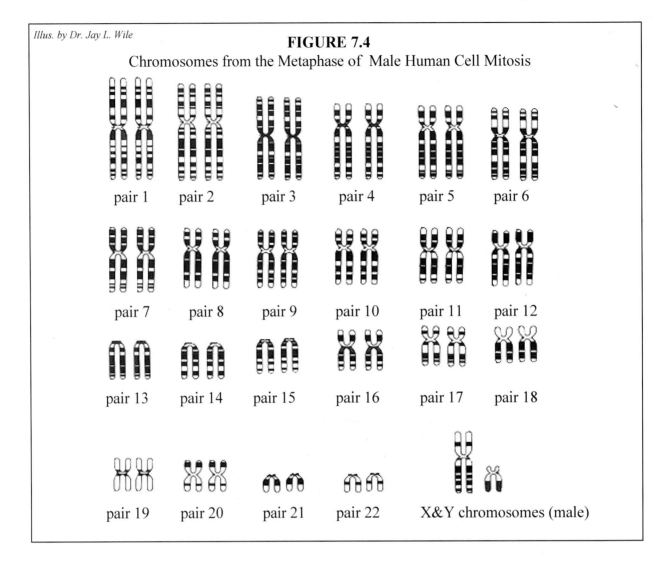

FIGURE 7.4

Chromosomes from the Metaphase of Male Human Cell Mitosis

pair 1   pair 2   pair 3   pair 4   pair 5   pair 6

pair 7   pair 8   pair 9   pair 10   pair 11   pair 12

pair 13   pair 14   pair 15   pair 16   pair 17   pair 18

pair 19   pair 20   pair 21   pair 22   X&Y chromosomes (male)

Notice that each chromosome is repeated. Now remember, this is not because the chromosomes have been duplicated for reproduction. Each chromosome is attached to its duplicate at the centromere. That's why they all (more or less) have an "X"-shape to them. So the "X"-shape is the result of duplication, not the fact that there are two of each chromosome. It turns out that human cells have two of every chromosome. Thus, each cell has two of chromosome number 1, two of chromosome number 2, and so on. The only exception to this is in the male human. Females have two X chromosomes (the last one in the figure), while males have an X and a Y chromosome. We will learn a great deal more about that in the next module.

As a point of terminology, a figure like that of Figure 7.4 is called a **karyotype** (kehr' ee uh tipe).

Karyotype - The figure produced when the chromosomes of a species during metaphase are arranged according to size

In this particular karyotype, you can see that there are 23 pairs of chromosomes (chromosomes 1-22 and an XY pair). With the exception of the XY or XX pair, each homologue within a pair has exactly the same number of genes as its partner. These genes also code for the same trait. For example, suppose that the gene for eye color is located on chromosome 1. Well, each homologue in the pair would have a gene for eye color, and they would be in exactly the same spot on each chromosome. In the next module, we will learn the implications of the fact that human cells have two genes for each trait.

When a cell's chromosomes come in pairs, we say that it is a **diploid** (dye' ployd) cell.

Diploid cell - A cell whose chromosomes come in homologous pairs

Many species have diploid cells, and as we will learn in the next module, this results in a great deal of diversity within the species. Cells whose chromosomes do not come in homologous pairs are called **haploid** (hap' loyd) cells.

Haploid cells - Cells that have only one of each chromosome

As we will see in the next section, even species that have diploid cells will have some haploid cells.

Before we leave this section, we need to make sure that you understand what we mean when we talk about the number of chromosomes in a cell. When we say that a horse has 64 chromosomes, this means that it has 32 pairs of chromosomes. The total count is 64, but they come in 32 different pairs. Since this terminology can be a little confusing at times, we sometimes make it a little clearer by referring to the **diploid chromosome number** (sometimes abbreviated as "2n") or the **haploid chromosome number** (sometimes abbreviated as "n").

Diploid chromosome number (2n) - The total number of chromosomes in a diploid cell

Haploid chromosome number (n) - The number of homologous pairs in a diploid cell

Based on these definitions, we could say that the diploid chromosome number (2n) of a human cell is 46 while the haploid number (n) is 23.

ON YOUR OWN

7.6 A pea plant has 7 pairs of homologous chromosomes. What is its haploid number? What is its diploid number?

7.7 In a scientist's notebook, you find notes regarding a new species that is being studied. The notes say that the species is diploid, with a chromosome number of 17. Is this the haploid or diploid number? If this is the haploid chromosome number, give the diploid number. If this is the diploid number, give the corresponding haploid number.

Meiosis: The Cellular Basis of Sexual Reproduction

Now that we have spent some time discussing asexual reproduction and chromosome number, it is time to move on to sexual reproduction. There are many facets of sexual reproduction, but we will concentrate on the cellular level, because it forms the basis of what happens during all of the other stages of sexual reproduction. Sexual reproduction begins with **meiosis** (my oh' sis).

Meiosis - The process by which a diploid (2n) cell forms four gametes (n)

Now this definition might seem a little mystifying, especially since you do not know what a gamete is. That's okay. This definition will become very clear to you as you continue to read.

You learned way back in Module #1 that in sexual reproduction, the offspring is not identical to the parents. Instead, the offspring might have some characteristics in common with the parents, but other characteristics which are quite different from the parents. This happens because in sexual reproduction, each parent contributes DNA to the offspring. As a result, the offspring has DNA from both parents.

If a human has 46 chromosomes and each parent contributes to the DNA of the offspring, then it should make sense that each parent contributes 23 chromosomes. How does this happen? After all, the cells in each parent have 46 chromosomes. How does the DNA of the parents get separated so that each parent contributes only 23 chromosomes to the offspring? That's what the process of meiosis is all about. In meiosis, diploid cells get split into haploid cells called **gametes** (gam' eats).

Gametes - Haploid cells (n) produced by diploid cells (2n) for the purpose of reproduction

In animals, the gamete produced by the mother is called an "egg" cell, or "ovum," while the gamete produced by the father is called the "sperm" cell.

A human gamete produced by meiosis, then, will have only one of each chromosome pair. This makes 23 chromosomes. When a gamete from one parent joins up with a gamete from the other parent, the result will be a diploid cell that has 23 pairs of chromosomes, for a total of 46 chromosomes. One member of each pair will come from the father and the other member of each pair will come from the mother. When gametes join together, the resulting diploid cell is called a **zygote**. We actually presented the definition of a zygote back in Module #4, but now at least you know how a zygote forms.

So, before sexual reproduction can begin, diploid cells must form gametes. This is accomplished through the process of meiosis, which is actually similar to mitosis in several ways. In fact, the phases of meiosis have the same names as the phases of mitosis. The details of these phases, however, are a bit different. In addition, meiosis involves a few more steps. In fact, the process of meiosis is split into two groups of phases, **meiosis I** and **meiosis II**. Each of these parts of meiosis involves prophase, metaphase, anaphase, and telophase.

In order to begin meiosis, a cell must duplicate its DNA, just as it must for mitosis. In addition, the centrioles must duplicate as well. At that point, the cell is ready for the first stage of meiosis, **prophase I**. During this phase, the centrioles move to opposite sides of the cell, just as in mitosis. Also, the aster begins to grow, just as is the case in mitosis.

By the time the cell has reached the next phase, **metaphase I**, the spindle has formed, and the microtubules attach to the centromeres so that there is a single microtubule for each chromosome and its duplicate. This is different from what happens in the metaphase of mitosis, because in mitosis, one microtubule attaches to a chromosome and another microtubule from the other centriole attaches to the chromosome's duplicate. Thus, in mitosis, the microtubules attach so as to pull the chromosomes away from their duplicates. In meiosis, however, the microtubules attach so as to pull the homologous pairs apart while leaving the chromosomes attached to their duplicates.

In **anaphase I**, then, the homologous pairs are separated, being pulled towards the centrioles on the opposite sides of the cell. Now think about how this is different than mitosis. In the anaphase of mitosis, each chromosome is separated from its duplicate. Thus, the "X"-shape goes away, because the chromosome goes one way and its duplicate goes the other. Both chromosome homologues go one way while both duplicate homologues go the other. In anaphase I of meiosis, however, the chromosomes stay with their duplicates. Instead, the homologues are separated from one another, breaking up the homologous pairs.

In **telophase I**, then, the plasma membrane constricts along the equatorial plane, forming two cells. Now if you think about it, these cells are haploid, because the homologues have been separated. However, each chromosome still has its duplicate attached to it at the centromere. Thus, each of these cells needs to split those duplicates up. Well, if you think about it, these haploid cells are ready for mitosis, because they have their chromosomes duplicated with the duplicates attached to the originals. Thus, each of the two haploid cells goes through a process very similar to mitosis, which is called meiosis II.

In meiosis II, both cells enter **prophase II** by having the centrioles duplicate and once again travel to opposite ends of the cell, increasing the size of the aster. In **metaphase II**, the spindle is fully formed, the chromosomes are lined up along the equatorial plane, and the microtubules attach to each chromosome and its duplicate at the centromere. In **anaphase II**, the microtubules of the spindle pull the chromosomes away from their duplicates, with the originals being moved towards one centriole and the duplicates moving towards another. Finally, in **telophase II**, the plasma membrane in each cell constricts along the equatorial plane, forming two pairs of cells where there were originally just two cells. The four cells formed, then, are still haploid, because no homologous pairs are present. In addition, there are no chromosome duplicates in any of the cells. The process of meiosis is summarized in Figure 7.5.

FIGURE 7.5
The Process of Meiosis in Animal Cells

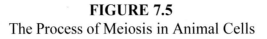

Meiosis I

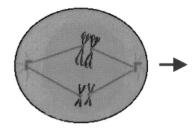

1. Early prophase I
The chromosomes and centriole duplicate and the centrioles begin to move.

2. Late prophase I
The chromosomes line up in homologous pairs and the spindle begins to form.

3. Metaphase I
The chromosomes line up on the equatorial plane. The microtubules attach to one homologue each.

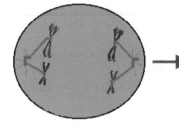
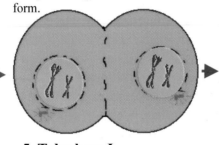
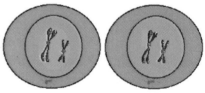

4. Anaphase I
The microtubules of the spindle pull the homologues apart.

5. Telophase I
The plasma membrane constricts, and the chromosomes become surrounded by nuclear material.

6. End of meiosis I
Two haploid cells result. They still have duplicated chromosomes, however, so they must enter meiosis II.

Meiosis II

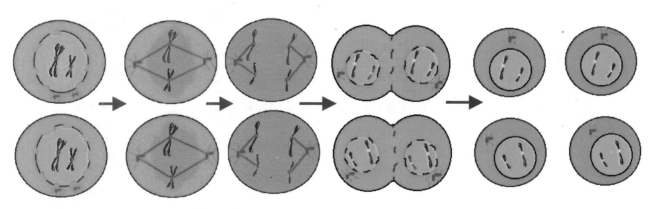

1. Prophase II
Both cells duplicate centrioles

2. Metaphase II
The chromosomes of each cell line up on the equatorial plane.

3. Anaphase II
The microtubules pull the duplicates from the originals.

4. Telophase II
The plasma membranes constrict, and nuclei form.

5. End of meiosis II
Four haploid cells (gametes), each with no duplicate chromosomes, result.

So if we compare mitosis and meiosis, we can see several differences. Mitosis takes one cell and makes an exact duplicate, resulting in two cells. If the original cell was diploid, then the copy will be diploid as well. In contrast, meiosis takes a single cell and produces four cells. The cells produced are quite different than the original, however. The original is diploid, but the four cells produced are haploid. These haploid cells are called gametes. In addition, meiosis has twice as many steps as mitosis. Finally, while the anaphase of mitosis keeps the homologous pairs of chromosomes together and separates the chromosome duplicates, anaphase I of meiosis separates the homologous pairs, keeping the chromosomes and their duplicates together. Although there are several differences between meiosis and mitosis, there are also similarities. The phases of each are remarkably similar. In fact, meiosis II is really just mitosis being performed on two cells.

Before we leave this section on meiosis, it is important for you to understand that although meiosis occurs in both the female of an animal species as well as the male, there are differences between the two. As we mentioned before, in males, the gametes produced by meiosis are called sperm, while in females, the gametes are called eggs. It turns out that these gametes are quite different. As a result, the process of meiosis II is different in males than in females.

In animal males, once telophase II ends, the gametes produced grow flagella. This is accomplished by the centriole, which moves to the plasma membrane and grows a microtubule through the membrane and out into the surroundings. The sperm can then use these flagella to move about in search of an egg. Male meiosis is shown in Figure 7.6.

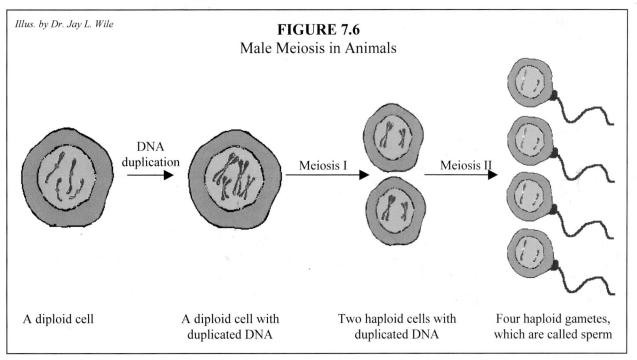

Illus. by Dr. Jay L. Wile

FIGURE 7.6
Male Meiosis in Animals

DNA duplication → Meiosis I → Meiosis II

A diploid cell | A diploid cell with duplicated DNA | Two haploid cells with duplicated DNA | Four haploid gametes, which are called sperm

In animal females, something quite different happens. For reasons unbeknownst to scientists at this time, one of the cells produced at the end of telophase I takes most of the cytoplasm as well as most of the organelles. Thus, this cell is much bigger than the other. Both cells go through meiosis II. However, since the small cell is so small, the two gametes it produces are quite small. In addition, for reasons we do not understand, when the big cell reaches telophase II, one of the two gametes once again takes most of the cytoplasm and organelles. As a result, meiosis II ends with three tiny gametes and one huge gamete. The three tiny gametes, often called **polar bodies** are useless and quickly degenerate and die. The only useful gamete is the large one, called the egg.

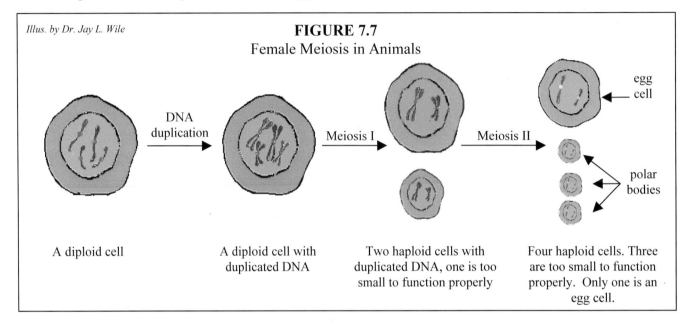

Illus. by Dr. Jay L. Wile

FIGURE 7.7
Female Meiosis in Animals

DNA duplication

Meiosis I

Meiosis II

egg cell

polar bodies

A diploid cell

A diploid cell with duplicated DNA

Two haploid cells with duplicated DNA, one is too small to function properly

Four haploid cells. Three are too small to function properly. Only one is an egg cell.

Since the functional egg produced by female meiosis has been formed by taking most of the cytoplasm and organelles during both meiosis I and meiosis II, it should make sense to you that the egg cell in a female is much bigger than the sperm cell in a male. In addition, since the sperm cells have a means of locomotion and the egg cell does not, the egg cell must sit still and wait for the sperm cell to travel to it. These facts are illustrated in Figure 7.8.

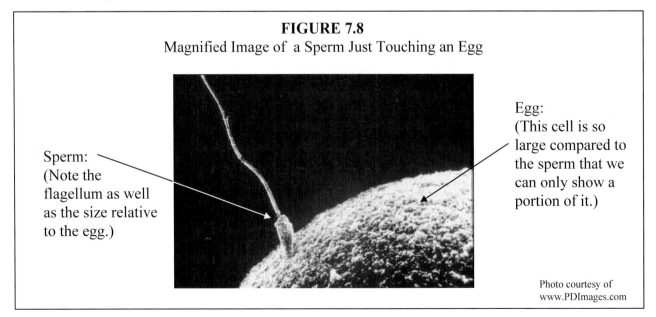

FIGURE 7.8
Magnified Image of a Sperm Just Touching an Egg

Sperm:
(Note the flagellum as well as the size relative to the egg.)

Egg:
(This cell is so large compared to the sperm that we can only show a portion of it.)

Photo courtesy of
www.PDImages.com

Once the egg and sperm meet, the sperm burrows into the egg and the haploid cells fuse to form a diploid cell. This diploid cell, called the zygote, can then begin to develop and grow. To do this, the zygote must begin producing many cells, which it does by mitosis.

ON YOUR OWN

7.8 Which phases of meiosis are essentially the same as the corresponding phases of mitosis? Which are different?

Meosis II & prophase I is about the same wheros metaphase anaphase I & telophase I is different

7.9 A cellular reproduction process results in four diploid cells. Is this mitosis or meiosis? How many cells underwent this process?

mitosis 2

7.10 A cellular reproduction process results in four haploid cells. Is this mitosis or meiosis? How many cells underwent this process?

meiosis 1

7.11 At the end of telophase I, there are two cells. Can these cells function on their own as if they were any other cell? Why or why not?

No they cant function on their own as normal cells

7.12 A sperm cell finds a polar body and attempts to fuse with it. Will a zygote develop?

No it wont

Viruses

In Modules #2 and #3, we spent considerable time discussing bacteria and protistans that cause disease. Despite all of that discussion, however, we have not mentioned perhaps the most common agent of disease: the **virus**. We decided to put off such a discussion until now. This is because viruses cause disease by overriding the normal cellular reproduction process. As a result, it seems only natural that a discussion of viruses should occur in the module that discusses reproduction.

The definition of a virus is rather specific.

> Virus - A non-cellular infectious agent that has two characteristics:
> (1) It has genetic material inside a protective protein coat
> (2) It cannot reproduce itself

Based on this definition, then, a virus is not alive. Since it cannot reproduce itself, it fails to meet one of life's basic criteria. A virus also has no means of taking in nutrients and converting them into useable energy. Thus, it fails to meet yet another criterion for life.

Well, if a virus is not alive, what is it? The best way to answer this question is to show you a couple of viruses.

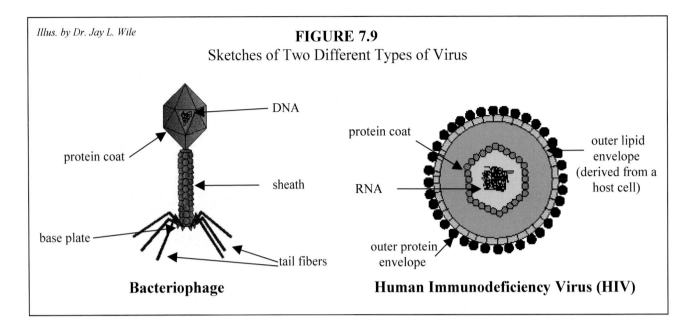

FIGURE 7.9
Sketches of Two Different Types of Virus

Illus. by Dr. Jay L. Wile

DNA

protein coat

sheath

base plate

tail fibers

Bacteriophage

protein coat

RNA

outer lipid
envelope
(derived from a
host cell)

outer protein
envelope

Human Immunodeficiency Virus (HIV)

Now don't let appearances deceive you here. Even though the bacteriophage looks like it has legs (the tail fibers), a body (the sheath), and a head (the protein coat), it is not a living organism. It is a highly organized mixture of chemicals that can perform some pretty specific tasks, nothing more. Notice from the figure that while the genetic material in the bacteriophage is DNA, the genetic material in the HIV is RNA. This is the case with viruses. Some have RNA as their genetic material, others have DNA.

The main thing you need to know about viruses is the way in which they infect their hosts. Since viruses cannot reproduce themselves, they rely on cells to do it for them. In order to do this, a virus will attach itself to a cell. The virus either enters the cell or injects its genetic material into the cell. The genetic material of the virus re-directs the cell's reproductive machinery to reproduce the DNA or RNA of the virus, as well as the proteins that make up the virus. The cell's biosynthetic machinery is then directed to assemble these pieces into new viruses. This continues until there are so many viruses that the cell ruptures, destroying the cell and releasing new viruses to infect other cells. This process is called the **lytic pathway**.

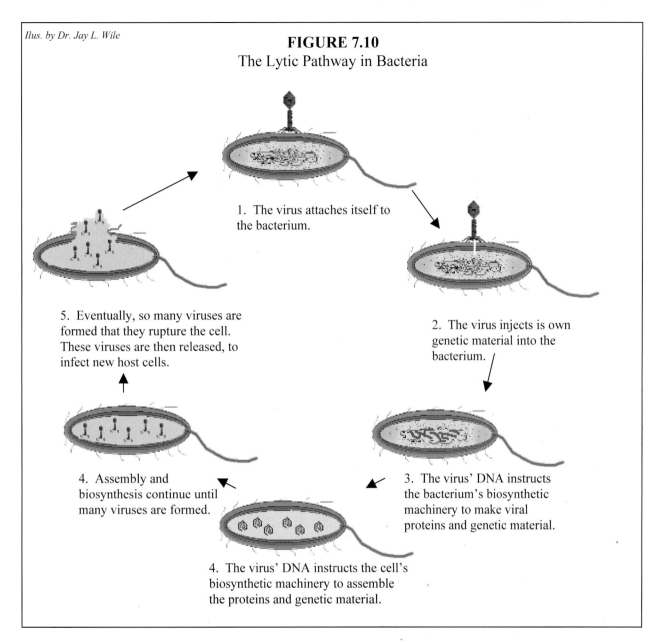

Ilus. by Dr. Jay L. Wile

FIGURE 7.10
The Lytic Pathway in Bacteria

1. The virus attaches itself to the bacterium.

5. Eventually, so many viruses are formed that they rupture the cell. These viruses are then released, to infect new host cells.

2. The virus injects is own genetic material into the bacterium.

4. Assembly and biosynthesis continue until many viruses are formed.

3. The virus' DNA instructs the bacterium's biosynthetic machinery to make viral proteins and genetic material.

4. The virus' DNA instructs the cell's biosynthetic machinery to assemble the proteins and genetic material.

Although Figure 7.10 is drawn for the specific case of a bacteriophage infecting a bacterium, all viruses reproduce and infect via the lytic pathway. Some viruses, however, can inject their genetic material into a cell so that it lies dormant for as long as several years before beginning the lytic pathway. Thus, the time between a host receiving the virus and manifesting the symptoms of the malady caused by that virus can be several years. This is the case with HIV, the virus currently thought to cause AIDS (acquired immune deficiency syndrome).

Viruses cause many diseases and afflictions. Warts, chicken pox, the common cold, influenza, several forms of cancer, mumps, measles, AIDS, and many other diseases are caused by viruses. Of course, viruses affect other organisms as well as humans. Plants can be killed by viral infections. As illustrated in Figure 7.10, even bacteria are subject to viruses!

In most organisms, there are infection-fighting agents that can destroy viruses and other pathogens. People, for example, have a host of infection fighting mechanisms in their bodies. **Phagocytic** (fag uh sih' tik) **cells** engulf and destroy viruses, bacteria, and other pathogens that invade the body. Remember from Module #6 that phagocytosis is the process by which cells engulf foreign substances or other cells. Well, there are certain cells that use this process to destroy pathogens, including viruses.

White blood cells, for example, are phagocytic cells. They typically live in the bloodstream but, when an infection strikes, they can leave the bloodstream and go to the point of infection in order to destroy the pathogen. Some phagocytic cells cannot move, however. When you have a cut that gets infected, it typically swells up, right? The reason that it swells is because extra body fluid is being added to that region. The extra body fluid tries to wash the pathogens into the **lymph** (limf) **nodes** in your body. These lymph nodes have phagocytic cells that cannot move. However, when a pathogen is carried to them by bodily fluid, then they can engulf the pathogen.

Besides phagocytic cells, humans also can produce **antibodies** to ward off pathogens.

Antibodies - Specialized proteins that aid in destroying infectious agents

While some antibodies can destroy a host of different pathogens, other antibodies are highly specific and can destroy only one type of pathogen. When your body is infected, the existing antibodies try to destroy the pathogen. If they cannot, then the body produces a specific antibody to try and ward off that specific pathogen. Once the body makes a specific antibody, it typically continues to produce the antibody, in case there is another infection.

This is the principle behind **vaccines**, the most common way that medical science can fight viruses.

Vaccine - A weakened or inactive version of a virus that stimulates the body's production of antibodies which can destroy the virus

When a person is infected by most viruses, the only way to rid the person of that virus is for the body to make a specific antibody which will destroy it. The body can produce antibodies against many, many different viruses. However, some virsues are so fast-acting that the body cannot produce antibodies before many viruses have completed the lytic pathway. By that time, the body is overwhelmed by the virus. As a result, the disease kills or permanently maims the person. When you are given a vaccine, a weakened or inactive version of the virus is injected into your body. This virus has been changed so that it cannot enter the lytic pathway. As a result, it simply "hangs out" until your body produces antibodies to destroy it. Once your body produces the antibodies, however, it never stops. Sometimes it slows down a bit, so for some vaccines you need a booster shot every now and again to keep your body actively making antibodies. That way, when you are exposed to the real virus, those antibodies specific to the virus can destroy it before it starts the lytic pathway, keeping you from contracting the disease.

So, in essence, a vaccine works by giving you a virus! Of course, since this altered form of the virus cannot enter the lytic pathway, it is obviously of no harm to you. However, your body doesn't know that, so it manufactures the antibodies to destroy the virus, thus protecting you if you come into contact with the real thing. This is why vaccines must be given *before* you get infected by the virus. Once infected, your body begins producing antibodies. The problem is, it might not do so fast enough to kill the viruses before they overwhelm the body. At that point, then, a vaccine will do you no good! Thus, vaccines act like a wall of defense. If you do not build the wall before the attackers come, you will be too late.

There is a movement afoot these days that says vaccines are bad for you. After all, proponents say, a vaccine injects a virus into your body! This is just like getting infected by the virus. In addition, the antibodies whose production it stimulates stay in your body forever. Thus, your body "remembers" the infection, and you will always have poorer health as a result. Nothing could be farther from the truth! Since the virus in a vaccine cannot enter the lytic pathway, it is completely safe. Thus, it is not like getting infected by the real thing. Also, there is nothing harmful about antibodies. Your body produces millions every day. Vaccines are truly one of the most wondrous inventions that God has allowed man to discover!

Now if you think about this for a moment, the concept of a vaccine is, in fact, a great testament to God's power and majesty. After 3,000 years of medical science, human beings cannot fight viruses as efficiently as God's creation (the human body) can. Thus, to protect ourselves, we simply stimulate the body to do what God designed it to do in the first place. In the case of vaccines then, human science gives the body a little push (infecting the body with an inactive or weakened virus), and the body does the rest (produces the antibodies). Three thousand years of medical science has confirmed that we are truly "fearfully and wonderfully made." (Psalm 139:14)

ON YOUR OWN

7.13 What is the principal difference between viruses and pathogenic bacteria?

Viruses aren't living

7.14 The human body can produce the antibodies which destroy smallpox. If this is the case, why did so many people die from it? Why didn't their bodies just kill the virus?

they couldn't produce antibodies fast enough

ANSWERS TO THE ON YOUR OWN QUESTIONS

7.1 <u>No, you would not expect them to be identical in every way. Because they have the same genes, they will have the same genetic tendencies and traits. Any characteristic that is determined completely by genetics (hair color, eye color, etc.) will be the same in each twin. However, any characteristics in which environmental and spiritual factors play a role (weight, personality, strength, etc.) will most likely be at least somewhat different in each, since they grew up in different environments.</u>

7.2 <u>The chromosome would unravel and fall apart.</u> Remember, the helix of DNA is wound around and supported by histones. Without the histones to wind up the DNA, the chromosome would not exist.

7.3 <u>Interphase is not a part of mitosis.</u> This phase of a cell's life is defined as the time between reproduction. As for the remaining phases, they occur in the following order: <u>prophase, metaphase, anaphase, telophase.</u>

7.4 <u>It must be a plant cell.</u> In animal cells, the two new cells are created when the plasma membrane constricts. This can't happen in a plant cell because of the rigid cell wall. Thus, plant cells use vesicles to carry the components of the plasma membrane and cell wall in between the two nuclei and build them there.

7.5 <u>Chromosomes are separated from their duplicates during anaphase.</u>

7.6 The haploid number is the number of homologous pairs. <u>Thus, the haploid number is 7</u>. The diploid number is the total number of chromosomes in the nucleus. Since there are 7 pairs and each pair has 2, there are a total of 14 chromosomes in the nucleus. <u>The diploid number, therefore, is 14</u>.

7.7 Since the number is odd, it cannot be a diploid number. Remember, diploid cells have chromosomes that come in pairs. Since there are two chromosomes in a pair, then the total number of chromosomes (the diploid number) will always be even. By virtue of the fact that this is an odd number, then, we know that <u>it is a haploid number</u>. Since the haploid number is 17, the diploid number will be twice that. Thus, <u>the diploid number is 34</u>.

7.8   Well, since meiosis II is essentially mitosis that occurs on two cells, the stages of meiosis II are all essentially the same as mitosis. Thus, <u>prophase II, metaphase II, anaphase II, and telophase II are essentially the same as the corresponding stages of mitosis. In addition, prophase I is essentially the same as the prophase of mitosis. Metaphase I is different than the metaphase of mitosis</u>, because the microtubules attach so as to separate homologues in metaphase II. In the metaphase of mitosis, the microtubules attach so as to pull the chromosome duplicates apart from their originals. <u>Anaphase I is also different than the anaphase of mitosis</u>, because in anaphase I the homologues get pulled apart, whereas in the anaphase of mitosis, the chromosome duplicates get separated from their originals. Finally, <u>telophase I is different than the telophase of mitosis because the former results in haploid cells while the latter results in diploid cells.</u>

7.9 <u>This must be mitosis</u>. Meiosis always results in haploid cells. Since mitosis makes two cells for every one that goes through the process, <u>two cells must have gone through mitosis to produce 4 cells in the end</u>.

7.10 <u>This was meiosis</u>. Mitosis results in diploid cells. Since meiosis forms four cells for every one that undergoes the process, <u>1 cell must have undergone meiosis to get 4 cells in the end</u>.

7.11 <u>These cells cannot function on their own as normal cells</u>. The cells are only haploid, so they contain only half of the necessary information to perform life functions.

7.12 <u>A zygote will not develop</u>. Polar bodies are so small that they cannot function properly.

7.13 <u>A pathogenic bacterium is alive; a virus is not</u>.

7.14 <u>Even though the body can produce the antibodies, it cannot do so fast enough to destroy the virus before many viruses reach the lytic pathway. As a result, the virus is reproduced so quickly that it overwhelms the body</u>. The smallpox vaccine defended so well against the virus, however, the smallpox virus has been wiped out. It now only exists in laboratories and, as a result, this vaccine is no longer necessary!

STUDY GUIDE FOR MODULE #7

1. Define the following terms:

    a. Genetics
    b. Genetic factors
    c. Environmental factors
    d. Spiritual factors
    e. Gene
    f. Chromosome
    g. Mitosis
    h. Interphase
    i. Centromere
    j. Mother cell
    k. Karyotype
    l. Diploid cell
    m. Haploid cells
    n. Diploid chromosome number
    o. Haploid chromosome number
    p. Meiosis
    q. Gametes
    r. Virus
    s. Antibodies
    t. Vaccine

2. Suppose scientists determine that a set of genes are significantly more prevalent in murderers than in the population at large. Would that mean that murderers are not at fault for what they do? Why or why not?

3. Why do proteins have to coat the DNA in chromosomes?

4. List (in order) the four stages of mitosis.

5. Identify the stage of mitosis for each of the following pictures :

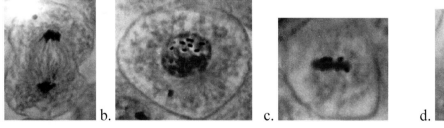

a.   b.   c.   d.

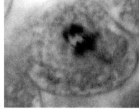

Photos by Kathleen J. Wile

6. The diploid number of a cell is 16. What is its haploid number?

7. The haploid number of a cell is 9. What is its diploid number?

8. What is the difference between a gamete and a normal cell?

9. List (in order) all of the stages of meiosis.

10. Which is closer to mitosis: meiosis I or meiosis II?

11. A single cell with 7 pairs of homologous chromosomes goes through meiosis I. How many cells result at the end of meiosis I? How many (total) chromosomes exist in each cell? Are the chromosomes in each cell duplicated or not?

12. Two cells that originally (prior to meiosis I) had 7 pairs of homologous chromosomes go through meiosis II. How many cells result? How many (total) chromosomes exist in each cell? Are the chromosomes in each cell duplicated or not?

13. What are gametes produced in animal males called? What are gametes produced in animal females called?

14. How many useful gametes are produced in the meiosis of animal males? What about animal females?

15. What is the difference between a polar body and an egg?

16. Which gamete can move on its own: the male gamete or the female gamete?

17. What is the purpose of the lytic pathway?

18. If a virus uses DNA as its genetic material, is it alive? Why or why not?

19. A person decides to wait until he contracts measles before getting the vaccine. What is wrong with this strategy?

20. What is it that makes the virus in a vaccine harmless?

Module #8: Genetics

Introduction

As we mentioned in the last module, the means by which an offspring receives traits and characteristics from its parents has always intrigued scientists. Now that you have learned how the genetic material of a cell gets split into gametes for the purpose of sexual reproduction, you are ready to learn the mechanisms of this fascinating process. Before we get into the nuts and bolts of genetics, however, we need to spend some time discussing the founder of modern genetic science: Gregor Mendel.

Gregor Mendel

Named Johann Mendel at birth, Gregor Mendel was born a peasant in Heinzendorf, Austria. His father was a farmer, and he taught his son the techniques of animal breeding and plant grafting. These pursuits intrigued young Johann, and later in life, he spent a great deal of time studying them as no one had done before. While still a child, his talent at learning impressed his school teacher so much that the teacher implored his parents to give him a higher education. At that time, a "higher education" really just meant our equivalent of high school. His parents agreed, but they were so poor that they could not properly afford an education. Johann struggled through his education, oftentimes nearly dying of starvation because he could not afford to eat. Eventually, his younger sister gave up her dowry so that he could properly finish his education.

Once he graduated, he spent some time trying to be a teacher. When he proved completely unsuccessful at that, he joined the Augustinian monastery of St. Thomas in Altbrun. You see, back then monasteries were a center for scholarly pursuits. This particular one had a great reputation for scientific achievement, so Mendel was happy to join. When a monk joins the Augustinian order, he must choose a new name. That was when Johann Mendel became Gregor Mendel.

While he was at the monastery, Gregor tried again to be a teacher. When he took the examinations to qualify himself, however, he failed. Nevertheless, one of his examiners, Andreas Baumgartner, was impressed by his originality of thought, and he pulled some strings to get Gregor admitted to the University of Vienna. While at the university, Mendel studied under Johann Christian Doppler, a physicist who became famous for an acoustical effect now known as the "Doppler effect." When you take physics, you will learn all about that. Doppler taught Mendel the proper way to conduct experiments, and that's all that Mendel needed.

After finishing his studies at the university, he once again tried to take the examination to qualify for a teaching position. His examiners failed him a second time, however. This time, he was failed for "showing too much original thought." Giving up the idea of being a teacher, he began his work as a scientist. For eight years, he conducted experiments on that which first intrigued him: breeding. He raised thousands of pea plants and carefully documented the results of breeding and cross-breeding them. We will learn much about these experiments in the next section. At the end of those eight years, he published a paper in which he presented a series of four conclusions that we now call Mendelian genetics. We will also learn about these conclusions in the next section.

Unfortunately, his paper went largely unnoticed among the scientific community. In addition, Mendel had to give up his scientific endeavors shortly after writing the paper, because he became embroiled in a political controversy. The government decided that it would tax the monasteries, and Mendel thought that this was an attack on religious freedom. As a deeply committed Christian, he bitterly opposed the government's idea and spent most of the rest of his life fighting against the taxation. Since he was so involved in this political struggle, he had no time for his scientific work. When he died in 1884, no one knew the significance of his experiments. Nevertheless, by 1935, his work was well-known throughout the scientific community, and today he is known as the father of modern genetics.

Gregor Mendel's life story is inspiring on several levels. It shows what can happen when a person has a true desire to learn. Mendel was willing to sacrifice *eating* in order to pursue an education. As a result, he unlocked one of the deep mysteries of God's Creation. Mendel's story also shows that when you fail, you should not give up. Mendel failed his teaching examination twice; nevertheless, he taught the worldwide scientific community about a process that had fascinated biologists since the beginning of scientific inquiry. Finally, his willingness to put all of that away in order to defend the faith against an attack from the government shows that Mendel had the proper set of priorities. He put his faith in Christ above everything!

Mendel's Experiments

During his eight years of scientific work, Mendel studied the breeding of pea plants. He noticed that pea plants had certain definable characteristics that seemed to change from plant to plant. For example, some pea plants were tall (about six feet in height) and others were short (about 2 feet high). In some pea plants, the flowers grew along the sides of the plant (these are called "axial flowers"), while in others, the flowers grew on top of the plant (these are called "terminal flowers"). Sometimes, the pea pods were green and other times they were yellow. In the same way, some pea plants produced yellow peas and some produced green ones. Finally, sometimes the peas were wrinkled and sometimes they were smooth.

Mendel noticed that some plants bred so as to produce offspring with the same characteristic. For example, some tall plants would always give rise to other tall plants. When this happens, we say that the tall plant has **bred true**.

<u>True breeding</u> - If an organism has a certain characteristic that is always passed on to all of its offspring, we say that this organism bred true with respect to that characteristic.

Mendel noticed that not all plants bred true. Some tall plants would produce a short plant every now and again. Mendel decided that the tendency for some plants to breed true while others do not was key to understanding the mysteries of reproduction.

With this thought in mind, Mendel came up with an ingenious set of experiments. He took a tall pea plant that always bred true and allowed it to sexually reproduce with a short pea plant that also always bred true. No matter how many times he did this, the offspring of such a union were always tall plants. Now most people might think that the sexual reproduction of a tall plant and a short plant would result in a medium-sized plant. This, however, was not the case. Mendel noticed that with the other definable characteristics listed above, the outcome was similar. Figure 8.1 compiles the results of this portion of Mendel's work.

Figure 8.1

Illus. by Dr. Jay L. Wile

Mendel's Initial Experiments

Plants Bred	Result
  Tall plants bred with Short plants.	 All offspring were tall plants.
  Axial-flowered plants bred with terminal-flowered plants.	 All offspring had axial flowers.
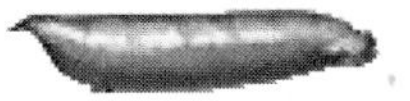  Green-pod plants bred with yellow-pod plants.	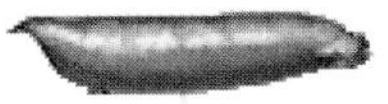 All offspring produced green pods.
 Yellow-pea plants bred with green-pea plants.	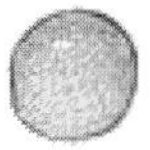 All offspring produced yellow peas.
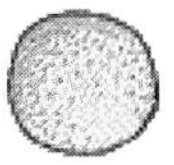  Smooth-pea plants bred with wrinkled-pea plants.	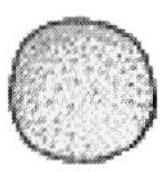 All offspring produced smooth peas.

Notice that in each experiment shown in Figure 8.1, the definable traits did not mix. A short plant bred with a tall plant did not produce a medium-sized plant. Instead, a tall plant was always produced. If a plant that produced smooth peas was bred with a plant that produced wrinkled peas, the result was not a plant that produced partly-smooth, partly-wrinkled peas. Instead, such a union would always result in a plant that produced smooth peas.

One interesting fact about pea plants (and many other plants as well) is that they can be self-bred. In other words, a pea plant can actually sexually reproduce with itself! Now the details of how this happens will be left to a discussion of plants in an upcoming module. For right now, just accept this startling revelation and realize that the process of a plant sexually reproducing with itself is called **self-pollination**. Mendel extended his studies by using this property of plants. He decided to see what would happen when the offspring produced in the experiments described in Figure 8.1 were self-pollinated. When he tried this, an amazing thing happened. The majority of the offspring produced had the same definable trait as the parent. However, a minority of the offspring had the other trait. Thus, if the tall plants produced in the first experiment were self-pollinated, 75% of the offspring produced were tall; however, 25% were short. These results are summarized in Figure 8.2.

FIGURE 8.2
Mendel's Second Set of Experiments

Illus. by Dr. Jay L. Wile

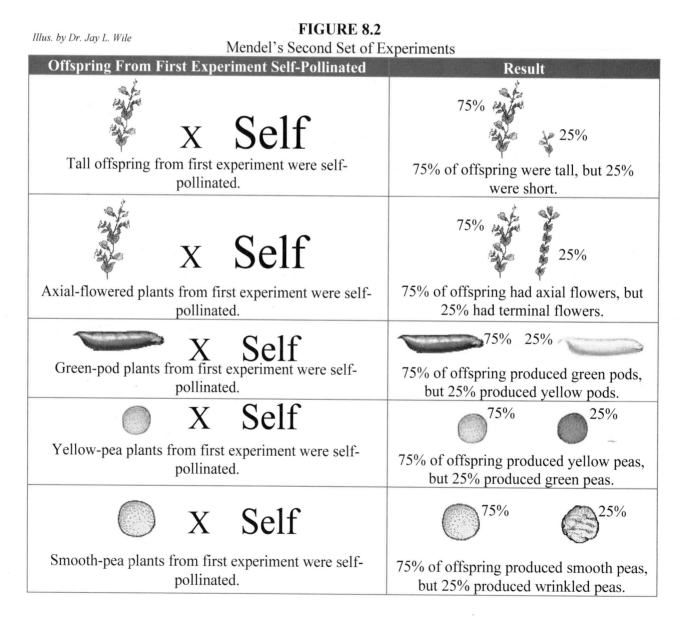

Offspring From First Experiment Self-Pollinated	Result
x Self Tall offspring from first experiment were self-pollinated.	75% 25% 75% of offspring were tall, but 25% were short.
x Self Axial-flowered plants from first experiment were self-pollinated.	75% 25% 75% of offspring had axial flowers, but 25% had terminal flowers.
X Self Green-pod plants from first experiment were self-pollinated.	75% 25% 75% of offspring produced green pods, but 25% produced yellow pods.
X Self Yellow-pea plants from first experiment were self-pollinated.	75% 25% 75% of offspring produced yellow peas, but 25% produced green peas.
X Self Smooth-pea plants from first experiment were self-pollinated.	75% 25% 75% of offspring produced smooth peas, but 25% produced wrinkled peas.

These two sets of experiments led Mendel to develop four principles of genetics.

1. **The traits of an organism are determined by packets of information called "factors."**

2. **Each organism has not one, but two factors that determine its traits.**

3. **In sexual reproduction, each parent contributes ONLY ONE of its factors to the offspring.**

4. **In each definable trait, there is a dominant factor. If it exists in an organism, the trait determined by that dominant factor will be expressed.**

How do these principles help explain the data that Mendel collected? Well, Mendel assumed that if a pea plant always bred true, then it must have two factors that were identical. In other words, if a tall pea plant always produced tall offspring, it must have two factors that correspond to the trait of being tall. A short plant that always produced short plants must have two factors that correspond to the trait of being short. Since each parent had two of the same factor, they always contributed that factor to the offspring. Thus, the offspring produced when a true-bred short plant was bred with a true-bred tall plant would always have one factor that corresponded to being tall and one factor that corresponded to being short. By further assuming that the factor corresponding to being tall was dominant, then each offspring would be tall, because each offspring had that dominant factor. This kind of reasoning explained all of Mendel's observations in his first set of experiments.

We are virtually certain that the previous paragraph was rather confusing to you, so we now want to explain it in a different way. Suppose we represented these factors with letters. Let's say that a capital "T" represents the factor that tells a plant it will grow tall. Further, let's say that a small "t" represents the factor that tells a plant it will stay short. Why did we choose a capital "T" for tall and a small "t" for short? Well, we are going to assume that the factor for tallness is dominant. The capital "T" will help us keep that in mind. Now, if we take a true-bred tall pea plant, it must have two factors that both tell it to grow tall. Thus, we could represent its factors as "TT." The true-bred short plant must have two factors telling it to stay short. Thus, we could represent its factors as "tt." Now suppose these two plants were bred and produced four offspring. What would the factors in their offspring look like? Study Figure 8.3.

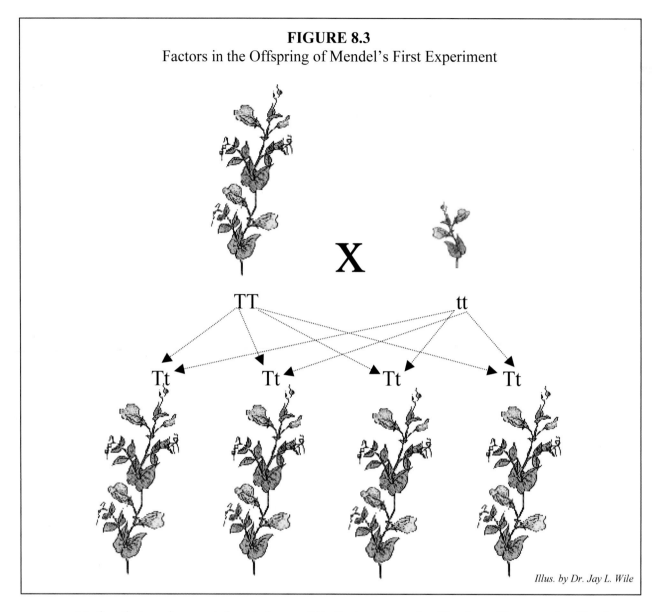

FIGURE 8.3
Factors in the Offspring of Mendel's First Experiment

Illus. by Dr. Jay L. Wile

Notice that each parent donated one of its factors to each offspring. Thus, each offspring received a "T" factor from the tall parent and a "t" factor from the short parent. Their factors, then, are all "Tt." Well, since the factor for tallness (T) is dominant, then that trait will be expressed in every offspring that has it. Since each offspring has a dominant factor, each offspring will be tall. Now you should see why Mendel's principles explain the data in the first experiment.

What about the data in the second experiment? Well, in that experiment, each of the offspring represented in Figure 8.3 was self-pollinated. If one of the offspring in Figure 8.3 is self-pollinated, it is like having two parents, each with the same set of factors ("Tt"). Thus, we now get the situation shown in Figure 8.4.

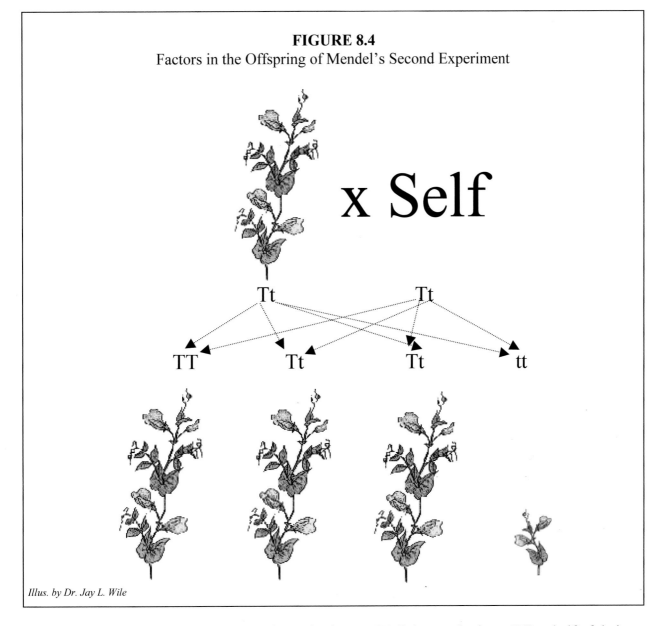

FIGURE 8.4
Factors in the Offspring of Mendel's Second Experiment

x Self

Illus. by Dr. Jay L. Wile

In this case, since both parents have the factors "Tt," they each give a "T" to half of their offspring and a "t" to the other half. As a result, one of their offspring has "TT" as its factors. Obviously, then, this plant will be tall. Two of the remaining offspring have "Tt" as their factors. Since the "T" is dominant, then they will be tall as well. The final offspring, however, inherited a "t" from each parent. As a result, it has the factors "tt." Thus, this plant will be short, because it has only the factors that tell it to stay short, and no dominant factor telling it to grow tall.

Now do you see how Mendel's fours principles can explain all of his experiments? Despite the fact that we've learned a great deal more about genetics since Mendel's time, these four principals are still the foundation of the science. This should give you a strong appreciation for Mendel's creativity and scientific ability. Without the aid of microscope, chemical analysis, or electronic instrumentation, he was able to devise and analyze experiments so as to form the foundation of a science that has been growing for more than one hundred years since his death!

ON YOUR OWN

8.1 The factor for producing smooth peas (we will call it "S") is dominant over the factor for producing wrinkled peas (which we will call "s"). Three plants (a, b, and c) have the following factors. Determine whether they will produce smooth peas or wrinkled peas.

          a. ss   b. Ss   c. SS

8.2 The factor for producing yellow peas ("Y") is dominant over the factor for producing green peas ("y"). Suppose a pea plant produces yellow peas. What possible combination(s) of factors can it have?

8.3 What possible combination(s) of factors are possible for a pea plant that produces green peas?

Updating the Terminology

        As we said before, Mendel's work is still used today as the foundation of genetics. Nevertheless, we have learned a great deal about this fascinating science since Mendel published his work, so we need to update the terminology a little bit. For example, biologists do not use the term "factor" any more. We now know that the packets of information that Mendel envisioned are genes. In addition, since microscopic studies have shown us that most animals have homologous pairs of chromosomes, we know genes come in pairs, with one gene on each homologous chromosome. Each gene that makes up one of these pairs is called an **allele** (uh leel').

Allele - One of a pair of genes that occupies the same position on homologous chromosomes

Thus, Mendel's "factors" are now called alleles. Each definable trait has two alleles that help determine it, and they can each be represented by a letter.

        When we put two alleles together ("Tt" for example), we say that we are describing a **genotype** (jee' nuh tipe).

Genotype - Two-letter set that represents the alleles an organism possesses for a certain trait

When we presented the offspring in Figure 8.3, the "Tt" we used to represent them is called their genotype. So by presenting the genotype of an organism, we are telling you what alleles an organism has.

        Now we learned in Figure 8.4 that different genotypes can result in apparently identical organisms. After all, in Figure 8.4, one of the offspring had "TT" as its genotype and two others had "Tt" as their genotype. Nevertheless, all three of those plants were of exactly the same height. We determined that was the case since the "T" allele was dominant. Thus, if a plant had just one "T" in its genotype, then it would be tall. Well, the *expression* of an organism's genotype is called the **phenotype** (fee' nuh tipe).

> Phenotype - The observable expression of an organism's genes

Thus, there are two phenotypes possible in Figure 8.4: tall and short. That's how the genes studied in Figure 8.4 are expressed. The genotypes "TT" and "Tt" end up resulting in the same phenotype: tall. The genotype "tt" results in the other phenotype: short. We can therefore say that a phenotype is the outward expression of an organism's genotype.

Since the genotype "TT" and "Tt" result in the same phenotype (tall), we have developed terminology to separate these two genotypes. When a genotype is composed of identical alleles ("TT" or "tt", for example), then we say that the genotype is **homozygous** (ho muh zy' gus). When a genotype has mixed alleles ("Tt", for example), we say that it is **heterozygous** (het uh roh zy' gus).

> Homozygous genotype - A genotype in which both alleles are identical

> Heterozygous genotype - A genotype with two different alleles

We need to clarify two things before we go any further. First of all, there is no such thing as a "tT" genotype. This is because we have no way of knowing which chromosome is which in a homologous pair of chromosomes. Thus, in a heterozygous genotype, we always write the dominant allele first. Secondly, if we call the "T" allele dominant, what do we call the "t" allele? We call it **recessive**.

> Dominant allele - An allele that will determine phenotype if even one is present in the genotype

> Recessive allele - An allele that will not determine the phenotype unless the genotype is homozygous with that allele

Now we realize that this is probably a lot of terminology to throw at you in one section, but it is important that you master it before you go any further. Therefore, look over the definitions again and make sure that you understand them. Then, to help you familiarize yourself with the terminology some more, read the following, which re-states Mendel's principles in modern terminology:

A RESTATEMENT OF MENDEL'S PRINCIPLES

1. **The traits of an organism are determined by its genes.**

2. **Each organism has two alleles that make up the genotype of a given trait.**

3. **In sexual reproduction, each parent contributes ONLY ONE of its alleles to the offspring.**

4. **In each genotype, there is a dominant allele. If it exists in an organism, the phenotype is determined by that allele.**

Now that you know Mendel's principals using updated terminology, you need not remember the ones using his older terminology. We will refer only to these from now on.

Before we finish this section, we need to make sure that you understand how the information you have learned in this module relates to meiosis, which you learned about in the last module. Remember, the reason animals have two alleles for each genetic trait is because they have diploid cells. Thus, they have homologous pairs of chromosomes. In sexual reproduction, however, each parent contributes only one allele to the offspring. This is why meiosis takes diploid cells and makes them haploid. The process of meiosis separates the homologous pairs, separating the alleles from each other. Thus, each gamete produced has only one allele for each trait. When the male gamete (sperm) fuses with the female gamete (egg) and fertilization takes place, the resulting zygote then has two alleles: one from the father and one from the mother.

So the process of meiosis separates the alleles in the parents, so that each parent contributes only one allele. If this is the case, then, how do we know which allele gets donated? Well, it all depends on which gamete fertilizes which. Remember, the male makes four functional sperm for each meiosis. Thus, two of the sperm have one of the male's two alleles and the other two have the other allele. Thus, if the male's genotype for a particular trait is "Tt," then two sperm will have the "T" allele and two will have the "t" allele. For the female, only one functional egg is produced. The allele that it has depends completely on which cell in meiosis I got the majority of the cytoplasm. For a heterozygous female, then, whether or not the single, functional egg has a "T" or "t" allele depends entirely on which cell got the most cytoplasm in meiosis I. The process which determines which cell gets the majority of cytoplasm is completely random; thus, half of all meiosis processes in a heterozygous female will produce an egg with a "T" allele while the other half will produce an egg with a "t" allele.

In the end, then, the genotype of the offspring is determined by which sperm fertilizes which egg. If a "T" sperm fertilizes a "t" egg, then the genotype of the offspring is "Tt." Similarly, if a "t" sperm fertilizes a "T" egg, the genotype of the offspring will once again be "Tt." On the other hand, if a "T" sperm fertilizes a "T" egg, the genotype of the offspring will be "TT," and a "t" sperm fertilizing a "t" egg will result in a "tt" offspring. Thus, the fact that each parent contributes only one allele is a result of meiosis. The final result of which allele gets contributed, however, is determined by which gamete participates in fertilization.

ON YOUR OWN

8.4 A student repeats one of Mendel's experiments. He self-pollinates a "Ss" pea plant, where "S" is the dominant allele for smooth peas and "s" is the recessive allele for wrinkled peas. The result is one "SS" offspring, two "Ss" offspring, and one "ss" offspring. For each of these three sets of offspring, list their genotype, whether they are homozygous or heterozygous, and their phenotype.

Punnett Squares

Since the genotype of an offspring depends on which gamete from the mother and which gamete from the father ends up participating in the fertilization process, the traits of the offspring can never be determined with absolute certainty. However, if we concentrate on a single, definable trait, we can determine the percent chance that the offspring has for possessing a certain genotype and corresponding phenotype. For example, if you look back at Figure 8.4, you will see that out of four offspring, 1 had the "TT" genotype (resulting in a tall phenotype), two had the "Tt" genotype (also resulting in a tall phenotype) and one had the "tt" genotype (resulting in the small phenotype). Thus, the chance of an offspring having the "TT" genotype was 25% (one in four). The chance of an offspring having a "Tt" genotype was 50% (two in four), and the chance for a "tt" genotype was 25% (one in four). Since these genotypes resulted in three tall plants and one small plant, the chance for a tall phenotype was 75% (3 in 4), and the chance for a small phenotype was 25% (one in four).

It turns out that for any single, definable trait, we can predict the likelihood of offspring with a given genotype and phenotype. We do this using **Punnett squares**. When we construct a Punnett square, we take the genotype of one parent and the genotype of another and place them on the top and side of a grid. This allows us to predict all possible genotypes of the offspring. For example, another experiment that Mendel performed dealt with the pea plant's flower color. In the pea plant, the allele for a flower color of purple is dominant (we will call it "P") and the allele for a flower color of white ("p") is recessive. In the first experiment, Mendel bred a homozygous purple-flowered pea plant ("PP") with a homozygous white-flowered pea plant ("pp"). Using Punnett squares, we can predict what possible genotypes and phenotypes will be produced. We illustrate this in Example 8.1.

EXAMPLE 8.1

A homozygous, purple-flowered pea plant is bred with a homozygous, white-flowered pea plant. If the allele for purple is dominant, what possible genotypes and phenotypes will be produced? What is the percent chance for each?

Since we are told that the allele for a purple flower is dominant, we can use "P" for that allele and "p" for the allele that causes white flowers. The genotypes of the parents (since they are homozygous), then, are "PP" and "pp." To construct a Punnett square, we take one parent's genotype and place each allele at the top of a 2x2 grid. We then take the other parent's genotype and place each allele to the left of the grid:

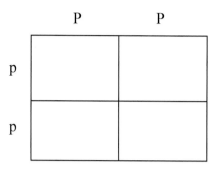

Now, since each parent can donate either allele to the offspring, all possible genotypes are determined by taking the allele to the left of a row and adding it to the allele on top of the column. That two-letter sequence goes in the appropriate box, giving you a possible genotype. in this case, the result is:

	P	P
p	Pp	Pp
p	Pp	Pp

Since there are four boxes, there are four possible genotypes. In this case, they are all "Pp," resulting in a purple phenotype. Thus, the chance for genotype is <u>100% "Pp"</u>, and the phenotype chance is <u>100% purple flowers.</u>

Now that was probably not very informative, since all offspring had the same genotype. Therefore, let's look at a more complex example. In this example, we use a term that you will see a lot in genetics. When we breed two individuals, we often call it a "cross." Thus, we will sometimes say that we are "crossing" two individuals instead of "breeding."

EXAMPLE 8.2

A heterozygous purple-flowered pea plant is crossed with a heterozygous purple-flowered pea plant. What are the possible phenotypes and genotypes, along with the percentage chance of each?

Since we have already been told that the allele for a purple flower is dominant, we can use "P" for that allele and "p" for the allele that causes white flowers. Since the parents are heterozygous, they must have one of each. Their genotypes, then, are both "Pp." The Punnett square we initially set up will therefore look like this:

Now, since each parent can donate either allele to the offspring, all possible genotypes are determined by taking the allele to the left of a row and adding it to the allele on top of the column. That two-letter sequence goes in the appropriate box, giving you a possible genotype. in this case, the result is:

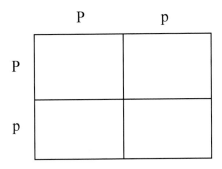

Based on this, then, we see three different genotypes. <u>The "PP" genotype has a 25% chance of occurring</u>, since it appears in one of the four boxes. <u>The "Pp" genotype has a 50% chance of occurring</u>, since it appears in two of the four boxes. <u>The "pp" genotype has a 25% chance of occurring</u>, since it appears in one of the four boxes. Since any offspring that has even one "P" allele will have purple flowers, and that happens in three of the four boxes, <u>the purple flower phenotype has a 75% chance of occurring</u>. The only white-flowered pea plant will be the one with genotype "pp," which happens in one of the four boxes. <u>The white-flowered phenotype, then, has a 25% chance of occurring</u>.

ON YOUR OWN

8.5 Using Punnett squares, predict the possible genotypes and phenotypes, along with the percentage chance for each, when a heterozygous, purple-flowered pea plant is crossed with a homozygous, white-flowered pea plant.

Pedigrees

Let's suppose we want to study an organism but we really don't know much about its genotype. Is there any way that we can find this out? Yes, it turns out that in many cases, you can determine the genotype of a organism if you construct a **pedigree**.

Pedigree - A diagram that follows a particular species' phenotype through several generations

Figure 8.5 is an example of a pedigree.

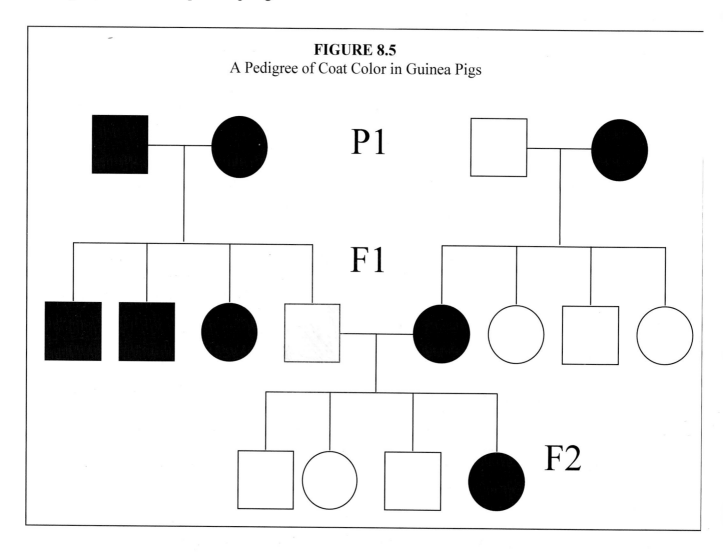

FIGURE 8.5
A Pedigree of Coat Color in Guinea Pigs

In this pedigree, squares represent male guinea pigs and circles represent females. In addition, a filled-in square or circle indicates that the guinea pig has a black coat and a hollow square or circle indicates a white coat. When a circle and square are linked by a horizontal line, it indicates that they have been bred. The symbols that "hang" off of that solid line are the offspring of the breeding. So, in Figure 8.5, on the left, a black-coated male and a black-coated female are bred and produce four offspring: two black coated males, a black-coated female, and a white-coated male. The white-coated male was then bred with a black-coated female from another breeding. The letters "P1," "F1," and "F2" refer to the generations. "P1" indicated the initial parents. "F1" represents their offspring, and "F2" represents their offspring's offspring. In this pedigree, then, we are following the phenotype of coat color through three generations. If more generations were present, they would be labeled "F3," "F4," etc. That's what we mean when we say that a pedigree follows a particular species' phenotype through several generations.

What can we learn through such a pedigree? The first thing we learn is that the allele which determines a black coat must be the dominant allele. Why? Think about it. In the cross between the black-coated male and black-coated female, a white-coated male was produced. This means that the allele for a white coat *must* be present in the parents. However, the parents are both black-coated. According to the principles of Mendelian genetics, the only way that a black-coated guinea pig can have an allele for a white coat is if the allele for white coat is recessive! We will therefore call it "b." This means that the allele for black coat is dominant. We will therefore call it "B." In addition, since we know that the allele for a white coat is recessive, then each parent must have one. After all, for a recessive allele to be expressed, the organism must have two of that allele. Thus, since one of the offspring has two white-coat alleles, one must have come from each parent. Each parent, therefore, has an allele for a white coat. Nevertheless, each parent has a black coat. This means that each parent also must have an allele for a black coat. Thus, by looking at this pedigree, we can already say that the black-coated female and black-coated male in the first cross both have a genotype of "Bb." By just looking at the pedigree, we have determined the genotype of the first two parents!

We can also determine the genotypes of the other parents in P1. After all, the white-coated male must be "bb," that's the only way it can be white. Also, since some of their offspring are white, this indicates that the female parent must also have the recessive allele. This indicates that the black-coated female must be "Bb." Do you see how that works? By just following the phenotype through a generation, we have already determined the genotype of the original parents!

Now you should notice something that looks wrong in the figure. If the P1 parents on the right-hand side of the figure (the white-coated male and black-coated female) have genotypes "bb" and "Bb" as we just determined, you can work out the Punnett square to determine that the probability of a black-coated phenotype is 50% and the probability of the white-coated phenotype is 50%. Thus, you might think that there should be two white-coated offspring and two black-coated offspring. Why are there three of one and one of the other?

Remember, just because the *probability* of having a white coat is 2 out of 4 (50%), it doesn't mean that two out of *every* 4 offspring will be white-coated. It only means that, if you have lots and lots of offspring, on average, 2 out of 4 will be white-coated. Nevertheless, for only a few offspring, the probabilities might not exactly work out. After all, if you flip a coin, the odds are 50% that you will get heads and 50% that you will get tails. This doesn't mean that *for every two times you flip* it one time will give you heads and the next will give you tails. Indeed, if you take a few minutes to flip a coin right now, you might get 2 or 3 heads in a row before getting a tail. If, however, you flipped the coin 1,000 times, you probably would get a total of 500 heads and 500 tails. The same thing holds true in pedigrees. You can't expect the probabilities to work out exactly each time!

EXAMPLE 8.3

The following pedigree represents experiments involving the eye color of a certain animal. The only possible eye colors for this animal are blue and brown. A shaded square or circle indicates the brown eye color, while a hollow circle or square indicates the blue eye color. What are the genotypes of individuals 1-4 in this pedigree?

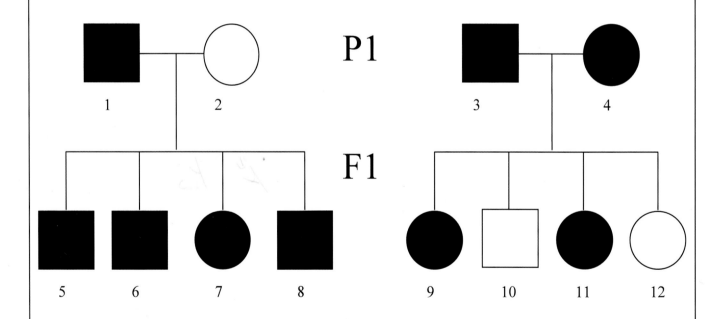

First of all, as we look at this pedigree, we realize that blue eye color (hollow circles and squares) must be the recessive allele. We know that because even though 3 and 4 both have brown eyes, some of their offspring are blue-eyed. Therefore, they both must have a blue-eyed allele. Since they must have a blue-eyed allele but are nevertheless brown-eyed, then the brown eye allele must be dominant. Therefore, we will call the brown eye allele "B" and the blue eye allele "b." Right away, then, we know that the genotype of 2 is "bb." After all, the only way you will express the phenotype of a recessive allele is if you have two of those alleles. Also, since 3 and 4 (who are both brown-eyed) must have a recessive allele to produce blue-eyed offspring, they must both be "Bb." What about 1? Well, when 1 and 2 breed, there are no blue-eyed offspring, despite the fact that 2 has both blue eye alleles. This indicates that 1 must always give the dominant allele. Thus, it is probably "BB." In the end, then, 1 is "BB," 2 is "bb," 3 is "Bb," and 4 is "Bb."

ON YOUR OWN

8.6 In the following pedigree, a biologist studies the presence of a tail on a certain species of animal. Some individuals have a tail (represented by the filled circles and squares) and others do not (represented by the hollow circles and squares). Which allele is dominant, the one for having a tail or the one for not having a tail? What are the genotypes of individuals 1-4?

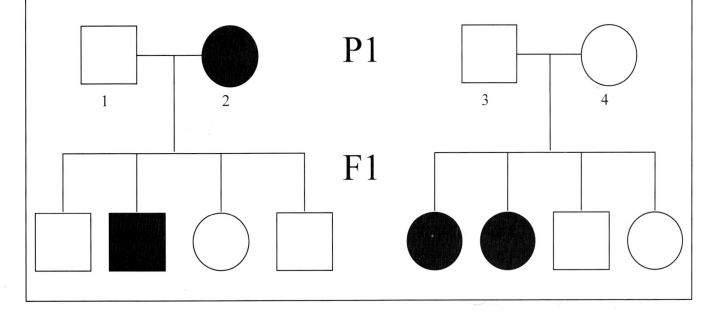

More Complex Genetic Crosses

The genetic crosses that you have just done are relatively simple because they concentrate on only one trait. When we are looking at the sexual reproduction of two individuals and concentrate on only one trait, we say that we are doing a **monohybrid cross**.

Monohybrid cross - A cross between two individuals concentrating on only one definable trait

More interesting situations develop, however, when we look at a **dihybrid cross**.

Dihybrid cross - A cross between two individuals concentrating on two definable traits

For example, in the pea plants that we discussed in Examples 8.1 and 8.2, we looked only at flower color. As we know, however, a pea plant has many more traits. For example, it can be either short or tall, it can produce smooth peas or wrinkled peas, it can produce green or yellow peas, etc. What if we were to look at two of these traits simultaneously? With a little more work, we can predict the possible results when a tall, purple-flowered pea plant is crossed with a short, white-flowered pea plant.

How do we do this? Well, it involves making a Punnett square, but the size of that Punnett square depends on the genotypes present in each parent. You see, when we make a Punnett square, we need to know the alleles that will be present in all gametes produced by each parent. The alleles in each possible gamete from one parent will be listed at the top of the columns in the square, and the alleles in each possible gamete from the other parent will be listed at the left of rows in the square. In order to determine the gametes from each parent, however, we need to think a little more about meiosis.

In meiosis, each allele is separated. Half of the gametes end up with one allele, and half end up with the other. When we are talking about an individual trait (flower color, for example), the possibilities of what allele goes into what gamete are limited. If an organism has the "PP" genotype, for example, there is only one possible allele for each gamete, the "P" allele. If an organism has the "Pp" genotype, then half of the gametes will have the "P" allele and the other half will have the "p" allele. Rather simple, huh?

When we start considering more traits, however, determining all possible gametes gets more complicated. Suppose, for example, a pea plant has the "Pp" genotype for flower color and the "Tt" genotype for height. What are the possible allele combinations for the gametes produced by meiosis? Well, when the alleles split in meiosis, the "P" allele for flower color might go into the same gamete as the "T" allele for height. Thus, the gamete would be *PT*. We italicize this combination of letters to make it clear that this is not a genotype. It is simply a list of two of the alleles that exist in a gamete. If the "P" allele for flower color went to the gamete that got the "t" allele for height, then the resulting gamete would be *Pt*. Since which allele goes where is completely random in meiosis, both of these combinations are possible! Furthermore, the "p" allele for flower color might go into the gamete that got the "T" allele for height. This would result in a gamete that has *pT*. Finally, if the "p" allele for flower color went to the gamete that got the "t" allele for height, the result would be a *pt* gamete.

Now we realize that this might be a little confusing, but stay with us here. Since all of these combinations of alleles are possible in a gamete produced by this plant, we must include all of them in our Punnett square. Thus, the Punnett square will have the following columns: *PT, Pt, pT, pt*. Notice that when writing the alleles in a gamete, we always keep the allele for one trait first and the allele for the other trait second. Unlike genotypes, we do not worry about whether an allele is dominant or recessive when determining where to put it. In the case of writing out the alleles in a gamete, we always keep them ordered by the trait that they affect.

In the end, then, when we are considering more than one trait, we have to determine all possible allele combinations in the gametes produced by each parent. That will help us build the Punnett square. Study the next example and then do the "on your own" problems that follow so that you can be prepared to do the Punnett squares afterwards.

EXAMPLE 8.4

A tall, white-flowered pea plant is heterozygous in the gene that determines height and homozygous in the gene that determines flower color. What will the possible allele combinations be for the gametes produced by this plant?

Since the plant is heterozygous in height, its genotype is "Tt." Note that we didn't really have to tell you that it is homozygous in flower color. The *only* way a pea plant can be white is if its genotype is "pp." So, given these two genotypes, how many possible allele combinations are there? Well, the "T" allele could go with a "p" allele, making a *pT* gamete. Also, the "t" allele could go with a "p" allele, making a *pt* gamete. Hmm. That's all of the possible combinations. Since the plant is homozygous in one trait, the number of possible combinations is realtively low. Thus, the possible gametes formed are just *pT* and *pt*.

ON YOUR OWN

8.7 Suppose we are going to set up a Punnett square that concentrated on the smooth/wrinkled nature of the peas produced in a pea plant ("S" for smooth, "s" for wrinkled) and the color of the pea produced ("Y" for yellow and "y" for green). What are all possible allele combinations for the gametes produced by a plant that is heterozygous in both traits?

Now that you have experience thinking about the possible allele combinations in gametes when more than one trait is considered, it is time to study an example of making the Punnett square for a dihybrid cross.

EXAMPLE 8.5

A pea plant that is tall and purple-flowered is homozygous in both corresponding genotypes. What are the possible genotypes and phenotypes, along with their percentage chances for occurring, when this pea plant is crossed with a short, white-flowered pea plant?

Since the tall, purple-flowered pea plant is homozygous in both corresponding genotypes, it has the "TT" and "PP" genotypes. What about the other plant? We are not told whether it is homozygous or heterozygous. Well, we are not told because we do not need to be told. As we already know, the allele for shortness is recessive. Thus, to be short, the plant must have the "tt" genotype. Similarly, since white flowers are recessive, it must have the "pp" genotype.

Now, to make a Punnett square for this cross, we need to think about what possible gametes will be produced by meiosis. To do that, we need to come up with all possible combinations of alleles in each parent. Well, for the "TT", "PP" parent, there is only one possible combination. No matter what, a "P" allele will be mixed with a "T" allele. The only possible gamete, then is *PT*. Likewise, since the other parent is homozygous in both traits, the only possible combination is *pt*. Thus, our Punnett square is rather simple:

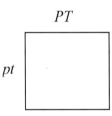

We can then determine the genotype of the offspring by just putting the alleles that affect the same trait together:

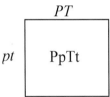

Since this is the only genotype possible, <u>100% of the genotypes will be "PpTt," with a resulting phenotype of tall with purple flowers</u>.

Now that example probably didn't seem too bad. Since both parents were homozygous, there was only one possible combination of alleles for each of their gametes. Thus, the Punnett square was rather simple. Notice that in the previous Punnett squares you drew, there were always two rows and two columns. This resulted in four boxes. We call such a Punnett square a "2x2" Punnett square, because there are 2 rows and 2 columns. In the example you just studied, there was only one row and one column, so we would call it a 1x1 Punnett square. In the next example, things are a bit more complicated, resulting in a 4x4 Punnett square.

EXAMPLE 8.6

A tall pea plant that produces purple flowers is heterozygous in both of the relevant genotypes. What are the genotypes and phenotypes, along with their percentages, for the offspring produced when this plant is self-pollinated?

Since this plant is heterozygous in both genotypes, we know that it is "Pp" and Tt." Self pollination is just like having two parents, each with the same genotype. So, we need to determine all possible combinations of alleles for the gametes produced by this genotype. Well, the "P" allele might go with the "T" allele, making a gamete that is *"PT."* Of course, the "P" allele might also go with the "t" allele, making a *"Pt"* gamete. The "p" allele could go with the "T" allele (for a *"pT"* gamete), or the "p" allele could go with the "t" allele, making a *"pt"* gamete. Thus, there are four possible gametes. Since self pollination is the same as having two parents with identical genotypes, then the Punnett square we draw will have these four gametes both on top of the columns as well as to the left of the rows, making a 4x4 Punnett square:

	PT	*Pt*	*pT*	*pt*
PT	*PPTT*	*PPTt*	*PpTT*	*PpTt*
Pt	*PPTt*	*PPtt*	*PpTt*	*Pptt*
pT	*PpTT*	*PpTt*	*ppTT*	*ppTt*
pt	*PpTt*	*Pptt*	*ppTt*	*pptt*

Since there are 16 boxes, we have 16 possibilities. The genotype "PPTT" appears once. Remember, to calculate the percentage chance of something happening, we take the number of times it occurs and then divide by the total number possible, and then we multiply by 100 to make it a percentage. Thus, the percentage chance of this genotype occurring is (1/16)x100 = 6.25%. The genotype "PPTt" happens twice, making its percentage chance (2/16)x100 = 12.5%. The genotype "PPtt" happens once, making its percentage chance (1/16)x100 = 6.25%. The genotype "PpTT" happens twice, making its percentage chance (2/16)x100 = 12.5%. The genotype "ppTT" happens once, making its percentage chance (1/16)x100 = 6.25%. The genotype "ppTt" happens twice, making its percentage chance (2/16)x100 = 12.5%. The genotype "PpTt" happens four times, making its percentage chance (4/16)x100 = 25%. The genotype "Pptt" happens twice, making its percentage chance (2/16)x100 = 12.5%. Finally, the genotype "pptt" happens once, making its percentage chance (1/16)x100 = 6.25%.

Now remember, many genotypes will result in the same phenotype. The genotype "PPTT" will result in a tall pea plant with purple flowers, but so will "PpTt," "PPTt," and PpTT." Adding the percentages of each genotype together, then, the probability of producing a tall pea plant with purple flowers is 6.25% + 25% + 12.5% + 12.5% = 56.25%. The genotype "ppTT" results in a tall pea plant with white flowers, but so does "ppTt." The probability for producing a tall pea plant with white flowers is therefore 6.25% + 12.5% = 18.75%. A short pea plant with purple flowers will result from genotypes "PPtt" and "Pptt." The percentage chance of making a short pea plant with purple flowers is 6.25% + 12.5% = 18.75%. Finally, the genotypes "pptt" will result in a short pea plant with white flowers, giving a probability of 6.25%.

That was pretty difficult, so we need to go back and review what we did. First of all, we recognized that we were dealing with two distinct traits. Thus, we knew that we had a dihybrid cross on our hands. We therefore determined all possible combinations of the 4 alleles present, to determine all of the different gametes that the parents could produce. We found 4 different gametes, so we made a 4x4 Punnett square. After filling in the boxes, we counted the number of times that each genotype occurred, and then divided by the total number of boxes (16) and multiplied by 100. This gave us the percentage chance of making each genotype. Then, we determined all of the genotypes that would lead to the same phenotype and added their percentages, giving us the total percent chance of producing that phenotype.

This is tough stuff; no question about it. Nevertheless, it is important for you to get some practice doing it yourself. Thus, perform the following "experiment." The answers to the experiment are given after the answers to the "on your own" problems, at the back of the module.

EXPERIMENT 8.1
A Dihybrid Cross

Material: Lab notebook

Object: To help understand how multiple traits are passed from one generation to another

Background: The ability of some people to taste PTC (a bitter substance to some people and tasteless to others) and the ability to roll their tongues into a U shape when extended from the mouth, are both dominant traits. We will designate them with a "T" and an "R," respectively. Persons who have these dominant alleles are called "tasters" and "rollers." The inability to taste PTC or to roll the tongue is caused by being homozygous in the recessive alleles, which we will designate as "t" and "r," respectively. Those who are homozygous with the recessive alleles are called "nontasters" and "nonrollers".

Procedure:
1. Draw a 1 by 1 Punnett Square.
2. Fill in the square with the following cross: a man that is a homozygous taster/roller and a woman that is a nontaster/nonroller.
3. What is the genotype and the phenotype of the children?
4. Now make a 4 by 4 Punnett square.
5. Cross a heterozygous taster/roller with another of the same genotype.
6. Note the percentage chance of producing each of the following: roller/taster(s), roller/nontaster(s), nonroller/taster(s) and nonroller/nontaster(s).

<u>Sex and Sex-Linked Genetic Traits</u>

As we learned back in Module #7, humans have 23 homologous pairs of chromosomes. The last pair, however, are a little different than the other 22. Thus, we call the first 22 pairs of chromosomes **autosomes** (aw' toh sohms) and the last pair **sex chromosomes**, because the last pair of chromosomes determines the sex of an individual.

<u>Autosomes</u> - Chromosomes that do not determine the sex of an individual

<u>Sex chromosomes</u> - Chromosomes that determine the sex of an individual

If a person's sex chromosomes are perfectly homologous, they are called an XX pair, and the person is female. In males, the sex chromosomes (called an XY pair) are not perfectly homologous. There are fewer genes on the second chromosome (called the "Y" chromosome) than there are on the X. This causes an interesting effect called **sex-linked genetic traits**, which we will discuss in a moment.

Before we do that, however, we just need to briefly make it clear how the sex of offspring is determined using these chromosomes. In meiosis I, the homologous pairs of chromosomes are split up. Thus, since females have two X chromosomes, each cell at the end of meiosis I has an X chromosome. As a result, all gametes produced by a female have an X chromosome. At the end of meiosis I in the male, however, one cell has an X chromosome and the other has a Y. As a result, half of the gametes produced by the male will have an X chromosome and half will have a Y. In the end, a Punnett square for the sex chromosomes would look like this:

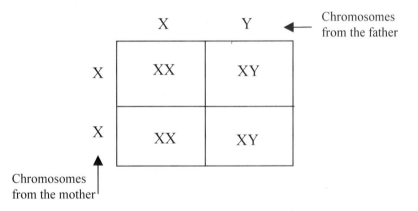

So you see that half of the offspring will have XX chromosomes and therefore be female and the other half will have XY chromosomes and will therefore be male. As a result, half of the babies born are male and half are female. Once again, in a given family, half of the children may not be male, because we are only discussing probabilities here. If you count *enough* children, half will be male and half will be female. It is not unusual, however, for a single family to "beat the odds" and have significantly more of one sex than the other.

Sex-linked genetic traits exist because the Y chromosome does not carry as many alleles as does the X chromosome. As a result, there are some traits for which alleles only exist on the

X chromosome and not the Y chromosome. In essence, this means that for certain genetic traits, males have only one allele-not two. For males, then, expressing the recessive phenotype for this trait is much more likely. After all, if you have two alleles for a given genetic trait, then both recessive alleles must be present for you to express the recessive phenotype. For sex-linked traits, however, the male has only one allele. Thus, the male needs only one recessive allele to express the recessive phenotype!

Consider, for example, the fruit fly. If you leave fruit exposed for very long, you will most likely attract these tiny flies from the genus *Drosophila*. If you examine them under a microscope, you will see that the vast majority have brick-red eyes. A few, however, have white eyes, and that phenotype is much more likely among male *Drosophila* than female *Drosophila*. The reason for this is that the eye color allele for *Drosophila* exists on the X chromosome but not on the Y chromosome. Thus, to have white eyes, a female must have two recessive alleles. A male, however, has only one allele to begin with. If that allele is recessive, then the male's eyes will be white.

In analyzing sex-linked characteristics, we can still use a Punnett square. The notation has to be a little more involved, however. Instead of just representing an allele with a capital letter when it is dominant and a small letter when it is recessive, we have to note the fact that it exists only on the X chromosome. We do this by using an X with a superscript. In the case of *Drosophila*, for example, we could say that the allele for a red eye can be noted with an "X^R." The X refers to the fact that it is carried on the X chromosome, and the capital "R" indicates that it is the dominant allele. The recessive, white allele, then, would be noted as "X^r." The Y chromosome from the male has no allele, so it will just be noted as a "Y." Let's see how this notation works in an example.

EXAMPLE 8.7

A heterozygous, red-eyed female is crossed with a red-eyed male. Will any of the offspring be white-eyed? What is the percentage chance of getting a white-eyed female? What is the percentage chance of getting a white-eyed male?

If the female is heterozygous, it has one of each allele. Thus, its gametes will be abbreviated as "X^R" and "X^r." The male has the allele only on the X chromosome, and, since he is red-eyed, it is the dominant allele. Thus, the male's gametes will be labeled as "X^R" and "Y." Remember, since the Y chromosome does not have this allele, it has no superscript. Now that we have the two gametes produced by the male and female, we can make a Punnett Square:

	X^R	Y
X^R	$X^R X^R$	$X^R Y$
X^r	$X^R X^r$	$X^r Y$

The two females (the ones with two X chromosomes) will both end up red-eyed, since they each have at least one of the dominant allele. The males (the ones with XY chromosomes), however, show a different story. One of the males (the one in the upper, right-hand corner) will be red-eyed, because he got the dominant allele from his heterozygous mother. The other one, however, will be white-eyed, since his only allele is the recessive one he got from his mother. Thus, there will be white-eyed offspring. The percent chance of a female offspring being white-eyed is 0%, but 50% of all males will be white-eyed.

Do you see the reasoning behind sex-linked characteristics? If we build the Punnett square showing that the allele exists on the X chromosome but not on the Y chromosome, then predicting the sex and phenotype of the offspring is a piece of cake! Try your hand at it with the following "experiment." Once again, the answers to the experiment come at the back of the module, after the answers to the "on your own" problems.

EXPERIMENT 8.2
Sex-linked Genetic Traits

Material: Lab notebook

Object: To help understand how sex-linked traits are passed from parents to offspring

Background: You should have read in your history books about the English and Russian royalty's problems with hemophilia, a disease that inhibits the blood's ability to form clots. People with this disease are prone to excessive bleeding. In fact, it is very possible for hemophiliacs to bleed to death as a result of minor cuts. This devastating disease is a sex-linked, recessive characteristic. Queen Victoria of England was heterozygous in this trait. Thus, she carried the recessive allele for hemophilia, but since she had the dominant allele on her other X-chromosome, she did not actually have the disease.

Procedure:

1. Since this trait is sex-linked, we designate the gene as X^h. We use a small h, because it is recessive.
2. If Queen Victoria was heterozygous, then she must be $X^H X^h$.
3. Queen Victoria's husband did not have hemophilia, so his genotype would had to have been $X^H Y$.
4. Form a Punnett that tells what the children's genotype might be. Also state all the possible phenotypes.
5. Their son Leopold was a hemophiliac. What was his genotype?
6. Hemophilia can also be present in females. Write a Punnett square to show how it is possible to produce hemophiliac females. Use whatever genotypes you need for the parents in order to get a female offspring with hemophilia.

Genetic Disorders and Diseases

Although most of the health maladies that exist are due to infectious agents such as bacteria, protozoa, and viruses, a few health problems can be directly linked to a person's genetic code. In general, there are five means by which genetic abnormalities occur: **autosomal inheritance, sex-linked inheritance, allele mutation, changes in the chromosome structure** and **changes in the chromosome number**. We will discuss each individually.

When a genetic disorder is inherited from one of the 22 pairs of autosomes, it is called **autosomal inheritance**.

<u>Autosomal inheritance</u> - Inheritance of a genetic trait not on a sex chromosome

In autosomal inheritance, a gene exists on an autosome that causes a particular malady. Sometimes the genetic disorder is recessive and thus the offspring have to have both recessive alleles in order to have the disease. In other cases, the genetic disorder comes from the dominant allele, and you must have both recessive alleles in order to not have the disease.

In Module #5, we discussed lactose intolerance, which causes people to get sick when they eat too many dairy products. This is because those who have the disorder cannot produce one of the enzymes needed to digest lactose, which is the sugar that gives dairy products their sweetness. This condition is caused by a recessive allele. If a person is homozygous in this allele, he or she is missing a key piece of information for building the enzyme, and as a result, cannot digest lactose.

Using a Punnett square, you can see that it is possible for two parents without this disorder to have a child with it. Suppose we call the allele for lactose intolerance "l." We use a small "l" because it is a recessive allele. The dominant allele, which keeps you from having the disorder, we will call "L." Suppose two heterozygous parents have children. Since they are heterozygous, neither will have the disorder. However, when they have children, the Punnett square will look like this:

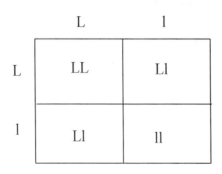

While there is a 75% percent chance of their children not having the disorder ("LL" and "Ll" will not have the disorder), there is a 25% chance of a child having it ("ll" will have the disorder).

People who are heterozygous in a genetic disorder are called **genetic disease carriers**.

Genetic disease carrier - A person who is heterozygous in a recessive genetic disorder

Since genetic disease carriers have the dominant allele that keeps them from expressing the genetic disorder, they themselves are healthy. However, as the Punnett square demonstrates, their children can end up with the disorder.

Some genetic disorders are actually caused by the dominant form of the allele. Probably the best known case is called **Huntington's disorder**. The vast majority of the people in the world are homozygous recessive in the allele that causes Huntington's disorder. If someone has the dominant allele, however, their nervous system deteriorates rapidly once they hit their forties. They eventually have trouble doing even the simplest of tasks, and they end up dying at a young age. The particularly bad thing about this disease is that anyone who has it is very likely to pass it on to their children, since the allele is dominant. Even if someone who is heterozygous in the allele has children with someone that does not have the disease at all, the children have a 50% chance of getting the disease. If the parent has the disease and is homozygous, all of the children will have the disease! Unfortunately, most people do not know that they have the disease until they are forty or so, and by that time, they have had their children.

As you saw doing Experiment 8.2, some genetic disorders are sex-linked.

Sex-linked inheritance - Inheritance of a genetic trait located on the sex chromosomes

Hemophilia is probably the best example of a sex-linked genetic disorder. Hemophilia was particularly well-known among the royal families of England and Russia. One reason that this was the case is that oftentimes members of the royal family married their own relatives, to keep the blood line "pure." Interestingly enough, this practice enhances the chance of your children having a genetic disease! Since members of a family have similar genetic codes, it is more likely for both parents to be disease carriers or to have the disease if they come from the same family. Thus, having children with the disease becomes much more likely as well. This is one reason that most countries have laws against people marrying others in their own family.

Sometimes, for reasons that we still do not understand, an allele will undergo a radical change during meiosis. When this happens, we say that the allele has mutated.

Mutation - A radical chemical change in one or more alleles

Sometimes when this happens, the results are benign. Most of the time, however, the results are quite bad. **Hutchinson-Gilford progeria syndrome**, for example, results when an allele on one of the human autosomes mutates during meiosis. A child with this allele ages rapidly. By age 3 or four, they have many of the same characteristics of a very old person: wrinkles, hair loss, arthritis. Because of their rapid aging, they die extremely young. The oldest living person with this disease died at age 18. Most die by the time they reach 10.

Another form of genetic disorder comes from a change in the structure of the chromosomes.

Change in chromosome structure - A situation in which the chromosome loses or gains genes during meiosis

When a chromosome has too many or too few genes, then the body does not understand all of the genetic information. As a result, disorders occur. For example, **cri-du-chat**, commonly called the "cat's-cry disease" is caused when the fifth chromosome is damaged during meiosis. The chromosome loses some of its genes, and an offspring produced by the damaged gamete will be mentally retarded. In addition, the child's larynx (the organ we use to make sounds) is abnormal, and the infant's cry sounds like the meow of a cat.

The last way a genetic disorder occurs is when a person receives too few or too many of a particular chromosome.

Change in chromosome number - A situation in which abnormal cellular events in meiosis lead to either none of a particular chromosome in the gamete or more than one chromosome in the gamete

If a gamete that has none of a particular chromosome goes through fertilization, then the resulting offspring will not be homozygous in that chromosome. Alternatively, if it has two of a particular chromosome, then after fertilization, the offspring will have three! The results of this can be quite devastating.

Down's syndrome, for example, occurs when a person has 3 of chromosome 21. This happens because either the father's sperm or the mother's egg has 2 of that chromosome rather than one. Those who have Down's syndrome (about 1 in 1,000 of the newborns in North America) are mentally retarded. Their skeleton also develops more slowly than it should; thus they are short and their muscles are slack. Great strides have been made in the treatment of those with this genetic order. Even though there is no cure, with the proper physical and mental therapy, most people with Down's syndrome can lead very happy and productive lives.

Summing Up

In the end, genetics is a very complex, interesting subject. As is the case with most of this course, however, we have just scratched the surface! For example, we spent a lot of time working with traits that were determined by dominant and recessive alleles. Well, it turns out that not all genetic traits have dominant and recessive alleles. Many alleles express **incomplete dominance**. For example, snapdragon flowers can be either white, red, or pink. This is because there is an allele for red flower color and another for white flower color, but neither is dominant. If a snapdragon is homozygous in the allele for red, then its flowers will be red. If it is homozygous in the allele for white, it will have white flowers. If a snapdragon is heterozygous, however, its flowers will be pink, because neither allele is completely dominant. Instead, the two flower colors "mix," forming a color in between the two. Incomplete dominance is common in genetics and adds yet another layer of complexity to this interesting subject!

Some of the most exciting careers in biology are in the field of genetics, and the research going on is truly remarkable. If you liked this subject at all, you might consider doing a little more investigation into genetics and genetic engineering. We will be taking another look at genetics next semester, when we discuss the creation/evolution controversy. The science of genetics has probably done more to illustrate the fallacies in the theory of evolution than any other single subject. You will see how in the next module!

Before we leave this module, however, we need to re-emphasize something you learned in Module #7. We told you then that an organism's characteristics are not wholly determined by genetics. There are environmental factors and, for humans, spiritual factors that influence the makeup of an organism. With all of the talk of Punnett squares and the emphasis on phenotypes that are completely determined by genetics, it is important to remind ourselves that genetics is not the only thing at play. Perform the following experiment to remind yourself of this important fact!

EXPERIMENT 8.3
The Environmental Factor and Its Effect on Radish Leaf Color

Materials:
- 60 radish seeds (purchase locally)
- 2 shallow pans or dishes
- Potting soil
- Clear plastic wrap
- Box to cover one dish
- Water
- Lab notebook
- Magnifying glass (if available)

Object: To observe the effect that the environment has on a genotype to make its phenotype change

Procedure: 1. Label one dish "light" and the other dish "dark".
               2. Place soil in each dish and then add water until the soil is wet.
               3. Spread 30 seeds across the top of each dish and cover with clear plastic wrap.
               4. Place both dishes in a warm place where there is sunlight, but not direct sunlight.
               5. Cover one dish with the box, so that no light will get in.
               6. Check the dishes daily.
               7. Add water with a dropper if soil begins to dry.

8. Watch for the seeds to sprout. When half of the seeds have sprouted, begin examining them each day. A magnifying glass will be useful in these inspections.
9. The two leaves, which are called **cotyledons**, will be either yellow or green.
10. In your notebook, set up a chart like the one below and fill in the results from your observations. Do this for the next **6** days.

	Dish - Light				Dish - Dark		
	# Green	# Yellow	% Yellow		# Green	# Yellow	% Yellow
Day 1							
Day 2							
Day 3							
Day 4							
Day 5							
Day 6							

11. Your observations should tell you that:
    a. Those in the light will have some yellow to begin with, but will turn green as exposed to sunlight.
    b. Those in the dark will continue to stay yellow and will increase in number due to lack of light.
12. Your results show you that the genetic traits can be influenced by environmental conditions.
13. After your sixth day of recording data, this experiment is completed. Clean up all materials and return them to the proper place.

ANSWERS TO THE ON YOUR OWN PROBLEMS

8.1 a. Since both factors indicate that the peas will be wrinkled, <u>the plant will produce wrinkled peas</u>.

b. This plant has one factor for smooth and one for wrinkled peas. We are told, however, that smooth is dominant. Thus, because it has one dominant factor, <u>the plant will produce smooth peas</u>.

c. Since both factors indicate that the peas will be smooth, <u>the plant will produce smooth peas</u>.

8.2 If a pea plant produces yellow peas, one possible set of factors would be "YY." However, since yellow peas are dominant, a plant needs to have only one yellow factor in order to produce yellow peas. Thus, "Yy" is also a possibility. The plant, therefore, can have <u>either "YY" or "Yy"</u> and produce yellow peas.

8.3 Since the green pea factor "y" is not dominant, the only way that a plant can produce green peas is if both of its factors are for green peas. Thus, the only possible way to produce green peas is to have the <u>"yy" factors</u>.

8.4 <u>The "SS" offspring's genotype is, simply, "SS."</u> Since it has the same alleles, <u>it is homozygous</u>. Since "S" means smooth peas, then <u>its phenotype is for smooth peas. The "Ss" offspring have the "Ss" genotype, they are heterozygous</u> because their alleles are different, <u>and since the smooth pea allele is dominant, they also have the phenotype for smooth peas. The "ss" offspring are homozygous; they have the genotype "ss," and they have the phenotype for wrinkled peas.</u>

8.5 We are using "P" to represent the purple-flower allele and "p" to represent the white-flower allele. Since one plant is heterozygous, its genotype is "Pp." Since the white-flowered plant is homozygous, its genotype is "pp." Note that we did not have to tell you that the white-flowered plant was homozygous. Since that allele is recessive, a white-flowered plant *must* be homozygous. The resulting Punnett square, then, is

	P	p
p	Pp	pp
p	Pp	pp

Note that it doesn't matter whether you put the "Pp" genotype on the top or to the left. The important thing is the genotypes produced. <u>Two offspring (50%) will have the "Pp" genotype while the other two (50%) will have the "pp" genotype.</u> Since the purple-flower allele is dominant, a "Pp" genotype leads to a purple flower. A "pp" genotype, however, leads to a white flower. Thus, <u>two (50%) will have the purple flower phenotype and two (50%) will have the white flower phenotype.</u>

8.6 The cross between individuals 1 and 2 does not tell us much, since they have different phenotypes. However, both 3 and 4 have the same phenotype (no tail). When crossed, they have some offspring with tails and some without. Thus, they must each have both alleles. As a result, we know that <u>the no tail allele is dominant</u>, because these two individuals have both alleles but express the no tail phenotype. We will therefore call the no tail allele "N" and the allele for a tail "n." This tells us that <u>3 and 4 have the "Nn" genotype.</u> Also, since the only way to have a tail is to have both recessive alleles, <u>we know that 2 has the "nn" genotype.</u> Finally, the only way that 1 and 2 can have children with tails is if 1 also has an allele for a tail. Since 1 expresses the no-tail phenotype, however, <u>1 has the "Nn" genotype.</u>

8.7 If the plant is heterozygous in both traits, its genotype is "Ss" and "Yy." In meiosis, the "S" could go with the "Y" or it could go with the "y." In the same way, the "s" could go with the "Y" or with the "y." Thus the possible gametes are: <u>*SY, Sy, sY, and sy.*</u>

EXPERIMENT 8.1 ANSWERS

1&2

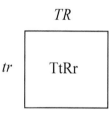

3. <u>The genotype is TtRr, making all children with the taster/roller phenotype.</u>

4&5.

	TR	*Tr*	*tR*	*tr*
TR	*TTRR*	*TTRr*	*TtRR*	*TtRr*
Tr	*TTRr*	*TTrr*	*TtRr*	*Ttrr*
tR	*TtRR*	*TtRr*	*ttRR*	*ttRr*
tr	*TtRr*	*Ttrr*	*ttRr*	*ttrr*

6. <u>roller / taster</u> (genotypes TTRR, TtRr, TTRr, TtRR) 9 of 16 or <u>56.25 %</u>
   <u>roller / nontaster</u> (genotypes ttRR, ttRr) 3 of 16 or <u>18.75 %</u>
   <u>nonroller / taster</u> (genotypes TTrr, Ttrr) 3 of 16 or <u>18.75 %</u>
   <u>nonroller / nontaster</u> (genotype ttrr) <u>1 of 16 or 6.25 %</u>

EXPERIMENT 8.2 ANSWERS

4.

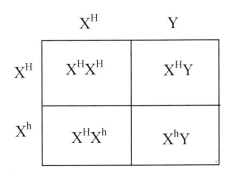

<u>The girls will never be hemophiliacs, but half of the boys will be</u>.

5. To have the disease, you must have only the recessive allele(s). For a son, that means <u>X^hY</u>

6. For a female to have the disease, she must have the recessive allele on both X chromosomes. Since one X always comes from the father, then the father must have a recessive allele on his X chromosome. Also, the mother needs to have at least one, but she could have 2 recessive alleles. In the end, then, there are two possibilities, either of which is correct:

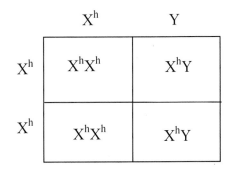

OR

STUDY GUIDE FOR MODULE #8

1. Define the following terms:

a. True breeding
b. Allele
c. Genotype
d. Phenotype
e. Homozygous genotype
f. Heterozygous genotype
g. Dominant allele
h. Recessive allele
i. Mendel's principles of genetics (use updated terminology)
j. Pedigree
k. Monohybrid cross
l. Dihybrid cross
m. Autosomes
n. Sex chromosomes
o. Autosomal inheritance
p. Genetic disease carrier
q. Sex-linked inheritance
r. Mutation
s. Change in chromosome structure
t. Change in chromosome number

2. Three pea plants have the following alleles for yellow ("Y") and green ("y") peas. What is the genotype and phenotype of each? Note whether they are homozygous or heterozygous.

a. YY   b. Yy   c. yy

3. What process causes gametes to have only one allele, since other human cells have two of each allele?

4. A pea plant which is homozygous in the dominant, axial flower allele ("A") is crossed with a pea plant that is heterozygous in that allele. What are the possible genotypes and phenotypes, along with their percentage chances, for the offspring?

5. A woman is heterozygous in the ability to roll her tongue when extended. If she marries a man who cannot roll his tongue, what percentage of their children will be able to roll their tongues? Remember, the allele for being able to roll your tongue is dominant.

6. Recall that in guinea pig coat color, black (filled circles and squares) is dominant and white (hollow circles and squares) is recessive. What is the genotype of the male parent in the cross below?

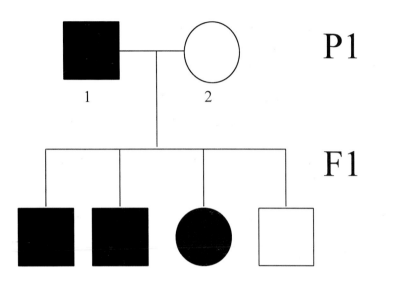

7. The following pedigree is for the presence or absence of wings on a certain insect. The hollow circles and squares represent insects without wings, while the filled circles and squares represent insects with wings. Which is the dominant allele? What are the genotypes of individuals 1-4?

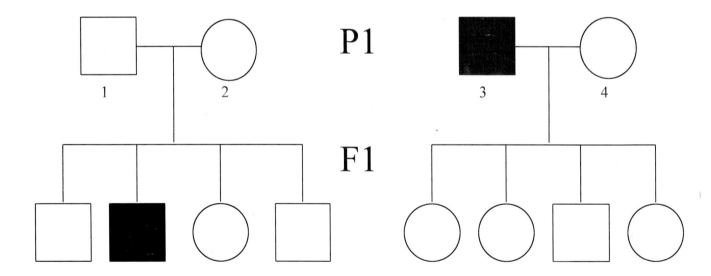

8. Give the possible phenotypes and the percentage chance for each in the dihybrid cross between a pea plant that is homozygous in producing smooth, yellow peas and a pea plant that produces wrinkled, green peas. The smooth and yellow alleles are dominant.

9. Give the possible phenotypes and the percentage chance for each in the dihybrid cross between a pea plant that is heterozygous in producing smooth, yellow peas and another with the same genotype.

10. In fruit flies, the color of the eye is a genetic trait that is sex-linked. What is the percentage of males that will have white-eyes when a heterozygous, red-eyed female is crossed with a white-eyed male? What is the percentage of females that will have white-eyes from the same cross?

11. In the case of fruit fly eye color again, what must be the genotype of a male fly if, when crossed with a heterozygous female, there is no possibility of having a female with white eyes?

12. If a gamete has two alleles for the same genetic trait, what type of genetic disorder will result in a zygote formed with this gamete?

13. A person carries a genetic disorder but does not have the disorder. How is that possible?

14. Do sex-linked genetic disorders affect men and women the same? If not, which sex is affected more and why?

15. Two individuals have the exact same genotype for a certain trait but they are not identical when it comes to that trait. How is this possible?

Module #9: Evolution - Part Scientific Theory, Part Unconfirmed Hypothesis

Introduction

In 1859, Charles R. Darwin published a book entitled *On The Origin of Species by Means of Natural Selection or the Preservation of Favoured Races in the Struggle for Life.* Typically abbreviated as *The Origin of Species*, Darwin's book caused a firestorm in the scientific community. In *The Origin of Species*, he proposed a theory that attempted to explain the diversity of life which exists on earth. This theory, now known as the theory of evolution, made no reference to God. Instead, it proposed that the same kinds of processes which we see occurring today are, in fact, responsible for all of the species on the planet. In effect, Darwin's book proposed to answer the age old question, "How did we get here?" without ever referring to a supernatural Creator.

This sent shockwaves throughout the scientific community because, at that time, science was inseparably linked to God. Most scientists were formally trained in the Bible. Many were, in fact, clergy of the church. Back then, it would have been very rare indeed to find a scientific publication that did not refer to God or the Bible throughout the entire work. The very idea, then, that Darwin proposed a theory which did not refer to or require God was astonishing! To some, it was horrendous. They could not imagine a scientist who didn't refer to God as a natural part of his scientific inquiries. To others, it was what they had wanted for a long time. They did not believe in God, and finally a theory had come along that told them how we came to be without referring to anything supernatural.

Sadly, Darwin's theory (with ample modifications) has become the standard explanation for the origin of life on this planet. As you will see in this module, however, it is not because of the scientific evidence. Based on the scientific method that we discussed in Module #1, the important part of Darwin's theory never really left the stage of being an hypothesis. Nevertheless, the majority of scientists today consider his theory to be a scientific law! Why is this the case? Hopefully, that is one of the questions which this module will answer for you.

Before we begin discussing Darwin and his theory, however, we need to point something out. There are many ways that people of faith have approached the theory of evolution. On one end of the spectrum, there are those who assume that it must be true, since the majority of scientists today believe in it. As a result, they try to make the scriptures of their faith consistent with it. In the case of Christians and Jews, for example, this means trying to interpret the first few chapters of Genesis in such a way as to allow for the process of evolution. On the other end of the spectrum, there are those who refuse to even consider the theory, because it runs counter to what they believe Scripture teaches. Christians and Jews on this end of the spectrum, for example, will not even consider the theory, because their interpretation of the first few chapters of Genesis flatly contradicts the theory of evolution. Of course, there are many people in between these two ends of the spectrum, attempting to either integrate the idea of evolution into their faith on one level or another, or rejecting part or all of the theory because it is considered incompatible with their faith.

Who is right? What is the proper way to view the theory of evolution in the light of faith? We will not answer those questions. Clearly these authors have very strong ideas on how Christians should approach the theory of evolution, but we do not think that such opinions belong in a science textbook. Questions of scripture interpretation and fundamental dogma should be left to other, more inspirational books. Instead, we choose to approach evolution strictly from a scientific perspective. We will examine the theory and see where it stands in the light of today's current scientific knowledge.

Charles Darwin

Because he is such an important figure in the field of biology, it is necessary to look at the life of Charles Darwin in some detail. This will help us gain insight into how he developed his theory, and, hopefully, you might learn a lesson or two from his story.

Charles Robert Darwin was born in the village of Shrewsbury, England on February 12, 1809.   On that same day, more that 4,000 miles away, Abraham Lincoln was born in Kentucky. Darwin lived in a relatively wealthy family where education and artistic enrichment were stressed. In 1825, he enrolled at the University of Edinburgh, where all the men in his family had been educated. His father wanted him to study medicine, but he was sickened at the first sight of surgery being performed without anesthesia. He also showed little aptitude for the subject, so after two years, he abandoned the study of medicine.

When he left the study of medicine, he transferred to Christ's College in Cambridge, England to study theology. Contrary to what you might have heard about Darwin, he was a deeply committed Christian at this point in his life. During this part of his life, he said that he did not "in the least doubt the strict and literal truth of every word of the Bible" (Julian Huxley and H.B.D. Kettlewell, 1965, *Charles Darwin And His World*, page 15). Since he was a committed Christian, the study of theology came quite natural to him, and he graduated with a BA in Theology, Euclid, and the Classics.

Although his degree was in Theology, Euclid, and the Classics, Darwin developed a keen interest in geology while at Cambridge. Thus, when he had the opportunity to accompany a Cambridge professor, Adam Sedgewick, on a geology field trip in the summer of 1831, he jumped at the chance. While on that field trip, he was offered the position of naturalist on the *HMS Beagle*, a ship that planned to circumnavigate the globe. Although you might think it unusual for a ship to employ a naturalist, nearly every ship had such a position available. Darwin eagerly accepted the assignment, and that voyage changed both Darwin's life and the face of science forever.

Before that voyage, Darwin had read a book written by Thomas Malthus entitled *An Essay on the Principles of Population*. In this book, Malthus said that all individuals within a population struggle against other individuals to obtain what is necessary (food, shelter, a mate, etc.) in order to survive. While on board the *HMS Beagle*, Darwin read some of the works of a controversial geologist named Sir Charles Lyell. Lyell was one of the first scientists who rejected the history of the world as told in the Old Testament and tried to show that the same processes which we see at work today could, given eons and eons of time, produce all of the geological features in the world. Lyell gave this concept the catch phrase, "The present is the key to the past."

Darwin voyaged on the *HMS Beagle* for five years and, during that time, he made many observations. Each time the ship dropped anchor, Darwin collected samples and made observations of the species native to whatever island or land mass he was on. These observations, some of which we will detail later, combined with the ideas of Malthus and Lyell, led Darwin to formulate his theory, which he called "natural selection." Although his theory was completely formulated by the time he had left the *HMS Beagle*, he did not publish his book for another 23 years. Part of the delay was due to Darwin trying to perfect his work, but most of it was due to his wife, who recognized the devastating effect that his work could have on the church. She pleaded with him not to publish, and he respected her wishes for some time, but in the end, he felt that he had to communicate his ideas to the scientific world. He therefore published *The Origin of Species* in 1859.

It is important to note that Darwin was a careful, meticulous scientist. He was not the anti-religion crusader that many have made him out to be. If you actually read his work, you will find that it is quite even-handed. Indeed, Darwin devoted more space to discussing the scientific data that seemed to contradict his main hypothesis than he did to the discussion of the data that supported it! You will not find that kind of even-handedness in the majority of scientific writing that occurs today. Indeed, modern scientists (especially evolutionary crusaders) could learn a lot from Darwin's style. Darwin's only real mistake was to allow his faith to erode as a result of the science he pursued on the *HMS Beagle*.

The best illustration of how Darwin's faith eroded while on the *HMS Beagle* and the years after can be found by comparing two statements he made. During the earliest part of his voyage, he wrote in his diary that he often bore the brunt of a good deal of laughter "...from several of the officers for quoting the Bible as final authority on some moral point" (Bern Dibner, 1964, *Darwin Of The Beagle*, page 82). Only a few years after his voyage, however, he stated "...that the Old Testament from its manifest false history of the world, with the Tower of Babel, the rainbow as a sign, etc., etc., and from its attributing to God the feelings of a revengeful tyrant, was no more to be trusted than the sacred books of the Hindoos [sic], or the beliefs of any barbarian." (Bern Dibner, 1964, *Darwin Of The Beagle*, pages 82-83) Clearly these are the statements of a man whose faith was once strong but eroded over time to nothing!

Charles Darwin died in 1882 as the result of a long illness. He died a celebrated naturalist whose views were said to usher in a new age of science. He was buried in Westminster Abbey along with such scientific greats as Sir Isaac Newton and Michael Faraday. There is a myth going around the Christian community that Darwin recanted his theory on his deathbed. This is a lie, started by the widow of Sir James Hope, fleet admiral for the Royal Navy. She claims to have visited Darwin shortly before his death and to have heard him recant his theory and ask to be told how he might be saved. Darwin's own daughter Henrietta, however, said, "...[the admiral's widow] was not present during his last illness, or any illness. I believe he never even saw her..." In addition, she states that "He never recanted any of his scientific views, either then or earlier. . . . The whole story has no foundation whatever." (Paul F. Boller and John George, 1989, *They Never Said It: A Book Of Fake Quotes, Misquotes, & Misleading Attributions*, pages 19-20). Although the story of a deathbed recantation by Darwin is appealing to Christians, it is almost certainly a lie and therefore such a story does not, in any way, honor God.

Although this biography was rather long and involved, it was necessary for three reasons. First, it is important that the phony story of Darwin's deathbed recantation not be spread any further. Second, it is important for you to realize that although Darwin's theory has had devastating effects on the faith of many people, Darwin himself was not an anti-religion crusader like most evolutionists are today. Darwin was a careful, dedicated scientist who started his career as a Bible-believing Christian. He had no intentions of harming the church. He was merely communicating what he thought were the obvious conclusions of science. Finally, a look at Darwin's life can show you how horrible the results are when you put your faith in science. As we stated in Module #1, science is limited and is constantly changing. What we thought were scientific laws less than a century ago are now shown to be wrong. Indeed, as you will see in the rest of this module, we now know that most of Darwin's ideas were very wrong! You simply cannot put your faith in something as flawed as science. Had Darwin realized that, he would not have allowed his faith in the Bible to be eroded, and he might never have produced this errant theory that has had such a devastating effect on the faith of others!

Darwin's Theory

During his time on the *HMS Beagle,* Darwin had a chance to investigate a small chain of islands called the "Galapagos archipelago." The 13 islands of the archipelago are the result of volcanic activity, and these islands still exist about 600 miles west of the South American nation of Ecuador (on the equator). Although Darwin was pleased to study a wide variety of plant and animal life on these islands, he concentrated on the finches that lived there. Now known as "Darwin's finches," many science historians credit these birds as inspiring Darwin's theory of evolution through natural selection.

You see, there were (and still are) many different species of finches living in the Galapagos. These species had several common characteristics (they were all of approximately the same size and shape, for example), but there were specific differences between each species that were of great interest to Darwin. In particular, Darwin concentrated on the beak of each individual species of finch. Some of the species had the typical small, delicate beak that you expect for a small-to-mid-size bird like a finch. Other species, however, had short, stout beaks much like those that you would expect to see on a parrot. Still others had straight, thin beaks that were ideal for boring through wood in search of food. Still others had specially curved beaks that seemed perfect for probing flowers.

These differences fascinated Darwin. You see, the scientists of Darwin's day would have looked at each of these species and assumed that God had designed each one individually. In other words, they believed that God uniquely and individually created the wood-boring finch, the finch with the parrot's beak, the finch with the curved beak, and the finch with the standard beak.

Darwin, however, imagined something else. He said that other than the beak and a few trivial differences (plumage color, for example), these finches were all remarkably similar. Since they were all so similar, he imagined that they all came from a common ancestor long ago. As the feeding needs of the finches changed, however, this common ancestor began to give rise to many different species of finch, each with unique beaks.

How did Darwin propose that this could happen? Well, he said, look at what happens when any species reproduces. When two people have a baby, for example, the baby has many characteristics in common with his parents. The baby's eye color might be the same as his mother's, and his hair color might be the same as his father's. Nevertheless, the baby usually has some characteristics that do not seem to come from either parent. Many tall professional basketball players, for example, have very short parents. Thus, although offspring do tend to resemble their parents, they also have a few characteristics that are quite different than the corresponding characteristics in their parents. It is these differences, Darwin thought, that could be responsible for all of the finches in the Galapagos.

Suppose, Darwin imagined, that there was only one species of finch living on the islands. If food supplies were to grow scarce, then these finches would compete with one another for these dwindling food supplies, as Malthus predicted. When this competition began occurring, any finch that had an advantage would be more likely to win the competition than one who didn't. Thus, suppose a finch was born that had a beak which was stronger than the typical finch. Well, that finch might be able to find a new source of food (hard nuts that other finches couldn't break open, for example). With this new source of food, this strong-beaked finch would certainly win the competition for survival. As it reproduced, then, it would pass on this new, strong beak to its offspring. Over many, many generations, each time one of the these finches was born with an even stronger beak, it would be more likely to survive, because it could continue to find more food than the finches with which it was competing. Thus, this competition, combined with the natural differences that arise between parent and offspring, could, over generations, produce a finch whose beak was short and stout, like that of a parrot.

During that same time (or perhaps later), another finch might have been born whose beak was curved a bit. This would make it easier for that finch to probe flowers for food, giving it an edge in the competition to survive. As time went on, this finch would survive and would pass on the curved beak to its offspring. Each time a finch was born whose beak was more ideally shaped for the task of probing flowers, it would have a better edge in the competition for survival, making it more likely to live and pass its new characteristic on to more and more offspring. Thus, as time went on, the original finch species that lived in the Galapagos would eventually give rise to two new species of finch: one with a short, stout beak and another with a curved beak.

This is the mechanism by which Darwin imagined that all species of finch he observed could have originated from a single type of finch long ago. He called this mechanism "natural selection," because he said that due to the fierce competition which occurred between members of a species, any individual that had a unique characteristic making it more likely to win the competition would be selected by nature to survive. As time went on, these unique characteristics would continue to "pile up" on one another until eventually, a new species was formed.

Darwin, of course, did not stop there. After all, he imagined, if such a mechanism could be responsible for causing finches to develop new beaks, why couldn't that same mechanism allow them to develop longer, stronger wings, longer, sharper talons, and keener eyesight, eventually forming an eagle! These ideas led Darwin to his overall theory of evolution. At one time, Darwin believed, there was a relatively simple (most likely aquatic) life form that existed on earth. Darwin made no speculations of how that life formed developed, but others who followed have constructed wild scenarios that try to explain the formation of this organism without the intervention of a supernatural creator. This life form, Darwin assumed, would begin to reproduce and, as is the case today, variations would occur in the reproduction. These variations, guided by the process of natural selection, would eventually "pile up" so as to form new species. These species would, in the same way, give rise to other species. Thus, over the course of eons of time, Darwin believed that this mechanism could explain the existence of all life forms on the planet.

Hopefully you can see how Malthus and Lyell influenced Darwin's thinking. After all, Malthus gave Darwin the idea that individuals within a species compete with one another in order to survive. This led to Darwin's idea of natural selection. Lyell's concept that the present is the key to the past allowed Darwin to speculate that the same variations which we see in reproduction today could, over vast ages of time, be responsible for all of the variations between the species that exist on the planet. In other words, Darwin did not dream up this theory on his own. He was influenced by the works of others.

If you remember our discussion of the scientific method, you will recognize that at this point in the story, Darwin's idea was really no more than an hypothesis. Darwin made a bunch of observations and then proceeded to develop an explanation for those observations. Did the concept of evolution through natural selection ever make it past the stage of being an hypothesis? Well, the answer to that is both yes and no. Hopefully you will see what we mean in the next section.

ON YOUR OWN

9.1 What two concepts promoted by other scientists influenced Darwin in developing his ideas?

9.2 The cheetah is the fastest land-dwelling animal on the planet. It has been observed to reach peak speeds of almost 80 miles per hour! It uses this speed to catch animals that are too fast to be killed by most other predators. According to Darwin's ideas, the cheetah did not always exist. Instead, it arose from slower predators. Use Darwin's reasoning to explain how natural selection could produce such a creature.

9.3 Why is Darwin's hypothesis sometimes called "the survival of the fittest?"

Microevolution and Macroevolution

As we learned in Module #1, once an hypothesis is formed, it is tested against experimental data. If the data continue to support the hypothesis, it eventually becomes a theory. If it does not, it must be altered or discarded. This is the next step in the scientific method, and it is where Darwin's hypothesis ran into some trouble.

After leaving the *HMS Beagle*, Darwin began experimenting with pigeons. He raised and bred them, trying to see if natural selection could result in new species of pigeon. To investigate further, he talked to other animal breeders. He interviewed those who bred dogs, horses, and pigeons, looking to their experience as a guide to whether or not his hypothesis could be correct. Indeed, he found much evidence which confirmed at least part of his hypothesis. He noted several cases of breeders who, over several generations, succeeded in producing pigeons which were so different from the species with which the breeder started that they could, indeed, be classified as a new species of pigeon. The same seemed to be the case with dogs and horses as well.

As another piece of evidence for his hypothesis of natural selection, Darwin compared the domesticated versions of many animals with their wild counterpart. Wild dogs, for example, looked and behaved quite differently than domesticated dogs. In fact, many breeds of domesticated dog can not reproduce with wild dogs. Thus, by the definition of species as laid out in Module #1, these domestic dogs would be considered a wholly different species than the species of wild dog. Despite these incredible differences, domestic dogs were, many generations ago, simply wild dogs that men began to train and domesticate. Over generations, however, dog breeders would selectively mate those dogs which had what the breeder considered the best traits for domestication. Thus, the more "wild" dogs were not allowed to reproduce, and the more tame dogs were. This "man-made" selection, Darwin realized, mimicked natural selection, allowing the small variations that occurred during reproduction to "pile up," leading to new species of dog.

With these observations, Darwin was able to do two things. First, he established as a valid scientific theory the idea that the natural variations which occur during reproduction could, when guided by natural (or man-made) selection, take one species and pile up so many changes that the result could be something reasonably classified as another species. In other words, he showed that his explanation for the many species of finches in the Galapagos archipelago was scientifically viable. Second, Darwin was able to destroy forever an idea which had been established for generations before him: **the immutability of species**.

The immutability of species - The idea that each individual species on the planet was specially created by God and could never fundamentally change

In other words, scientists of Darwin's day believed that every breed of dog was created during the time of Creation and has existed, essentially the same, ever since that time. In the case of dogs, for example, those who held to the idea of the immutability of the species would say that in the

Garden of Eden, there were doberman pinchers, Saint Bernards, dachshunds, and chihuahuas. Each of these breeds of dog continued, essentially unchanged, up to the present. Darwin masterfully showed that this just wasn't true. He showed that all of these breeds of dog came from some original dog ancestor, and the natural variations that occurred in reproduction, guided by natural (or man-made) selection resulted in the many different breeds of dog that exist to this day.

Although it sounds like Darwin had remarkable success in testing his hypothesis, you need to realize that what Darwin showed to be true was only a small part of his theory of evolution. The idea that one ancestral finch could, over generations, give rise to many different species of finch was revolutionary, but it was not where Darwin wanted to stop. Once he had destroyed the idea of the immutability of the species, he wanted to go much farther. He wanted to show that this same process, over millions (or perhaps billions) of years could, eventually, cause the finch to give rise to the eagle. This is where Darwin ran into all sorts of trouble when comparing his hypothesis to the data.

Although it was rather easy to show that a species of wild dog could, over time, give rise to several breeds of domestic dog, it was quite another to show that a dog could give rise to a radically different species, such as a horse or a cow! In fact, Darwin found *some* evidence for this idea, but it was inconclusive at best. We will be looking at this data in-depth in the next section, but for right now it is enough to say that there was so much data contradicting this hypothesis, he spent the majority of his book discussing it.

So, in the end, Darwin found ample evidence that starting with a basic life form (a finch, for example), many other specialized species of this life form (many species of finch) can arise as a result of variation guided by natural selection. However, when it came to showing that a basic life form (once again, a finch) could evolve into a completely different life form (like an eagle) by natural selection, there was precious little evidence for his hypothesis and plenty of evidence against it. This has led scientists to divide Darwin's theory of evolution into two parts: the theory of **microevolution** and the hypothesis of **macroevolution**.

Microevolution - The theory that natural selection can, over time, take an organism and transform it into a more specialized species of that organism

Macroevolution - The hypothesis that the same processes which work in microevolution can, over eons of time, transform an organism into a completely different kind of organism

The distinction between macroevolution and microevolution cannot be over-emphasized. There is so much evidence to support the idea of microevolution that it is a well-documented scientific theory. There is so little evidence for macroevolution and there is so much evidence against it that it is considered, at best, an unconfirmed hypothesis.

Well, if Darwin could not find much evidence in support of macroevolution and found a lot of evidence against it, how did it become so popular among scientists? There are several answers to that question, but one of the most important ones is that at the time, scientists were rather ignorant about a great many things which we take for granted. As a result, Darwin could argue his point rather convincingly.

Darwin basically said that since microevolution is so clearly apparent from a scientific point of view, then macroevolution should also be rather obvious. After all, if finches can change a little over a small amount of time, shouldn't they be able to change a lot over a long period of time? Assuming that the amount of change a given species can experience is essentially limitless, it will just take a little longer for microevolution to slowly lead to macroevolution.

To scientists of Darwin's day, this sounded like a reasonable argument. You see, they didn't know what we know about genetics. They didn't know that the genetic code is responsible for determining the range of characteristics that a species has. Thus, they didn't know that the natural variation which we see in reproduction today is simply the result of different alleles being expressed in different individuals. Since we know that the number of alleles in the genetic code of any species is limited, then we also know that the natural variation which occurs as a part of reproduction is limited as well. Thus, unlike Darwin argued, the variation which a species can experience is NOT unlimited. It is limited by the number and type of alleles in the species' genetic code. Thus, today we know that macroevolution cannot occur the same way that microevolution occurs.

You see, microevolution is simple to explain. When God created the animals and plants, he built into their genetic code a great amount of variability. As these plants and animals began reproducing, this variability began manifesting itself. This built-in variation was then acted upon by natural selection to create the variations that we see within a particular kind of creature. Thus, God probably created a "typical" dog during the creation week and then the process of microevolution produced the many variations of dog that we see today. Microevolution, then, is a testament to God's foresight. As the Creator, God knew that the creatures of His world would have to adapt in order to survive. Thus, He built in their genetic codes the ability to change, and microevolution is simply the theory that describes how that change takes place.

Macroevolution, however, is something quite different. The hypothesis of macroevolution assumes that a given life form has an unlimited ability to change. This means that some process must exist to *add information to the creature's genetic code*. After all, a creature's ability to change is limited by the information in the genetic code. There are only a certain number of genes and alleles of those genes. There is therefore only a certain number of possible variations in genotype and therefore a limited number of possible phenotypes. Thus, in order to get an unlimited amount of change, a creature must somehow find a way to add genes and alleles to its genetic code! This is something altogether different than microevolution and, as we will see in the next few sections, there is precious little data supporting such an hypothesis and quite a lot of data contradicting it.

In the end, then, we can distinguish between microevolution and macroevolution by referring to genetics. If we are talking about a species varying *within* its genetic code, then we are talking about microevolution. This is how wild dogs became domestic dogs and how the many varieties of finch formed in the Galapagos archipelago. On the other hand, if we are talking about a species suddenly *adding information to* its genetic code, we are talking about macroevolution. The distinction is quite important, because the former is a well-established scientific theory while the latter is a very shaky hypothesis!

ON YOUR OWN

9.4 In the woods outside London before the industrial revolution, the trees were mostly white with small black specks. The peppered moths that lived in those woods were also mostly white with small black specks. Because the moths blended in well with the trees, they were able to hide from predators. During the industrial revolution, the people of London burned so much coal that the trees in these woods actually turned black due to pollution. In just a few years, the peppered moths became mostly black with small white specks, once again blending in very well with the blackened trees. Is this an example of microevolution or macroevolution? Why?

9.5 Some biologists believe that the whale once had a cow-like ancestor that lived on land. This ancestor was very heavy, so it started spending a great deal of time in the water. The water helped buoy it up, making it easier for the animal to walk. As time went on, the animal began adapting to the water, slowly changing its legs into fins and its skin into a soft, rubbery substance ideal for swimming in the water. Eventually, the cow-like creature gave rise to the whale. Is this an example of microevolution or macroevolution?

Inconclusive Evidence: The Geological Column

As we said in the previous section, when Darwin was collecting data, he found some evidence for his hypothesis of macroevolution. For example, he looked at the data uncovered by geology and found an amazing fact. He found that as geologists dug deeper and deeper into the earth, the rock that they dug up was laid down in distinct layers which geologists called **strata**.

Strata - Distinct layers of rock

In these strata were the preserved remains of once-living organisms, which are called **fossils**.

Fossils - Preserved remains of once-living organisms

Now none of this was really that amazing. Geologists had known this for some time. The fact that Darwin found amazing was that as one encountered strata deeper and deeper in the earth, the fossils found in those strata were of simpler and simpler life forms. This fact is illustrated in Figure 9.1.

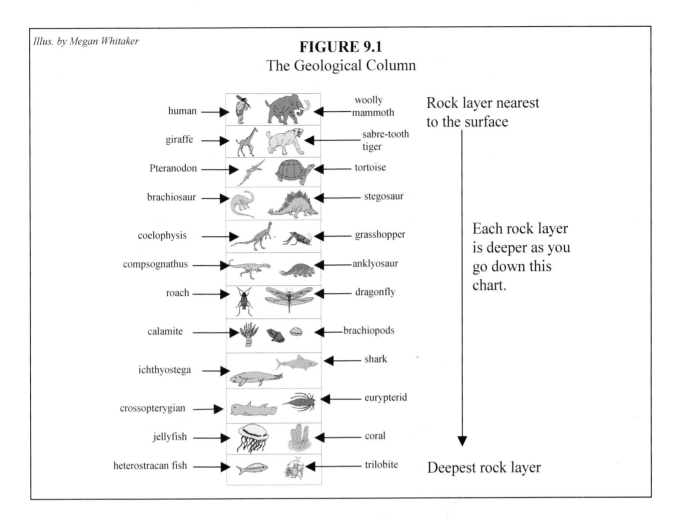

Illus. by Megan Whitaker

FIGURE 9.1
The Geological Column

human — woolly mammoth
giraffe — sabre-tooth tiger
Pteranodon — tortoise
brachiosaur — stegosaur
coelophysis — grasshopper
compsognathus — anklyosaur
roach — dragonfly
calamite — brachiopods
ichthyostega — shark
crossopterygian — eurypterid
jellyfish — coral
heterostracan fish — trilobite

Rock layer nearest to the surface

Each rock layer is deeper as you go down this chart.

Deepest rock layer

Now we have already clearly seen that there is no such thing as a simple life form, but Darwin did not have the benefit of microscopic analysis and detailed explanations of how organisms work. Instead, he saw that the strata which were near the surface of the earth contained fossils of "complex" animals like horses, lions, and humans. As they dug deeper, however, they found that these "complex" fossils disappeared and gave way to fossils of "simpler" life forms like rats, rabbits, and snakes. Even deeper, these fossils disappeared and gave rise to even "simpler" fossils such as fish, squid, and crabs.

Before we discuss how Darwin interpreted this geological column, there is an important point that must be made. You may have already seen this kind of representation of the geological column, and you most likely will see it again. The problem is, this is not at all what the geological column actually looks like! You see, 95% of all fossils that we recover are of clams. There are, quite literally, fossilized clams in every region of the earth in nearly every layer of rock. So what we are showing in this representation of the geological column is really only about 5% of the fossil record. Thus, it is not a realistic representation of what the geological column actually looks like. Nevertheless, since the geological column is usually discussed in reference to macroevolution, the clams simply are ignored. Therefore, geologists ignore the vast majority of fossils and concentrate only on fossils like those pictured here. Thus, any conclusions you make based on the geological column are, in fact, based on a tiny minority of the available data!

Now we can discuss how Darwin interpreted the geological column. To do that, we must discuss how the geologist Lyell (who influenced Darwin tremendously) interpreted it. Lyell said that the strata shown in Figure 9.1 were laid down sequentially over vast eons of time. Using his idea that the present is the key to the past, Lyell said that the strata seen by geologists were formed when dirt and sediment accumulated slowly over time. We see this happening today, he said, and it results in layers of dirt and sediment. Eventually, Lyell postulated, pressure and heat from the sun would take this pile of dirt and sediment and harden it into rock. This would result in a single layer of rock. As time went on, another layer of dirt and sediment would accumulate on top of this layer of rock, eventually forming another layer of rock on top of the previous layer. This would happen over and over again, eventually forming the layers of rock seen in the geological column.

Now, of course, all of this is wild speculation, but it is pretty much accepted by geologists today as the way in which the rock strata in the geological column were formed. Well, if you accept this speculation as fact, then you are left to conclude something rather obvious. The deeper a rock layer is in the geological column, the farther back in earth's past it was formed. After all, since Lyell's process requires the lower rock layers to form before the higher ones can, then the rock layers on top should be younger than the rock layers on the bottom.

So, from the concrete scientific fact that the geological column is formed of many rock strata, Lyell's speculation led Darwin to conclude that the lower the strata in the geological column, the older the rock. Well, thought Darwin, if this is the case, then fossils found in strata that are low in the geological column are the remains of creatures that lived long ago. In the same way, fossils in strata near the top of the geological column must be the remains of creatures that lived in the more recent past.

Based on Lyell's speculation, then, Darwin argued that the geological column shows us that long ago, only "simple" life forms existed. That is why we see fossils of only "simple" forms in the lowest geological strata. As you look up the geological column, however, the fossils become more and more complex. This indicated to Darwin that as time went on, life forms got more and more complex. Well, Darwin argued, this is great evidence for macroevolution. After all, macroevolution predicts that life started out simple and, over eons of time guided by natural selection, more complex forms of life emerged.

Now you have to realize that Darwin's conclusion is based on two big assumptions. He assumed that Lyell was right in the speculation that the lower a rock layer is in the geological column, the older the rock is. In addition, he had to believe Lyell's mechanism for how rock layers were formed. He had to believe that layers of rock are formed by the slow accumulation of sediment over eons of time. Darwin's conclusion regarding the fact that the geological column is evidence for macroevolution will only be right if those two assumptions are right.

So the big question is, are those two assumptions right? The answer, from a scientific standpoint, is that we don't know. Since this is not a geology course, we do not want to spend a lot of time on the specifics of how rock strata form. Instead, we will simply say that scientists have seen rock layers form slowly as a result of a process much like that suggested by Lyell. Scientists have also seen that natural catastrophes like floods and volcanic eruptions can lay down many layers of rock virtually overnight. American geologists who have spent time studying the results of the eruption of Mount St. Helens in the southern part of the state of Washington, for example, have documented the formation of rock layers, canyons, and fossils all in the span of less than 24 hours after the main eruption.

Science tells us, then, that rock strata like those in the geological column can be formed either slowly in a process like that suggested by Lyell or quickly in natural catastrophes. So, if you *want* the geological column to provide evidence for macroevolution, it does. You simply have to assume that it was formed much in the way Lyell suggested that it was formed. If, on the other hand, you *don't want* the geological column to provide evidence for macroevolution, you can say that it was not formed that way. Instead, it was formed quickly as the result of natural catastrophe. These authors, for example, consider the major parts of the geological column to be the result of Noah's flood. An excellent book written by Dr. Henry Morris entitled *The Genesis Flood* gives convincing evidence that this is, indeed, the case. Of course, those who believe that the geological column was formed according to the speculations of Lyell also have evidence of their own, so the final answer is not clear.

Since, from a scientific point of view, we really don't know whether the geological column was formed according to the speculations of Lyell or by natural catastrophe, the data from the geological column is inconclusive. IF the geological column was formed according to the speculations of Lyell, then it is excellent evidence for macroevolution. IF, on the other hand, it was formed by natural catastrophe, then it is excellent evidence against macroevolution, because that would mean all life forms fossilized in the geological column existed at the same time (the time of the catastrophe). Creation scientists believe that there was a great natural catastrophe (Noah's Flood) that could easily explain the geological column. Thus, they believe that the geological column is evidence against macroevolution. Evolutionists, on the other hand, do not believe in Noah's Flood, so they believe that the geological column is evidence for macroevolution.

ON YOUR OWN

9.6 What is the big assumption that Darwin had to make to interpret the geological column as evidence for macroevolution?

9.7 Why is the geological column not conclusive evidence for or against macroevolution?

The Details of the Fossil Record: Evidence Against Macroevolution

Well, if the geological column holds no real evidence for or against macroevolution, where must we look next? We can look the same place Darwin did. After examining the geological column, Darwin looked at the details of the fossil record. If macroevolution did occur, geologists should be able to find series of fossils that demonstrated how one species slowly evolved into another. If wild dogs, for example, did eventually give rise to horses, then there should be fossils of animals that were somewhere in between a dog and a horse. Darwin called these life forms (which he assumed must have existed) **intermediate links**, because they represented a link between one species and another. Unfortunately for Darwin, there were only a few examples of fossils that might be interpreted to be intermediate links, and even for those fossils, their status as intermediate links was quite questionable.

This lack of intermediate links was the most vexing problem that Darwin had with his hypothesis. In fact, in his book, he stated:

"Geological research, though it has added numerous species to existing and extinct genera, and has made the intervals between some few groups less wide than they otherwise would have been, yet has done scarcely anything in breaking the distinction between species, by connecting them together by numerous, fine, intermediate varieties; and this not having been affected, is probably the gravest and most obvious of all the many objections which can be raised against my views." (*The Origin of Species*, 6th ed, 1962, Collier Books, NY, p.462.)

Notice what he says here. Darwin's hypothesis says that one species eventually led to another. Thus, there should be "fine, intermediate varieties" of fossils in between species. The fact that there weren't was a problem that he called "grave." Additionally, note that he admits there are "many objections" which can be raised against his views. As we stated before, Darwin was very open about the flaws that he saw in his macroevolutionary hypothesis. In fact, he devoted more pages of his book detailing the evidence against macroevolution than he did detailing the evidence for it!

Although Darwin could not find any good examples of intermediate links in the fossil record, he had a hope. He figured that geology was still in its infant stage as a science, and that geologists just hadn't found the intermediate link fossils yet. He was convinced that as time went on, geologists would find these intermediate links. Thus, he said that the intermediate links were currently just "missing" from the fossil record, but they would be found in time. Critics of macroevolution quickly coined the phrase "missing link" to emphasize that the fossil record was devoid of any evidence for macroevolution.

Well, what of these missing links? Has geology uncovered them? The answer to that is an unequivocal NO. Read, for example, the words of Dr. David Raup, the curator of the Chicago Field Museum of Natural History and an expert on the fossil record.

"Well, we are now about 120 years after Darwin, and knowledge of the fossil
record has been greatly expanded…ironically, we have *even fewer examples* of
evolutionary transition than we had in Darwin's time. By this I mean that some of
the classic cases of Darwinian change in the fossil record, such as the evolution of
the horse in North America, have had to be discarded or modified as the result of
more detailed information." (*Field Museum Bulletin* January, 1979)

So Dr. Raup says (and all experts on the fossil record agree) that the missing links are still
missing! Darwin saw this fact as strong evidence against macroevolution, and Dr. Raup says that
the situation is worse now than ever! The fossil record, then, is clearly strong evidence against
macroevolution.

Now think about this for a minute. The hypothesis of macroevolution tries to explain
something about earth's past. Since no one was around back then to tell us whether or not
macroevolution actually happened, it is necessary to look for data that either supports or
contradicts the hypothesis. Well, if you're looking for data about earth's history, where is the
first place you would look? You would look in the fossil record! What does the fossil record
say? It says that macroevolution never happened! Do you see what we mean when we say that
scientists don't believe in macroevolution today because of the evidence? If the fossil record (the
main place you look for information about earth's past) says that macroevolution did not occur,
then scientists simply should not believe that it did occur. Unfortunately, many scientists look
past the fossil record, because they *want* to believe in macroevolution!

At this point, you might be thinking, "Wait a minute, haven't I seen museum exhibits,
television programs, and books with detailed, macroevolutionary sequences? Where have they
come from if the fossil record is so devoid of evolutionary intermediate links?" Well, they come
either from misinterpretations of the fossil records or from the imaginations of those who really
want to believe in evolution.

For example, take another look at the quote from Dr. David Raup. He mentions "…the
evolution of the horse in North America" as an example of an evolutionary sequence that had to
be "…discarded or modified as the result of more detailed information." You see, a set of fossils
that supposedly showed the evolution of the horse from a small, dog-like creature was once
considered excellent evidence for evolution. If you examine Figure 9.2, you will see why.

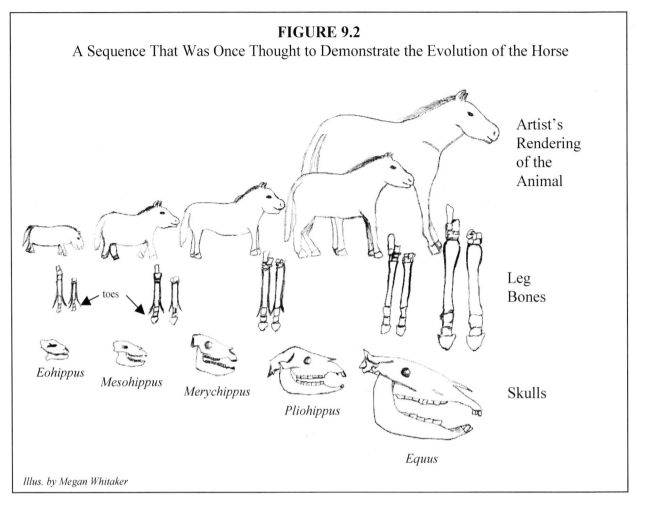

FIGURE 9.2

A Sequence That Was Once Thought to Demonstrate the Evolution of the Horse

Artist's Rendering of the Animal

Leg Bones

toes

Eohippus   *Mesohippus*

Merychippus

Pliohippus

Skulls

Equus

Illus. by Megan Whitaker

It's pretty obvious why these fossils used to be thought of as excellent evidence for macroevolution. This is just what macroevolution predicts the fossil record should be full of. One species (the *Eohippus*) slowly changing into several other species, gradually developing into something completely different (the modern horse, *Equus*). Look at the gradual change in the leg bones. *Eohippus* starts out with toes, but as you go across the sequence, the toes begin to disappear and the final animal (*Equus*) has no toes, just a hoof. Excellent evidence for macroevolution, right?

Well, it would be, except for a couple of inconvenient facts (the "more detailed information" that Dr. Raup spoke of). The first problem is geography. There is no site in the world where this supposed macroevolutionary succession can be seen in the fossil record. Instead, these fossils have been gathered from several different continents and then simply "thrown together" to form a macroevolutionary sequence. Clearly, however, if each member of the sequence evolved from the previous member, they would have had to be at least on the same continent!

Of course, geography is only the first problem with the sequence. Another problem is time. You see, macroevolution demands that the first member of the sequence (*Eohippus*) must have lived *before* the others. Then slowly, over time, the next member of the sequence (*Mesohippus*) should appear, followed by the next (*Merychippus*), and so on. The plain fact is that *Eohippus* has been found in rock strata just as near to the surface of the earth as has *Equus*. In fact, there seems to be no pattern between the depth at which you find the fossils of these creatures and their position in the supposed macroevolutionary sequence. Instead, geologists admit that all of these creatures lived at the same time. Based on that fact alone, then, this cannot be a macroevolutionary sequence.

There is still one more problem with this sequence! Although the leg bones and skulls seem to show a gradual macroevolutionary sequence, much of the rest of the skeletons of these creatures do not. The number of ribs on each creature varies from 15 to 19 with no discernible pattern. Also, the number of lumbar vertebrae (bone segments in the backbone) vary from 6 to 8 with no discernible pattern. If this really were a macroevolutionary sequence, the ribs and vertebrae should vary as smoothly as the leg bones seem to.

So, in the end, even those who believe in macroevolution admit that this is not a macroevolutionary sequence. Instead, these are just creatures whose similar structures make them *appear* to form a macroevolutionary sequence. Once all of the facts are known, this just becomes another in a long list of "evidences" for macroevolution that have been overturned "…as the result of more detailed information."

We have spent so much time on this figure because it brings up a very important point about how the hypothesis of macroevolution is treated in the media, in textbooks, and in museums. Even though evolutionary scientists have admitted for more than 20 years that the data in Figure 9.2 is not a macroevolutionary sequence, it is still presented in the media, in textbooks, and in museums as great evidence for macroevolution! You see, as Dr. Raup admitted earlier, there are really no intermediate links in the fossil record. Thus, in order to convince people to believe in macroevolution, textbooks, museums, and the media *simply tell lies*. Dr. Niles Eldredge, curator of the American Museum of Natural History, and a believer in macroevolution, admits this.

> "I admit that an awful lot of that has gotten into the textbooks as though it were true. For instance, the most famous example still on exhibit downstairs [in *his* museum] is the exhibit on horse evolution prepared perhaps 50 years ago. That has been presented as literal truth in textbook after textbook. Now I think that that is lamentable..." (As quoted in *Darwin's Enigma: Fossils and Other Problems*, Master Books 1988, p.78)

Sadly, this is not the only example of macroevolutionists lying in order to make their case believable. The history of paleontology (the study of fossils) yields many more examples. Also, molecular biology (which we will study in an upcoming section) provides still more!

So, after everything is said and done, are there any fossils of creatures that are intermediate links? The answer to that question depends on how you look at the fossils. For example, one of the fossils that macroevolutionists (currently) think represents an intermediate link between monkey and man is *Australopithecus* (aw stray' low pih thih cus) *afarensis* (ah fuh' ren sis). A sketch of the most complete skeleton of this creature, nicknamed "Lucy," is shown in Figure 9.3.

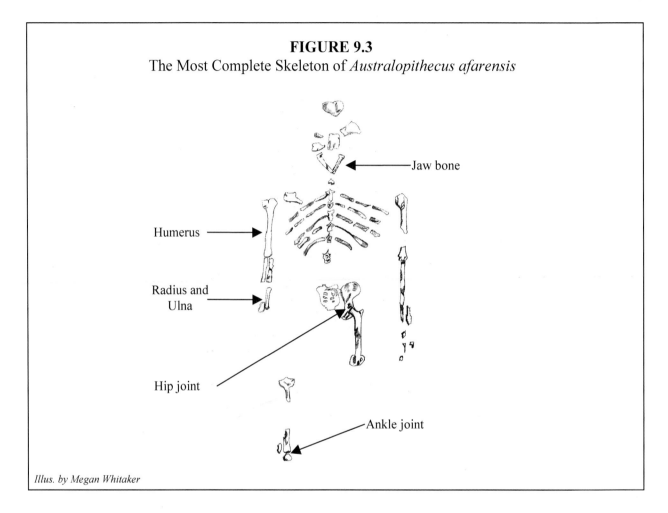

FIGURE 9.3
The Most Complete Skeleton of *Australopithecus afarensis*

Jaw bone

Humerus

Radius and Ulna

Hip joint

Ankle joint

Illus. by Megan Whitaker

Although this might not look like a complete skeleton to you, it is actually quite a find. One of the problems with studying fossils is the fact that they tend to be very incomplete. As a result, scientists are left to speculate on the nature of a creature based often on far too little evidence. Many fossils were proclaimed intermediate links based on just a few tiny fragments of bone. Perhaps the greatest example was Nebraska Man. Scientists found a single tooth among the remains of some ancient tools. Based on *that single tooth*, scientists proclaimed that they had found one of the intermediate links between man and ape. They had artists draw full pictures of Nebraska Man, based only on a tooth! The pictures looked very convincing - a hulking, brutish figure that indeed looked to be a link between man and ape. Later on, it was conclusively shown that the tooth actually belonged to a peccary, which is a certain type of pig! Thus, you should never believe the drawings that you see in books. Find out what *fossils* exist for the supposed creature, you will most likely be amazed at how little there is!

This specimen of *Australopithecus afarensis*, however, is quite a nice specimen, so we can learn a lot from it. When scientists examine such a specimen, they typically look at a few "key" bones that are characteristic of a certain kind of creature. For example, jaw bones are often very distinctive among most animals. Notice how the jaw bone of *Australopithecus afarensis* is shaped like a "V." This is very distinctive of an ape. Also, the ratio between the size of the humerus and the size of the radius or ulna is also helpful in determining the kind of creature that *Australopithecus afarensis* is. Once again, the ratio of these bones indicates that *Australopithecus afarensis* is an ape. In fact, each bone in this entire skeleton indicates that *Australopithecus afarensis* is an ape.

If the bones indicate that it is an ape, why is *Australopithecus afarensis* considered an intermediate link? Well the hip joint and ankle joint can be constructed in such a way as to make *Australopithecus afarensis* stand upright in a relatively comfortable manner. This is unusual in apes. Most apes tend to be comfortable on all fours, whereas humans tend to be comfortable standing upright. Thus, because it is *possible* that *Australopithecus afarensis* might have stood upright, it is considered an intermediate link between man and ape.

Of course, not even all evolutionists agree that *Australopithecus afarensis* stood upright. There is still a great deal of debate in the literature on this point. Nevertheless, most evolutionists do agree that *Australopithecus afarensis* is an intermediate link between ape and man. Of course, creation scientists contend that since *every bone in its body* indicates that it is an ape, it is safest to assume that *Australopithecus afarensis* is an ape. So, whether or not *Australopithecus afarensis* is an intermediate link depends on your point of view. If you want to believe in macroevolution, it is. If you do not want to believe in macroevolution, it is not.

There are only a few other examples of possible intermediate links. Perhaps the most famous is *Archaeopteryx* (ark ee' op ter icks), which evolutionists want to believe is an intermediate link between reptiles and birds. Once again, however, the fossil seems to indicate that it is a fully-functional bird. Because of some small differences between that fossil and the "typical" bird, however, macroevolutionists want to believe it is an intermediate link. The point, however, is quite clear. If macroevolution happened, then the fossil record should be littered with intermediate links, as Darwin predicted. Instead, macroevolutionists can only present a few highly questionable ones. These supposed intermediate links always closely resemble one of the two species they are supposed to link together. If intermediate links existed anywhere except in the imaginations of committed macroevolutionists, you would think that at least one unambiguous intermediate link could be found somewhere in the fossil record!

ON YOUR OWN

9.8 If macroevolution really occurred, would you expect to find more fossils of individual species or of intermediate links?

9.9 Why is the supposed macroevolutionary sequence for the horse wrong?

9.10 Why do macroevolutionists consider *Australopithecus afarensis* an intermediate link between man and ape? Why do creation scientists think that it is not?

Structural Homology: Formerly Evidence for Macroevolution, Now Evidence Against It

Another evidence that Darwin detailed in his book came from comparing the skeletal structures of vastly different species. He looked at data like that given in Figure 9.4, and he thought it gave excellent evidence for macroevolution.

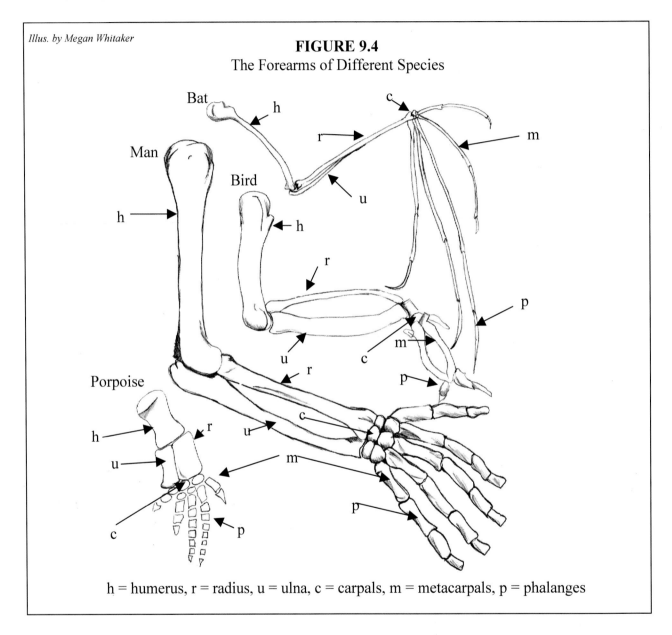

FIGURE 9.4
The Forearms of Different Species

Illus. by Megan Whitaker

h = humerus, r = radius, u = ulna, c = carpals, m = metacarpals, p = phalanges

Notice in the figure that although these arm bones come from very different species, they have an amazing amount in common. They all are jointed about halfway down. They all consist of a single, upper arm bone, the humerus (hyou' mer us), and two lower arm bones, the radius and ulna (uhl' nah). They all end in a palm consisting of carpals (car' puhls) , metacarpals (met uh car' puhls), and phalanges (flan' jeez). Indeed, with the exception of the bird, they all have five digits!

Darwin said that these incredible similarities between such vastly different species was excellent evidence that they all had a common ancestor. After all, he said, that common ancestor probably had a forearm which was jointed in the middle, had a humerus, followed by a radius and an ulna, followed by carpals, metacarpals, and phalanges. This general forearm structure, then, over eons of time and guided by natural selection, simply adapted to the needs of each individual species as it arose. Thus, much like you can tell that a man and his grandson are related because of their striking similarities, you can assume that man and the bat are related by a common ancestor because of the striking similarities in their forearms.

The study of similar structures in different species is called **structural homology** (hum awl' uh gee).

<u>Structural homology</u> - The study of similar structures (bones or organs, for example) in different species

In Darwin's time, structural homology was very strong evidence for macroevolution. How could vastly different species have such similar characteristics unless they were all related by a common ancestor? If they all had a common ancestor, then clearly macroevolution would have to have occurred in order to turn this common ancestor into these vastly different species, right?

That sounded like a great argument in Darwin's time because scientists back then had no idea how traits were passed on from generation to generation. With the advent of Mendelian genetics, however, scientists finally began to understand how this happens. As scientists began to understand genetics and DNA better, they developed technology to actually determine the sequence of nucleotide bases in an organism's DNA. This spelled the end of structural homology as evidence of macroevolution.

You see, if structural homology was the result of common ancestry, then that fact should show up in the genetic codes of the organisms which possess similar structures. Take, for example, the forearms shown in Figure 9.4. If the forearms of the bat, bird, man, and porpoise look so similar because they all inherited their forearms from a common ancestor, then the parts of their DNA that contain the information regarding the forearms should be similar. After all, traits are passed from parent to offspring through DNA. If each one of these creatures inherited their forearm structure from a common ancestor, then the portions of DNA which contain information about the forearm would all have come from that same, common ancestor. As a result, those portions of the DNA should be similar from organism to organism.

Is this the case? Is structural homology the result of similar DNA sequences? No, it is not! Read the words of Dr. Michael Denton, an evolutionist:

"The evolutionary basis of homology is perhaps even more severely damaged by the discovery that apparently homologous structures are specified by quite different genes in different species...With the demise of any sort of straightforward explanation for homology one of the major pillars of evolution theory has become so weakened that its value as evidence for evolution is greatly diminished." (*Evolution: A Theory in Crisis*, Michael Denton, Adler and Adler 1985, pp.149 and 151)

Notice what Dr. Denton says here. Apparently homologous structures are specified by quite different genes in different species. Well, if they are specified by different genes, then there is *no way* that the homologous structures could have been inherited from a common ancestor. As basic Mendelian genetics tells us, the only way to inherit something from an ancestor is through the genetic code!

To creation scientists, Figure 9.4 actually offers excellent evidence for a Creator. After all, any good engineer, once he or she finds a design that works, tends to stick with that design and simply adapt it from situation to situation. Thus, structural homology is, to creation scientists, evidence of common design, not common ancestry.

Think about what we have learned so far. Darwin's two greatest evidences for macroevolution were the geological column and structural homology. As we have seen, based on today's understanding of science, the geological column provides inconclusive evidence, and structural homology now provides *evidence against macroevolution*. In addition, the fossil record provides evidence against macroevolution. Is this the end of the story? Actually, no. It turns out that we haven't even studied the data that speaks out against macroevolution the strongest. We save that for the next section!

ON YOUR OWN

9.11 What major scientific breakthrough led to structural homology changing from evidence for macroevolution to evidence against it?

Molecular Biology: The Nail in Macroevolution's Coffin

Of all the scientific data that provides evidence against evolution, perhaps the most important comes from the field of molecular biology, which studies the properties and structures of the molecules important to biology. Think back to what you learned in the three previous modules. Aside from DNA, what is the most important type of molecule in the chemistry of life? Proteins! As a result, a large amount of the research effort in molecular biology centers around understanding proteins.

Early on, molecular biologists noticed something rather amazing. There are certain proteins that are common to many species. Most animals, for example, have the protein hemoglobin. This protein transports oxygen through the bloodstream to the cells. In addition, most organisms have the protein cytochrome C, which takes part in cellular metabolism. Interestingly enough, these proteins are not identical from species to species. In other words, the cytochrome C that you find in a bacterium is different than the cytochrome C that you find in a human.

How do these proteins vary from species to species? Well, what determines the structure and function of a protein? In Module #5, we learned that the sequence of amino acids within a protein determines its structure and function. Thus, if you were to examine the amino acids in the cytochrome C of different species, you would find slightly different sequences. For example, Table 9.1 lists the amino acid sequences of the same portion of cytochrome C for several different types of organisms.

TABLE 9.1
Partial Cytochrome C Amino Acid Sequences

Organism	Amino Acid Sequence (partial)
Horse	Gly-Leu-Phe-Gly-Arg-Lys-Thu-Gly-Glu(NH_2)-Ala-Pro
Kangaroo	Gly-Ile-Phe-Gly-Arg-Lys-Thu-Gly-Glu(NH_2)-Ala-Pro
Human	Gly-Leu-Phe-Gly-Arg-Lys-Thr-Gly-Glu(NH_2)-Ala-Pro
Chicken	Gly-Leu-Phe-Gly-Arg-Lys-Thr-Gly-Glu(NH_2)-Ala-Glu[*]
Tuna	Gly-Leu-Phe-Gly-Arg-Lys-Thr-Gly-Glu(NH_2)-Ala-Glu[*]
Moth	Gly-Phe-Gly-Arg-His-Thr-Gly-Glu(NH_2)-Ala-Pro-Gly
Yeast	Gly-Ile-Phe-Gly-Arg-His-Ser-Gly-Glu(NH_2)-Ala-Glu(NH_2)

[*]NOTE: The chicken and tuna Cytochrome C sequences are the same for this *partial* amino acid sequence. Cytochrome C has many, many amino acids. We are looking only at one section here. The differences between chicken and tuna Cytochrome C occur in other sections of the protein.

Now remember what we are looking at here. Each three-letter abbreviation represents a specific amino acid. Thus, "Gly" stands for the amino acid glycine, "Leu" stands for the amino acid leucine, and so on. Don't worry about the NH_2. That's an amine functional group that really has no bearing on our discussion.

What do we see in studying Table 9.1? Well, first notice that all of these sequences are very similar. That's not at all surprising, because the protein is the same in each case: cytochrome C. It performs the same basic function in each organism, but in order to be able to work with the specific chemistry of that organism, it is slightly different in each specific case. That's where the difference in the sequences come from. The cytochrome C section shown here, for example, is nearly identical between the horse and the kangaroo. The only difference is the second amino acid, which is leucine (Leu) in the horse and isoleucine (Ile) in the kangaroo. Because of this one difference, the cytochrome C of a horse will not work in a kangaroo or vice-versa.

What does all of this tell us? Well, think about how proteins are made. They are made in the cells *according to the instructions of DNA.* Thus, by looking at the amino acid sequences in a protein that is common among many species, you are actually looking at the differences between a specific part of those organisms' genetic code: the part that determines the makeup of that protein. If macroevolution is true, then that portion of the genetic code should reflect how "closely-related" two species are. If two species are closely related, then their DNA that codes for a common protein should be closely related. If they are only distantly related, then their DNA sequences that code for that same protein should have more significant differences between them. Looking at the difference between the amino acid sequences of a common protein is a way to determine just how many differences exist between the DNA of the organisms in question.

For example, the portion of the amino acid sequence for cytochrome C shown in the table is 11 amino acids long. Of those amino acids, there is only one difference between the horse and the kangaroo. We can therefore calculate the percentage difference between the cytochrome C amino acid sequence in a horse and the cytochrome C amino acid sequence in a kangaroo.

$$\text{percent difference} = \frac{1}{11} \times 100 = 9.1\%$$

Comparing the amino acid sequences in cytochrome C for the yeast and the horse, however, there are 4 differences. Thus, the percent difference is

$$\text{percent difference} = \frac{4}{11} \times 100 = 36.4\%$$

These two comparisons, then, tell us that the portion of a horse's genetic code that determines the makeup of the cytochrome C protein is much closer to the kangaroo's than it is to a yeast's. From a macroevolutionary point of view, this would tell us that the horse is more closely related to the kangaroo than to the yeast. In other words, in some macroevolutionary scheme in which one life form evolves into another, the kangaroo would be closer to the horse than would the yeast.

That makes sense, doesn't it? After all, a yeast is considered, by macroevolutionists, to be a rather "simple" life form, whereas kangaroos and horses are rather complex. As a result, it makes sense that a horse is more closely related to a kangaroo than a yeast. When data like this first came out, it was hailed as great evidence for macroevolution. The macroevolutionists' hopes were high, because they thought that they had finally found a way to specifically determine exactly how closely related species were. Unfortunately for evolutionists, as more and more of these amino acid sequences were mapped out, the evidence quickly turned *against* macroevolution.

Consider, for example, the bacterium *Rhodospirillum* (rho doe spuh ril' ee um) *rubrum* (rube' rum). When its cytochrome C amino acid sequence was mapped out and compared to vastly different organisms, nothing made sense in terms of the macroevolutionary hypothesis. Table 9.2 shows the percentage difference between the amino acid sequence in a *Rhodospirillum rubrum*'s cytochrome C and the amino acid sequence of other organisms' cytochrome C.

TABLE 9.2
Percent Differences Between a Bacterium's Cytochrome C and That of Other Organisms

Organism	Percentage Difference
Horse	64%
Pigeon	64%
Tuna	65%
Silkworm moth	65%
Wheat	66%
Yeast	69%

Now remember what macroevolution says. It says that "complex" life forms evolved from "simple" ones. Well, the "simplest" life form on the planet is a bacterium. Of the organisms listed in the table, the yeast (a single-celled fungus) is probably the next "simplest" life form. Increasing in complexity then comes the silkworm moth, followed by the tuna, followed by the pigeon, followed by the horse. Thus, macroevolution would assume that the bacterium eventually evolved into the yeast, which eventually evolved into the silkworm moth, etc., etc., all the way up to the horse. As a result, then, the yeast should be more closely related to the bacterium, the silkworm moth the next most closely related, and so on. According to the sequence of amino acids in their cytochrome C, however, each organism in the table is essentially as closely related to the bacterium as any other organism on the table! If anything, the bacterium is more closely related to the *most complex* organisms, not the least complex ones!

In other words, the data presented in Table 9.2 show none of the evolutionary relationships that *should* exist if macroevolution really occurred. Instead, this data seems to indicate that the bacterium is just as different from the horse as it is from the yeast! As more and more data like this was collected, it became increasingly obvious that this was the "normal" pattern to emerge. Regardless of the protein studied, the amino acid sequences seemed to indicate that each individual type of organism was just as different from one type of organism as it was from another. Just to make it clear that the data is really overwhelming on this point, we present a few more tables like Table 9.2.

TABLE 9.3
More Percentage Differences Between Cytochrome C Amino Acid Sequences

The cytochrome C of a lamprey eel compared to other organisms.		The cytochrome C of a carp compared to other organisms.		The cytochrome C of a pigeon compared to other organisms.		The cytochrome C of a horse compared to other organisms.	
Organism	% diff	Organism	% diff	Organism	% diff	Organism	% diff
Horse	15%	Horse	13%	Horse	11%	Pigeon	11%
Pigeon	18%	Pigeon	14%	Carp	14%	Turtle	11%
Turtle	18%	Turtle	13%	Turtle	8%	Carp	13%
Carp	12%	Lamprey	12%	Lamprey	18%	Lamprey	15%

Now think about what this data tells us. We are comparing the cytochrome C's of a mammal (horse), a bird (pigeon), a reptile (turtle), a fish (carp), and an eel (lamprey). According to macroevolution, the eel would have come first, then later evolved into the fish, which later evolved into an amphibian, which later evolved into the reptile, which later evolved into the bird and the horse. Thus, the eel should be most similar to the fish, then the reptile, and then the bird, and it should have the least in common with the horse. Instead, the data tell us that the cytochrome C in the lamprey eel is most like the carp, *but then next most like the horse, and has essentially the same differences between the turtle and the pigeon*!

If you look at the data presented in Tables 9.2 and 9.3, it is clear that you can establish no evolutionary trends. This is the case with the vast majority of the data collected from molecular biology. If you map the amino acid sequences of virtually any protein and compare the differences between organisms that have that protein, you will generally find no macroevolutionary trends. Instead, each kind of organism seems to be equally or nearly equally different from every other kind of organism. As is the case with all of science, there are exceptions to this general rule, but those exceptions are quite rare.

Even though the exceptions are rare, many disreputable macroevolutionists actually highlight those exceptions as evidence for evolution. You see, if you pick and choose your data very carefully, you *can* find examples of molecular biological data that *seem* to indicate a macroevolutionary trend. For example, look at Table 9.4.

TABLE 9.4
A Highly Selective View of Some Cytochrome C Sequences Compared to Human Cytochrome C Sequences

Organism	% Difference Compared to Human
Rhesus monkey	1%
Rabbit	9%
Horse	12%
Tuna	20%
Screw Worm	25%
Wheat	38%
Yeast	41%
Bacterium	65%

Now this table does seem to indicate a macroevolutionary trend, doesn't it? After all, as you go down the table, you are getting to "simpler" and "simpler" life forms. As the life forms get "simpler," their cytochrome C seems to get more and more different from human, doesn't it?

This table, or something like it is often presented as strong evidence for macroevolution. There's only one problem. Much like the macroevolutionary sequence of the horse, evolutionists know that this is a big lie! In order to construct this table, macroevolutionists must *ignore 99%*

of the data from molecular biology and choose only the 1% of the data that agrees with their hypothesis. Then, they usually proceed to tell the reader that this kind of data is typical of the data from molecular biology! Of course, if you read the scientific literature concerning the data from molecular biology, there is a lot of hand-wringing among macroevolutionists. They just don't understand why 99% of the data disagree with their hypothesis. Instead of admitting this in textbooks and the like, however, they simply sweep that 99% of the data under the rug, so as not to "confuse" students!

Once again, we see that those who believe in macroevolution are forced to lie in order to make their case look stronger than it is. The data from molecular biology clearly wipes away any hope of being able to relate species in some sort of macroevolutionary framework. Those who really want to believe, however, simply discard the data that disagrees with what they want to believe and concentrate on the tiny subset of data that agrees with their preconceptions. Clearly, this is not real science!

ON YOUR OWN

9.12 A molecular biologist details the amino acid sequences in a common protein for the following creatures: a human, a rat, an amoeba, a fish, and a frog. Assuming macroevolution did occur, list these creatures in terms of increasing similarity between their protein and the human protein.

9.13 Why is the comparison of amino acid sequences in common proteins such a useful tool in determining whether or not macroevolution occurred?

Macroevolution Today

Because there is virtually no data to support the hypothesis of macroevolution, the hypothesis has changed greatly since Darwin's time. Mostly to try and get around the fact that there seem to be no intermediate links in the fossil record, macroevolutionists have come up with two major revisions in Darwin's hypothesis. The first revision is typically called **neo-Darwinism** and the second is usually called **punctuated equilibrium**.

Neo-Darwinism was the first revision to occur in Darwin's theory. Remember, Darwin pretty much established the theory of microevolution, but when it came around to providing evidence for macroevolution, he was stuck. When scientists began to understand genetics, they understood why he got stuck. As we said before, microevolution is simply the variation of a type of organism *within* its genetic code. Thus, microevolution is understandable in terms of Mendelian genetics. Macroevolution, however, is a completely different story. In order for macroevolution to occur, a species would have to *add* information to its genetic code. How could that happen? Well, neo-Darwinists said that it could happen by mutation. You see, Darwin thought that all of macroevolution could occur by the normal changes that happen during the reproductive process. When scientists came to understand genetics, however, it became clear that those kinds of changes only explained microevolution. Something else had to happen to add information to the genetic code. That something else was mutation.

Remember, mutation is a mistake that occurs during meiosis or mitosis. When it occurs, the genetic code is altered. Now as we discussed previously, that alteration always seems to be bad for the organism, but that didn't seem to bother the neo-Darwinists. Suppose, they speculated, that a few *good* mutations could occur in a species. They might, by blind chance, produce a new trait in the organism that made it more fit to survive. This, trait, then, would be passed on, adding new information into the genetic code of the species!

What really excited neo-Darwinists about this idea was not only did it provide a means by which an organism could add information into its genetic code, but it also allowed them to explain away at least some of the discrepancy between their hypothesis and the fossil record. After all, mutations result in dramatic changes between parent and offspring. If macroevolution occurred by these dramatic changes, then perhaps there shouldn't be as many intermediate links as Darwin first suspected. This modification, then, seemed to kill two birds with one stone.

Of course, one big problem with this idea is that there has never, ever been a documented case of a mutation that made an organism more fit to survive. Most mutations are detrimental to an organism. Some are only mildly harmful, and even fewer seem to be benign. Sometimes, a mutation will occur that helps an organism in one way and hurts it in another. For example, sickle-cell anemia is a disease in humans that is caused by mutation. The result is somewhat beneficial, because those who have the disease are much less likely to die from malaria. However, sickle-cell anemia makes your blood much less efficient at sending oxygen to your cells. As a result 25% of all who have this disease die prematurely. Is that a beneficial mutation? Of course not. Nevertheless, if you ask a macroevolutionist for a beneficial mutation, this is *the only one* he or she will give you. Why? Because it's the closest thing to a beneficial mutation that we have ever seen!

Now the fact that no truly beneficial mutations exist has never bothered neo-Darwinists much. After all, they can always say that sometime, way back when, there were beneficial mutations. The problem for neo-Darwinists, however, is still in the fossil record. Even though neo-Darwinism reduced the number of intermediate links needed for evolution, it did not reduce the number to near zero, which is what exists in the fossil record. Thus, another revision had to be made.

The latest revision is called punctuated equilibrium, and is just a spin-off of neo-Darwinism. It says that macroevolutionary steps occurred over very short time intervals, followed by long time periods of no macroevolution occurring. The idea is that a group of organisms might be suddenly exposed to high levels of toxic chemicals or radiation. This would speed up the mutation rates among the offspring. Most of these offspring would die, but a few lucky ones would get several good mutations all at once, and become more fit to survive. As the high levels of toxic chemicals or radiation continued, these few lucky offspring would produce more offspring with more mutations. Once again, most would die, but a few lucky ones might get several mutations that were good, making them more fit to survive. After just a few generations, the radiation or chemicals would be gone, and a new species would have emerged. That species would exist, essentially unchanged, for millions of years until another episode of high mutation rates occurred.

Notice what punctuated equilibrium does. It still counts on mutations to add information to the genetic code of the species that is evolving, but it expects those mutations to occur over only a few generations. Thus, the intermediate links would not exist for very long. What does that accomplish? Well, if the intermediate links didn't live for very long, then their chance of being fossilized is very small compared to the final product, which would live for millions of years essentially unchanged. Thus, by altering the hypothesis, we can now account for the fact that no intermediate links appear in the fossil record.

Now of course, this still doesn't explain why structural homology is not echoed in the genetic code, and it still doesn't tell us why the vast majority of molecular biology data indicates no macroevolutionary trends, but at least the problem with the fossil record is fixed! This is pretty much the current state of macroevolutionary thinking. The hypothesis has become so void of factual evidence that its proponents must cook up wild scenarios to explain away the lack of data in support of it.

ON YOUR OWN

9.14 What is the main difference between Darwin's hypothesis of macroevolution and the neo-Darwinist hypothesis?

9.15 What does punctuated equilibrium explain that neo-Darwinism and Darwin's original hypothesis cannot?

9.16 The three graphs below are hypothetical graphs that plot macroevolutionary change (on the y-axis) versus time (on the x-axis). Which graph represents Darwin's original hypothesis, which represents neo-Darwinism, and which represents punctuated equilibrium?

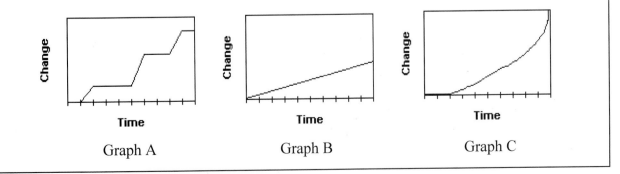

Graph A   Graph B   Graph C

Why Do So Many Scientists Believe in Macroevolution?

In this module, we have laid down some of the basic data relevant to evolution. We have shown that while microevolution is a well-established scientific theory, macroevolution is, at best, an unconfirmed hypothesis. There is no convincing data in favor of the hypothesis and plenty of data contradicting it. Given these facts, then, why do so many scientists believe in macroevolution? There isn't a simple answer to this question. Instead, there are a host of reasons that tend to lead to this situation.

The first reason is that scientists are simply indoctrinated at a very early age to become evolutionists. Since alternatives to evolution are not allowed in most classrooms (high school *or* college), students are left with no other explanation to one of life's most critical questions. If you think about it for a moment, you will realize how absurd such a situation is. The whole idea of intellectual pursuit is based on considering alternative ideas. Nevertheless, most schools will not even allow the consideration of any idea for life's origin other than macroevolution. One is forced to wonder why macroevolutionists are so afraid to allow their idea to compete with others!

The next reason is a consequence of the first. Those of us who do not believe in macroevolution are regularly ridiculed on college campuses. The best way to illustrate this is to look at the personal experiences of one of these authors, Dr. Jay L. Wile.

"Personally, I began experiencing the ridicule heaped on creationists when I attended graduate school at the University of Rochester. At that time, I had to choose a professor who would be my research advisor. His role was to direct my research and provide me with the training necessary to become a scientist in my chosen field. I picked my research advisor based on the kind of work that he did. The man I chose did what I considered to be the most interesting chemical research that went on at the University of Rochester.

While he was, in fact, an incredibly gifted scientist, my chosen research advisor was a committed macroevolutionist. As my creationist views became apparent to him, he ridiculed them. Of course, he would not discuss the data relevant to macroevolution with me. Instead, he simply made fun of me. In addition, one of the other members of our research team, a man I still consider to be the greatest scientist with whom I have ever worked, also constantly made fun of my creationist stand. Such a situation was very difficult for me. After all, I admired these two men. I aspired to become the caliber of scientist that they both were. Nevertheless, they made fun of my most strongly held scientific beliefs! Throughout my graduate school career, I concluded that if I was not as strongly convinced as I was by the data, I would have left my creationist stand behind to escape the ridicule of these scientific role models."

Sadly, Dr. Wile's story is very common among creation scientists.

The third reason that most scientists still believe in macroevolution stems from the fact that most of the scientists have never investigated the data. There is, of course, a very good reason for this. Any time that a scientist spends researching something that is not directly related to his or her field, the less likely that person is to become a great scientist. Today's science is so intricate and so specialized, that a person must devote himself completely to his field, or he will be left behind. Thus, since scientists have been indoctrinated over the years to believe that macroevolution is the only scientifically sound belief system, they are not about to waste time and energy researching the facts surrounding the issue. The wasted time would damage their careers too severely.

In addition to the time constraints that inhibit scientists in examining questions outside of their field, there is a definite pressure in the university setting to either believe in macroevolution or simply stay quiet about it! For example, a professor of biology at San Francisco State University was forbidden to teach the introductory biology course that he had been teaching for more than a decade because he began stressing the design elements that are prevalent in the world around us. The biology department felt that this would "confuse" students when they later reached a course on evolution, so they forbade him to teach that course! Creationist professors regularly feel great pressure from their university to keep quiet about their creationist views, and their colleagues constantly ridicule them for their beliefs.

In the end, then, scientists are indoctrinated at a very early age to be macroevolutionists; they are ridiculed throughout their careers if they believe otherwise; they usually cannot spare the time necessary to look into the facts regarding macroevolution; and they are often punished for believing anything else. These issues keep scientists from learning the data necessary to help them understand the fact that macroevolution is no more than an unconfirmed hypothesis. Although things seem to be changing (a lower percentage of scientists believe in macroevolution today than any time over the last 50 years), the pace of change will stay very slow unless schools (both at the secondary and college level) allow more academic freedom to simply discuss the data relevant to macroevolution and allow the presentation of other alternatives.

ANSWERS TO THE ON YOUR OWN PROBLEMS

9.1 Darwin was influenced by <u>the idea of a struggle for survival</u> as first proposed by Thomas Malthus. In addition, Sir Charles Lyell's idea that <u>the present is the key to the past</u> influenced him.

9.2 <u>Darwin would say that in the beginning, only slow predators existed. As food got scarce, however, any predator born that was slightly faster than the others would be able to get to more food; thus, this predator would be naturally selected to survive and would pass on its speed to its offspring. As generation after generation passed, each time a predator was born that was slightly faster than its peers, it would be naturally selected to survive. Thus, the extra speed would "pile up" generation after generation until, eventually, a cheetah was formed.</u>

9.3 Darwin believed that any variation which gave an organism an advantage in the struggle for life would make that organism more likely to live. <u>If the organism has an advantage, we could say that it was more "fit" for survival than those without that advantage. Thus, the "fittest" organisms will tend to survive.</u>

9.4 <u>This is an example of microevolution.</u> The moths did not become some other insect. They simply changed color. This is variation within the moth's genetic code, which is microevolution.

9.5 <u>This wild scenario is an example of macroevolution.</u> In order for something like a cow to turn into a whale, radical changes must be made to the genetic code. This is what must happen in macroevolution.

9.6 In order for the geological column to be interpreted as evidence for macroevolution, <u>Darwin had to assume that each strata of rock was laid down individually over long periods of time, according to the speculations of Lyell.</u> That way, the lower rock strata could be interpreted as older than the upper rock strata.

9.7 <u>It is not conclusive because whether the geological column supports or contradicts macroevolution depends on an assumption that cannot be checked.</u> Since scientists have seen rock strata form both as the result of a Lyell-like process and as the result of natural catastrophe, we have no idea which process is responsible for forming the geological column. Without knowing how the geological column was formed, we have no idea whether it gives evidence for or against macroevolution.

9.8 <u>You would expect more intermediate links than individual species.</u> After all, to get from one species to another there must have been several steps along the way. This would result in many intermediate links for each new species.

9.9 <u>It is wrong because the fossils have no common geographic location, they all lived at the same time, and the other parts of the creatures' skeletons do not show the trends that the leg bones and skulls do.</u>

9.10 Evolutionists think that since it is possible that this creature stood upright, it represents a link between man and ape. Creation scientists disagree because every bone in its body is characteristic of an ape. Thus, it seems most scientific to classify it as an ape.

9.11 Structural homology turned from evidence for macroevolution to evidence against it when scientists learned about genetics and how to map out nucleotide sequences.

9.12 According to a macroevolutionist, the order of creatures in terms of *increasing* protein similarity to human protein would be:

amoeba, fish, frog, rat

Since macroevolutinists believe that life evolved from the simple to the complex, then the more complex the organism is, the more similar its protein will be to the corresponding human protein. As we pointed out in the text, this is precisely what we *don't* see in the data!

9.13 Since DNA codes for the amino acid sequences in a protein, tracking the amino acid sequences is like comparing equivalent strands of DNA in different creatures. We assume the strands are equivalent because they code for the same protein.

9.14 The real difference is mutations. Neo-Darwinists use mutations to add information to the genetic code. Darwin thought that the normal variations that are the result of reproduction could drive macroevolution.

9.15 Punctuated equilibrium explains the lack of intermediate links in the fossil record. By constructing the scenario, punctuated equilibrium says that intermediate links only live during a brief period in time. This makes them very unlikely to be fossilized.

9.16 Darwin's original hypothesis assumed that the natural variation as a result of reproduction was responsible for macroevolution. This would result in very slow change. Thus, the graph that changes the slowest (B) represents Darwin's original hypothesis. Neo-Darwinism allowed for faster change using mutations, so (C) represents neo-Darwinism. In punctuated equilibrium, there are brief periods of change followed by long periods of no change. Thus, (A) represents punctuated equilibrium.

STUDY GUIDE FOR MODULE #9

1. Define the following terms:

   a The immutability of species

   b. Microevolution

   c. Macroevolution

   d. Strata

   e. Fossils

   f. Structural homology

2. Where did Darwin do most of the work which led to his hypothesis of evolution?

3. Did Darwin ever recant his scientific beliefs?

4. What was the main idea that Thomas Malthus' work gave to Darwin?

5. What was the main idea that Sir Charles Lyell's work gave to Darwin?

6. What age-old concept was Darwin able to dispel with his research?

7. Suppose a herd of horses were living in an area where food near the ground was scarce but there was plenty of food in the trees. If, after several generations, the horses gave rise to giraffes that could easily reach the food in the trees, would this be an example of microevolution or macroevolution?

8. Consider a fish population that is trying to survive under conditions of extremely cold water. If, over several generations, the fish develop thicker fat layers under their skin for better insulation, is this an example of microevolution or macroevolution?

9. From a genetic point of view, what is the main difference between microevolution and macroevolution?

10. In this module, we studied four main sets of data: the geological column, the fossil record, structural homology, and molecular biology. For each set of data, indicate whether it is evidence for or against macroevolution or if it is inconclusive. Briefly explain why.

11. Name two creatures that macroevolutionists claim are intermediate links and why they are not really intermediate links.

12. Consider the following amino acid sequences that make up a small portion of a protein:

   a. Gly-Ile-Gly-Gly-Arg-His-Gly-Gly-Glu(NH$_2$)-Glu-Glu(NH$_2$)-Lys-Lys-Lys

   b. Gly-Leu-Phe-Gly-Arg-Lys-Ser-Gly-Glu(NH$_2$)-Gly-Glu(NH$_2$)-Ala-Arg-Lys

   c. Leu-Ile-Gly-Gly-Arg-His-Ser-Gly-Glu(NH$_2$)-Ala-Glu(NH$_2$)-Arg-Arg-Arg

Which protein would you expect to be the most similar to a protein with the following subset of amino acids?

   Gly-Ile-Phe-Gly-Arg-His-Ser-Gly-Glu(NH$_2$)-Ala-Glu(NH$_2$)-Arg-Arg-Lys

13. Based on macroevolutionary assumptions, which organism's cytochrome C should most resemble that of a yeast: a kangaroo or a bacterium?

14. What problem with Darwin's hypothesis did neo-Darwinism hope to solve?

15. What problem with Darwin's hypothesis did punctuated equilibrium attempt to solve?

16. How would an adherent to punctuated equilibrium explain the lack of intermediate links in the fossil record?

17. What problems still exist for those who believe in punctuated equilibrium?

MODULE #10: Ecosystems

Introduction

Have you ever wondered why certain creatures can only be found in certain places on the earth? For example, where can you find polar bears? You can either find them in zoos or in the Arctic. Why? If polar bears can live in zoos, then they must be able to live in climates other than that found in the Arctic. Why, then, do they only live in the Arctic? Well, they live there because God has designed them to be a part of the Arctic's **ecosystem**.

<u>Ecosystem</u> - An association of living organisms and their physical environment

In order to assure that life can continue on any part of the earth, there must be living creatures (and a few non-living chemical and physical processes) that all play critical roles in the support of life. If any one of these creatures or factors were to disappear or to cease functioning properly, the ecosystem would fall out of balance, creating havoc. In addition, if a creature or process that isn't supposed to be there is suddenly added, similar havoc would ensue.

One of the best examples of what happens when an ecosystem is thrown out of balance comes from the country of Australia. In 1859, an Australian sheep rancher named Thomas Austin decided that he wanted to hunt rabbits on his ranch. There was only one problem: rabbits do not live in Australia. So he decided to import a few (24 to be exact) from England, where he had grown up. Well, in only six years, his land was home to more than 10,000 rabbits! He had hunted to his heart's content, claiming to have killed more than 20,000 rabbits, but they continued to overrun his ranch. They destroyed his grasslands, making it impossible for his sheep to graze. In essence, his sheep were starving, all because he had imported 24 rabbits six years earlier.

What happened? Well, Australia has its own, unique ecosystem. Several specific species of animals and plants live in that region of the earth, and they all have particular roles to play in order to make sure that life can exist there. When the rabbits were introduced into Australia, it upset the ecosystem. You see, in England, rabbits have several natural predators. They are, in essence, the main food supply for many creatures in England. Thus, God has designed them to reproduce rapidly, knowing that many, many rabbits are needed each year to meet the food needs of several creatures in England. None of the creatures in Australia, however, ate rabbits. With no predators, the rabbits were able to reproduce rapidly, with nothing to hold their numbers in check. In just a few years, Australia was overrun by rabbits!

By 1883, the Australian government decided this was a major problem. The rabbits had spread out more than 1,000 miles from Austin's ranch and were destroying some of the major grazing lands in the country. As a result, the government decided that something had to be done. What would you do to fix this problem? Would you import a bunch of natural predators (such as hawks or foxes) to start killing the rabbits? Well, that might take care of the rabbits, but then what would happen when the rabbits were gone? You would still have one creature (the fox or hawk) that didn't belong in the ecosystem. No one knows what damage that would cause!

Well, the first thing that the government tried was to hire exterminators to try and kill the rabbits. They spent thousands of pounds (the Australian unit of currency at that time) on poisons, rabbit traps, and hunters. When that didn't work, they even built a huge fence (42 inches high and almost 1200 miles across) to try and contain the rabbits. None of this worked. Rabbits were still overrunning Australia, destroying thousands of acres of prime grazing land.

By 1937, scientists had discovered and tested a virus that was harmful only to the English strain of rabbit that was imported to Australia. After extensive research, they decided that the virus would harm no other creature in Australia except the foreign rabbit. Thus, they infected many rabbits with the virus and released them into Australia. For a while, nothing happened, and scientists thought something had gone wrong. In 1950, however, a huge number of dead rabbits were found in a certain part of Australia. They had died of the viral disease. Soon, many more were dying. It turns out that mosquitoes who had bit the diseased rabbits were slowly infecting non-diseased rabbits. Soon, the rabbit population was under control.

Does that mean rabbits are gone from Australia? No, they are not. You see, as the rabbit population dwindled, so did the rate of infection. After all, the fewer rabbits there were, the fewer mosquitoes would get the virus by biting diseased rabbits. With fewer mosquitoes that carried the virus, then, even fewer rabbits would get newly infected. Thus, the rabbit population will probably never die in Australia. However, by finding a way to kill large numbers off, scientists were able to control an immense threat to the grazing lands of Australia.

This story helps to illustrate the way in which ecosystems across the world are finely balanced. If one creature is removed or added to an ecosystem, the results could be disastrous. Thus, an entire science has developed around the study of ecosystems. It is called **ecology.**

<u>Ecology</u> - The study of relationships among organisms in ecosystems

Generally, we hear about ecology in the context of pollution. When the media needs to hear scientific opinions regarding pollution and its effects, they usually interview ecologists. Ecology, however, is a much broader science. There is a lot more to be learned about ecosystems than what pollution does to them. This module will give you a glimpse of this fascinating subfield of biology.

Energy and Ecosystems

In order to survive, all organisms need energy, and lots of it. Where does that energy come from? Ultimately, almost all of it comes from the sun. As we learned in Module #1, autotrophic organisms convert the energy from the sun into food for themselves. This food gives them the energy to sustain their life functions. We say that "almost all" of the energy used in ecosystems comes from the sun because, as we learned in Modules #2 and #3, not all autotrophs use the sun's energy to make their food. Some autotrophs (chemosynthetic ones, for example) use alternative forms of energy (chemosynthetic autotrophs use the energy stored in chemicals). Since these types of autotrophs are rare, only a small fraction of the energy used in ecosystems comes from a source other than the sunlight, so it is safe to say that "almost all" ecosystem energy comes from the sun.

No matter what autotrophs use as their energy source, we already learned that they are called **producers**, because they produce food for themselves and for other creatures. The creatures that use this food (one way or another) we called **consumers**. In this module, we will be a little more specific. **Herbivores**, as you should recall, eat only plants. Since they get their food directly from the producers, we call them **primary consumers**.

Primary consumer - An organism that eats producers

Of course, God has designed His creatures to get food not only directly from the producers, but also from other consumers. **Carnivores**, for example, eat other organisms that are not producers. Since carnivores do not get their food directly from the producers, we call some of them **secondary consumers**.

Secondary consumer - An organism that eats primary consumers

Finally, carnivores that eat other carnivores are called **tertiary** (ter' she air ee) **consumers**.

Tertiary consumer - An organism that eats secondary consumers

These relationships, called **trophic** (troh' fick) **levels**, are shown in Figure 10.1.

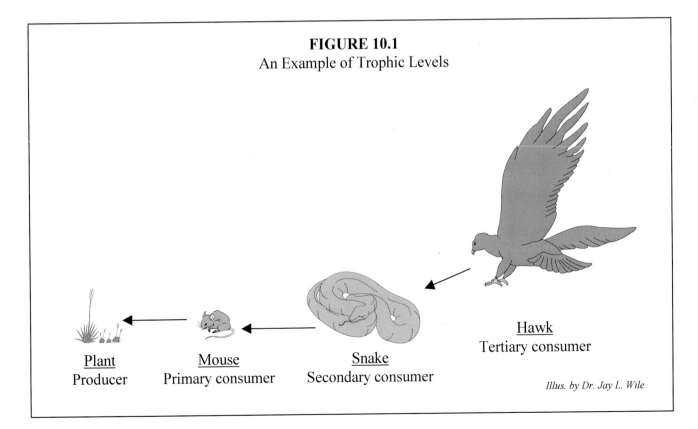

FIGURE 10.1
An Example of Trophic Levels

Plant
Producer

Mouse
Primary consumer

Snake
Secondary consumer

Hawk
Tertiary consumer

Illus. by Dr. Jay L. Wile

A diagram like the one shown in Figure 10.1 is often called a **food chain**. Since the plant is first, we say that it is lowest on the food chain. Likewise, the hawk is last, so we say that it is highest on the food chain. Although such diagrams are useful, they tend to be a little simplistic. For example, a hawk can change its trophic level. In the food chain diagrammed above, the trophic level of the hawk is tertiary consumer, because it is eating the snake. Sometimes, however, hawks will eat mice. When they eat mice, they are secondary consumers. Thus, the trophic levels of some organisms can change. As a result, a more complex diagram, called a **food web**, is often used to provide a more accurate description of the feeding relationships between organisms in an ecosystem.

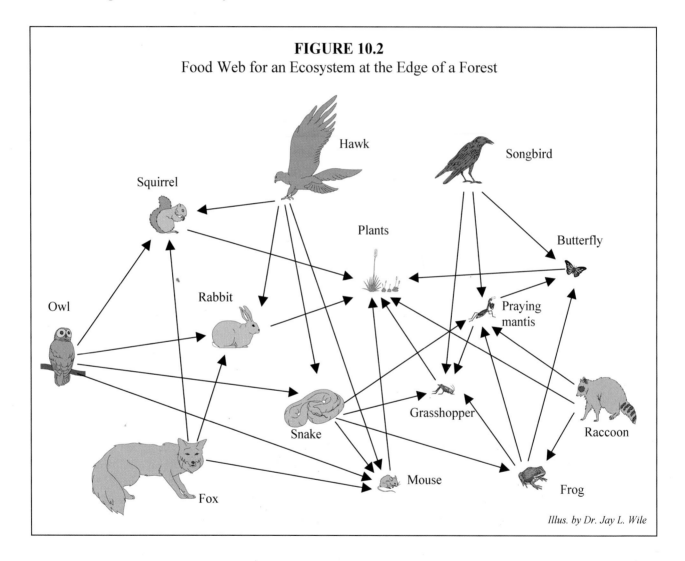

FIGURE 10.2
Food Web for an Ecosystem at the Edge of a Forest

Illus. by Dr. Jay L. Wile

Using the food web, you can easily see how some creatures remain at the same trophic level while others vary theirs. The fox, for example, is shown as always being a secondary consumer, because it is shown eating only the mouse, rabbit, or squirrel. All three of these creatures are primary consumers, so the fox is always a secondary consumer. Look at the raccoon, however. It is shown eating plants, a praying mantis, and a frog. When the raccoon eats plants, it is a primary consumer. When it eats a praying mantis or a frog, it is a tertiary consumer. This is why a food web is a much more realistic representation of trophic levels in an ecosystem than is a food chain.

So why are trophic levels so important in an ecosystem? First of all, as we learned by the story of the rabbits that were imported into Australia, every living organism must have some sort of predator, or the organism will overrun its ecosystem. Food webs show you the predator/prey relationships in an ecosystem. What about the creatures at the edge of the food web in Figure 10.2? Do the fox, owl, hawk, songbird, and raccoon have predators? Of course they do! If they did not, then they would overrun the ecosystem in no time at all. Why aren't their predators pictured in the figure? Well, at some point, you have to limit your diagram of a food web because it is just too complex when you try to include all predator/prey relationships in it. Also, some predator/prey relationships are rather difficult to diagram. For example, some species of songbird actually prey on the hawk. They don't eat adult hawks, but they eat hawk eggs before the eggs have had a chance to hatch. A food web showing a songbird eating a hawk would look a little silly, but nevertheless, songbirds do prey on hawk eggs. So Figure 10.2, as is the case with most attempts to diagram Creation, is a highly simplified picture.

The second reason that trophic levels are important is that they actually track energy as it moves through the ecosystem. When energy comes into the ecosystem, it is usually through the sun. The energy of the sunlight is used by producers to make food for themselves. When the producers are eaten by primary consumers, the energy moves to the next trophic level. As secondary consumers eat primary consumers, energy moves up to the next trophic level again. Well, each time energy moves up a trophic level, *most of that energy is wasted.* This is a very important point to remember:

Energy is wasted each time it moves up a trophic level in an ecosystem

How is that energy wasted? Well, think about it. When a fox eats a rabbit, it doesn't eat the whole thing, does it? It only eats the meat and fat. Thus, the skin, fur, skeleton, etc. of the rabbit is completely unused by the fox. Nevertheless, it took energy to form and maintain the skin, hair, and skeleton of the rabbit. All of that energy is wasted!

What does this mean? Well, it means that the lower an organism's trophic level, the less energy the ecosystem must invest in it. Plants, for example, require the least amount of energy. When a primary consumer eats the plant, however, up to 90% of that energy is wasted. As a result, it takes a LOT of plants to satisfy the feeding needs of only one primary consumer. Once again, when a secondary consumer eats a primary consumer, most of the energy contained in the primary consumer is wasted. Thus, it takes a LOT of primary consumers to meet the feeding needs of one secondary consumer. Likewise, it takes a LOT of secondary consumers to meet the feeding needs of one tertiary consumer.

One of the best ways to illustrate the consequence of energy loss in an ecosystem is via the **ecological pyramid**.

<u>Ecological pyramid</u> - A diagram that shows the biomass of organisms at each trophic level

Of course, this definition does us little good if we are not familiar with the term **biomass**!

<u>Biomass</u> - A measure of the mass of organisms within a region divided by the area of that region

For example, suppose you went to a pond that was a perfect square measuring 4 feet by 4 feet. If you caught all of the fish in that pond, you could calculate the biomass of the fish by measuring the mass of all of the fish that you caught and dividing that by the area of the pond (16 square feet).

Biomass, then, tells us something about how much biological matter exists in a certain region. The ecological pyramid separates biomass out according to trophic level so that you can get a feel for how much energy must exist at each level. For example, ecologists have extensively studied the aquatic ecosystem in Silver Springs, Florida. In a given year, the ecological pyramid for this ecosystem is displayed in Figure 10.3.

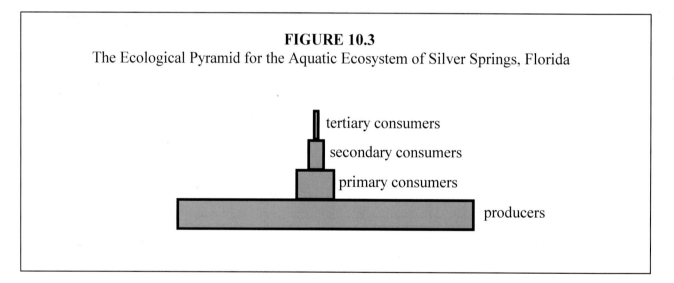

FIGURE 10.3
The Ecological Pyramid for the Aquatic Ecosystem of Silver Springs, Florida

In this figure, the length of the rectangle is an indication of the amount of biomass at each trophic level. As you can see, the biomass of producers is huge compared to the biomass of primary consumers, which is large compared to the biomass of secondary consumers, which is large compared to the biomass of tertiary consumers. If no energy were wasted from trophic level to trophic level, then the biomass of each trophic level would be the same. However, since energy is lost at each trophic level, there must be more and more creatures as you go to the lower trophic levels. It is important to note that you should look at the differences in trophic levels in terms of the *percentage* change, not the absolute change in the length of the rectangles. In other words, it is irrelevant what the actual size of each rectangle is. Instead, the only thing that is important is the percentage change between each rectangle.

Now before we leave this section, we hope that you have noticed an entire class of organism which we have simply ignored in this discussion. We have ignored the decomposers. We have done this on purpose, because it is hard to fit them into this kind of framework. Since decomposers are rather indiscriminate, they tend to feed at all trophic levels. They feed on dead plants, dead primary consumers, dead secondary consumers, and dead tertiary consumers. As a result, they have their own, unique trophic level. It is hard, therefore, to include them in a food chain or a food web. In addition, the biomass of decomposers is remarkably variable depending on the ecosystem studied. As a result, they do not fit nicely into ecological pyramids either. Therefore, we simply ignored them as a part of our discussion. As mentioned previously, however, decomposers are vitally important to an ecosystem. They are the last to use the energy, and they recycle matter so that it can be used again. Thus, do not think that they are unimportant just because we ignored them in this discussion!

ON YOUR OWN

10.1 Looking at Figure 10.2, identify all possible trophic levels of each organism in the figure.

10.2 Based on Figure 10.2 and the idea of an ecological pyramid, would you expect more owls or more mice at the edge of a forest?

10.3 Based on the ecological pyramid in Figure 10.3, between which two trophic levels is the greatest amount of energy wasted?

Symbiosis

Although the previous section concentrated on the predator/prey relationship, there are other relationships which exist between organisms in an ecosystem. For example, we have already discussed the phenomenon of symbiosis, where two or more organisms live in a mutually beneficial relationship. In Module #3, for example, we discussed the termite and the protozoa from genus *Trichonympha*. The termite lives on wood, but cannot digest one of the principal components of wood, cellulose. Since the termite cannot digest cellulose, it would continue to build up in the termite's system until it became toxic. To keep this from happening, protozoa from genus *Trichonympha* live in the gut of termites, eating the cellulose that the termites cannot digest. This keeps the termites alive while providing an ample food supply for the protozoa. In Module #4, we also learned about the symbiotic relationships between algae and fungi, which we call lichens, as well as the symbiotic relationships between plants and fungi, which we call mycorrhizae.

It turns out that there are many, many examples of symbiosis in God's Creation. The three that you have already learned about are examples from the land, but there are also incredible symbiotic relationships that exist between organisms underwater. Consider, for example, the amazing life of the clownfish. Shown in Figure 10.4, these fish get their name from their bright orange and white markings. What makes them unique, however, is that they live their lives among the stinging tentacles of sea anemones (uh nem' uh neez), which are also shown in the figure.

FIGURE 10.4
A Clownfish in the Tentacles of a Sea Anemone

Photos from the MasterClips collection

Sea anemones are normally deadly to fish. Their tentacles are full of stingers, each of which delivers a paralyzing poison to any fish that they touch. Once the fish is paralyzed, the anemone eats it. This doesn't happen to clownfish, however. When a clownfish touches the tentacles of a sea anemone, it does not get stung. This is because the clownfish is missing something that nearly every other fish has. You see, most fish are covered with a thin coat of mucus, which makes them feel slimy to the touch. If you compare the mucus coating of a clownfish to that of nearly any other fish, you will find one particular amino acid missing. This amino acid is the chemical that triggers the stingers in a sea anemone. Since the clownfish doesn't have that amino acid, the sea anemone does not sting the clownfish!

So, the clownfish can swim in, out, and among the tentacles of the sea anemone without any adverse effects. As a result, the clownfish is well-protected from predators. After all, any predator that goes into the tentacles of the sea anemone will immediately be killed. Thus, the clownfish benefits from the relationship by being protected from predators. The sea anemone benefits because clownfish tend to swim near, but not in the anemone. This tantalizes predator fish. When the predator fish tries to attack the clownfish, the clownfish darts back into the tentacles of the sea anemone. Often, the predator fish, which normally knows to steer clear of the anemone, will be overcome with the desire to eat the clownfish and will not realize that it is being led into the tentacles of death. Thus, the sea anemone benefits because the clownfish tends to attract food to it.

Another example of symbiosis can be found in the relationship between the blind shrimp and the goby (goh' bee), pictured in Figure 10.5.

FIGURE 10.5
The Blind Shrimp and the Goby

Photo © Gary Bell

These two organisms exist in most tropical seas, and they are always found together. They live in a hole on the ocean floor that has been dug by the blind shrimp. If you've ever been to the beach and tried to dig a hole in the sand under the water, you will have some idea about how hard it is for the blind shrimp to maintain its home. Ocean currents are constantly throwing debris into the hole, so the shrimp has an almost never-ending job of clearing the debris away from the hole. This puts the blind shrimp in danger, because (as its name implies) it cannot see. It therefore has no way of knowing whether or not it is in danger of predators. This is where the goby comes in.

The goby, a fish, has excellent vision. When the shrimp needs to clear the hole of debris, the goby goes out with it. While the shrimp digs, the goby keeps watch. The entire time that the shrimp is digging, it keeps one of its feelers on the goby. If the goby sees a predator, it signals the blind shrimp by flicking its tail, and they both head down the hole at lightning speed. Since the hole is too small for predators to follow, they are protected from danger.

Once again, then, this is a mutually beneficial relationship. The blind shrimp benefits because it is warned about the presence of predators. The goby benefits because it has access to a hole in which it can hide from predators. Since the goby has no means by which to dig a hole, it would not have such protection were it not for the blind shrimp. Even if it were able to find a hole that was unoccupied, the hole would fill up quickly from the debris carried by ocean currents. With the help of the blind shrimp, however, the goby gets a permanent home in which it can hide from predators.

The last example of symbiosis that we want to cover is rather striking. It is the relationship that exists between two fish: the Oriental sweetlips (the big fish in Figure 10.6) and the blue-streak wrasse (the small fish in Figure 10.6).

FIGURE 10.6
The Oriental Sweetlips and the Blue-Streak Wrasse

Photo © Gary Bell

Now if you were to just glance at this figure, you might think that it depicts a big fish eating a little fish. Although it looks like this is the case, the figure actually depicts a big fish (the Oriental sweetlips) *getting its teeth cleaned* by a little fish (the blue-streak wrasse).

You see, some fish have to have teeth. If they didn't, then there would be some underwater creatures without predators. As we have already seen, this would have disastrous effects on the ecosystem. The Oriental sweetlips is one of those fish that have teeth. Now although teeth are useful, they actually are hard to maintain. If you do not clean them, they rot and fall out. Thus, the Oriental sweetlips must have some means by which it cleans its teeth. How does a fish clean its teeth? A person cleans his or her teeth by brushing them. A dog doesn't brush its teeth to clean them. Instead, it chews on hard, edible substances (like bones). This tends to flake the crud off of its teeth. A fish could, in principle, do that, but there are very few hard, edible substances underwater. Thus, the Oriental sweetlips must find some other way of getting its teeth clean.

The way the Oriental sweetlips does this is quite surprising. After spending its day feeding on little fish, the Oriental sweetlips decides that it is time to get its teeth cleaned. To do this, it looks for a particular color of coral. When it finds that coral, it swims up to the coral with its mouth wide open. Then, the little blue-streak wrasses dart out of the coral and *swim right into the mouth or up the gills of the Oriental sweetlips*! These little fish then proceed to eat all of the crud off of the Oriental sweetlips' teeth. This is the main food source of the blue-streak wrasse! If all of this isn't amazing enough, the Oriental sweetlips also knows not to chomp down on the blue-streak wrasse when it finishes its work. It allows the little fish to swim back out of its mouth so that it can have its teeth cleaned tomorrow. Once again, this is a mutually beneficial relationship. Without the blue-streak wrasse, the Oriental sweetlips would lose its teeth. Without the Oriental sweetlips, the blue-streak wrasse would starve.

Studying these incredible relationships should give you a great appreciation for the power and ingenuity of God. It should also tell you something about the hypothesis of macroevolution, which we studied in the last module. Remember, macroevolution hopes to explain the origin of all species on this planet through competition and natural selection. Think about this hypothesis in light of the symbiosis that we have just studied. According to macroevolution, species should compete with one another for survival. The many, many examples of symbiosis in Creation tell us that species do not necessarily compete. They often help each other out. This flies in the face of a fundamental assumption in macroevolution.

Symbiosis presents a much greater challenge to the hypothesis of macroevolution, however. Macroevolution can never hope to explain many of the examples of symbiosis in Creation. For example, think about how macroevolution would have to explain the relationship between the Oriental sweetlips and the blue-streak wrasse. At some point in time, macroevolutionists would say, the sweetlips' ancestors had no teeth. In a number of generations, however, teeth began to form in a few of the ancestor's offspring. Now, in order for these teeth to avoid falling out, this new fish would have to develop the instinct for seeking out the wrasse, allowing the little fish to swim into its mouth, and not eating the blue-streak wrasse when the little fish is done with its work. This instinct, of course, would have to evolve *at exactly the same time* that the Oriental sweetlips' teeth evolved. That's not enough, however. *At the exactly same time that the teeth and instincts evolved in the Oriental sweetlips, the blue-streak wrasse would have to develop the instinct to swim right into the Oriental sweetlips' mouth without fear of being eaten!* Remember, if all of these things didn't happen in the same exact generation, then the system would not work.

Obviously it is ridiculous to believe in all of these chance coincidences occurring at the same time. Over and over again, however, this is what the scientist who believes in macroevolution *must* believe. When one, as a scientist, begins to look at the world around him, he sees far too many of these "happy coincidences." Between all of the precisely balanced properties of atoms and subatomic particles, the incredibly complex web of interactions that makes life on earth possible, the complicated nature of life itself, and the amazing design features of the animals we see in the world around us, it becomes intuitively obvious that this fantastic world around us could never have appeared by chance. It must have been designed by a powerful and intelligent designer. As Sir Frederick Hoyle (arguably England's foremost astrophysicist) says,

> "...A common sense interpretation of the facts suggests that a superintellect has monkeyed with physics, as well as with chemistry and biology, and that there are no blind forces worth speaking about in nature. The numbers one calculates from the facts seem to me so overwhelming as to put this conclusion almost beyond question." (Frederick Hoyle, *Engineering and Science* November, 1981, p. 12)

Symbiosis is just one more example of Creation providing witness for its Creator.

ON YOUR OWN

10.4 Of the three new examples of symbiosis that you learned about in this section, which are vital to the existence of each organism? Can any of the organisms discussed exist without their symbiotic partners?

10.5 Although macroevolution can never hope to explain most of the examples of symbiosis in Creation, it can explain some. Which of the three examples of symbiosis that you read about in this section would be easiest to explain in terms of macroevolution? How would macroevolution attempt to explain the relationship?

The Physical Environment

So far, our discussion of ecosystems has revolved around the associations between organisms. Specifically, we have discussed the feeding relationships that exist between organisms and the phenomenon of symbiosis. However, the definition of ecosystem includes something besides the associations that exist between organisms; it includes the physical environment. The physical environment is composed of all non-living things in the ecosystem. This includes, but is not limited to, such things as weather, temperature, the chemicals in the ecosystem, the air that the organisms breathe, the availability of water, and so on. These are the things we want to talk about in this section.

Now if you think about it, in order to begin to study the physical environment of an ecosystem, you first have to determine what its limits are. For example, although you could say that the whole earth is a single ecosystem, it would be awfully hard to study. After all, the physical environment changes dramatically from place to place around the earth. In deserts, water is scare, whereas in the ocean, water is obviously quite plentiful. In the Arctic, the temperature is always rather low, whereas near the equator, the temperature is high. Thus, in order to make an ecosystem understandable, we must establish limits. The more you restrict the region that you are considering, the easier it is to understand the ecosystem.

In Figures 10.1 and 10.2, for example, we highlighted the trophic levels in an ecosystem at the edge of a forest. By restricting ourselves to the creatures in a region like that, our food web was at least manageable. We could, however, define an even easier to understand ecosystem. If you were to walk into your backyard or a park and measure out a region of grass that is 2 feet by 2 feet, you would also have an ecosystem. This ecosystem would have the benefit of being very easy to study. Primarily, there would be producers (the grass and plants) and primary consumers (the insects and an occasional squirrel or rabbit). If your cat came into that region and ate a mouse, you would have a secondary consumer. Of course, the drawback of such a narrowly-defined ecosystem is that you would not learn an enormous amount by studying it, because there is not a great diversity of life in it. In the end, then, an ecosystem really consists of a region that we choose to define. If we define it narrowly, it is an easy-to-understand ecosystem, but we will not learn an enormous amount from it. On the other hand, if it is large, it will be quite complex, but we can learn a great deal from it.

Once we have defined an ecosystem, we can begin to look at its physical environment. Now, as we have already discussed, there are a whole host of things that make up the physical environment of an ecosystem, and it is simply far too complex to study even the majority of these things. As a result, we will concentrate on just a few of the more important substances that make up the physical environment of an ecosystem: **water**, **oxygen**, and **carbon**. By looking at how these substances are distributed in an ecosystem, you will get some idea how incredibly complex and wonderful God's Creation is!

ON YOUR OWN

10.6 A biologist is studying two different ecosystems. One is defined as a 30-yard stretch of ocean shore. The other is a forest that runs one mile across and half of a mile wide. Which ecosystem will be the easiest to study? Which will most likely reveal more information?

10.7 The following is a list of things found near a pond. Which are a part of the ecosystem's physical environment?

     a. water   b. rocks   c. fish   d. algae   e. grass   f. sand

---

The Water Cycle

Probably the single most important substance for life on earth is water. Almost all creatures on earth need water to survive. In fact, there are many more organisms that do not need oxygen than those that do not need water. Thus, water is vital to all ecosystems. It is important, then, to understand where water comes from and where it goes to in an ecosystem. Consider, for example, what happens to water in an ocean shore ecosystem.

FIGURE 10.7
The Water Cycle at the Ocean's Shore

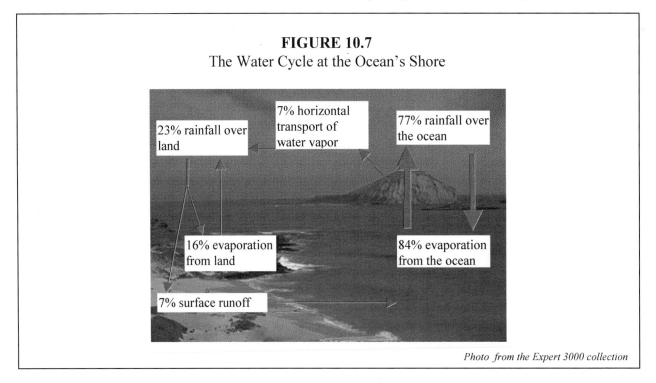

23% rainfall over land

7% horizontal transport of water vapor

77% rainfall over the ocean

16% evaporation from land

84% evaporation from the ocean

7% surface runoff

Photo from the Expert 3000 collection

This diagram represents what we call the **water cycle**. It attempts to detail where water comes from and where it goes to in an ecosystem.

Looking at Figure 10.7, you will see that water seems to cycle from the ocean to the shore and back again. Water evaporates from the ocean and travels into the air, where it eventually forms clouds. Clouds are simply water droplets suspended on fine dust particles in the air. The clouds eventually cause rain, which brings water back into the ocean. More water evaporates from the ocean, however, than what falls back in the form of rain. If that were the end of the story, then, the oceans would slowly run dry. Luckily for us, however, that's not what happens.

The water vapor that evaporates from the ocean forms clouds which travel through the air. If they travel far enough, those clouds can cause rain that falls on the land. As you can see from the figure, more water falls on the land than does the water that evaporates from it. Thus, the land has excess water. This water trickles back into the sea, as surface runoff. So you see that while 84% of all evaporation in the ecosystem comes from the ocean, the rest of it (16%) comes from land. The fact that only 77% of the rain occurs over the ocean and the rest of it (23%) occurs over land seems to form a 7% deficit, meaning that the ocean gives up 7% of all that evaporates to the land. However, the land gives that 7% back to the ocean as surface runoff, assuring that the amount of water on land and the amount of water in the ocean stay constant.

It turns out that the water cycle does more than make sure that water stays where it should be in an ecosystem. It is also responsible for transporting nutrients from one part of an ecosystem to another, or even *between one ecosystem and another*. For example, in the ocean shore ecosystem depicted in Figure 10.7, the surface runoff water carries with it nutrients that were contained in the soil and sand on and near the shore. These nutrients include chemicals that are the result of decomposers feeding off of dead creatures and minerals that were held in the soil and sand.

The role of the water cycle in transporting nutrients is best seen through a study of a **watershed**.

Watershed - An ecosystem where all water runoff drains into a single river or stream

Figure 10.8 is an illustration of a watershed.

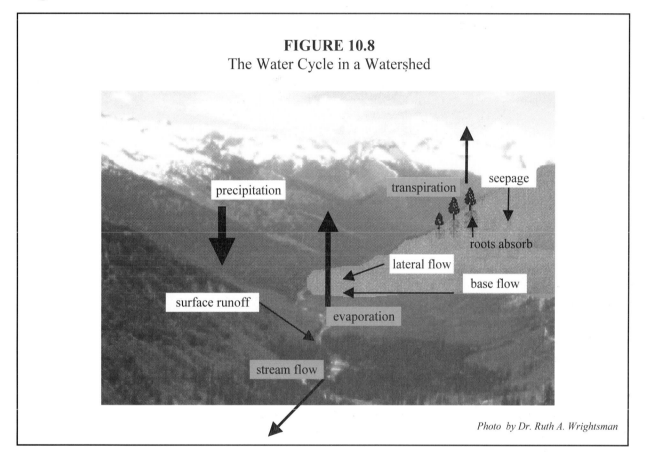

FIGURE 10.8
The Water Cycle in a Watershed

Photo by Dr. Ruth A. Wrightsman

Looking at Figure 10.8, we see that water comes into the watershed in one way: through rainfall (which we call **precipitation**). It leaves in one of three ways: through **transpiration** (tran spuh ray' shun), **evaporation**, or **stream flow**.

Transpiration - Evaporation of water from the leaves of a plant

Along the way, a lot of other things happen to the water. Water can flow over the land and into the stream as surface runoff; or it can be absorbed by the soil and flow laterally into the stream; or it can seep deep into the ground until it becomes part of the groundwater, which eventually flows along the base of the soil until it reaches the stream.

Watershed ecosystems play a vital role in the distribution of water throughout the earth. The plants in the ecosystem moderate the amount of nutrients carried away by the overland flow, and the plants' roots play the same role for the lateral and base flow. Without these plants, too many nutrients would flow into the water, causing algal blooms that can destroy the water ecosystem as well as depriving the soil of nutrients that would promote plant growth. This is why ecologists are worried about deforestation. If too many forests are cut down, then watersheds will not be able to control nutrient flow in the water cycle. This could result in disastrous effects to local ecosystems.

Despite the obvious need for strong forest and plant life in watershed ecosystems, many third-world countries find that the economic benefit derived from cutting down the trees is too great to worry about long-term environmental effects. As a result, forests are being leveled in places like India, Mexico, and Brazil, destroying the local watersheds. In the United States, environmental regulations have forced logging companies to replant trees at a *faster* rate than they cut them down. As a result, there is more forest coverage in the United States today than there has been in the past 100 years. If other countries can force their people to do the same, then watersheds across the world would be better protected.

Although this gives you an idea of how the water cycle works, it should be obvious to you that the water cycle is different in each ecosystem. Just as the water cycle in an ocean shore ecosystem is different than the water cycle in a watershed, it is different in just about every other ecosystem as well. Thus, if you want a complete picture of the water cycle around the world, you would need to study the water cycle in a variety of different ecosystems.

ON YOUR OWN

10.8 There are at least three different means by which water can potentially leave an ecosystem. Which of these means exists in a watershed but not an ocean shore environment?

10.9 In a watershed, water continually leaves the ecosystem because it flows out via the stream or river. Based on this fact, which is greater: the amount of water that participates in evaporation and transpiration combined or the amount of water that participates in precipitation?

The Oxygen Cycle

If water is the most important substance for life, then oxygen comes in a close second. Although some organisms do exist without oxygen, they are, by far, the exception. The vast majority of organisms are aerobic, requiring oxygen for respiration. Thus, the oxygen cycle is obviously an important part of the physical environment in any ecosystem.

The air which makes up our atmosphere is about 21% oxygen. This often surprises students, because they think that since organisms breathe air specifically for oxygen, then the air must be made up of mostly oxygen. In fact, less than one fourth of the air is oxygen. Almost all of the rest of it is nitrogen, and a small part of it is made up of several other gases including argon, carbon dioxide, and ozone. It turns out that this mixture of gases is simply ideal for supporting life. When you take chemistry, you will find out why this is the case.

Well, if the mixture of gases that makes up the air we breathe is ideal for the purpose of supporting life, then that mixture should never be allowed to change. This is a problem, since organisms are constantly breathing oxygen from the air, using up the oxygen. If this used oxygen were not continually being replaced by new oxygen, then the amount of oxygen in the air would drop to a percentage much lower than 21%. Eventually, there wouldn't be enough oxygen in the atmosphere to sustain life. Luckily, however, this is not the case. The Creator has designed His Creation to continually replenish the oxygen used by organisms. The physical processes which cause this to happen are collectively called **the oxygen cycle**.

The oxygen cycle is depicted in Figure 10.9.

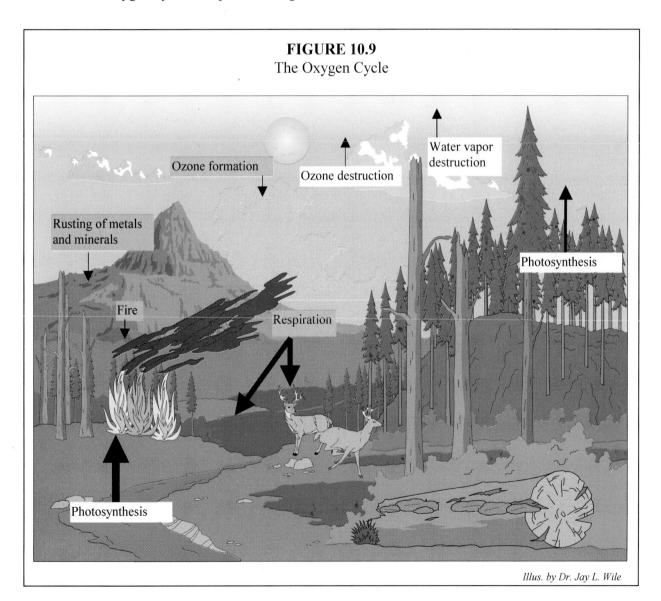

FIGURE 10.9
The Oxygen Cycle

Illus. by Dr. Jay L. Wile

In the oxygen cycle, oxygen is used up in many different ways, as illustrated by the arrows pointing downwards. The thicker the arrow, the more oxygen that is used up in the process. Organisms (both aquatic and land-based) use up oxygen as a part of their respiration. Natural and human-made fires use up oxygen. The rusting of metals is actually a chemical process that takes oxygen. Finally, some oxygen is converted into ozone by the energy of the sun. It actually turns out to be a good thing that oxygen is used up to make ozone, because ozone forms a protective shield (the ozone layer) that protects us from certain harmful rays which come from the sun. When you take chemistry, you will learn more about this fascinating substance.

Although oxygen is used up by all of these processes, the amount of oxygen in the air does not diminish because there are also means by which oxygen is restored to the air. The major way this happens is by photosynthesis. Green plants on land perform photosynthesis, as do phytoplankton in the water. As we learned in Module #3, phytoplankton are responsible for the vast majority (almost 75%) of all photosynthesis on earth. In order to perform photosynthesis, the organisms require carbon dioxide (CO_2) and water, as well as energy from the sun. The amount of water in the air, of course, is regulated by the water cycle. As we will learn in the next section, the amount of carbon dioxide in the air is regulated by the carbon cycle. Thus, the cycles that we are discussing here are not isolated from one another. Instead, they all work together to regulate the physical environment of the many ecosystems on earth. We simply separate them in order to make them each a little more understandable.

Even though photosynthesis is the main way that oxygen is replenished, it is not the only way. When certain rays from the sun hit ozone molecules, oxygen is formed. Also, when these rays hit water vapor in the air, the water can decompose into hydrogen and oxygen. This, then, is another way that oxygen is replenished in the air around us. The amount of water vapor in the air is controlled by the water cycle, once again demonstrating that these cycles are not independent of one another.

ON YOUR OWN

10.10 In Biosphere II, an attempt at engineering an artificial ecosystem, scientists could not develop a well-regulated oxygen cycle. Typically, the biosphere kept running out of oxygen, despite the fact that green plants were plentiful in the ecosystem. What, most likely, did the biosphere lack?

The Carbon Cycle

As you learned in Module #5, the chemistry of life is built on the chemical element carbon. Indeed, in order for a molecule to be considered organic, it must have carbon atoms in it. The carbon cycle regulates the amount of carbon in ecosystems, principally by keeping careful track of carbon dioxide (CO_2). The carbon cycle is illustrated in Figure 10.10.

FIGURE 10.10
The Carbon Cycle

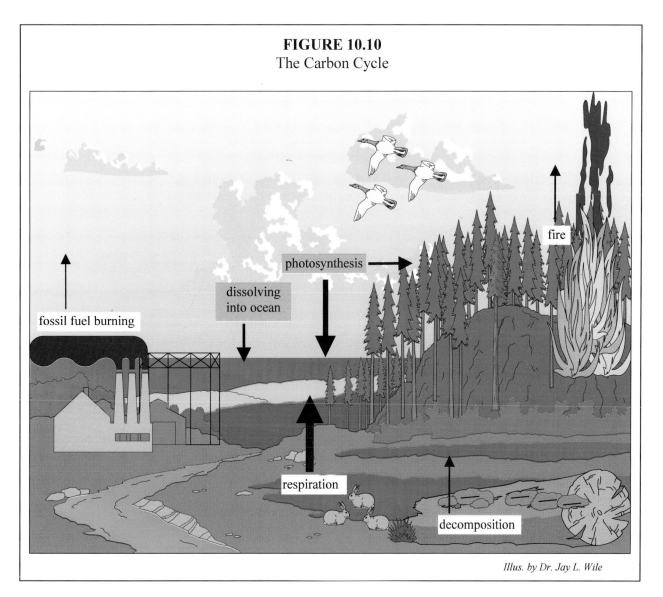

Illus. by Dr. Jay L. Wile

The amount of carbon in an ecosystem can be tracked using carbon dioxide because, at one point or another, all carbon from organic molecules passes through a stage in which it is part of a CO_2 molecule. Carbon that makes up a living organism's body, for example, will be converted to carbon dioxide by decomposers after the organism dies. When organisms eat food (either from producers or consumers), the carbon in that food is eventually converted to carbon dioxide and is then exhaled. Thus, by seeing where carbon dioxide goes to and comes from in an ecosystem, we are actually tracking all of the carbon in that ecosystem.

We see, then, that carbon dioxide enters the air through a variety of different processes. When decomposers do their job, the dead organic matter is converted to carbon dioxide. The respiration of all aerobic organisms results in carbon dioxide. When something burns (either by natural or human-made means), carbon dioxide is produced. All of these processes tend to increase the amount of carbon dioxide in the air.

Although carbon dioxide is constantly being added to the air, it is also constantly being taken away. Since photosynthesis requires carbon dioxide, an enormous amount of carbon dioxide is taken out of the air by this process. In addition, carbon dioxide also dissolves into the ocean, where shellfish react it with lime in order to form their shells. Some of the dissolved carbon dioxide also reacts with other chemicals in the ocean to form minerals, which get stored in the sediments at the bottom of the sea.

It turns out that the amount of carbon in the atmosphere is important to track for reasons other than trying to see where the carbon is in an ecosystem. Carbon dioxide itself has a very important task for keeping the earth hospitable for life. It helps keep the planet warm. How does it do that? Well, carbon dioxide is one of the principal gases involved in a process known as the **greenhouse effect**.

Greenhouse effect - The process by which certain gases (principally water, carbon dioxide, and methane) trap heat that would otherwise escape the earth and radiate into space

You see, the light shining on the earth from the sun tends to warm the planet; however, the earth tends to reflect a lot of that light rather than absorbing it. When that light is reflected by the earth, it does not warm the planet at all. If that were the end of the story, then the earth would be a VERY cold place. So cold, in fact, that life could not exist anywhere on its surface.

Fortunately, however, that is not the end of the story. It turns out that certain gases (principally water vapor, methane, and carbon dioxide) actually trap much of the light that reflects off of the earth. As a result, the energy from that light is not lost. Instead, it is held in the air, and the net effect is to warm up the planet. Because of the greenhouse effect, the earth is warm enough to support life. To become more familiar with the concept of the greenhouse effect, perform the following experiment.

EXPERIMENT 10.1
Carbon Dioxide and the Greenhouse Effect

Supplies:

- Thermometer (It can be a cooking thermometer or a weather thermometer. It just needs to be able to read temperatures from room temperature to about 100 degrees Fahrenheit.)
- Two clear ZIPLOC® sandwich bags
- Sunny windowsill (If it's not sunny today, just wait until it is.)
- Plastic, 2-liter soda bottle
- Vinegar
- Baking soda
- Teaspoon

Object: To observe the ability of carbon dioxide to absorb energy from sunlight.

Procedure:

A. Running the experiment with air

1. Open your ZIPLOC® bag. Now you want to fill your bag with air. Do this by holding the bag wide open, with the open side facing down. Then, raise the bag as high as you can and quickly lower it. This should fill the bag with air. Now zip it locked. You should now have a ZIPLOC® bag that is inflated with air.

2. Place your thermometer on the sunny windowsill and lay the inflated bag on top of it. Arrange the bag so that it completely covers the thermometer, but so you can still look through the bag and read the thermometer. If the bag is too small to cover the thermometer completely, arrange it so that the bulb of the thermometer is completely covered.

3. Allow the bag and thermometer to sit for 15 minutes and then read the temperature from the thermometer.

B. Running the experiment with carbon dioxide

1. Take the thermometer and ZIPLOC® bag off of the windowsill. Open the ZIPLOC® bag and throw it away (or recycle it!). Place the thermometer in a cool area while you prepare the next part of the experiment.

2. Take the 2-liter soda bottle and fill it about one-third of the way full with vinegar.

3. Measure out 1 teaspoon of baking soda and add it to the vinegar. The contents of the bottle will begin to bubble. Those bubbles tell you that a gas is being formed. The gas is carbon dioxide. Wait for a while. This will allow the carbon dioxide to push the air out of the bottle.

4. Repeat step 3. By doing that step twice, you are ensuring that there is only carbon dioxide in the bottle above the vinegar and baking soda.

5. Measure out another teaspoon of baking soda, but this time keep it in the spoon.

6. Open the ZIPLOC® bag and press it flat to remove any air in it. Keep it handy.

7. This has to be done quickly. Add the baking soda to the vinegar and then quickly hold the ZIPLOC® bag over the opening of the bottle. Don't worry about making sure all of the baking soda lands in the bottle. In this step, speed is very important. Immediately close the bag around the bottle opening, so that the carbon dioxide

coming from the bottle inflates the bag. Allow this to continue until the bag is as inflated as the previous bag that you used. If you can't fully inflate the bag the first time, carefully lift it off of the bottle and try again by adding more baking soda. If you are careful enough, you won't lose much carbon dioxide when you lift the bag. Once it is inflated, carefully remove the bag from the bottle and quickly zip it closed. You should now have a bag filled with carbon dioxide.

8. Go to the windowsill and place the thermometer and your new bag in the same position that you did in part "A" of the experiment. Allow them to sit for 15 minutes again. After the 15 minutes, read the temperature

C. Analyzing the data

Look at the two temperatures. They should be different. The temperature that you got in part "A" should be higher than the one you got in part "B." Why? Well, carbon dioxide absorbs energy from sunlight. This is what causes the greenhouse effect. In your experiment, because carbon dioxide absorbed the energy from the sunlight, there was less energy available to the thermometer. As a result, the temperature that the thermometer read was lower.

Since carbon dioxide is one of the principal gases that participate in the greenhouse effect, the amount of carbon dioxide in the atmosphere must be regulated very carefully by the carbon cycle. After all, if the amount of carbon dioxide in the air decreased significantly, then the earth would get colder and colder, perhaps turning it into a frigid wasteland. If, instead, the amount of carbon dioxide in the atmosphere increased significantly, the greenhouse effect would get greater and greater, leading to a warmer and warmer planet. If the planet got too warm, disastrous things could happen to certain ecosystems. The north and south poles, for example, are filled with miles and miles of ice. If those regions of the earth were to get too warm, that ice could melt, causing global flooding. Also, most ecosystems are very sensitive to temperature. A significant change in the temperature of some ecosystems could destroy much of the life contained in them.

Some people are very worried that this very thing is occurring today. Because we burn a large amount of coal, oil, and wood as a part of normal life, people have been adding carbon dioxide to the atmosphere. As a result, the amount of carbon dioxide in the atmosphere has been rising steadily for the past 100 years or so. Some people are worried that the air is getting so rich in carbon dioxide that the greenhouse effect is already becoming enhanced and the earth is already getting abnormally warm. They call this phenomenon **global warming**, and they are convinced that the earth will simply get too hot if something isn't done to stop the buildup of carbon dioxide in the air.

Although the fear that too much carbon dioxide in the air could lead to global warming is based on sound scientific reasoning, reality is just a bit more complex than that. To illustrate what we mean, take a look at the data presented in Figure 10.11. This data shows both the amount of carbon dioxide in the air and the change in the average temperature of the earth as a function of the year in which those measurements were taken.

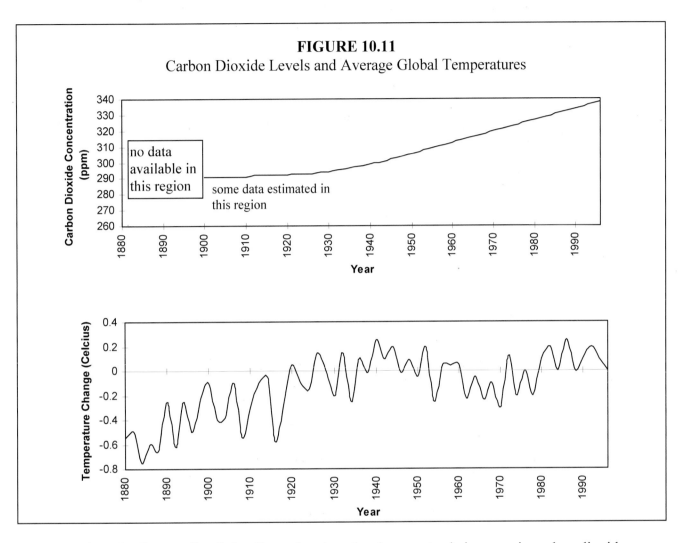

FIGURE 10.11
Carbon Dioxide Levels and Average Global Temperatures

What does the figure tell us? It tells us that there has been a steady increase in carbon dioxide levels in the air since about 1920 (see the top graph). If you look at the bottom graph, however, you do not see a corresponding increase in the temperature of the earth. Instead, from about 1880 to 1920, the average global temperature increased in a very shaky pattern by about 0.5 degrees Celsius (0.9 degrees Fahrenheit). After that, however, the temperature change varies up and down quite a bit, but continues to hover around zero. In other words, over the time that the amount of carbon dioxide in the air increased steadily, the average temperature of the earth, on average, did not change!

Does this mean that the amount of carbon dioxide in the air does not effect the temperature of the earth? No, of course not. We know that the greenhouse effect is real, or we wouldn't be here. What these data tell us is that reality is always more complex than theory. Remember, in order to see a change in the greenhouse effect, there must be a significant change in the amount of carbon dioxide in the air. Unfortunately, scientists do not know what a significant change would be. If the amount of carbon dioxide in the air doubled, would that be significant? From the standpoint of the greenhouse effect, we really do not know. What we know for sure, however, is that the sum total of all carbon dioxide produced by human activity is approximately 3% of the carbon dioxide produced by the other processes which make up the carbon cycle. One could argue, then, that this amount of added carbon dioxide is simply not significant compared to all of the other processes that add carbon dioxide to the air.

Also, the way in which carbon dioxide is added to the air is very important in the greenhouse effect. When humans burn fossil fuels, carbon dioxide is not the only gas that is released. Many other gases are released as well. Some of these gases tend to reflect light rather than absorb it. This actually *reduces* the amount of energy absorbed by the earth, causing a net cooling effect. It could be that any increase in the greenhouse effect due to human-produced carbon dioxide is offset by the cooling caused by the other chemicals associated with human activity.

We know for sure, however, that right now all measured data indicates that the greenhouse effect is not being enhanced enough to cause global warming. As a result, the majority of atmospheric scientists (scientists who study the air) do not consider global warming a problem. Now that little fact might surprise you, because you might have heard the television or radio news say that there is a "scientific consensus" that global warming is a big problem today. The scientific consensus, however, is precisely the opposite. In a 1993 Gallup poll conducted for the Institute of Science, Technology, and the Media, only 17% of all atmospheric scientists believe that global warming is a problem. *The majority (53%) said that global warming is not a problem and the rest (30%) said that there was not enough information to make that decision.* These results are typical of any global warming poll that is performed on atmospheric scientists.

If the data seems to indicate that no global warming has occurred, and if the majority of scientists in the field say that global warming is not a problem, then why does the media seem to say the exact opposite? We have no idea. Perhaps it has something to do with a desire to push an agenda rather than to communicate the truth. Regardless of why, it is very clear that the popular media are ignoring the facts about global warming. Hopefully, the people will be able to learn the facts through some other means.

ON YOUR OWN

10.11 If a tree dies and slowly rots away, does this add or remove carbon dioxide from the air?

10.12 If you were able to see carbon dioxide as it interacted with the surface of the ocean, you would see the ocean absorb an enormous amount of carbon dioxide. Only a small portion of that carbon dioxide actually dissolves in the ocean, however. The majority of the carbon dioxide goes into the ocean but is not dissolved. If it isn't dissolved, what is it used for?

Summing Up

In this module, we have only scratched the surface of the science of ecology. In studying ecosystems, we have only concentrated on two relationships between organisms (trophic and symbiotic relationships) and three processes related to the physical environment (the water, oxygen, and carbon cycles). However, even this cursory glance at ecosystems should tell you something rather profound: ecosystems are intricately balanced, complex systems. If even one of the many, many processes that regulate the physical environment were to change significantly, life would cease to exist. If even one organism is introduced to an ecosystem in which it does not belong, a disaster can occur!

This intricacy and complexity, of course, is a testament to the power and knowledge of God. Just think about it! God has placed all of the organisms in just the right ecosystems so that every organism has enough predators to keep its population in check but not so many that the population dies out. He has designed some organisms to work together in order to survive. In addition, He has set up literally millions and millions of chemical and physical processes that regulate the physical environments of all ecosystems on the earth. Is there any way that all of this could just have happened by chance? Of course not! Even the most basic ecosystem in the world is already more intricate and complex than the most sophisticated machine that human science has ever developed. Clearly, systems of that complexity require an incredibly knowledgeable and powerful Designer. That Designer has left His fingerprints all over His Creation. It is up to us to recognize that and give Him the praise that He deserves.

ANSWERS TO THE ON YOUR OWN PROBLEMS

10.1

Organism	Possible Trophic Levels
Hawk	secondary consumer, tertiary consumer
Songbird	secondary consumer, tertiary consumer
Squirrel	primary consumer
Plants	producers
Butterfly	primary consumer
Owl	secondary consumer, tertiary consumer
Rabbit	primary consumer
Praying mantis	secondary consumer
Snake	secondary consumer, tertiary consumer
Grasshopper	primary consumer
Raccoon	primary consumer, tertiary consumer
Fox	secondary consumer
Mouse	primary consumer
Frog	secondary consumer, tertiary consumer

10.2 There would be more mice. Since energy is wasted at each trophic level, there should be more primary consumers than secondary consumers.

10.3 Remember, we do not look at the absolute change in the length of each rectangle. Instead, we look at the percentage change. For example, the rectangle for primary consumers is 1/8 (12.5%) the size of the one for producers. Thus, there was a 7/8 (87.5%) change in the step from producers to primary consumers. The secondary consumer rectangle, however, is about 1/3 (33%) the size of the primary consumer rectangle. Thus, there was only a 67% change going from primary consumers to secondary consumers. Likewise, it looks like there is about a 67% change going from secondary consumers to tertiary consumers.   This indicates that an inordinately large amount of producers are needed to support the feeding needs of the primary consumers. Thus, the largest amount of energy is wasted between the producers and the primary consumers.

10.4 The clownfish/anemone symbiosis is not absolutely necessary for the mutual survival of the two organisms. After all, many anemones exist without clownfish. Clownfish just make it easier for the anemone to trap prey. In the same way, clownish can still swim away and hide from predators like any other fish. The other two relationships are necessary, however. The blind shrimp could not live without the eyes of the goby, and the Oriental sweetlips' teeth would rot and fall out without the blue-streak wrasse.

10.5 The clownfish/anemone symbiosis could be explained by macroevolution. According to the hypothesis, a clownfish happened to be born without the one amino acid that triggers the anemone's sting. When the clownfish inadvertently swam into the anemone, it did not die and was protected from its predators. That strain of clownfish was naturally selected to survive, because it was so well protected from predators. Thus, it passed on its trait to its offspring.

10.6 The 30-yard stretch of ocean shore is the simplest ecosystem, because it is small. The biologist will probably learn more from the other ecosystem, however.

10.7 The physical environment includes all non-living surroundings. Thus, a, b, and f are part of the physical environment.

10.8 Stream flow does not exist in an ocean shore ecosystem. You might be tempted to say transpiration here, simply because transpiration is not in Figure 10.7. Remember, however, that transpiration is from plants, which do exist in ocean shore ecosystems. In the figure, transpiration is lumped together as evaporation from land.

10.9 Think about it. Since water continually leaves the watershed by way of the stream, then in order to keep the stream from running dry, a lot more water has to enter the ecosystem than that which evaporates and transpires. As a result, the amount of precipitation is the largest.

10.10 The vast majority of photosynthesis is done by phytoplankton. The Biosphere II probably didn't have enough phytoplankton in its bodies of water.

10.11 When the tree rots, it is decomposed by decomposers. This adds carbon dioxide to the air.

10.12 The majority of the carbon dioxide is going into the ocean so that it can be a part of the photosynthesis of phytoplankton.

STUDY GUIDE FOR MODULE #10

1 Define the following terms:

a. Ecosystem

f.   Ecological pyramid

b. Ecology

g.   Biomass

c. Primary consumer

h.   Watershed

d. Secondary consumer

i.   Transpiration

e. Tertiary consumer

j.   Greenhouse effect

2. When fruits or vegetables are imported into the U.S. from a foreign country, they are always very closely inspected for insects, even though the vast majority of insects are not really harmful. Why is the inspection done?

3. In the following abbreviated food chain:

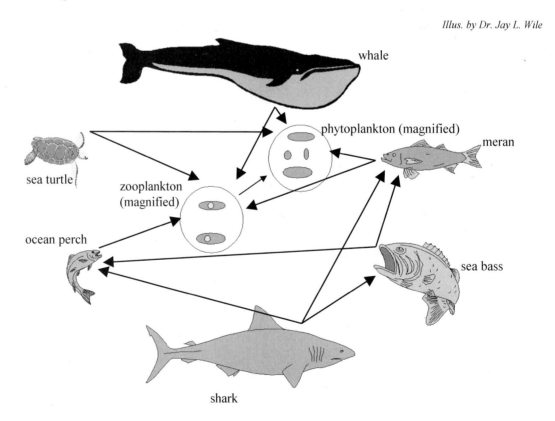

Illus. by Dr. Jay L. Wile

List all possible trophic levels for each organism.

4. Consider the following ecological pyramid

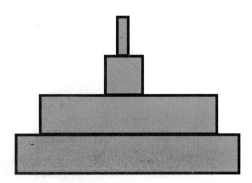

a. Which two trophic levels have the greatest disparity in biomass?

b. Between which two trophic levels is the smallest amount of energy wasted?

5. Name the participants in the three new symbiotic relationships that we learned in this module. Briefly describe the roles of each participant.

6. What fundamental assumption of macroevolution does symbiosis seem to contradict?

7. In the water cycle of an ocean shore ecosystem, more water evaporates from the ocean than falls back into the ocean in the form of rain. Why doesn't the ocean lose water?

8. What does the water cycle accomplish besides balancing out the water in an ecosystem?

9. What is the possible consequence if deforestation occurs in a watershed?

10. What is the principal means by which oxygen is taken from the air? What is the principal means by which it is restored to the air?

11. Name the other ways that oxygen is removed from the air.

12. Name the other ways that oxygen is replenished in the air.

13. Name the ways in which carbon dioxide is removed from the air.

14. Name the ways in which carbon dioxide is replenished in the air.

15. What human activity worries those who think that global warming is a problem?

16. Is global warming occurring now?

Module #11: The Invertebrates of Kingdom Animalia

Introduction

We've taken a break from looking at individual groups within biology's classification scheme and have, instead, looked at more global issues such as chemistry, genetics, and ecosystems. A biologist's chief work, however, is to study the creatures that God has placed in His marvelous Creation, so it is time to start doing that again. We have two kingdoms left to discuss in this course: kingdom Plantae and kingdom Animalia. We will actually jump back and forth between these kingdoms over the rest of the course, but for right now, we will start our discussion with kingdom Animalia.

In the classification scheme that we use here, there are 19 phyla in kingdom Animalia. Now before you get too worried, we assure you that we will not cover every one of them! Instead, we will pick a few representative phyla and learn some basic characteristics of each. When we get to phylum Chordata (kor dah' tah), however, we will spend a great deal of time on it. Why? Well, that's because most of the creatures which students consider to be "animals" (birds, fish, and mammals) belong in that phylum.

Kingdom Animalia is often split into two groups. Although they are not official classification groups, biologists use the terms often, so it is good to know them.

Invertebrates (in vur' tuh brates) - Animals that lack a backbone

Vertebrates (vur' tuh brates) - Animals that possess a backbone

Invertebrates make up all of the phyla in kingdom Animalia except for phylum Chordata. As you might imagine, then, there are far more invertebrates in kingdom Animalia than there are vertebrates. However, vertebrates are the organisms with which we are most familiar. As a result, we will spend an undue amount of time on them in a later module. For now, we will begin with invertebrates. We will start off with some of the more exotic invertebrates and then, in the next module, focus on a single phylum of invertebrates, phylum Arthropoda (are throw' pah duh). Before we do that, however, we need to discuss something that is used in our classification scheme: symmetry (sim' uh tree).

Symmetry

When looking at an organism, it is often possible to split that organism into two identical halves. For example, a human being has a left side and a right side. Each side seems to be identical in every way: This kind of symmetry is illustrated in Figure 11.1

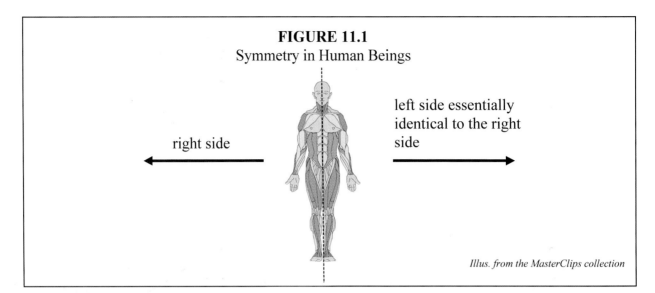

FIGURE 11.1
Symmetry in Human Beings

right side

left side essentially identical to the right side

Illus. from the MasterClips collection

Although this is the most familiar kind of symmetry in the animal kingdom, there are actually a total of three different kinds of symmetry that we can find.

Spherical symmetry - An organism possesses spherical symmetry if it can be cut into two identical halves by any cut that runs through the organism's center.

Radial symmetry - An organism possesses radial symmetry if it can be cut into two identical halves by any longitudinal cut through its center.

Bilateral (bye lat' uh ruhl) symmetry - An organism possesses bilateral symmetry if it can only be cut into two identical halves by a single longitudinal cut along its center which divides it into right and left halves.

Hopefully, you can tell from these definitions that humans have bilateral symmetry. If, however, these definitions are not clear, take a look at Figure 11.2, which illustrates them all.

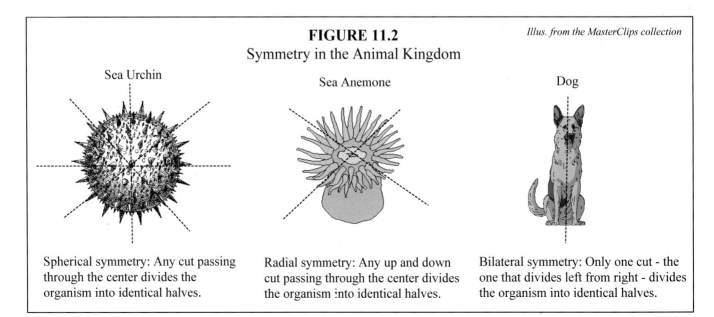

FIGURE 11.2   *Illus. from the MasterClips collection*
Symmetry in the Animal Kingdom

Sea Urchin   Sea Anemone   Dog

Spherical symmetry: Any cut passing through the center divides the organism into identical halves.

Radial symmetry: Any up and down cut passing through the center divides the organism into identical halves.

Bilateral symmetry: Only one cut - the one that divides left from right - divides the organism into identical halves.

Now realize that not all animals possess any kind of symmetry. If they do, however, it is going to be one of these three types. Be sure that you understand how to recognize symmetry by performing the following "on your own" problem.

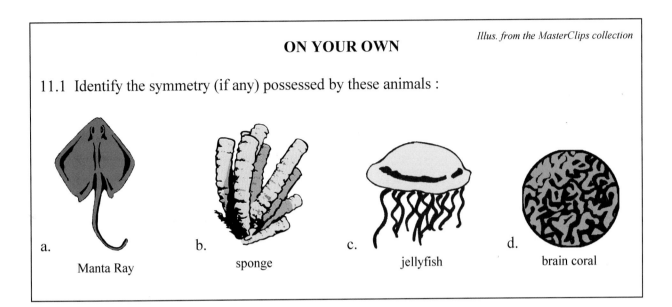

ON YOUR OWN

Illus. from the MasterClips collection

11.1 Identify the symmetry (if any) possessed by these animals :

a. Manta Ray   b. sponge   c. jellyfish   d. brain coral

Now that we've looked at the symmetry that exists in the animal kingdom, it is time to study some of the major phyla of invertebrates, some of which exhibit these symmetries.

Phylum Porifera: The Sponges

Have you ever washed dishes or cleaned up a mess with a sponge? Most likely, the sponge that you used was synthetic, but its structure is based on a whole group of creatures commonly called "sponges," the members of phylum Porifera (poor if' uh ruh).

Although there are a handful of freshwater species, most sponges live in marine environments. They are often mistaken for plants, because they are anchored to an immobile object and are therefore unable to move. Nevertheless, they are classified in the animal kingdom because of their cellular structure and the way they feed.

Sponges are amazing creatures. They possess no symmetry and take on a variety of shapes. These shapes include flat, tubular, branched, cuplike, and vaselike. Some sponges are no larger than the size of a fingernail, whereas others are large enough for you to sit in! A few examples of these incredible creatures are shown in Figure 11.3.

FIGURE 11.3
Some Members of Phylum Porifera

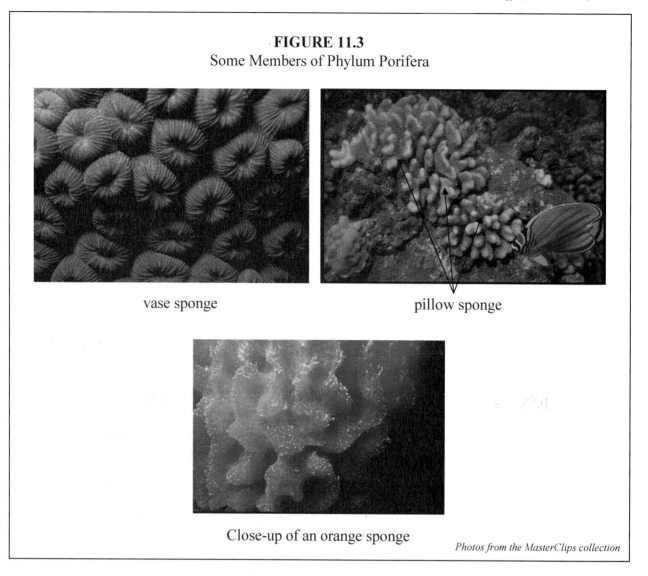

vase sponge   pillow sponge

Close-up of an orange sponge

Photos from the MasterClips collection

These creatures have an interesting anatomy. Sponges have two layers of cells which are separated by a thin, jelly-like substance. The outer layer of cells is called the **epidermis** (ep uh dur' miss), and the jelly-like substance is called the **mesenchyme** (mes' uhn kime).

Epidermis - An outer layer of cells designed to provide protection

Mesenchyme - The jelly-like substance that separates the epidermis from the inner cells in a
             sponge

Although the mesenchyme is not made up of cellular material, it is necessary to the life of the sponge, as you will learn in a minute.

A sponge is supported by a network of **spicules** that are found mostly in the mesenchyme. They weave throughout the mesenchyme, providing a framework that supports the sponge. The spicules are made of lime (calcium carbonate) or silica, depending on the species of the sponge. In some sponges, these spicules actually extend through the epidermis, giving the sponge a spiny or velvety look. Some sponges do not have spicules, however. These sponges are supported by a tough web of protein called **spongin**. Typically, sponges that have spicules feel hard and spiny, whereas sponges that have spongin are soft. Whether or not a sponge has spicules and what those spicules (if they exist) are made of help biologists further classify sponges.

Although sponges have interesting anatomy, their feeding habits are what really makes the sponge amazing! Since they cannot move, they cannot seek out prey. Instead, they must force their prey to come to them. How do they do this? Well, it all starts with the canals and cavities that exist in the body of the sponge. These canals or cavities are lined with **collar cells**.

Collar cells - Flagellated cells that pump water into a sponge

These cells have flagella that beat constantly, pulling water into the sponge. When the water is drawn in, the algae, bacteria, and organic debris contained in the water are drawn in as well. The sponge extracts these things from the water and eats them.

The sponge has no organs, so it must have specialized cells that take care of digestion. These cells are called **amebocytes** (uh mee' buh sites).

Amebocytes - Cells in a sponge that perform digestion and transport functions

These amebocytes travel freely in the mesenchyme. They digest the food that the sponge has extracted from the water and transport the food to the parts of the sponge that need it. In addition, they take in waste products from the inner cells and travel to the epidermis where the waste products are released. Amebocytes also exchange gases with the surroundings, bringing respiratory gases (such as oxygen) to the inner cell layers. On top of all this, these useful little cells also produce the lime or silica that make up the spicules.

Once food has been extracted and digested, the water must be expelled to allow fresh water to be drawn in. This is accomplished by more collar cells. Thus, sponges are continually pulling water in and then pumping it back out. During that process, the water is "cleaned" of algae, bacteria, and organic debris. Now you see why sponges have so many holes in them. Water must be continually pumped in and out of the sponge in order to provide it with food. Thus, it must be filled with many intertwined canals to allow for the passage of so much water.

Because of these canals, as well as the properties of spongin, sponges that contain spongin are very useful for humankind. As we mentioned before, the sponges you use for cleaning are usually synthetic, but for a long time, people used natural sponges for cleaning, because spongin is a soft, absorbent material. They did not use living sponges, because living sponges usually have a strong, foul odor. Instead, people used the "skeletons" of long dead sponges. You see, the protein spongin does not decay very quickly. Thus, much like the bones of a fossil, the spongin of a sponge stays around long after the sponge has died.

Sponges are used for many purposes other than just cleaning. During Roman times, they were used to pad the armor of soldiers. They are still used today as painting tools, and certain surgical swabs are made from sponges. Because sponges are so useful, there is a very profitable "fishing" industry built around them. Most of this "sponge fishing" occurs in either the Mediterranean Sea or the Gulf of Mexico, where the spongin-containing species of sponge are more plentiful.

Sponges have several modes of reproduction at their disposal. Their standard form of reproduction is asexual budding. They can also regenerate; if a portion of a sponge is cut away, it can grow into a new sponge. Also, during periods of freezing temperatures, sponges can produce a **gemmule** (jem' yool).

Gemmule - A cluster of cells encased in a hard, spicule-reinforced shell

A gemmule, much like a cyst, can survive through a long period of inclement weather. Once conditions become favorable again, the gemmule will break open and a sponge will grow from the encased cells.

Sponges also have a sexual mode of reproduction. Under certain conditions, collar cells can produce either eggs or sperm. These gametes are then released into the flow of water that the collar cells are maintaining. If a sperm cell and an egg cell meet, fertilization occurs and a zygote is formed.

If you have a microscope, you can learn more about sponges by performing the following experiment.

EXPERIMENT 11.1

Observation of the Spicules of a Sponge From Genus *Grantia*

Supplies:

- Microscope
- Prepared slide of *Grantia* Spicules
- Lab notebook
- Colored pencils
- Other sponges (optional)

Object: To observe a specimen from the phylum Porifera and note the simplicity , yet the complexity, of this animal's support structure. There are approximately 5,000 species of sponges. You will observe a species from genus *Grantia*.

Note: You can observe other sponges at an aquarium near you. Sponges can be iridescent, can encrust, and can be round shaped or tubular. Crabs, shrimp and brittle stars sometimes seek shelter within the maze of pores in a sponge, converting the sponge into a living apartment house. Natural sea sponges can also be purchased at art supply stores and hobby shops.

Procedure:

    A. Set up the microscope as instructed in previous experiments.

    B. Place the prepared slide under the microscope.

    C. Observe under low power and draw a section in your notebook. This slide shows you the spicules of the support system of the sponge. These are produced by amebocytes and come in a variety of shapes: needle, multipronged "jack", hooked or barbed. The shape of the spicules is used to classify sponges. Identify the shapes that you see.

    D. Observe under high power and draw one microscope field.

    E. (Optional) If you have purchased sponges from an art store, slice off a small section and make a wet mount by wetting the slice and covering it with a coverslip. The thinner the slice, the more you will see.

    F. Observe under low power and high power and sketch a section of the sponge in your notebook, one for each power.

    G. Clean up and return equipment to proper place.

ON YOUR OWN

11.2 One biology book calls sponges "tireless, natural pumps." Why is this a good description of sponges?

11.3 If a sponge has plenty of food but cannot distribute it to all of its cells, what is the sponge missing?

11.4 A sponge feels hard and prickly. Does it contain spicules or spongin?

Phylum Cnidaria

    The next set of invertebrates we want to discuss are the members of phylum Cnidaria (nih dahr' ee uh). Members of this phylum, which include jellyfish, sea anemones, and hydra, have two basic forms: the **polyp** and the **medusa**.

Polyp - A sessile, tubular cnidarian with a mouth and tentacles at one end and a basal disk at the other

Medusa - A free-swimming cnidarian with a bell-shaped body and tentacles

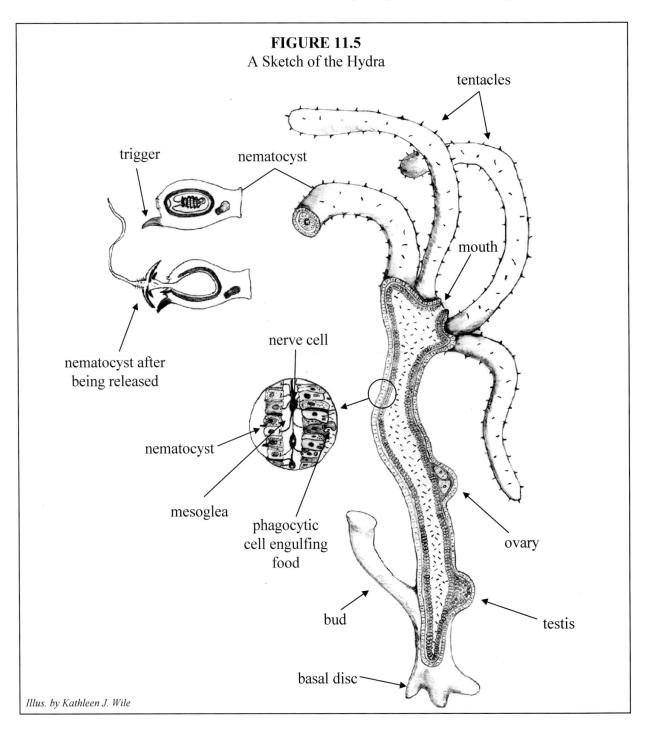

FIGURE 11.5
A Sketch of the Hydra

Illus. by Kathleen J. Wile

Notice that the mouth of the hydra is surrounded by tentacles. These tentacles are covered with nematocysts, as is the hydra's outer epithelial layer. The nematocysts actually lie within special structures that contain a pressure-sensitive trigger. When something brushes up against the trigger, the nematocyst is unleashed, stinging whatever set off the trigger. Isn't that neat? This little creature has a natural "booby trap" that is designed to be released when something touches it. The triggers on these cells are very sensitive and very reliable. Can you imagine such a thing forming by chance? Of course not. The kind of engineering needed to design such a system is far too complex to have arisen by chance. This is just one more tangible evidence of God's handiwork in His Creation!

In the drawing, you can also see how the extracellular/intracellular digestion of the hydra works. When prey is subdued by the nematocysts, the tentacles push it into the mouth so that it enters the hydra's gut. There, the prey is broken down until it is in small enough pieces for phagocytic cells to engulf the pieces and finish off the digestion. Waste products are then released into the surroundings right through the body of the hydra.

Hydra reproduce by budding, which is an asexual form of reproduction. The bud starts as a small bump on the side of the hydra. It grows tentacles and elongates until it actually looks like a miniature hydra. It then separates from the adult, anchors itself to something, and is a new hydra.

Although budding is one means of reproduction in the hydra, there is a sexual mode of reproduction as well. Egg or sperm gametes are formed in the hydra and pack together in little bumps in the animal's body wall. The bumps in which sperm cells form are called **testes** (test' ez), while the bumps that contain the egg cells are called **ovaries** (oh'vuh reez).

Testes - The organ that produces sperm

Ovaries - The organ that produces eggs

Although most hydra produce either eggs or sperm, a few hydra have been observed that actually form both! Egg and sperm are released in the water, and when a sperm meets an egg, fertilization occurs.

When the hydra was first discovered, many biologists thought that it was one of the "missing links" that Charles Darwin had spoken of. Remember from Module #9 that Darwin acknowledged the appalling lack of transitional forms in the fossil record, but he thought that maybe they had not been discovered yet. Thus, they were termed "missing links." When the hydra was first discovered, it was hailed as a macroevolutionary link between plants and animals. After all, here was a creature that stayed in one place (like a plant) but ate food (like an animal). As with all such "missing links," however, the more biologists learned about the creature, the more they realized that it could not be a missing link but was, instead, a unique form of life. To learn more about this interesting little creature, perform the following experiment. This is a microscope experiment, so if you do not have a microscope, please skip it.

EXPERIMENT 11.2
Observation of a Hydra

Supplies:

- Microscope
- Prepared slide of *Hydra*

- Lab notebook
- Colored pencils

Object: To observe the hydra as a typical member of phylum Cnidaria.

Note: Other forms of Cnidaria might be observed at your local zoo or aquarium. For those who wish to study this phylum more intensely, live or preserved specimens may be purchased from a biological supply house.

Procedure:

A. Set up the microscope as instructed in previous experiments.

B. Place the prepared hydra slide under the microscope and observe under low power.

C. Draw the entire hydra in your notebook. Things to note and label: tentacles, mouth, bud, basal disc, nematocyst and any reproductive structures that might be present. (Not all specimens will have reproductive organs present.)

D. Now turn to high power and observe several fields. Note the shape of the cells in the tentacles and on the body of the hydra.

E. Draw a portion of the wall of the body of the hydra and a portion of the tentacle wall. See if you can recognize the neomatocysts and any evidence of the poisonous barb they contain.

F. Clean up and put your equipment away.

Although we spent a great deal of time on the hydra, there are many other interesting organisms that make up phylum Cnidaria as well. A sea anemone is pictured in Figure 11.4b. Remember from Module #10 that the sea anemone has a symbiotic relationship with the clownfish. The clownfish can brush up against the tentacles of the sea anemone without getting stung by its nematocysts. How is that possible? Well, if you thought that the nematocysts in the hydra were well-engineered, get a load of this.

The nematocysts of a sea anemone are not triggered by a pressure-sensitive device like the hydra's are. Instead, the sea anemone has a *chemical recognition system* that looks for a specific amino acid. This amino acid exists in the slimy covering of most fish, but not the clownfish. Thus, the clownfish never sets off the sea anemone's nematocysts, and can nestle comfortably among its tentacles. Now think about that for a moment. A chemical recognition system for releasing nematocysts is much more complicated than a pressure-sensitive booby trap. Both such systems are present in the members of phylum Cnidaria. Furthermore, the organisms in this phylum are *the only ones* that possess nematocysts. The design apparent in this phylum should make you once again marvel at the power and wisdom of the Creator!

Another set of cnidarians are collectively called **corals**, an example of which is illustrated in Figure 11.4c. These organisms are very tiny polyps that live in self-made stone structures. Much like the amebocytes in sponges, these creatures have cells that produce a stone-like substance which is used to form a cup-like "house" in which the polyp lives. Although the substance that this residence is made of is much like stone, when the coral is alive, it stays flexible enough to open and close. When the coral feeds, the cup is open, and when it is not feeding, the cup stays closed. Corals are important because they gather together and form huge colonies, actually attaching their cup-like structures to one another. As corals begin to die, more corals attach themselves to the top of the dead ones, enlarging the size of the colony. These colonies get so large that they form great reefs (coral reefs) that can stretch for miles. These coral reefs form important ecosystems in warm, shallow marine waters and are home to thousands of other organisms.

A jellyfish is shown in Figure 11.4d. Although the jellyfish is usually recognized in the medusa form, some jellyfish (the ones in genus *Aurelia*, for example) actually go through the polyp form before they turn into the medusa form! A sketch of the lifecycle of an *Aurelia* (aw reel' yuh) is shown in Figure 11.6.

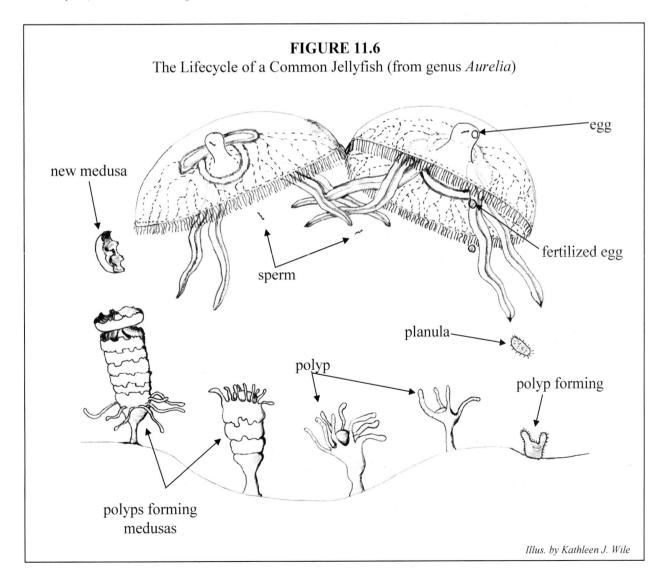

FIGURE 11.6
The Lifecycle of a Common Jellyfish (from genus *Aurelia*)

new medusa

egg

sperm

fertilized egg

planula

polyp

polyp forming

polyps forming
medusas

Illus. by Kathleen J. Wile

Unlike many cnidarians, the common jellyfish has definite males and females during the medusa stage. The male releases sperm which swim in the water. Some of these sperm will enter the gut of the female, where an egg is waiting to be fertilized. When fertilization takes place, the zygote leaves the gut through the mouth and clings to the female's tentacles for a while. After some development, the zygote has become a **planula** (plan' yuh luh), which swims away and attaches itself to an underwater base. It then grows tentacles and a mouth, forming a polyp. During its time as a polyp, the jellyfish might asexually reproduce by budding. Eventually, however, the polyp undergoes a dramatic change. It becomes almost cylindrical, with stacks of rings. These rings eventually separate, each forming a medusa.

If you are ever in the ocean swimming and you see a jellyfish, just remember that this recognizable form only represents one part of this fascinating creature's lifecycle. If you were to examine the bottom of the ocean, you would probably see the polyp form as well. Of course, if you see a jellyfish, you should immediately leave the water! Why? Well, like all cnidarians, the jellyfish has nematocysts in its tentacles. If you are unfortunate enough to brush up against those tentacles, the jellyfish will release its nematocysts into you! Now, you are far too big for the jellyfish to paralyze, but the poison released into your system will cause great swelling and irritation. Some people also have an allergic reaction to the poison, and that allergic reaction has been known to result in death. So even though jellyfish are quite beautiful and very interesting, it is best to stay away from them!

ON YOUR OWN

11.5 An organism has a mouth, tentacles, and a gut. It also has bilateral symmetry. Can it be placed in phylum Cnidaria?

11.6 If a clownfish were to brush up against a hydra, would the hydra release its nematocysts into the clownfish?

11.7 Nearly every organism in Creation needs some sort of respiratory and excretion systems. Why don't cnidarians need these?

11.8 If a jellyfish reproduces asexually, is it in polyp form or medusa form?

Phylum Annelida

It is now time to discuss one of the more familiar invertebrates, the members of phylum Annelida (ann uh lee' duh). Most likely, you have fished with members of this phylum, found them under rocks, or seen them crawling along the sidewalk after a strong rain. Most likely, you have even squished them underfoot. You see, phylum Annelida is made up of worms. It turns out, however, that there are several different types of worms in the animal kingdom, far too many for just one phylum. As a result, phylum Annelida contains only one particular type of worm, the **segmented worm**.

The most familiar type of segmented worm is the common earthworm, a member of genus *Lumbricus* (loom brih' cus). The basic anatomy of the common earthworm is illustrated in Figure 11.7 .

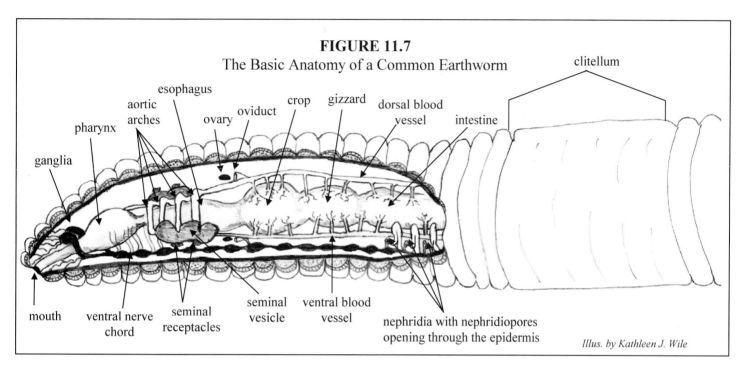

FIGURE 11.7
The Basic Anatomy of a Common Earthworm

Illus. by Kathleen J. Wile

These worms inhabit almost every bit of moist soil in the world. They possess bilateral symmetry and are made up of segments that look like little rings stacked on top of one another. Every earthworm also has a **clitellum** (klye tel' uhm), which is a barrel-shaped swelling that usually starts at the thirty-second segment and covers all segments up to thirty-seven. This structure aids in reproduction (as we will see later), and it helps us to distinguish the head of the worm from its tail.

That last statement might surprise you a bit. When you look at an earthworm, it certainly doesn't seem to have a head and a tail. Nevertheless, it does. In the animal kingdom, we usually distinguish the region of the organism that contains the head from the region that contains the tail with the terms **anterior end** and **posterior end**.

Anterior end - The end of an animal that contains its head

Posterior end - The end of an animal that contains the tail

The anterior end of the earthworm is usually a little more pointed than the posterior, and it is also usually darker than the rest of the body. The earthworm's mouth is at the tip of the anterior end, and it has another opening at the tip of its posterior end called the **anus**.

The earthworm moves with a combination of two muscle layers and small bristles called **setae**. The worm has a **circular layer** of muscles that, when contracted, makes the earthworm long and thin. When the other layer (the **longitudinal layer**) contracts, the earthworm gets short and thick. In order to move, the earthworm uses its posterior setae to anchor its posterior end while it contracts its circular muscles. This stretches the earthworm out. Since the posterior end is anchored, however, the net effect is to simply push the anterior end forward. Once the circular muscles have contracted completely, the earthworm then uses it anterior setae to anchor its anterior end and at the same time releases its posterior setae. Then, the longitudinal muscles contract. Since the anterior is anchored and the posterior is not, this causes the posterior to move forward as the longitudinal muscles cause the earthworm's body to become shorter. This is how the earthworm crawls along a surface and tunnels through dirt.

Feeding Habits of the Earthworm

The earthworm is the first animal that we have studied which has a complete digestive system. It ingests soil into its mouth by using its powerful **pharynx** (fare' inks) to create a sucking action. The soil is then passed into the **crop**, where it is stored for a while. Eventually, the soil makes it to the **gizzard**, where muscular contractions grind the soil, breaking into small pieces any vegetation, refuse, or decaying organic matter that happens to be in the soil. These pieces are what the worm actually digests as food. The rest of the soil is passed down through the **intestine** and out through the anus. Digestible materials are digested by enzymes in the intestine. The digested food is then absorbed by blood that circulates through the walls of the intestine. The blood then transports the digested food to cells throughout the worm's body.

Although most waste is excreted through the anus, some of the waste products of metabolism (water, for example) are gathered in small organs called **nephridia** (nuh frid' ee uh). These organs, which function like "simple" kidneys, are in every segment of the earthworm's body except the first three and the very last one. They push the waste products out of the earthworm through tiny holes called **nephridiopores** (nuh frid' ee uh poors).

The feeding habits of the earthworm actually produce two beneficial results for the earthworm and one for the ecosystem in which it resides. First, of course, the earthworm gets the food it needs to survive. Secondly, however, it also loosens the soil, allowing the earthworm to tunnel though it. In other words, in order to tunnel through soil, the earthworm actually eats it! Finally, the earthworm's feeding habits make the soil in which it resides very fertile. By loosening the soil, it helps oxygen and water filter into the soil, which makes it easier for plants to absorb them. Also, because the earthworm is constantly tunneling up and down in the soil, it takes the nutrient-rich soil at the surface and mixes it with the mineral-rich soil below the surface, producing a fertile layer of topsoil. Soil cannot stay fertile year after year without the help of earthworms!

This is, yet again, a great example of the design that is apparent in Creation. Since plants constantly extract minerals from the topsoil in which they grow, there would be no minerals left in only a few short years. In addition, when plants die, they decay on the surface. Thus, without the aid of earthworms, there would be a nutrient-rich layer of soil at the surface, but those nutrients could not be absorbed by the plants' roots. At the same time, there would be no minerals for the roots to absorb, because they would be too low for the plants' root systems to reach. The worm, however, was created to continually mix the lower soil with the upper soil, recycling the nutrients and bringing up more minerals. Now you know why all moist soil contains earthworms. They are the caretakers that God has made for the plants, allowing them to continue to grow!

The Respiratory and Circulatory Systems in an Earthworm

Looking back at Figure 11.7, you can see that there are still some parts of the earthworm's anatomy that we have not discussed. For example, the earthworm is the first animal we have discussed which has a full-fledged **circulatory system**.

Circulatory system - A system designed to transport food and other necessary substances throughout a creature's body

Composed of a complex series of blood vessels, this circulatory system allows vital substances such as food and oxygen to travel to all cells within the earthworm's body. Although the earthworm has a full-fledged circulatory system, it does not have a heart. Instead, the two main blood vessels, the **dorsal blood vessel** and the **ventral blood vessel**, pump the blood, causing it to move throughout the earthworm's body. The dorsal vessel pumps blood to the anterior end of the earthworm, while the ventral vessel pumps blood towards the posterior end. All along the way, smaller blood vessels branch off of these two main vessels, allowing blood to flow to all regions of the earthworm's body. The dorsal and ventral blood vessels are linked together by a series of strongly-muscled vessels known as **aortic** (ay or' tik) **arches**, which help regulate the blood flow.

The blood picks up nutrients as it flows through the vessels that are a part of the intestine. However, the blood must also take oxygen to the cells, and pick up carbon dioxide from the cells so that it can be released as waste. This happens at the surface of the earthworm's body, through its thin epidermis. As the blood flows through the epidermis, it absorbs oxygen that has diffused in from the surroundings and releases the carbon dioxide that it has picked up from the cells, which diffuses out into the surroundings. In order to be able to absorb oxygen and release carbon dioxide in this way, the earthworm's epidermis is covered in a moist layer called a **cuticle**.

If you have ever seen a shriveled up earthworm lying on the sidewalk, you now know why it died. If the earthworm is out in the sun too long, the cuticle begins to dry. This stops the exchange of gases with the surroundings, suffocating the worm. Under normal circumstances, the earthworm can tunnel underground and allow the moist soil to wet its cuticle. If, however, the worm is unlucky enough to be caught on a sidewalk, it cannot tunnel underground, and it therefore suffocates. An earthworm can also suffocate during a rainstorm. If the soil in which it is tunneling gets too wet, the water takes the place of oxygen in the dirt. As a result, there is little oxygen to absorb. Thus, the earthworm quickly tunnels to the surface. While on the surface, of course, it is exposed to birds and other predators, so it will only stay there while the soil remains too wet to inhabit. That's why the best time to look for worms is right after a heavy rain.

How does the earthworm know when to leave the protective soil and expose itself to predators? The earthworm has a complex **nervous system** that allows it to respond to outside stimuli.

Nervous system - A system of sensitive cells that respond to stimuli such as sound, touch, and
taste

The nervous system is controlled by a small brain composed of two masses of nerve cell bodies that scientists call **ganglia** (gan' glee uh).

Ganglia - (singular: ganglion) Masses of nerve cell bodies

Although not as sophisticated as a real brain, these two ganglia do control the nervous system and coordinate the response of the earthworm. A large **ventral nerve cord** travels down the bottom of the earthworm, and the cord has a small ganglion at each segment of the worm. Nerves stretch out from these ganglia, sensing touch, light, and certain chemicals. If these nerves sense too much water, for example, then they send a message to the ganglion from which they extend, which then sends a message to the main ganglia. The main ganglia then tell the earthworm to head for the surface.

The Earthworm's Reproductive System

Perhaps the most interesting aspect of the earthworm is its method of reproduction. Earthworms are **hermaphroditic** (hur maf ruh dit' ik)

Hermaphroditic - Possessing both the male and the female reproductive organs

In other words, a single worm produces both eggs and sperm. Unlike a hydra which can, on occasion, do the same thing, the earthworm cannot mate with itself. Instead, it must find a partner with which to reproduce.

Here's how that works. Sperm produced by the testes of an earthworm are stored in **seminal vesicles** until the earthworm mates. Eggs produced in the ovaries are likewise stored until mating in tubular **oviducts**. When an earthworm mates, it finds another earthworm pointing in the opposite direction, and the two worms attach themselves together with a **slime tube**, exchanging their sperm. They each empty their seminal vesicles into small pouches called seminal receptacles in the other worm. In this way, the sperm of one worm is exchanged for the sperm of another. Once the sperm is exchanged, the worms separate, but their anterior ends remain wrapped in a slime tube.

A few days later, a **cocoon** forms around the clitellum of each worm. Each worm then backs out of the cocoon, leaving the slime tube behind as well. As the cocoon passes over the oviducts, eggs are released into it. Then, as the cocoon passes over the seminal receptacles, sperm are released as well. When the worm is completely free of the cocoon, the cocoon seals, and fertilization takes place. In two to three weeks, a young earthworm will break out of the cocoon.

Now we realize that we have thrown a lot of information at you. By far, this is the most complex organism that we have studied so far, so it is only natural for your head to be swimming right now with all of these facts. They are important facts, however, and you need to remember them. Specifically, you need to be able to label all of the organs shown in Figure 11.7 as well as be able to explain the use of each. This is a lot of information, so you will have to study this section of the text a few more times in order to retain it all. The earthworm, however, is the only organism in this module that you will have to know in such a detailed fashion.

Other Segmented Worms

Earthworms, of course, are not the only kind of segmented worms. The leech is another kind of segmented worm that you might have heard about. They live in marine, freshwater, and moist soil environments and move about by using two suckers, one on the posterior end and the other on the anterior end. These same suckers are used to latch onto their prey and suck out its blood. Although most leeches actually feed on small prey, killing them before they begin ingesting their blood, some leeches simply attach to their prey and suck out its blood while the prey stays alive. This particular type of leech was used in medicine at one time. Before we knew much about medicine, it was believed that many diseases were the result of "bad blood." To get rid of the blood, leeches were applied to the sick person. We know now, of course, that such a practice is harmful to the patient, but it was common medical practice for quite some time!

Other, more exotic worms also exist in this phylum. For example, Figure 11.8 shows two animals that you might not, at first glance, think of as worms. Nevertheless, they both belong to class Polychaeta (pol ee kay' tuh), which is a class in this phylum.

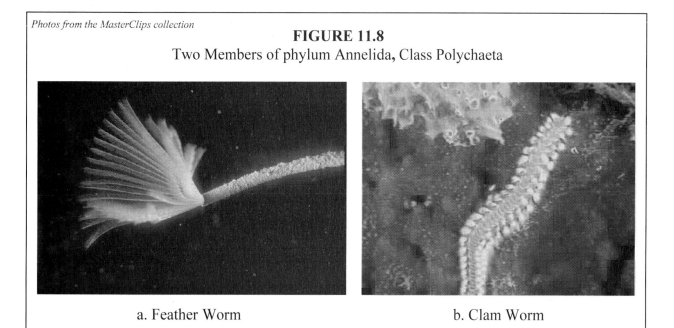

Photos from the MasterClips collection

FIGURE 11.8
Two Members of phylum Annelida, Class Polychaeta

a. Feather Worm   b. Clam Worm

The polychaete pictured in Figure 11.8a is called the "feather worm." It lives in a tube which it constructs, and has tentacles that stick out the end of the tube. The tentacles are covered with cilia and mucus, and when algae or organic matter in the ocean water stick to the mucus, the cilia move it down into the mouth of the worm. Its name comes from the fact that the tube and cilia-covered tentacles combine to look like a feather-duster. The polychaete pictured in Figure 11.8b is another marine-dweller, the "clam worm." This worm is a powerful predator, catching its prey by extending its pharynx out of its mouth and attaching it to the creature. When it pulls the pharynx back into its mouth, the prey is helplessly drawn inside as well!

ON YOUR OWN

11.9 If an earthworm's mouth is fully functional but it cannot ingest soil, what organ is malfunctioning?

11.10 What process will stop if the earthworm's cuticle dries up?

11.11 In a dissected earthworm, you see a nerve chord with ganglia at each segment right next to a long blood vessel. Is blood flowing towards the anterior or the posterior in this blood vessel?

11.12 An earthworm's seminal vesicle is empty but its oviducts are full. Has the earthworm mated yet?

If you purchased the dissection kit described in the beginning of the book, you should perform the following earthworm dissection so that you will be more familiar with the earthworm's anatomy. If you did not, please skip the experiment.

Optional Dissection Experiment #1
The Earthworm

Be careful! Dissection tools are SHARP!

Supplies:

- Dissecting tools and tray that came with your dissection kit
- Earthworm specimen
- Magnifying glass
- Laboratory Notebook

Object: To become more familiar with the earthworm's anatomy through dissection

Procedure:

A. Examine your earthworm specimen carefully. Rub your fingers lightly across the surface until you feel bristles. Those bristles are the **setae**. Write in your laboratory notebook how many setae you find on each of the worm's segments.

B. Using a magnifying glass, try to find the **nephridiopores**.

C. Examine the **clitellum**. In your laboratory notebook, write down how many segments there are in it.

D. Now you are ready to begin the dissection. Place the specimen ventral side (the side with the setae) down on the tray. Pin the anterior and posterior ends as illustrated in the drawing below. Try not to pierce any organs in the process!

Illus. by Kathleen Wile

E. Use your scissors to cut through the body wall along the line in the drawing above. Start about an inch posterior of the clitellum and just to the left of center. Being careful not to cut anything but the body wall, extend your cut all the way to the anterior end.

F. Pull apart the edges of the cut and peer in. You can probably see the **intestine**. The space in between the body wall and the intestine is called the **body cavity**.

G. Notice that the body cavity is separated by partitions that run from the body wall to the intestine. These are called **septa**. Using forceps (tweezers) and your probe, break the septa as shown in the drawing below:

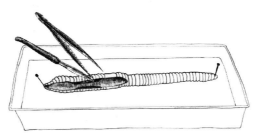

Illus. by Kathleen Wile

H. Peel back the body wall on both sides of the cut and pin it down, as shown in the drawing below:

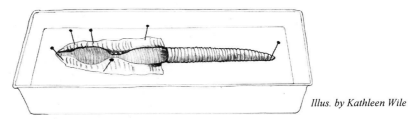

Illus. by Kathleen Wile

I. Now the internal structures should be visible. Make a drawing of your dissected earthworm in your laboratory notebook. As you identify the structures listed below, label them in your drawing. Note any structures you could not see as well as any organs which you saw but could not identify.

J. Using Figure 11.7 as a guide, identify the following digestive structures:

   Pharynx - A thick-walled structure in the area of segments 4-7
   Esophagus - The structure that extends from the pharynx to about segment 14

Crop - A bulge just posterior of the esophagus
Gizzard - The structure posterior to the crop
Intestine - The structure which extends from the gizzard to the anus

Review the functions of these structures as described in the text.

K. Once again, using Figure 11.7 as a guide, identify the following circulatory system structures:

Dorsal blood vessel - A dark, brownish vessel running along the dorsal side of the intestine. It might actually lie on the intestine.
Aortic arches - You will have to remove the seminal vesicles (see Figure 11.7) and septa to see the arches clearly. Remove them only from the left side of the earthworm and examine the aortic arches that are revealed. They will look like large tubes.
Ventral blood vessel - Use your probe to move aside (*do not remove*) the intestine near the posterior end of your cut. This should reveal the ventral blood vessel, which looks very similar to the dorsal blood vessel.

Review the functions of these structures as described in the text.

L. Locate the **nephridia**. The best way to do this is to extend your cut another two inches to the posterior. Without tearing the septa, remove the intestine from this region (and *only* this region) and then use your magnifying glass to find the nephridia. They will be in all segments except the first three and the last one, so there should be plenty to see. If you cannot find them, don't worry. They are, perhaps, the most difficult of the earthworm's internal structures to find.

M. If your dissection has been a bit sloppy, this exercise might not turn out too well. Using Figure 11.7 as a guide, try to find the **ganglia** that form the earthworm's "brain." They should be just anterior of the pharynx. Follow the **ventral nerve cord** from the ganglia. Note the small ganglion (which looks like a bulge) that appears at each segment.

N. To get a look at the reproductive structures, remove a portion of the digestive track. Do so by cutting across the intestine near the clitellum. Make a similar cut just posterior to the pharynx. You have now cut out a section of the digestive tract. Gently pull that section (that portion of the intestine, the gizzard, and the crop), out of the earthworm.

O. If you were careful in the previous step, you will now see white structures in segments 9-13. Those are the **seminal vesicles.** Directly under the seminal vesicles are the small **ovaries**. In addition, segments 9 and 10 contain the **seminal receptacles**, which are pairs of small, white, round structures.

P. Be sure that you have labeled all of the structures in your drawing and have listed all of those that you could not find. Dispose of your specimen. Clean and dry your dissection tools, tray, and pins. Put everything back into its proper place.

Phylum Platyhelminthes

Although the members of phylum Annelida are the worms with which you are mostly familiar, there are other kinds of worms that are just far too different to be contained in the same phylum. The members of phylum Platyhelminthes (plat ee hel min' theez) are more commonly called "flatworms," because compared to the cylindrical segmented worms, members of this phylum are rather flat. A common example of the organisms in this phylum is the planarian, pictured in Figure 11.9.

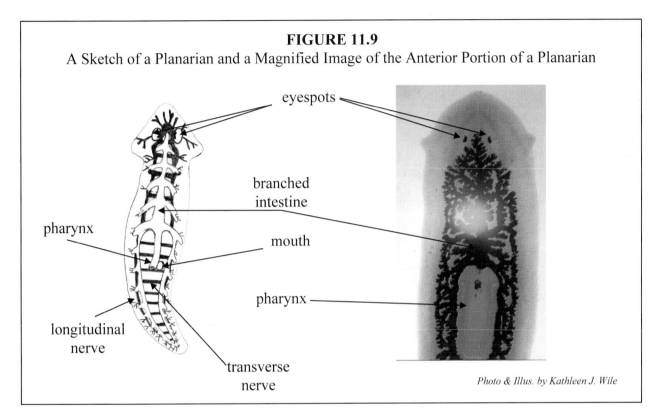

FIGURE 11.9

A Sketch of a Planarian and a Magnified Image of the Anterior Portion of a Planarian

eyespots

branched intestine

pharynx

mouth

pharynx

longitudinal nerve

transverse nerve

Photo & Illus. by Kathleen J. Wile

Planarians live in fresh water and swim there freely. Although planarians live in fresh water, there are many flatworms that live in marine environments as well.

Planarians usually eat small organisms that they grab with their mouth. Digestive enzymes are then secreted through the mouth, and, after the organism is broken down into small particles, a tubelike pharynx is extended out of the mouth, sucking up the small particles. These particles then go into the intestine, where more digestive enzymes continue to break them down. The intestine itself is highly branched and runs throughout most of the body. Unlike the members of phylum Annelida, planarians do not need a circulatory system, because the intestine branches extend throughout most of the body. This allows all cells to be close enough to the intestine that digested food can get to them simply by diffusion.

The nervous system of planarians is a little more complex than the one you studied in earthworms. At the head of the organism lies a mass of nerve tissue that is best described as a brain. This brain has nerve branchings that go throughout the head and also attach to the worm's eyespots. Although planarians do not really "see" with these eyespots, they can sense light and move towards it. Planarians move towards light in order to seek out prey such as photosynthetic organisms. Two longitudinal nerves extend from the brain to the posterior of the body, and those nerves are connected to each other by a series of transverse nerves that go across the body. This complex nervous system allows the planarian to have senses of taste, smell, and touch, as well as the ability to sense light.

Planarians sexually reproduce just like earthworms. They are hermaphroditic, but can only reproduce upon the exchange of sperm with a mate. One interesting way that planarians can asexually reproduce is by **regeneration**.

Regeneration - The ability to re-grow a missing part of the body

If a planarian decides to asexually reproduce, it simply tears itself in half. Then, the two halves each regenerate their missing half, producing two planarians where there was only one before. Although some earthworms have the ability to regenerate, they never use it for reproductive purposes. They only use it in case of an accident. While planarians can use regeneration to heal from an accident, they also use it deliberately, as a means of asexual reproduction!

If you have a microscope, you can learn more about planarians by performing the following experiment.

EXPERIMENT 11.3
Observation of a Planarian

Supplies:
- Microscope
- Slide of *Planaria* (injected whole mount)
- Lab notebook
- Colored pencils

Object: To observe a planarian as an example of organisms from phylum Platyhelminthes

Note: For those wishing to study the flat worms more thoroughly, a local veterinarian or medical doctor might have preserved specimens of tapeworms and/or flukes. Also, live planarians may be purchased from a biological house. Live planarians are extremely interesting. They react to stimuli like air (use an eye dropper), movement, touch and chemicals including salt water. If split in half longitudinally, they will regenerate.

Procedure:

A. Set up the microscope as instructed in previous experiments.

B. Place the prepared slide under the microscope.

C. Observe under low power and draw the whole planaria. Your injected whole mount should help you see the two eyespots, the intestine, the pharynx, the mouth and maybe the longitudinal and transverse nerves. Note that the mouth is more than halfway down the body, at the end of the pharynx. Looking closely, you might see the tubules and flame cells that are part of the excretory system. Be sure to include all organs that you can see in your drawing.

D. Clean up and put your equipment away.

Other Members of Phylum Platyhelminthes

Although planarians make a good case study in phylum Platyhelminthes, there are many other types of flatworms as well. For example, there are a host of parasitic flatworms called **tapeworms** and **flukes**. These two classes of flatworm are quite different from each other, but they have much in common. Mostly, they do not have elaborate nervous or digestive systems as do the planarians. After all, as parasites, they need not look for prey, and their host typically does a great deal of digestion for them. Instead, the main features of parasitic flatworms are mechanisms that protect them from the digestive juices and infection-fighting mechanisms of their host as well as suckers, hooks, or both to help them hold their position within the host.

ON YOUR OWN

11.13 Compare the digestion of a planarian with a fungus. What are the similarities?

11.14 A flatworm has very complex digestive and nervous systems. Is this flatworm likely to be parasitic? Why or why not?

Phylum Nematoda

Members of phylum Nematoda (nem uh toh' duh), often called roundworms, are tiny, cylindrical worms which possess bilateral symmetry. Like their common name implies, members of this phylum are round, rather than flat. Thus, they look more like members of phylum Annelida than members of phylum Platyhelminthes. Typically, however, they are much smaller than the members of phylum Annelida.

Roundworms are particularly hardy. They live in virtually every environment in Creation. You can find them in dirt, in the snow of the arctic tundra, the heat of hot springs, and the terribly hot and high-pressure environment of hydrothermal vents at the bottom of the ocean. Their bodies are essentially comprised of a tube within a tube. The outer tube, which is tapered on both ends, is made up of the epidermis and a cuticle that the epidermis secretes. The inner tube is a digestive canal which is open on both ends.

Many members of this phylum are parasitic. People, for example, can be infected by about 30 species of roundworms. A roundworm's cuticle protects it from the digestive juices of its host, allowing the worm to live in the host's intestine. Millions of people in third world countries are infected by parasitic roundworms. A common example of a parasitic roundworm is the trichina worm, *Trichinella* (try' kin el uh) *spiralis* (spuh' ral us). Adult trichina worms live in the intestines of pigs and certain game animals. When the female reproduces, the young make their way to the host's blood vessels, and from there they go to the host's muscle tissue. Once in the muscle tissue, the young develop for about 14 days and then form cysts, which allow them to survive there for a long time.

When a person eats pork or other game animals that are infected with trichina worms, the digestive juices in the person's intestine free the juvenile worms from their cysts and they begin to develop and mature, reaching adulthood in about two days. As they mature, they cause severe irritation of the intestinal tract, causing abdominal pain, nausea, vomiting, and watery stools. When the trichina worms mature, they mate. The young produced by this mating migrate to the muscle tissue, and cause muscle damage which often results in muscle pain, swelling, and joint pain. Such infection is called **trichinosis**. Although most people who contract trichinosis recover in about 6 months, the disease often results in permanent heart or eye damage. Nearly five percent of trichinosis cases are fatal.

Trichinella spiralis infection in people is usually caused by eating poorly-cooked pork. Despite rigid standards of meat inspection in the US, infected pork can sometimes pass inspection because the trichina worm cysts are very difficult to detect. If infected pork is not cooked thoroughly, then the cysts can survive, resulting in the infection discussed above. Many Old Testament scholars consider some of the dietary laws of the Israelites (see Lev 11:7, for example) God's way of protecting His people from trichinosis. To completely kill the trichina worm cysts in pork, the meat must be frozen at 5 °F for 21 days or -22 °F for 25 hours. Clearly, such processing was not available to the people of Old Testament times; thus, God decided to protect His people from trichinosis by simply forbidding them to eat pork.

This is an important point. The Israelites in Old Testament times didn't know anything about *Trichinella spiralis*. In order to protect His people, then, God simply forbade them to eat pork. Now to someone who was alive back then, such a rule might have seemed arbitrary and unfair. After all, they might have asked themselves, why can't we eat pork? Is God simply trying to keep us from enjoying ourselves? No, of course not. God was simply taking care of His people by forbidding them to eat something that was unhealthy for them! If you ever think that one of God's rules is designed simply to make life more difficult for you, just remember trichinosis. God's rules protect us from dangers that we do not understand, just like God's Law forbidding pork protected the children of Israel from a disease that they did not understand.

Phylum Mollusca

Phylum Mollusca (muh lus' kuh) contains many organisms, including clams, oysters, snails, and squid. Although these organisms might seem rather different to you, they actually share several features. Most members of phylum Mollusca, have the following features in common:

Mantle - A sheath of tissue that encloses the vital organs of a mollusk, makes the mollusk's shell, and performs respiration

Shell - A tough, multilayered structure secreted by the mantle. It is usually used for protection, but sometimes for body support

Visceral (vis' ur ul) hump - A hump that contains a mollusk's heart, digestive, and excretory organs

Foot - A muscular organ that is used for locomotion and takes a variety of forms depending on the animal

Radula - A organ covered with teeth that mollusks use to scrape food into their mouths

The common snail (shown in Figure 11.10) is a good example of the members of this phylum.

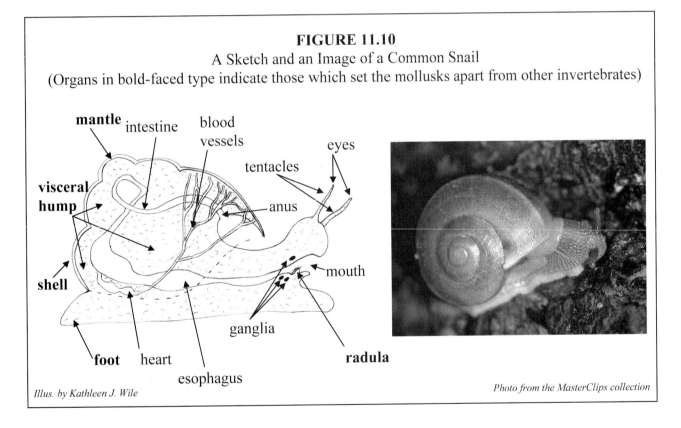

FIGURE 11.10
A Sketch and an Image of a Common Snail
(Organs in bold-faced type indicate those which set the mollusks apart from other invertebrates)

Illus. by Kathleen J. Wile

Photo from the MasterClips collection

The snail is a **univalve**, meaning it has only one shell. Organisms like clams are **bivalve**, because they have two shells, one on the top and one on the bottom.

Univalve - An organism with a single shell

<u>Bivalve</u> - An organism with two shells

In a snail, the shell exists for protection. When threatened, the snail can retreat into it shell, and, in the case of most snails, a "door" can be shut, sealing the shell.

The snail moves by laying down a thin layer of slime upon which it glides by rhythmically contracting its foot. This method of locomotion allows the snail to move at the "incredible" rate of 3 meters (about 9 feet) per hour! While moving at this speed, the snail grazes on plant material by extending its radula and grating the plant material. The food is digested in the intestine and nondigestible substances are expelled through the anus.

The foot is directly below the visceral hump, which is contained in the shell. Surrounding the visceral hump, under the shell, is the mantle. The mantle is rich in blood vessels, as it is the organ of respiration. The snail's head contains its main sensory organs: two tentacles that end in eyes. The nerves in the tentacles provide the sense of touch, while nerves in the head provide a sense of smell. The eyes, of course, provide the snail with a primitive sense of sight.

ON YOUR OWN

11.15 If you were to observe a patch of grass that a snail had just traveled across, what would you expect to find?

11.16 What does a clam use to get food into its mouth?

<u>Summing Up The Invertebrates</u>

Since there are 18 phyla of invertebrates in kingdom Animalia, it is simply impossible to discuss them all in a first-year biology course. We have tried to discuss the major ones, but please realize that there are several phyla that we have left out completely. Phylum Echinodermata, for example, contains sea urchins and starfish. There are several phyla of worms that we did not even mention, and even in the phyla that we did discuss, we did not talk about all of the major classes. Nevertheless, the organisms that we did discuss should give you a good idea about the general characteristics of the invertebrate animals. In order to try and give you a more complete view of at least one phylum of invertebrates, we will spend all of next module on one phylum: Arthropoda.

Before you begin the study guide, we need to remind you to use the study guide and "on your own" problems as a guide for what you need to know on the test. Clearly, there is no way for you to remember all of the information that we provided in this module. As a result, we will only require you to know the intricate details of the earthworm's anatomy. Although we discussed the anatomy of other organisms in some detail, you will only need to know the general principles that we discussed. Thus, use the study guide and "on your own" problems to help you decide what to concentrate on as you prepare for the test.

ANSWERS TO THE ON YOUR OWN PROBLEMS

11.1 a. The manta can only be cut in identical pieces if the cut separates left from right. This means it possesses <u>bilateral symmetry</u>.

b. There is really no way to cut the sponge to get identical halves. It therefore has <u>no symmetry</u>.

c. The jellyfish can be cut in identical halves with any cut so long as it is up and down. This gives it <u>radial symmetry</u>.

d. The brain coral is the same everywhere, so it has <u>spherical symmetry</u>.

11.2 <u>Sponges can be described as pumps because they continually pump water into themselves in order to extract their food and then pump it back out again</u>.

11.3 <u>The sponge is missing its amebocytes</u>. Sponges use amebocytes to distribute food amongst its cells. Without them, there is no way for the food to get there.

11.4 Only spongin has the soft, absorbent feel we are used to in a sponge. Spicules, on the other hand, are hard and prickly because they are made of lime or silica. Thus, <u>this sponge contains spicules</u>.

11.5 Although the mouth, tentacles, and gut are common to all cnidarians, the bilateral symmetry is not. Cnidarians must have radial symmetry, so <u>it is not in phylum Cnidaria</u>.

11.6 <u>Yes, the hydra would release nematocysts into the clownfish</u>. Unlike the sea anemone, the hydra's mechanism for releasing nematocysts is pressure-sensitive. Thus, *anything* that brushes up against the hydra will be stung.

11.7 <u>Cnidarians do not need respiratory and excretion systems because their bodies are so thin that gases can easily diffuse in and out directly through the body</u>. As a result, there is no need for an elaborate system to perform the function of taking in or releasing gases.

11.8 When in medusa form, jellyfish can only reproduce sexually. Thus, <u>the jellyfish must be in polyp form</u>.

11.9 The pharynx sucks soil into the mouth. Thus, <u>the pharynx must be malfunctioning</u>.

11.10 When an earthworm's cuticle dries up, its <u>respiration stops</u>.

11.11 If the nerve cord has ganglia, we are looking at the ventral nerve cord. This means that the blood vessel next to it is the ventral blood vessel. Since the ventral blood vessel pumps blood to the posterior of the earthworm, <u>the blood is flowing to the posterior</u>.

11.12 If the oviducts are full, then it is at or past the stage where it is ready to mate. If its seminal vesicles are empty, however, then it must have already exchanged sperm. Thus, <u>the earthworm must have already mated</u>, but it must still be waiting for the cocoon to finish forming.

11.13 <u>The digestion of a planarian and fungus are somewhat similar because the fungus digests its food completely before taking it in, and the planarian digests its food somewhat before taking it in</u>. Thus, although the process is not identical, there is at least some digestion before ingestion in both creatures.

11.14 <u>The flatworm is most likely not parasitic</u>. Remember, parasitic flatworms don't need much of a digestive system because their host digests most of their food for them. Also, they do not need much of a nervous system because they do not need to seek out prey.

11.15 <u>You would expect to find a trail of slime</u>. This is because the snail excretes a layer of slime upon which it glides.

11.16 Even though we did not study clams explicitly, they are members of phylum Mollusca, and all mollusks use their <u>radula</u> to pull food into their mouths.

STUDY GUIDE FOR MODULE #11

⚹ 1. Define the following terms:

⚹a. Invertebrates

⚹b. Vertebrates

⚹c. Spherical symmetry

⚹d. Radial symmetry

⚹e. Bilateral symmetry

⚹f. Epidermis

g. Mesenchyme

h. Collar cells

i. Amebocytes

j. Gemmule

k. Polyp

⚹l. Medusa

⚹m. Epithelium

n. Mesoglea

⚹o. Nematocysts

⚹p. Testes

⚹q. Ovaries

⚹r. Anterior end

⚹s. Posterior end

⚹t. Circulatory system

⚹u. Nervous system

⚹v. Ganglia

⚹w. Hermaphroditic

⚹x. Regeneration

y. Mantle

z. Shell

aa. Visceral hump

bb. Foot

cc. Radula

dd. Univalve

ee. Bivalve

⚹ 2. Do the vast majority of animals have backbones?

⚹ 3. Determine the symmetry of the following organisms:

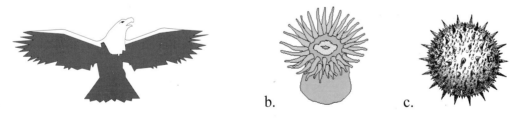

a.   b.   c.

Illus. from the Masterclips collection

4. How do sponges get their prey?

⚹ 5. If a sponge is soft, does it contain spicules or spongin? What purpose do these substances serve in a sponge?

6. What is the predominant mode of asexual reproduction in a sponge?

7. What roles do amebocytes play in the anatomy of a sponge?

8. When does a sponge produce gemmules?

9. What is the difference between the nematocysts of a hydra and those of a sea anemone?

✳10. Why do cnidarians not need respiratory or excretory systems?

11. Some biology books say that jellyfish live "dual lives." Why?

12. If a jellyfish reproduces sexually, what form is it in?

13. What is another name for a coral colony?

14. Name all of the structures in the diagram below:

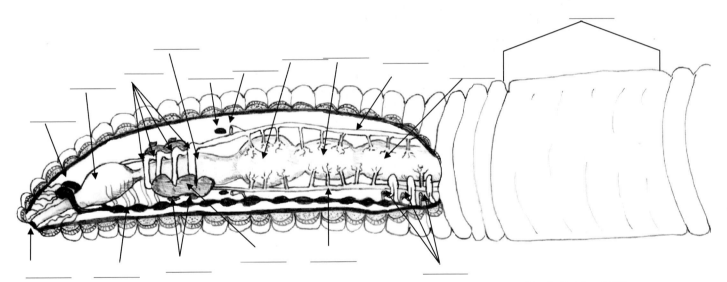

Illus. by Kathleen J. Wile

✳15. What benefits do earthworms give the plants in the soil that they inhabit?

✳16. If you pick up two earthworms and the first feels very slimy near the clitellum and the second does not, what can you conclude about the first earthworm?

17. What similarities exist between the hydra's sexual reproduction and the earthworm's? What differences exist?

✳18. What will happen to an earthworm if its cuticle gets dry?

19. Why don't planarians need circulatory systems?

20. If a flatworm has no complex nervous or digestive systems, is it most likely free-living or parasitic?

21. What is the main mode of asexual reproduction in a planarian?

* 22. Place each organism in one of the following phyla: Porifera, Cnidaria, Annelida, Mollusca, Platyhelminthes

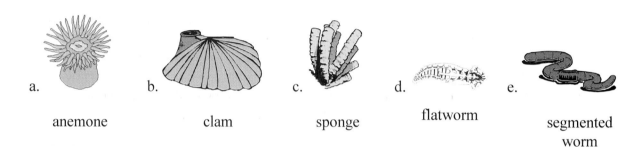

a. anemone   b. clam   c. sponge   d. flatworm   e. segmented worm

a-c,e. *Illus. from the MasterClips collection*
d. *Illus. by Kathleen J. Wile*

Module #12: Phylum Arthropoda

Introduction

As you have learned already, invertebrates make up the vast majority of the animal kingdom. In this module, we study the most populous phylum of invertebrates, phylum Arthropoda (ar thrah' poh duh). This phylum contains crayfish, lobsters, spiders, scorpions, and insects. In fact, it contains more species than *all of the other phyla in kingdom Animalia combined!* Obviously, then, this important phylum deserves an in-depth look.

Arthropods are all around us. They crawl on the ground, fly in the air, and skim across the water. They have an amazing effect on the environment. Arthropods, for example, help plants reproduce by carrying pollen from one plant to another. They also produce many useful items such as silk, wax, honey, and drugs. Although arthropods are very necessary for the earth's ecosystem, they can also be quite dangerous. Some arthropods transmit deadly diseases, while others have been responsible for the destruction of millions of acres of crops.

General Characteristics of Arthropods

Although this phylum is vast and diverse, there are many common characteristics that unite arthropods. These characteristics are important to know and understand.

An Exoskeleton

All arthropods have an **exoskeleton**.

Exoskeleton - A body covering, typically made of chitin, that provides support and protection

As the definition states, the exoskeleton is generally made of chitin. This chemical has the useful property of being both tough and flexible. In addition to chitin, there is usually a mineral substance in the exoskeleton of an arthropod that makes it hard. The hard, tough, exoskeleton can be thought of as a suit of armor that an arthropod wears. It is flexible enough to move with the creature, but it is tough enough to provide a good measure of protection.

Unlike a suit of armor, however, exoskeletons serve another purpose. The invertebrates that we have studied so far have not needed any support. They either float in the water (like medusae), attach themselves to an object (like sponges), or build themselves a container (like clams). The arthropods, however, must be able to move about or fly. As a result, their fleshy bodies must have support. Humans (and all vertebrates) get their support from their skeleton, a network of bones that runs throughout the body. Arthropods get their support from their exoskeleton. Thus, you could say that while humans (and all vertebrates) have their skeletons on the inside, arthropods have their skeleton on the outside. That's where the term "exoskeleton" comes from.

Although the exoskeleton is necessary for the existence of arthropods, it comes at a cost. You see, the exoskeleton is heavy. It is so heavy, in fact, that it limits the growth potential of an arthropod. As an arthropod increases in size, the amount of exoskeleton must increase as well. This causes the arthropod to get quite heavy. For each arthropod, there comes a point at which the creature's muscles just aren't strong enough to carry around the weight of the exoskeleton. Thus, each arthropod is limited as to how big it can get. Class Crustacea (kruh stay' shuh) contains the largest arthropods, some of which can grow to 12 feet across. The other classes of arthropods contain species that rarely get much larger than 11 inches across.

Not only does the exoskeleton limit the growth *potential* of an arthropod, it also makes it hard for the arthropod to grow during its lifecycle. After all, the exoskeleton is secreted by the arthropod's epidermis and forms around the body. It cannot grow, however. Thus, as the body gets bigger, the exoskeleton gets more and more constricting. As a result, an arthropod must **molt** several times throughout the course of its lifetime.

Molt - To shed an old exoskeleton so that it can be replaced with a new one

Most arthropods molt by secreting enzymes that eat away at the exoskeleton, weakening it. Then, they take in water, swelling the exoskeleton until it breaks away. This is done while a new exoskeleton is being produced under the old one. Once the old one is gone, the arthropod's body can grow again, until it gets too constricted, at which time it will molt again.

Body Segmentation

Like the organisms in phylum Annelida, arthropods are segmented. This segmentation is quite different than the annelids' segmentation, however. In arthropods, the body is divided into three major divisions: the **head**, the **thorax** (thor' aks), and the **abdomen**. These divisions can sometimes be further segmented. In addition, some arthropods have the thorax and head united in a single segment called the **cephalothorax** (sef uh loh thor' aks).

Thorax - The body region between the head and the abdomen

Abdomen - The body region posterior to the thorax

Cephalothorax - A body region comprised of a head and a thorax together

This segmentation is necessary in order to allow the exoskeleton to shift with the movements of the body. The segments can move back and forth, forming "joints" in the "armor."

Jointed Appendages

The term "arthropoda" actually means "joint-footed." Needless to say, then, one of the common features to all arthropods is that their appendages are jointed. This, of course, is not unusual. Almost all vertebrates have jointed appendages as well. In a vertebrate (such as humans, for example), the muscles form over the joint and control the joint from outside. In arthropods, because the joints are in the exoskeleton, the muscles actually control the joints from the inside.

A Ventral Nervous System

In order to react to stimuli, seek out prey, and seek protection from predators, arthropods have a nervous system. Two ganglia form a brain, much like that of an earthworm only more complex. Again, like an earthworm, a ventral nerve chord runs from the ganglia to the posterior. The fact that it is placed at the bottom of the body (that's why we call it "ventral") is no accident. This placement provides maximum protection. It is not only protected by the exoskeleton, but also by the bulk of the body. Instinctively, arthropods do everything they can to avoid exposing their undersides, because instinct tells them that this negates the body's ability to protect the nerve chord.

The nervous system is fed with information through various sensory organs. **Antennae** in the head region provide touch, taste, and smell sensations to the nervous system. In addition, all arthropods have some sort of eyes. There are two different types of eye in phylum Arthropoda: **compound eyes** and **simple eyes**.

Compound eye - An eye made of many lenses, each with a very limited scope

Simple eye - An eye with only one lens

Now don't be fooled by their names. First of all, no eye is simple. It takes an enormous amount of engineering to come up with a system that can detect light, turn that light energy into electrical signals, send the electrical signals to the brain (or ganglia), and have the brain convert those signals into an image! Only God can create such a marvel. Also, a "simple" eye is not necessarily less desirable than a compound eye. For example, the human eye has only one lens; thus it is a simple eye. Nevertheless, it is the most marvelously engineered organ on the planet and provides better sight than the eyes of any other species.

Spiders have simple eyes. Because there is only one lens, and because that lens is small, the eye does not cover much area. If you want to get an idea of what a spider sees, take a look through a thin straw. That's the kind of area that a spider's eye covers. Flies, on the other hand, have compound eyes. As a result, their sight covers a greater area, but, since the lenses are individuals, the image is rather strange. A fly will see many versions of the same image, each slightly tilted with respect to the other, because the lenses are slightly tilted relative to one another. Thus, the fly gets a "mosaic" view of the world, whereas a spider gets a "tunnel" view.

Open Circulatory System

Arthropods have quite an unusual circulatory system. In order to bring vital substances to every cell in the body, a heart in the dorsal (upper) region of the body pumps blood into short vessels that *empty out into different cavities of the body*! This allows blood to flow right over all of the cells in that cavity. In a sense, then, arthropods are always internally bleeding.

Open circulatory system - A circulatory system that allows the blood to flow out of the blood
vessels and into various body cavities so that the cells are in direct
contact with the blood

Of course, once released to flow throughout the body, the blood has to be collected again and
then recycled back into the heart.

ON YOUR OWN

12.1 Suppose an arthropod lived in an ecosystem which contained no predators. The arthropod
would never, ever be threatened. Would it need an exoskeleton in this situation?

12.2 If you count the cephalothorax as one body segment, how many segments does an
arthropod with a cephalothorax have?

12.3 An organism moves its joints with muscles that lay on top of the joint. Is the organism an
arthropod?

12.4 If an arthropod cannot taste something, but the rest of the nervous system is operational,
what sensory organ is malfunctioning?

12.5 Why do we say that arthropods are constantly bleeding internally?

Class Crustacea

The first arthropods that we will study come from class Crustacea (kruh stay' shuh).
Crustaceans mostly live in either freshwater or marine environments (like the crayfish or the
lobster), although some species (like the pill bug, which is often called the "roly-poly") are land
dwellers. Since the crayfish is the "standard" crustacean to study, we will start there.

The Crayfish

The crayfish, often called the "crawdad," lives in fresh water. If you go to virtually any
river in the United States and look at the shoreline, you will probably see these crustaceans busily
moving up and down the shallow waters of the shore, looking for food. Fishermen will tell you
that crawdad meat is one of the best baits that you can use. In the south, they are even cooked for
food. Of course, you have to exercise a bit of caution when you try to catch them, because their
claws can give you a pretty good pinch!

A sketch of the crayfish, illustrating most of its major exterior features, is given in Figure
12.1.

FIGURE 12.1
The Crayfish

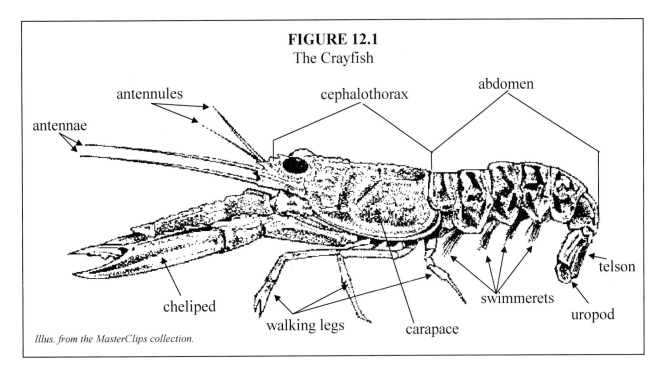

Illus. from the MasterClips collection.

The first thing that you should notice is that the crayfish is one of those arthropods with a cephalothorax. The cephalothorax is covered in a single plate called the **carapace** (kehr' uh pace). The abdomen is separated into six segments, each with its own protective plate of exoskeleton.

Crayfish have six sets of appendages that all perform different functions:

- **Walking legs** - These appendages are used for locomotion when the crayfish is on land or moving on the bottom of the lake or river in which it lives.

- **Swimmerets** - These aid in swimming as well as reproduction. In male crayfish, the first and second swimmerets transfer sperm to the female during mating. In females, the swimmerets carry both the eggs and the offspring.

- **Uropods and telson** - These appendages form the flipper-shaped tail that the crayfish uses for swimming.

- **Chelipeds** - The chelipeds (usually called "claws") are used for defense as well as to grab onto prey.

- **Antennules** - These small antennae aid the creature in balance, as well as provide taste and touch sensations.

- **Antennae** - These longer appendages are much more sensitive than the antennules, providing the crayfish with strong senses of taste and touch.

These appendages are all controlled by the nervous system, giving you some idea of how complex it is.

The Crayfish's Respiratory System

The crayfish gets its oxygen from the water through two sets of **gills** that are located in the cephalothorax. Gills are amazing organs contained in many organisms that live underwater. As you know, all aerobic organisms need oxygen to survive. For land-dwelling creatures, the oxygen is readily available from the surrounding air. But how do organisms that live underwater get their oxygen?

Well, oxygen gas is dissolved in the water. The problem is how to get the oxygen *out* of the water and *into* the creature. In the case of hydra, jellyfish, and other invertebrates that we studied in the previous module, the organism is designed to allow oxygen to diffuse into the body, while waste gases diffuse out. Although this system works for these creatures, it will not work for the rest of the underwater dwellers. Crayfish, for example, are protected and supported by an exoskeleton. No oxygen is going to diffuse through that! Even underwater dwellers that do not have exoskeletons (the vertebrate fish, for example) still have scales or some other coating that would prevent such an exchange of gas with the environment. How then, do these creatures get the oxygen that they need? They use gills.

The two sets of gills in a crayfish reside in two gill chambers that are slightly posterior to the head on each side of the crayfish. One of these sets of gills is illustrated in Figure 12.2.

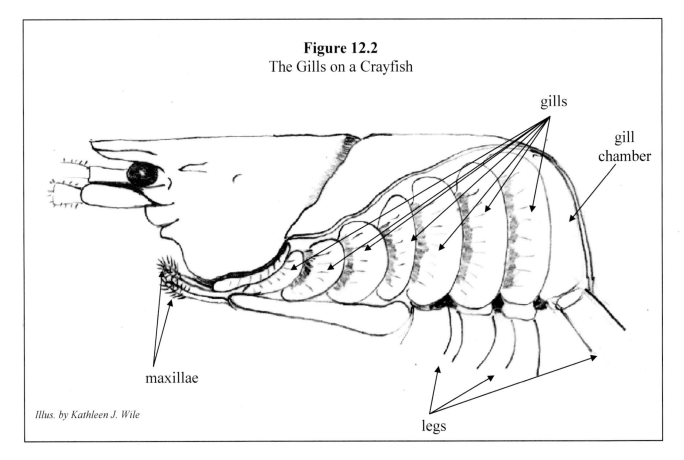

Figure 12.2
The Gills on a Crayfish

gills

gill chamber

maxillae

Illus. by Kathleen J. Wile

legs

The crayfish has small openings on the ventral side of the **gill chambers** that allow water from the surroundings to flow inside to the gills. Blood traveling through the gills can then release carbon dioxide into the water and absorb oxygen. Do you see what's happening in the gills? Since the exoskeleton cannot allow gases to be exchanged with the surroundings, God has designed gills that can do what the entire body of a hydra or jellyfish does. Of course, to make this system work, there have to be mechanisms by which water continually circulates through the gill chambers. This is accomplished by the motion of tiny appendages called **maxillae** (mak sil' ay) that continually push water through the gill chambers.

Interestingly enough, many crustaceans (including the crayfish) can actually store up a lot of water in their gill chambers so that they can make brief excursions out of the water. The gills continue to exchange oxygen and carbon dioxide with the stored water, until all of the oxygen is used up. After that, the crayfish must return to the water or it will suffocate. This situation is very similar to a land-dwelling animal that holds its breath while underwater. The lungs can continue to absorb oxygen from the stored air, but eventually, the animal must return to the surface of the water and breathe again. Crustaceans that have this ability typically use it to seek out new food sources near the shore of the lake, river, stream, or ocean in which they live. Some use it to move from one body of water to another.

Before we leave this section, it is important to dispel a myth that seems to be very popular in school today. As you might already know, water is composed of two hydrogen atoms connected to an oxygen atom. Thus, in a way, there is oxygen in every water molecule. For some reason, this makes students (and teachers, unfortunately) think that gills somehow decompose water, separating the hydrogen and oxygen atoms. Then, they use the oxygen atoms to perform respiration. *This is completely untrue.* The only way that gills can get oxygen is to extract the *oxygen that is dissolved* in the water. If a crayfish (or any organism that needs oxygen) were put in oxygen-poor water, it would suffocate immediately. The only way these creatures can get oxygen is if that oxygen has been dissolved in water. The oxygen atoms that are in a water molecule provide nothing for respiration!

The Crayfish's Circulatory System

Circulation is closely tied to respiration. After all, if the gills obtained oxygen but could not send it anywhere, the oxygen would not do much good, would it? As we mentioned in the previous section, arthropods have open circulatory systems. These circulatory systems are very interesting and deserve a little bit of study. Figure 12.3 illustrates many different systems in the crayfish. For right now, we will focus on the circulatory system.

FIGURE 12.3
The Major Systems of A Crayfish

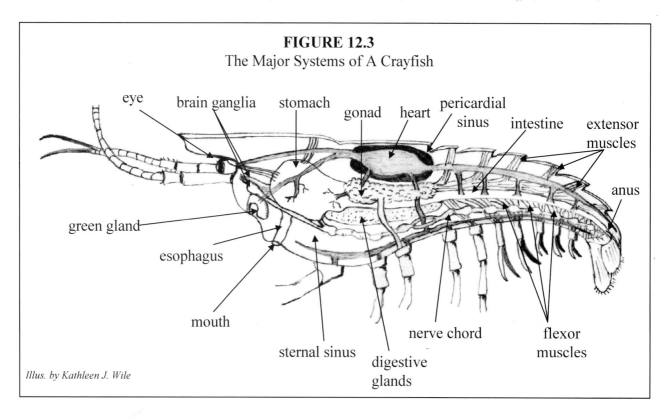

Illus. by Kathleen J. Wile

A crayfish has a heart in the dorsal (upper) part of its body. It rests in a cavity called the **pericardial** (pehr uh kar' dee uhl) **sinus** (sigh' nus). Blood collects in this cavity, and it enters the heart through one of three openings in the heart's surface. Each opening has a valve that closes when the heart is ready to pump. Once it absorbs the blood and closes these valves, the heart pumps blood through a series of blood vessels that are open at the other end. These vessels dump the blood directly into various body cavities. This allows blood to bathe every cell in the area, giving up the oxygen it is carrying and absorbing the carbon dioxide that the cell needs to give up.

Gravity causes the blood to fall into the **sternal** (stir' nuhl) **sinus**, where it is collected by blood vessels that are open at one end. Unlike the blood vessels that dump the blood into the body cavities, these vessels carry the blood back towards the pericardial sinus. On its way there, the blood is passed through the gills where it can release the carbon dioxide it has collected and pick up a fresh supply of oxygen. The blood also passes through a **green gland**, which cleans it of impurities and dumps those impurities back into the surroundings. Once the blood has passed through the gills and the green gland, it then makes its way back to the pericardial sinus to begin the trip all over again.

Why do arthropods have this unusual form of circulatory system? Well, it turns out to be a very efficient means by which gases and other vital substances can be sent to the cells. After all, the blood is in direct contact with the cells that it needs to supply. This obviously makes exchanging gases and other substances quite easy. If the open circulatory system is so efficient, why don't other organisms have it? We'll give you a hint. Mollusks have an open circulatory system as well. What do mollusks and arthropods have in common? They are both enclosed by a hard substance. Mollusks are covered in shells whereas arthropods are covered in exoskeletons. Why does that matter? Since there is a nice, hard covering that encloses the body, blood can flow right through the body without escaping. Thus, the design of the arthropods and mollusks allows them to have the most efficient circulatory systems in Creation.

Now before we go any further, we want you to stop a moment and think about what you have been reading. The crayfish has a means by which it can extract dissolved oxygen from the water and transport it throughout its body. The system that does the transporting is very efficient, allowing direct contact between the blood and the cells. In addition, the chemical nature of the blood allows it to pick up oxygen where it is supposed to (in the gills) and release it where it is supposed to (in the cells). At the same time, the blood can pick up carbon dioxide where it is supposed to (in the cells), and release it where it is supposed to (in the gills). Now this whole system could not work without an exoskeleton (or a shell in the case of mollusks) that keeps the open circulatory system from leaking blood. Isn't that amazing?

As amazing as all of that is, we are not done exploring the wonders of the crayfish's circulatory system. Remember, the exoskeleton allows the crayfish to have an open circulatory system. However, this exoskeleton can be damaged. Crayfish (and most other crustaceans) can lead pretty violent lives. When attacked by a predator, a crustacean can usually defend itself fiercely. As we already mentioned, the claws of a crustacean can be used to battle off enemies. Well, in the midst of such battles, it is not uncommon for a crustacean to lose one of its appendages. Its antennae could be broken off; it could lose a leg; it could even lose a claw! You would think that such a disaster would cause all of the crayfish's blood to flow out into the surroundings, killing the creature.

That's what you would think, but it's not what happens! You see, the crayfish and most other crustaceans have a double membrane in each appendage. When the crayfish loses an appendage, the membrane seals the resulting hole, keeping the blood in the body! If this isn't incredible enough, once the membrane seals the hole, the crayfish can *actually regenerate the missing appendage*. This isn't a means of asexual reproduction, as is the case with the planarian. It is a repair mechanism built in to most crustaceans! Crustaceans seem to know all about this marvelous mechanism, because if a crayfish is caught by one of its appendages, it will willingly break it off to escape!

Clearly there is a lot of detailed engineering going on here. Not only is there an enormous amount of design evident in just the circulatory system of crustaceans, but there are fail-safe mechanisms as well! In the case of disaster, built in systems react to minimize the damage and then rebuild whatever appendage was lost. That engineering is just one more example of the fact that God's Creation is a continuing testament to his majesty!

The Crayfish's Digestive System

Crayfish are scavengers, eating virtually anything that can be digested. In order to eat something, the crayfish first uses its mouth to break the food into small chunks. The food then enters a short esophagus and goes into a stomach that has essentially two regions. The first region, which is on the anterior side of the stomach, grinds the food up into fine particles. These fine particles are then sent to the other region of the stomach, which is on the posterior side. This region sorts the particles. If they are small enough, they are sent directly to digestive glands which secrete enzymes, completing the digestion process. If the particles are too large to be digested immediately, they are sent to the intestine. As they travel through the intestine, they are exposed to digestive enzymes, which digest what they can. Anything that remains at the other end of the intestine is considered indigestible and is expelled out the anus.

The Crayfish's Nervous System

A crayfish's brain is comprised of two **ganglia** that each have a **nerve chord**. These nerve chords join together posterior to the stomach and run along the ventral side of the crayfish. At regular intervals, there are ganglia that continually process the signals which run down the nerve chord. This system is fed information from various appendages throughout the body. The crayfish has compound eyes that send sight information to the brain. In addition to sight, the antennules and antennae send taste and touch information to the nervous system. Also, tiny bristles are found all over the crayfish's body, providing a sense of touch. These bristles are necessary because the exoskeleton is so hard that it is not sensitive to touch. The bristles make up for that, providing touch sensations to parts of the body that otherwise would not have them.

One rather interesting feature of the crayfish's nervous system can be found at the base of the antennules. As we mentioned before, antennules help the crayfish keep its balance. Here's how. At the base of each antennule is a **statocyst** (stat' uh sist).

<u>Statocyst</u> - The organ of balance in a crustacean

These statocysts are little containers that are lined with tiny hairs, each providing a sense of touch. Inside these hair-lined containers are a few grains of sand. These grains of sand shift when the crayfish is knocked off balance. The hairs detect that shift and send a message to the brain. The brain then sends signals to the flexor muscles and extensor muscles, which work the abdomen, swimmerets, telson, and uropods until the crayfish has righted itself again. This is how the crayfish knows which way is up.

The Crayfish's Reproductive System

Reproduction in the crayfish begins in the **gonad**, which produces gametes.

<u>Gonad</u> - A general term for the organ that produces gametes

In males, the gonad is called the testis (plural is testes), and in females the gonad is called the ovary. In male crayfish, sperm are formed in the gonad and transferred to the first and second pairs of swimmerets. Usually in the fall, crayfish will mate. Male crayfish will deposit their sperm into special containers that the female has. The female will then store the sperm until spring.

In the spring, the female produces eggs. When the eggs travel through the oviduct, they are fertilized by the sperm and go to the swimmerets. The fertilized eggs attach themselves to the swimmerets and develop for approximately six weeks, at which point the eggs hatch. The newborns look like miniature versions of the parents, and they tend to cling to the mother for many weeks after they hatch. During the first year of life, the average crayfish molts seven times. This is because the body experiences rapid growth, and the exoskeleton gets too restrictive. After a year is passed, crayfish tend to molt only twice per year. The average crayfish lives 4 to 8 years.

Other crustaceans

We have concentrated on the crayfish in this module because it is a good representative of the organisms from class Crustacea. Of course, there are many, many different kinds of crustaceans, a few of which are presented in Figure 12.4.

FIGURE 12.4
Some Other Organisms From Class Crustacea

Mantena Lobster

Photo by Kathleen J. Wile

Shrimp

Photo from the MasterClips collection

Spiny Crab

Photo from the Expert 3000 collection

Barnacles

Photo from the MasterClips collection

Lobsters, shrimp, and crabs live in virtually every marine environment and are popular dishes for many seafood lovers. Another marine crustacean that you see in nearly every marine environment is the barnacle (bar' nuh cuhl). Many students mistake barnacles for mollusks, because they appear to have a shell. Well, that's not really a shell. It's an exoskeleton formed out of calcium rather than chitin. Constantly moving appendages bring water into the exoskeleton, allowing the barnacle to breathe and take in food. Barnacles will stick to virtually anything. Although they usually attach themselves to rocks, barnacles have been found on boats, lobsters, large clams, and even whales' teeth!

An Important Note

As we did with the earthworm, we threw a lot of material at you in regards to the crayfish. Since this creature is such a good example of an arthropod, we want you to remember all of it. In other words, like the earthworm, you need to be able to point out all of the major organs and systems in a crayfish and know how they work. Thus, you will need to commit Figures 12.1 and 12.3 to memory. In addition, you will need to have a solid working knowledge about all of the systems discussed in this section. This may seem like a lot, but it is really necessary. The good news is that although we will study two more classes and several more examples of arthropods, you will not need to have such a detailed knowledge of them. Instead, you will just need to know the basics.

ON YOUR OWN

12.6 Although the gills of a certain crayfish seem to be working fine, the crayfish suffocates because it cannot get fresh water into the gill chambers. What organ or appendage is not working properly?

12.7 If open circulatory systems are more efficient than closed circulatory systems, why don't more creatures have them?

12.8 A crayfish loses its claw in a fight. What happens?

12.9 If a crayfish cannot stay upright in the water, what organ is most likely not working?

If you purchased the dissection kit, perform the following dissection lab. If not, go on to the next section.

Optional Dissection Experiment #2
The Crayfish

Be careful! Dissection tools are SHARP!

Supplies :

- Dissecting tools and tray that came with your dissection kit
- Crayfish specimen

- Magnifying glass
- Laboratory Notebook

Object: To become more familiar with the crayfish's anatomy through dissection

Procedure:

A. Examine your crayfish specimen. Identify all of the exterior features labeled in Figure 12.1.
B. Turn the crayfish ventral side up on the dissection tray and examine the mouthparts.

> **Mandibles** - One pair of mandibles can be found just posterior to the antennae. These are considered the jaws of the crayfish. Notice that, unlike human jaws, these mandibles open and close horizontally.
> **Maxillae** - These can be found just posterior to the mandibles. You should find 2 pairs.
> **Maxillipeds** - These can be found posterior to the maxillae. They are often called "jaw feet," because they appear to be tiny legs or feet. There are 3 pairs of them, and they are used by the crayfish to hold food in place.

C. Once you have located each of these mouthparts, use your forceps to remove them all from one side of the crayfish. You can remove them by simply grasping them with your forceps and plucking them out. Lay them on the dissecting tray side by side to examine their relative size. Note their relative size in your notebook.
D. Determine the sex of your crayfish. You can do this by closely examining the swimmerets. In males, the swimmerets closest to the anterior portion of the crayfish are modified for sperm transfer. In females, the anterior swimmerets are small in size relative to the others. Based on these criteria, record the sex of your crayfish in your notebook.
E. Now you are ready to look inside the crayfish. Place your specimen dorsal side up and use your scissors to cut the crayfish's carapace from the posterior end of the carapace to just behind the eyes. Then make a transverse cut just behind the eyes. These two cuts are illustrated by the white line in the drawing below:

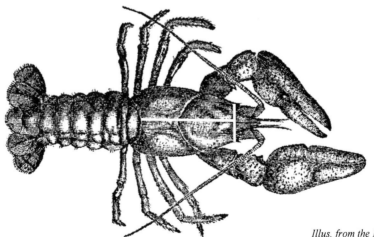

Illus. from the MasterClips collection

F. Carefully remove the carapace in the two pieces determined by your cuts. This should expose the **gills** (see Figure 12.2). Examine the structure of the gills. Count the gills and record the number in your notebook.

G. To make things easier, pull off the walking legs. Carefully remove the internal tissue on the dorsal side of the crayfish. Using Figure 12.3 as a guide, locate the following structures:

> **Heart** - The heart should be the structure closest to the top. Notice the main blood vessels attached to it.
> **Gonad** - Depending on the sex, you will see ovaries (female) or testes (male). The ovaries are darkly-colored and should look like a mass of eggs. The testes are small and white with coiled ducts attached.
> **Digestive Glands** - These are two lightly-colored masses on both sides of the body cavity.

H. Now you need to cut the abdomen open to reveal the intestine. Cut the dorsal side of the abdomen exoskeleton from the anterior end to the telson, and then open the abdomen. Look for a tube that runs its length. That's the **intestine**. You actually should see two tubes. The smaller, darker one is the **dorsal blood vessel**. Notice also the tissue beneath. These are the **abdominal muscles**, which is the part of the shrimp, lobster, or crayfish that we eat.

I. Trace the intestine forward to find the **stomach**.

J. Make a drawing of your crayfish as it looks now. Label all of the structures.

K. Next, you need to remove some of the internal organs. Do this by cutting the muscles just behind the eyes that lead to the stomach. Pull the stomach so that you can reach under it with your knife or a probe and cut the **esophagus**. Once you have done that, you should be able to pull the stomach and intestine out. This will bring a lot of other organs with it!

L. Now that you have cleared out the digestive organs, you should be able to see the **green glands** just posterior to and below the antennules. You may have to remove some tissue in the head in order to see them. They are green, but only slightly.

M. Looking between the eyes, you should see a mass of white tissue. This is the crayfish's **brain**. Once again, you might need to clean out some tissue to see it.

N. Try to trace the nerve cord from the brain to the abdomen. There is some hard tissue near the bottom that you will need to cut through in order to do this.

O. Make a drawing of the crayfish as it appears now. Label the structures you were able to find, and note in your laboratory notebook any structures you were unable to find.

P. Clean up everything. Wash and dry your tray and tools. Put everything away.

Class Arachnida

Are you ready for chills to run up and down your spine? Well, you had better be, because in this section, we are going to discuss class Arachnida (uh rak' nih duh), the class that contains spiders. For some reason, most people have a revulsion against spiders. This is probably because spiders typically live in the dark and attack their prey quickly with no warning. This seems to strike to the very core of some people's greatest fears.

Although spiders are greatly feared, there are only a few that are actually harmful. Many spider bites can hurt, but only a few spiders, like the black widow or the brown recluse, are poisonous to human beings. In fact, many spiders are actually quite beneficial to humans and to the ecosystem in general. Spiders help keep the population of insects in check. If it were not for these fearsome little arthropods, insects would overrun most ecosystems!

Organisms in class Arachnida have five common characteristics:

- They all possess four pairs of walking legs.

- They all have a cephalothorax instead of a separate head and thorax.

- They usually have four pairs of simple eyes.

- They have no antennae.

- Their respiration is done through organs known as "book lungs."

You probably know the first characteristic. It is usually the first thing taught to students as a means by which to separate arachnids from insects.

Since spiders are the most representative arachnids, we will take an in-depth look at them. As we mentioned at the close of the previous section, we will present a lot of information in this section for which you will not be held responsible on the test. Instead, we just want you to get a good appreciation for the organisms in this class. So read through this section, and then use the "on your own" questions and the study guide to help you determine what information is necessary for the test. For example, although we will present a figure that details the major systems and organs in an arachnid, you need not memorize them. We just want you to get an idea of what the creatures are like on the inside. As a result, the figure does not appear in the "on your own" problems or the study guide. This, then, tells you that it will not be on the test.

The Spider

Figure 12.5 shows two different spiders. They probably illustrate the most striking difference between the various species of spiders: some spiders spin webs, others do not.

FIGURE 12.5
The Tarantula and the Golden Garden Spider

Tarantula

Golden Garden Spider

Spiders are cunning predators that use many different means to catch their prey. Although the most common means of catching prey involves weaving a web (we will discuss this in a moment), many spiders do not weave webs at all. Tarantulas, for example, stalk their prey, sneaking up and pouncing on the creature unawares. Although the tarantula strikes fear into the heart of most who see it, the common species of this spider is not poisonous to humans. Of those that are, the poison usually results only in a stinging rash accompanied by swelling. Nevertheless, partly because of their size, and partly because of their reputation, tarantulas are one of the most feared groups of organisms on the planet. Some people use this to their advantage. There is a jewelry store in New York City (we are not allowed to mention its name) that had a terrible problem with theft. They fixed this by allowing tarantulas to walk freely in the display cases. The employees were taught that these particular tarantulas were harmless. Even if they were to bite, it would be of no more consequence than a wasp sting. Burglars, of course, do not know this, and since the day that the tarantulas began their vigil, the store has not suffered a *single* theft!

Of course, spiders are best known for weaving webs, because this is the most common way that they capture prey. They weave these webs out of spider silk, one of the most incredible substances in all of Creation. Spider silk is a very flexible substance, but at the same time it is very strong. Most man-made materials are either strong or flexible. It is very difficult to engineer a substance that is both. You've probably played with an abandoned web at one time, plucking the silk and watching it vibrate. This is a great illustration of spider silk's flexibility. What you probably don't know, however, is the strength of spider silk. It is rather easy to break the strands of a web, because the silk strands are so thin. However, if you were to weave a rope out of spider silk, *it would be stronger than the strongest steel pipe*, yet it would be almost as flexible as a rope. Scientists cannot make a material that has anything close to the properties of spider silk. If they could, however, think about the structures that it could make!

To illustrate just how incredible spider silk is, we want you to read about one application that scientists use it for today. In physics research, scientists are trying to develop a process called "laser-induced nuclear fusion." If it could ever be developed, it would result in a limitless, safe, and completely clean source of energy. This process involves taking a laser and inducing a nuclear reaction that turns hydrogen into helium. This is a violent reaction but, when done on a tiny sample, it is controllable, provided the sample is held in a very strong, very flexible container. Do you know what scientists use to make such a container? They use *spider silk*. Every laser-induced nuclear fusion facility has its own spider breeding ground. They continually harvest the spider silk from the spiders, making new fusion reaction containers. This is an expensive, time-consuming process. If modern science could produce a material that would do the job, they certainly wouldn't bother using spider silk. To this day, however, human science cannot produce what the spider has had since it was first created!

There are a host of different ways that spiders use their silk to catch their prey. Most weave webs. Some species of spider build a **sheet web**, which is a single, flat sheet of sticky silk. The spider hangs on the underside of the sheet and, when an insect gets caught, it paralyzes the creature and pulls it through the web. Some spiders spin tangles of webs that have no real discernible pattern. These webs are typically called **tangle webs**. The most geometrically stunning web is the **orb web**, like the one pictured in Figure 12.5. This web consists of concentric circles of sticky silk that are supported by "spokes" of non-sticky silk. These webs can span great distances, making them very efficient traps. The spider sits at the center of the web with its legs on the spokes. When an insect gets caught in the web, it begins to struggle. The spider can tell which part of the web that the insect is in because the spoke nearest the insect will shake more than the others. When the spider senses which spoke is shaking the most, it follows the spoke until it finds the insect.

Of course, there are spiders that produce silk but do not weave webs. The **trap door spider** digs a shallow hole in the ground and then weaves a "trap door" out of silk. The trap door is attached to the ground at one end but is free to move everywhere else. The spider holds on to tiny "handles" that it weaves into the trap door and keeps it slightly ajar. When an insect or crustacean walks by, the spider jumps out of the trap door, startling and killing its prey. It then drags it back into the hole and closes the trap door, feasting on its catch. Other spiders spin single strands of silk with very sticky ends. They launch the silk at their prey, catching them and "reeling" them in.

Now think about this for a moment. Consider the fact that spider silk is a marvelous material which 3,000 years of human science cannot even come close to manufacturing. Then realize that spiders use this to create such engineering marvels as an orb web, a trap door, or a projectile that can be reeled back in! Is there anything more marvelous in all of Creation? Well, believe it or not, there is! As you study more and more science, you will see that these kinds of engineering marvels exist all over Creation, and there are some even more awesome than the wonder of spider silk and how it is used. Through examples such as this one, we know that life is no accident. Such incredible systems cannot develop by chance. Clearly, Creation tells us what an awesome God we have!

The Major Points of Interest in Spider Anatomy

Figure 12.6 is a sketch of the major systems in a spider. We will not discuss all of them, because many organs in the spider are pretty much the same as those in the crayfish.

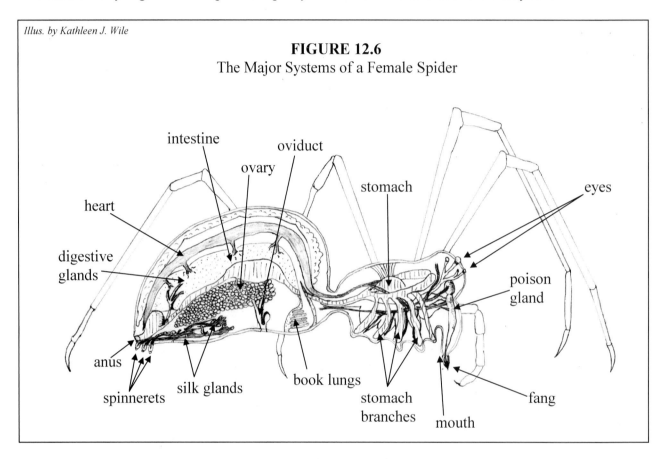

Illus. by Kathleen J. Wile

FIGURE 12.6
The Major Systems of a Female Spider

Of course, the most striking difference between the organs of a spider and those of a crayfish is the presence of **silk glands** and **spinnerets**. If a spider spins silk, it produces that marvelous substance in its silk glands and uses its spinnerets to spin the silk.

Regardless of how a spider catches its prey, once the prey is caught, they behave much the same. Spiders sink their **fangs** into the prey and, use their **poison glands** to inject a paralyzing poison into the prey. This immobilizes the creature, and the spider then secretes digestive enzymes into its body. The tissues that these enzymes partially digest are then sucked through the mouth and into the stomach. The poison of a spider is rarely harmful to humans. It is usually a very weak poison, designed to paralyze tiny prey. There are, however, a few species (the black widow and brown recluse spiders, for example) whose bite can be deadly to humans.

The **book lungs** of a spider do the same job that the gills do in a crayfish. Air enters the exoskeleton through a slit in the abdomen. There, the air encounters an organ that has several thin layers, almost like the pages of a book. As the air mingles with these "pages," oxygen is absorbed by the blood and carbon dioxide is released.

If you are interested in learning more about spiders, Moody Science Adventures produces a tape called *Treasure Hunt*, which includes a section called "Eight-Legged Engineer," an interesting program about the spider.

ON YOUR OWN

12.10 Do tarantulas have spinnerets and silk glands?

12.11 Where is the abdomen in Figure 12.6?

Classes Chilopoda and Diplopoda

It seems natural to discuss classes Chilopoda (kye lahp' uh duh) and Diplopoda (duh plahp' uh duh) right after class Arachnida, because the members of these classes are likely to make your skin crawl as well. Class Chilopoda is home to the arthropods normally called "centipedes," and class Diplopoda contains those arthropods usually called "millipedes." Although often used interchangeably, these two terms refer to two completely different creatures.

The term "centipede," which in Latin means "hundred legs," refers to the members of class Chilopoda. These arthropods have flat bodies that are divided into several segments, each of which contains a pair of legs. Their common name is misleading, however, because centipedes do not have anywhere near 100 legs. The head of a centipede contains antennae for sensory perception and several mouth parts. The body segment directly behind the head contains the first pair of legs, which have poisonous claws. These claws immobilize the insects and small animals that the centipede eats. The common centipedes with which you are familiar are rather small, but certain tropical species can reach lengths of up to one foot long. Although the bite of a centipede is painful to humans, it is rarely dangerous.

The term "millipede," which in Latin means "thousand legs," refers to the members of class Diplopoda. Once again, unlike their common name implies, these arthropods do not have anywhere near 1000 legs. They do have many more legs than centipedes, however, because each of their body segments contains two pairs of legs instead of just one. This is not the only difference between centipedes and millipedes, however. Millipedes have bodies that are rounded, rather than flat. Typically, their antennae are shorter than those of centipedes, and they do not have the poisonous claws that centipedes have. In fact, while centipedes are fierce predators, millipedes are typically docile. They move along the ground slowly, eating vegetation and organic debris. Often, when threatened, millipedes will simply roll into a ball, hoping that their strong exoskeletons will protect them.

Class Insecta

We now come to the largest class in Creation, class Insecta. There are more than 850,000 species of insects, and they represent almost 80% of the species in the animal kingdom. Insects have the following characteristics that separate them from other members of their phylum:

- They have three pairs of walking (or jumping) legs.

- They almost always have wings at some stage of their life.

- They have one pair of antennae.

- Their bodies have all three segments: head, thorax, and abdomen.

Although these characteristics are common to all insects, there is amazing variability in these and other characteristics within this huge class.

Insect Legs

Most insects use their legs for walking. As you would expect for arthropods, insect legs are jointed. They typically connect to the thorax of the insect, and the variety throughout class Insecta can be illustrated by the many different kinds of insect legs. For example, the front pair of legs on a praying mantis isn't just for walking. Each leg has a powerful claw that the mantis uses to capture prey. Flies, on the other hand, use all of their legs for walking, but the legs come equipped with sticky pads that allow them to walk up walls and even upside down on ceilings! The grasshopper uses its two most powerful legs for making great leaps, while the water strider has legs that are bristled at the end, allowing it to actually walk on water. Other insects use their legs to make sounds, while still others use them as a means to store food. This should give you a glimpse of the variety that exists in this class.

Insect Wings

Although most insects have 2 pairs of wings, some have only one pair and a very few have none at all. There are four basic types of wings in class Insecta:

- Membranous wings

- Scaled wings

- Leather-like wings

- Horny wings

Membranous wings are thin, transparent, and have a detailed network of veins that are visible. A fly's wings, for example, are membranous. **Scaled wings** have delicate scales that cover the wings. As a result, they are not transparent. The scales are easy to rub off, as if they were made of powder. Moths and butterflies have scaled wings. **Leather-like wings** are wings that appear to be a part of the exoskeleton. They are typically laid over a second, membranous pair of wings so as to protect them. Grasshoppers have one pair of membranous wings and one pair of leather-like wings. **Horny wings** are often hard to distinguish from leather-like wings. They also are used to cover and protect membranous wings. They are tougher than leather-like wings, however, and they typically cover almost the entire insect, rather than just the membranous wings, as in the case of leather-like wings.

The Basic Anatomy of an Insect

Since there is so much variety in class Insecta, it is hard to pull out a "representative" insect to study. Nevertheless, we will go ahead and use the grasshopper as our representative insect. Figure 12.7 contains a sketch of the outside and inside of a grasshopper.

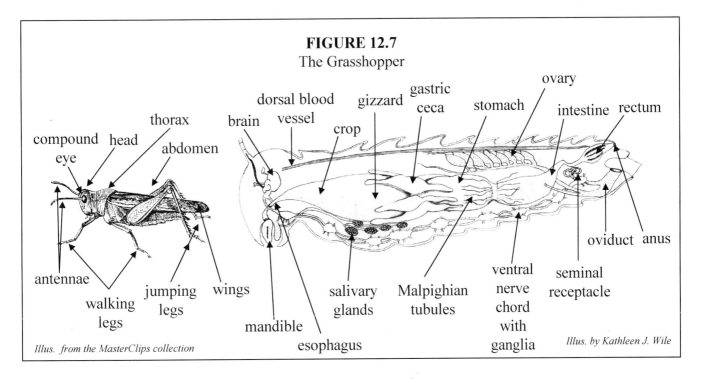

FIGURE 12.7
The Grasshopper

Illus. from the MasterClips collection

Illus. by Kathleen J. Wile

Respiration and Circulation in Insects

One of the most interesting aspects of insects is the fact that even though they have *no respiratory system*, they get plenty of oxygen. How is this possible? Well, insects have an elaborate system of interconnecting tubes called **tracheas** (tray key' uhs). These tubes are connected to the outside through a series of small holes in the exoskeleton called **spiracles** (spear' uh kuhls). The network of tracheae is so complex and thorough that air runs throughout the body, providing oxygen to all tissues! That's why you see no lungs in Figure 12.7. Air goes directly to the tissues, where oxygen and carbon dioxide are directly exchanged with the cells.

If insects have no respiratory system, why do they have a circulatory system, as is evidenced by a dorsal blood vessel? Well, cells need more than just oxygen to survive, and they need to expel more than just carbon dioxide. Thus, as the blood flows over the tissue (remember, this is an open circulatory system), it picks up other cellular waste products and delivers vital substances other than oxygen. Where do the waste products go? Well, insects don't have green glands, but they do have **Malpighian** (mal pig' hee ahn) **tubules**, which reside in the vicinity of the intersection between the stomach and the intestine. When blood flows over the Malpighian tubules, it is cleaned of the those waste products, and they are put into the intestine for elimination through the anus.

Another interesting aspect of insects is that many of them *have both simple and compound eyes* as a part of their nervous system. Typically, the compound eyes actually provide vision, while the simple eyes just look for the presence of light. The signals from both sets of eyes are fed into the brain, which is attached to the ventral nerve chord. This makes up the "backbone" of the nervous system. Besides the eyes, the nervous system also gets information from tactile hairs that provide touch, taste, and smell information. Some insects even have a sense of hearing, but the mechanism for detecting sound is not where you would expect. In a grasshopper, for example, there is a vibrating membrane (called the tympanic membrane) attached to the ventral nerve chord. The vibrations of this membrane provide a sense of hearing. Rather than being on the head, however, it is actually in the first segment of the abdomen!

The Feeding Habits of Insects

The grasshopper has a **mandible** which is designed for chewing. Although this is the case for the majority of insects, there are many different mouth structures in the insect world. This is because there are many different feeding habits apparent in class Insecta. Mosquitoes, for example, do not have mandibles. Instead, they have mouthparts designed to puncture and then suck. Flies, on the other hand, have an almost sponge-like mouth that is used to absorb food. Finally, butterflies have a mouth designed to siphon.

Once food has entered the mouth, however, most of the differences between insects disappear. The food is mixed with the secretions from the salivary glands. Enzymes and water in these secretions begin breaking down the starches that the insect has ingested. The food then passes through the esophagus and goes on to the crop where it might be stored for later use. Once the food is needed, it is sent to the gizzard where it is ground into fine particles. The gizzard empties into the stomach, where it is mixed with digestive enzymes which come from the **gastric ceca** (see' cuh). Undigested food then passes through the intestine, to the rectum, and out the anus.

Reproduction and Development in Insects

Perhaps the most extraordinary thing about insects is the means by which they develop from a fertilized egg into an adult. As is the case for almost all arthropods, the female and male sexes are completely separate in class Insecta. Females receive sperm from males during mating but, like the crayfish, store it in seminal receptacles for a time. When the female lays her eggs, they are fertilized by the stored sperm. We call this stage of the insect's development the **egg stage**. Once the insect hatches, things get really interesting.

When they hatch, most insect young are in a stage called the **larva stage**. In this stage, the young insect, regardless of its species, tends to resemble a segmented worm. They eat and molt over and over again, eventually entering the **pupa stage**. In this stage, the insect forms some sort of case around itself. The case might be formed of exoskeleton, or it might be woven from filaments. During this stage, *everything* changes. The organs are re-arranged and re-shaped, body structures are dismantled and re-formed, and an amazing transformation occurs. When the transformation is complete, the insect breaks out of its case and enters the **adult stage** of life. In the adult stage, the insect has all of the features and organs that are normally associated with its species.

Now, of course, you should have heard about this process before. It is called "metamorphosis" (met uh more' fuh sis) and is usually discussed in terms of the butterfly. When a butterfly's eggs hatch, the young are called caterpillars. They resemble segmented worms at this point in their life. This, then, is the larva stage for the butterfly. Eventually, the caterpillar weaves a cocoon or a chrysalis. This is the casing that forms in the pupa stage. When the insect emerges from the cocoon or chrysalis, it is an adult butterfly. What you might not have been aware of until now is that the *vast majority* of insects go through this transformation process. Flies, for example, go through the same process. When they are in their larva stage, they are called maggots, and when they are in their pupa stage they do not use a cocoon but rather a shell of exoskeleton. Nearly 90% of all insects develop in the same way!

In fact, if an insect does not go through the metamorphosis process described above, it goes through a similar one that involves only three stages. Thus, biologists classify the development of an insect as either **complete metamorphosis** or **incomplete metamorphosis**.

Complete metamorphosis - Insect development consisting of 4 stages: egg, larva, pupa, and adult

Incomplete metamorphosis - Insect development consisting of 3 stages: egg, nymph, and adult

When an insect develops through incomplete metamorphosis, it hatches from its egg stage into its **nymph stage**. In this stage, the insect looks like a miniature version of its adult form, but the proportions seem wrong. It lacks wings and reproductive organs. It molts several times during the nymph stage and, when it finally develops wings and reproductive organs, it is considered an adult.

ON YOUR OWN

12.12 An insect's outer wings are incredibly tough. Most likely, what kind of wings are they?

12.13 You can suffocate an insect by wrapping up its body, except for the head, in plastic wrap. Why, if the mouth is exposed to air, does the insect still suffocate?

12.14 An insect cannot digest food in its stomach due to a lack of digestive enzymes. Which organ is most likely not working?

12.15 An insect goes through a nymph stage in its development. Does it undergo complete or incomplete metamorphosis?

A Few Orders In Class Insecta

Class Insecta contains 17 different orders. Obviously, then, there is simply no way to go through them all. Nevertheless, to give you some feel for the diversity of life in this class, we need to at least touch on some of the major orders.

Order Lepidoptera: The Butterflies and Moths

Since we were just discussing the developmental process in insects, and since that process is usually associated with butterflies, we might as well start our insect order discussion with order Lepidoptera (lep uh dahp' tur uh). As we mentioned before, butterflies and moths (in their adult stage) have scaled wings. This is the major characteristic that sets them apart from most other orders in class Insecta.

Do you know the difference between a butterfly and a moth? Well, there are differences, but they are rather subtle. In its adult stage, the antennae of a moth usually look feathery, while the antennae of a butterfly look like straight stalks with knobs on the end. Also, the body of the adult is much slimmer in butterflies than it is in moths. Finally, when butterflies rest on the ground or sit on top of a plant, they typically hold their wings up vertically. Moths, on the other hand, tend to hold their wings out horizontally when resting.

Although butterflies and moths can be quite beautiful, they are not always the most desirable things to have around. Many butterflies eat ravenously while in their larva stage. Some have been known to cause millions of dollars worth of crop damage every year. Also, if certain types of moth lay their eggs in your closet, the larvae that hatch can destroy the clothes stored there. That's why many people put mothballs in their closets. Mothballs are small samples of a chemical called napthalene. They emit an odor which repels moths, so that when they are placed in a closet, moths will not enter the area and lay eggs.

Order Hymenoptera: Ants, Bees, and Wasps

Scaled wings, of course, are not the only kinds of wings in class Insecta. Members of order Hymenoptera (hi muh nahp' tur uh) have four membranous wings. Now that statement might confuse you in light of the fact that the title of this subsection includes ants as being a part of order Hymenoptera. Ants don't have wings, do they? Well, it turns out that they do, but you rarely see them. You'll see why in a moment. Members of this order usually have stingers, too. This is another characteristic that sets them apart from other organisms in class Insecta.

All species within order Hymenoptera are what biologists call **social insects**. They exist within a society in which they have very particular functions. The best example of this is bees. In a hive, bees belong to one of three groups: queen, drone, or worker. The vast majority of bees in the hive are worker bees. They are actually female bees that do not have reproductive capabilities. Instead, their egg-laying organ, the ovipositor, is a barbed stinger. When they sting an enemy, they release a poison with the stinger. The barb in the stinger, however, lodges in the victim, making it impossible for a bee to remove its stinger. Thus, the bee will rip off its last abdominal segment, leaving the stinger behind. This increases the effect of the poison, but it kills the bee. As their name implies, worker bees do all of the work. They build and maintain the hive, care for the eggs, care for the larvae once the eggs hatch, and protect the queen.

The queen bee is the only female with reproductive capabilities. Because she can lay eggs, however, she does not have a barbed stinger. A given hive has only one queen, so workers protect her at all times. If the queen dies the hive is out of luck, right? Wrong. You see, while the queen is alive, she produces a chemical which biologists call "queen substance." This substance is transferred to all workers, and it attracts workers to her. When the queen dies, the workers notice that they are no longer being given that substance, so they feed some of the developing larvae a special high protein food that they secrete. This causes those larvae to develop reproductive organs. The first one to enter adult stage is the new queen, and the rest are killed.

The drone bees are the only males in the hive. They are useless for all situations except mating. Their mouth is not designed to gather nectar from flowers, they cannot make honey or help maintain the hive. They simply wait their turn to mate with the queen. Since they are good only for reproductive purposes, worker bees have been observed killing drones or pushing them out of the hive when food becomes scarce!

As you no doubt are aware, bees make honey. What is honey and why do they make it? Honey is partially-digested nectar that the bees have sucked out of flowers. It serves as food for the queen and the drones, and it is also a backup supply of food for the workers when they cannot find nectar. The honey is stored in honeycombs, perfectly hexagonal storage chambers made out of wax that is secreted by (you guessed it) the workers. It turns out that no engineer could have designed a better system for the storage of honey. According to everyone who has studied the construction of a honeycomb, it provides the ideal balance between strength, durability, and high storage volume! Bees are so industrious that they make far more honey than they need under normal conditions, so humans (and some animals) take advantage of this by using the honey for themselves.

Now stop and think about this for a moment. The queen controls her "subjects" with a special chemical. When that chemical is no longer supplied, the workers "know" that they need a new queen, so they produce another substance that, when fed to the developing larvae, forms a new queen. At the same time, these workers store honey in a structure that science tells us is *perfectly* engineered for the task. If honey runs short, the workers also somehow "know" who among them (the drones) are expendable. Isn't that amazing? This is just one more of the incredibly designed systems in Creation that simply shouts out the glory of God!

We still haven't told you why you don't see wings on ants, have we? Well, you needed to understand the concept of social groups in insects first. The ants that you see most of the time are worker ants, which do not have reproductive capabilities. These ants do not have wings. Only ants that can reproduce have wings. They never venture out of the anthill, so you never see them. Even if you were to see them, you might still not see wings. The male ant dies shortly after mating, and the female ants loses her wings once she mates. Thus, only a small portion of any given ant species has wings, and those individuals either don't live long or don't keep their wings long. No wonder you never see them!

Order Coleoptera: The Beetles

No, we're not going to discuss the "Fab Four" here. Instead, we are going to discuss the largest order within class Insecta, order Coleoptera (ko lee ahp' tur uh). The name of this order literally means "sheath wing," and that is a very adequate description of what sets it apart from most other orders. All members of this order have horny wings. These wings are thick, sometimes colorful, and they typically cover the creature's entire body, protecting the membranous wings underneath. The horny wings make beetles look like little war machines, and that look is often enhanced by horns on their head or extended mandibles that can viciously cut through their enemies.

Beetles have a voracious appetite. Sometimes that is a good thing, and sometimes it is a bad thing. Many beetles feed on insects that damage crops. The ladybug, for example, feeds exclusively upon aphids and other crop-destroying insects. The fact that the ladybug has a huge appetite works out all the better for farmers. There are some beetles, however, (the Japanese beetle and the rice weevil, for example) that destroy crops themselves. The fact that they have a huge appetite is not good for farmers!

One particularly interesting beetle that we want to discuss is called the bombardier beetle. This ugly little beetle has one of the most beautiful defense mechanisms in all of Creation: a fully-equipped chemical weapon. This weapon begins with storage vessels that contain a mixture of two chemicals: hydroquinone and hydrogen peroxide. Under normal conditions, these chemicals would react but, while stored in the bombardier beetle's vessels, they are kept from reacting by the presence of a third chemical which inhibits the reaction. When the bombardier beetle feels threatened, however, it fills an empty reaction chamber in its body with the chemicals. Two other chemicals, catalase and peroxidase, are then added. These chemicals cause the hydrogen peroxide and hydroquinone to react violently. The violent reaction produces a great amount of heat and pressure in the reaction vessel. The beetle then points its tail in the direction of trouble and opens a valve that exists between the reaction vessel and the tail. A jet of steam which has a temperature of roughly 200°F shoots out the tail in the direction of danger. Any potential predator is immediately burned and frightened away! The bombardier beetle can perform this feat up to twenty times per day! The bombardier beetle is just another in the long list of organisms on this planet that tell us over and over again that this world was *designed*.

Order Diptera: Flies, Gnats, and Mosquitoes

We now come to the order with the most annoying creatures, order Diptera (dip' tur uh). All organisms in this order have a pair of membranous wings that they use to fly. Their second pair of wings are used as balancers, to keep them level in flight. Members of this order have no other defense mechanism except the ability to get away quickly. Their exoskeleton is rather weak; they have no stingers; and their mouths are designed to either pierce or suck, not to bite. Thus the only way that these creatures stay alive is to fly away. Given the fact that they are so prevalent on warm, sticky days, they obviously are good at it!

Flies are pests, but they can actually be quite dangerous. The average housefly carries about 30 million bacteria and viruses in its intestinal tract and another 500 million on its body. When a fly lands on a food source, it eats by excreting digestive juices onto the food and then sponging it back up. Many of the microorganisms and viruses it carries are transferred to the food in that process. Thus, if a fly lands on your food and begins to eat it, you run the risk of being infected by those microorganisms and viruses, some of which are pathogenic to humans.

Mosquitoes also carry potentially pathogenic microorganisms and viruses. In Module #3 we studied how the *Plasmodium* sporozoan, which causes malaria, is carried by the mosquito. Mosquitoes also carry some types of parasitic worms and the yellow fever virus. These creatures are quite effective at infecting you, because they feed on your blood. Their mouthparts are perfectly designed to taper into four needles that easily pierce the skin. Since your blood tries to clot in order to prevent bleeding, the mosquito then injects a chemical that counteracts the blood's clotting mechanism. This keeps the blood flowing freely, allowing the mosquito to get all that it wants. The problem is, of course, that any pathogens which the mosquito carries get mixed into your bloodstream along with the chemical that keeps the blood from clotting!

Order Orthoptera: Grasshoppers, Crickets, Cockroaches, and Praying Mantis

The last order that we want to discuss is order Orthoptera (or thahp' tur uh). We will not spend much time on it, though, because we have already made an in-depth study of the grasshopper, one of its members. Members of this order have one pair of leather-like wings that cover and protect their membranous wings. The praying mantis is an interesting member of this order. Its front two legs appear to be folded in prayer, but are actually powerful claws designed to capture prey. The female of the species actually uses her claws to tear the head off of the male once they have mated.

To help you review what you have learned in this section, perform the following experiment.

EXPERIMENT 12.1
Insect Classification

Supplies:

- Laboratory notebook
- Specimens in the pictures below
- Specimens collected from another source (In order to get more practice at recognizing the orders discussed in this book, you should try to find more examples of insects. Depending on the time of year and the area of the country that you live in, you may be able to collect some on your own. If you cannot, a library or museum might have an insect collection that you could study. If you know anyone in 4-H, they most likely have access to an insect collection. If an insect doesn't fit into one of the 5 orders that we have discussed here, do some research and see what order it does belong in!)

Object: To become familiar with classifying insects

Procedure: In your notebook, write down the letter of each specimen and the order to which it belongs. You can check your work at the end of the answers to the "on your own" problems.

a.

Photo from the MasterClips collection

b.

Photo from the MasterClips collection

c.

Photo from the MasterClips collection

d.

Photo from the MasterClips collection

e.

Photo from the MasterClips collection

f.

Photo by Rebecca Noelle Durnell

ANSWERS TO THE ON YOUR OWN PROBLEMS

12.1 <u>Yes, it would still need an exoskeleton.</u> Exoskeletons provide support for the body. Without it, the arthropod would be nothing more than a mass of tissue and organs.

12.2 Since the cephalothorax is a combination of the head and the thorax, only the abdomen is left. Thus, there are <u>two segments</u>.

12.3 <u>The organism is not an arthropod.</u> The muscles in an arthropod are inside the exoskeleton. Thus, the joints are controlled from the inside.

12.4 Since antennae provide the taste sensation, <u>the antennae must not be working</u>.

12.5 <u>Since arthropods have an open circulatory system, blood is constantly leaving the arteries and flowing directly through the tissue.</u>

12.6 In order to move fresh, oxygen-rich water into the gill chambers, the maxillae move back and forth. Thus, <u>the maxillae must not be working</u>.

12.7 <u>In order to have an open circulatory system, you need something like an exoskeleton to keep the blood from leaking out of the body.</u> Most organisms don't have such structures!

12.8 <u>When the claw falls off, a membrane closes to prevent bleeding. Then, the claw grows back over time</u>.

12.9 Since the statocyst keeps the crayfish's balance and tells it "which way is up," <u>the statocyst must be malfunctioning</u>.

12.10 <u>Tarantulas have neither spinnerets nor silk glands</u> because they do not spin silk. However, there are always exceptions to the rules in biology! It turns out that *some* tarantulas do spin silk. Some females spin webs, for example, in which they lay their eggs. These tarantulas obviously have both spinnerets and silk glands.

12.11 Since arachnids have a cephalothorax, the head and thorax are one segment. Notice in the figure that there is a "pinched" region of the body near the center. That marks the change from the cephalothorax to the abdomen. Thus, <u>the abdomen is posterior to the pinched region</u>.

12.12 Horny wings are the toughest. Thus, <u>they are probably horny wings</u>.

12.13 Insects get air through tiny pores in their exoskeleton called spiracles, not through their mouth. Thus, it is irrelevant that the mouth is free. <u>The insect suffocated because it could not take in air through the spiracles, which were blocked by the plastic</u>.

12.14 It is the gastric ceca's job to secrete digestive enzymes. Thus, <u>the gastric ceca are not working</u>.

12.15 The nymph stage is only present in <u>incomplete metamorphosis</u>.

ANSWERS TO EXPERIMENT 12.1

a. If the armor-like wings and the war-machine appearance didn't give it away, the horn on the head should have. This is from <u>order Coleoptera</u>.

b. The two membranous wings with two smaller membranous wings (balancers) should be a hint. Also, it looks a lot like a fly. This is from <u>order Diptera</u>.

c. This belongs in <u>order Lepidoptera</u> whether it is a butterfly or a moth.

d. This is a grasshopper, belonging to <u>order Orthoptera</u>.

e. Bees belong in <u>order Hymenoptera</u>.

f. This is a praying mantis, belonging to <u>order Orthoptera</u>.

STUDY GUIDE TO MODULE #12

1. Define the following terms:

a. Exoskeleton
b. Molt
c. Thorax
d. Abdomen
e. Cephalothorax
f. Compound eye
g. Simple eye
h. Open circulatory system
i. Statocyst
j. Gonad

✳ 2. Name the five common characteristics among the arthropods.

3. Identify the structures in the following diagram:

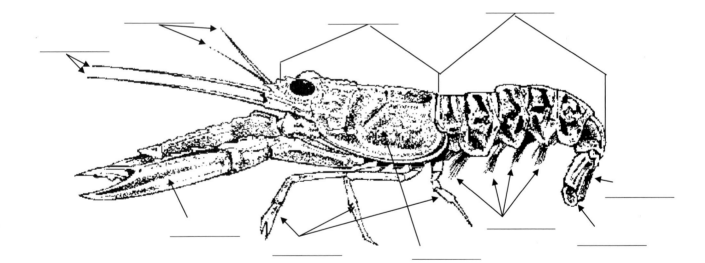

Illus. from the MasterClips collection

4. Identify the organs in the following diagram:

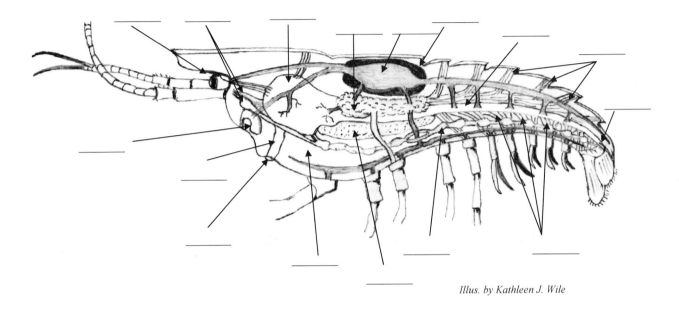

Illus. by Kathleen J. Wile

5. Explain the flow of blood in a crayfish, starting from the pericardial sinus.

✳ 6. What purpose does the green gland serve?

7. What structure (besides the gills and gill chamber) is vitally important for respiration in a crayfish?

✳ 8. What happens when a crayfish loses a limb?

✳ 9. Where do the fertilized eggs of a crayfish go?

✳ 10. Why do arthropods molt?

✳ 11. What two appendages are responsible for taste and touch in a crayfish?

✳ 12. What five characteristics set arachnids apart from the other arthropods?

13. Do all spiders make silk?

14. Do all spiders that make silk spin webs?

✳ 15. Why are the spider's lungs called book lungs?

✳ 16. What four characteristics set insects apart from the other arthropods?

17. Why don't insects have respiratory systems?

✷18. If an insect goes through a pupa stage, does it perform complete metamorphosis or incomplete metamorphosis?

19. What four types of wings exist among insects?

✷20. Match the orders listed below with the wings that they have:

Hymenoptera, Lepidoptera, Coleoptera, Diptera, Orthoptera

a. Two leather-like wings, two membranous wings

b. Four membranous wings

c. Two membranous wings, two membranous balancers

d. Two horny wings, two membranous wings

e. Scaled wings

Module #13: Phylum Chordata

Introduction

After spending two modules concentrating on the invertebrates of Kingdom Animalia, it is now time to start talking about the vertebrates in that same kingdom. Although there are many, many more invertebrates than vertebrates, most of the animals with which you are familiar are vertebrates. Thus, we should spend some time discussing them.

The vertebrates of Kingdom Animalia are all found in one phylum: phylum Chordata (kor dah' tuh). From fish to reptiles to mammals, all animals that have some kind of "backbone" are a part of this phylum. What do we mean by "some kind of backbone?" Well, most members of phylum Chordata have **vertebrae** (ver' tuh bray)

Vertebrae (singular: vertebra) - Segments of bone or some other hard substance that are arranged
                  into a backbone

When you think of a backbone, this is typically what you think about. As a result, the term "vertebrate" is generally used to refer only to creatures that have vertebrae. Not all members of phylum Chordata have such an obvious backbone, however. Some creatures in this phylum have a **notochord** (no' tuh kord) instead.

Notochord - A rod of tough, flexible material that runs the length of a creature's body, providing
                  the majority of its support

In fact, all members of phylum Chordata have a notochord at some point in their development. While some have a notochord throughout their entire life, others have it only in their larval stage. Many chordates have notochords only in the earliest stages of their development. For these creatures, the notochord turns into vertebrae before the creature is born or hatched.

The fate of the notochord in an organism actually provides the first level of classification in phylum Chordata. You see, this phylum contains so much diversity that it must be split into three subphlya. Subphylum Urochordata (yoor' uh kor dah' tuh) contains those creatures that have a notochord through the larva stage but then the notochord actually disappears in the adult stage of their life. Subphylum Cephalochordata (sef' uh loh kor dah' tuh) is comprised of those organisms who have notochords throughout their entire life. Finally, Subphylum Vertebrata (vurt uh bräht' uh) is made up of those vertebrates that develop vertebrae before they are born or hatched. This last subphylum is the group usually referred to as "vertebrates."

Some kind of backbone is not the only common characteristic among organisms within phylum Chordata. All members of this phylum also have a **dorsal nerve chord**. Remember, the arthropods we studied in the previous module had a ventral nerve chord. Thus, chordates have their nerve chord in precisely the opposite location. In arthropods, the nerve chord is placed on the lower portion of the body so that, in addition to the exoskeleton, the entire upper part of the body protects it. As a result, arthropods are always reluctant to expose their undersides, because the nerve chord has the least protection from that side of the body. Chordates, on the other hand, have their nerve chord on the dorsal side of the body. This works fine for most chordates because, in the case of vertebrates, the vertebrae actually encase the nerve chord, providing optimum protection.

Subphylum Urochordata

Subphylum Urochordata is comprised of those chordates that have a notochord only during the larval stage of their development. The best example of such a creature is the strange little **sea squirt**. This interesting marine-dwelling creature gets its name from the fact that in its adult stage, it squirts water through a siphon when it is disturbed. Some biologists refer to sea squirts as "tunicates," because in their adult stage, they cover themselves in a leathery "tunic" that they secrete. The rather interesting lifecycle of a sea squirt is illustrated in Figure 13.1.

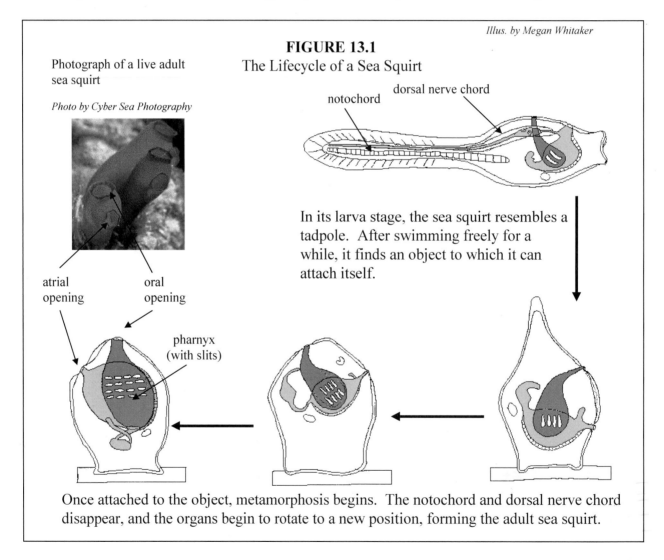

Illus. by Megan Whitaker

FIGURE 13.1
The Lifecycle of a Sea Squirt

Photograph of a live adult sea squirt

Photo by Cyber Sea Photography

atrial opening

oral opening

pharnyx (with slits)

notochord

dorsal nerve chord

In its larva stage, the sea squirt resembles a tadpole. After swimming freely for a while, it finds an object to which it can attach itself.

Once attached to the object, metamorphosis begins. The notochord and dorsal nerve chord disappear, and the organs begin to rotate to a new position, forming the adult sea squirt.

In its larval stage, the sea squirt resembles a tadpole. It has both a dorsal nerve chord and a notochord. It swims around for a brief period, and then it attaches itself to something on the bottom of the sea. Once attached, the sea squirt begins to develop into its adult form. The nerve chord and notochord are recycled to form new tissues, the organs rotate so that the oral and atrial openings point upwards, and a leathery "tunic" is formed around the animal's exterior for protection and support.

Much like a sponge, the adult sea squirt obtains food by siphoning water into its body through the atrial opening and expelling it through the oral opening. Phytoplankton and zooplankton that come in with this water are filtered out by the slits in the pharynx and then digested as food. Once again, like a sponge, the adult sea squirt does not move. It is firmly attached to whatever structure the tadpole chose.

ON YOUR OWN

13.1 If the sea squirt feeds and stays fixed like a sponge, why isn't it classified in phylum Porifera?

13.2 What structure in the adult sea squirt performs the function that the notochord performs in the larva?

Subphylum Cephalochordata

Unlike the members of Subphylum Urochordata, the organisms in Subphylum Cephalochordata retain their notochord throughout their entire life. A representative organism of this subphylum is the **lancelet**. This marine-living creature, also called "amphioxus" (am fee ox' us), is tapered at both ends, looking something like an eel. A sketch of a lancelet is shown in Figure 13.2.

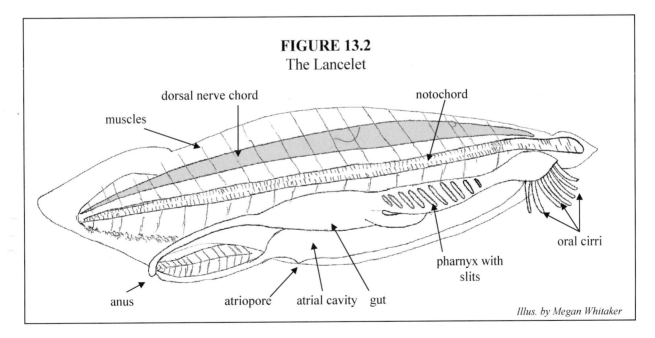

FIGURE 13.2
The Lancelet

Illus. by Megan Whitaker

This interesting little creature possesses a dorsal nerve chord along with the notochord that makes it a part of this subphylum. Its sensory organs are not as sophisticated as those found in most other creatures in this phylum. As a result, it tends to spend the majority of its time buried in the sand. This allows it to hide from predators without hampering its ability to obtain food. Like the sea squirt, the lancelet takes in food by filtering out the phytoplankton and zooplankton in the water.

Water is drawn into the creature by the beating of cilia located just inside the mouth. This water is first filtered through the **oral cirri** (sear' eye), which clean it of large, indigestible debris (such as the sand in which the lancelet is buried). As the water passes into the pharynx, it travels through slits like those of the sea squirt. The food particles in the water are trapped in the slits, and the water flows out into the **atrial** (ay' tree uhl) **cavity**, eventually leaving the body through the **atriopore** (ay' tree oh poor). The food particles that have been trapped in the slits of the pharynx get sent into the gut, where they are digested. Any undigested remains leave through the anus.

Although a relatively unknown organism in the United States, the lancelet is very common in tropical, marine environments. At Discovery Bay, Jamaica, for example, biologists have reported populations of up to 5,000 lancelets per square yard of sand! In many parts of the world, particularly in certain regions of Asia, the lancelet is actually a very important food item. There are some fishermen who make their living solely by harvesting lancelets.

ON YOUR OWN

13.3 A dead lancelet is found to have large particles of sand lodged in the slits of its pharynx. What, most likely, was not functioning properly prior to the lancelet's death?

Subphylum Vertebrata

We now turn our attention to the subphylum that contains the "true" vertebrates. These organisms have a backbone formed by vertebrae from the day they are born or hatched. There are many, many different creatures in this subphylum. There are so many, in fact, that we cannot discuss them all in what is left of this module. As a result, we will revisit this subphylum again in Module #16, where we will discuss reptiles, birds and mammals. In what remains of this module, we will talk about fish and amphibians.

Before we talk about the individual kinds of organisms in Subphylum Vertebrata, however, it is important to discuss the common features that you find among the many classes of organisms which compose this diverse subphylum. Although creatures like fish, amphibians, reptiles, birds, and mammals may seem quite different, they actually have many similarities. It is these similarities that allow us to classify them all in one subphylum.

The Endoskeleton

Unlike the arthropods who support their bodies with an exoskeleton, vertebrates get their bodily support from an **endoskeleton** (en doh skel' uh tuhn).

Endoskeleton - A skeleton on the inside of a creature's body, typically composed of bone or cartilage

Now this definition does us little good if we aren't clear on what **bone** and **cartilage** (kar' tuh lij) are.

When you think about bone, what comes to mind? You probably think about a hard, rock-like substance. Believe it or not, however, bone is *living tissue!* Figure 13.3 is a sketch of a typical vertebrate bone, showing its main structures.

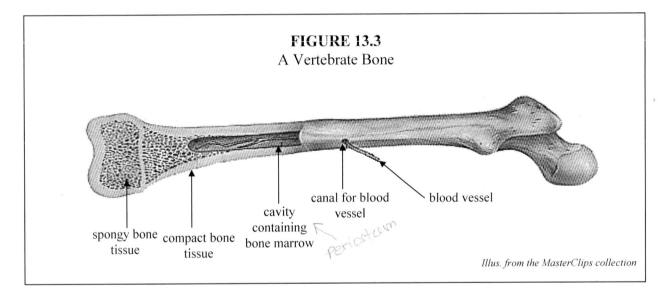

FIGURE 13.3
A Vertebrate Bone

spongy bone tissue   compact bone tissue   cavity containing bone marrow   canal for blood vessel   blood vessel   Periosteum

Illus. from the MasterClips collection

Bones in vertebrates come in all shapes and sizes. They are mostly composed of two different kinds of tissue: **compact bone tissue** and **spongy bone tissue**. Both of these tissues are composed of proteins called **collagen**. The collagen fibers are mixed with calcium-containing salts which harden them. This gives bone its hard, rock-like feel. Within the woven, hardened, collagen fibers, living cells are housed. The main difference between compact bone tissue and spongy bone tissue is how these calcium-hardened fibers are packed together. In compact bone tissue, they are packed together tightly, forming a hard, tough structure that can withstand strong shocks. In spongy bone tissue, on the other hand, the fibers are packed loosely, with a lot of space in between. This gives the tissue a spongy look. Although the tissue looks spongy, it is quite hard. It provides support to the bone without adding much weight.

In order to take nutrients and other necessary substances to the living cells in bone tissue, a series of interconnecting canals that contain blood vessels are woven throughout the bone. In the very center of the bone is a cavity that holds **bone marrow**.

Bone marrow - A soft tissue inside the bone that produces blood cells

So we see that the bones of a vertebrate are, in fact, composed of living tissue. They also do more than just support the body. Without bone marrow, vertebrates would have no blood cells! In addition, bone tissue is a repository for excess minerals which are vital to the chemistry of vertebrates. If a vertebrate runs low on these minerals, they can be extracted from the bone tissue as a short term fix for the shortage.

Now that you know what bone is, what is cartilage? Well, cartilage is a kind of bone precursor. It is composed of collagen fibers just like bone tissue, but it is not reinforced with calcium salts. This makes it more flexible than bone, but it also is much weaker. When most vertebrates develop prior to birth or hatching, the beginnings of their endoskeleton are formed as cartilage. Later on, this cartilage develops into bone tissue. Some vertebrates, however, never convert their endoskeleton from cartilage to bone. It simply stays as cartilage. This gives them more flexibility in their endoskeleton, but it also provides less protection and support. Even vertebrates who convert their cartilage to bone still have some cartilage. It functions as a connective tissue, helping keep the various bones of the skeleton connected.

Now that we know a little bit about bone and cartilage structure, we need to look at the endoskeleton a little more globally. In Figure 13.4, the skeleton of a typical human being is shown.

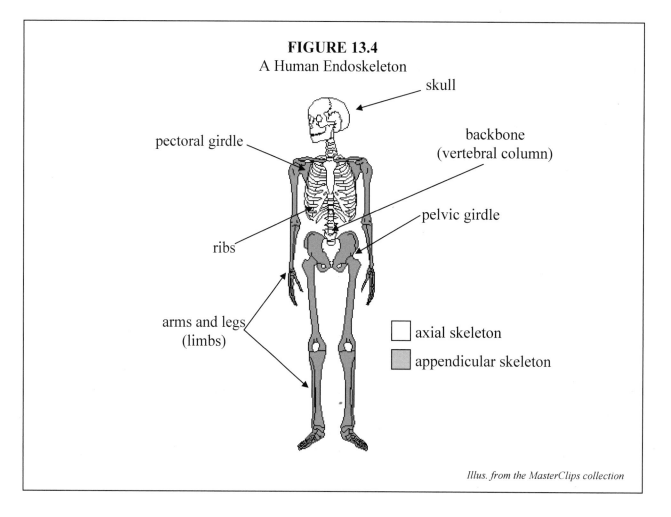

FIGURE 13.4
A Human Endoskeleton

Illus. from the MasterClips collection

Like most vertebrates, humans have endoskeletons which can be split into two major sections: an **axial skeleton** and an **appendicular** (ah pen dihk' you luhr) **skeleton**.

Axial skeleton - The portion of the skeleton that supports and protects the head, neck, and trunk

Appendicular skeleton - The portion of the skeleton that attaches to the axial skeleton and has the limbs attached to it

As you can see from the figure, the backbone (often called the **vertebral column**) , ribs, neck, and skull make up the axial skeleton. On the other hand, the **pectoral girdle** and the arms that attach to it as well as the **pelvic girdle** along with the legs that attach to it make up the appendicular skeleton (shaded gray in the figure). For creatures that have four limbs rather than two arms and two legs, the pectoral girdle is the portion of the skeleton to which the anterior limbs attach while the pelvic girdle is the portion of the skeleton to which the posterior limbs attach.

The Circulatory System

Unlike the open circulatory system that is found in arthropods, vertebrates have a **closed circulatory system**.

Closed circulatory system - A circulatory system in which the oxygen-carrying blood cells never leave the blood vessels

A vertebrate's circulatory system begins with a heart that is composed of two, three, or four chambers. When we study some individual organisms later on in this module, we will learn more about heart chambers. The heart pumps blood into **arteries**, which carry it away from the heart. These arteries branch to all parts of the body, sending the blood to **capillaries** (kap' uh lehr eez). These capillaries are tiny, thin-walled blood vessels that allow oxygen to leave the blood and flow out into the tissue. At the same time, the capillaries allow carbon dioxide to enter the blood, so that it can be carried away. The carbon dioxide-laden blood is then sent back to the heart in **veins**.

Arteries - Blood vessels that carry blood away from the heart

Capillaries - Tiny, thin-walled blood vessels that allow the exchange of gases and nutrients between the blood and cells

Veins - Blood vessels that carry blood back to the heart

The blood of all vertebrates is red. This is because the cells which carry the oxygen in the blood are full of a red protein called **hemoglobin** (hee' muh glo' bun). It is the hemoglobin in these cells that actually holds on to the oxygen and allows it to be transported by the blood. These hemoglobin-containing cells are called **red blood cells**, and they give blood its color. When red blood cells carry oxygen, they have a bright red color. When they lack oxygen, they have a darker red color. Because of this, when we look through our skin at our veins, the blood actually appears to be blue. Even though it appears to be blue, it is not. The blue color is the result of the fact that dark red blood does not reflect light through tissue as well as bright red blood does. As a result, the blood in our veins appears to be a different color than it really is.

The Nervous System

Vertebrates have the most complex nervous systems of all animals in Creation. Figure 13.5 illustrates the nervous system of a "typical" vertebrate.

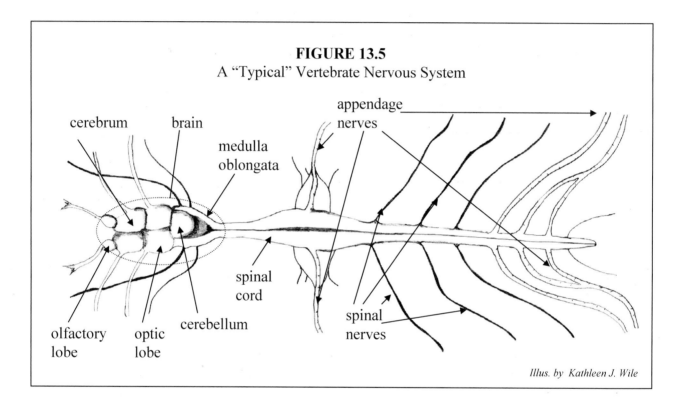

FIGURE 13.5
A "Typical" Vertebrate Nervous System

Illus. by Kathleen J. Wile

Although this is considered the layout of a "typical" vertebrate nervous system, you must realize that there is a great amount of diversity in Subphylum Vertebrata. Thus, many vertebrates have nervous systems that are far more complicated than that shown here. Nevertheless, this is a good starting point for understanding how vertebrates sense and respond to stimuli.

The nervous system is controlled by the **brain**. Most vertebrate brains are segmented into five different types of **lobes**.

<u>Olfactory lobes</u> - The regions of the brain that receive signals from the receptors in the nose

<u>Cerebrum</u> (suh ree' bruhm) - The lobes of the brain that integrate sensory information and coordinate the creature's response to that information

<u>Optic lobes</u> - The regions of the brain that receive signals from the receptors in the eyes

<u>Cerebellum</u> - The lobe that controls involuntary actions and refines muscle movement

<u>Medulla oblongata</u> (muh dul' uh ahb lawn gah' tuh) - The lobes that coordinate vital functions, such as those of the circulatory and respiratory systems, and transport signals from the brain to the **spinal cord**

Notice from the figure that with the exception of one type of lobe, the **cerebellum** (sehr uh bell' uhm), the types of lobes come in pairs, each with a right lobe and a left lobe. When this was first discovered, biologists thought that each lobe simply processed information for that side of the body. Thus, the right cerebrum lobe coordinated responses to stimuli for the right side of the creature, and the left lobe did so for the left side. As our knowledge of brain function has increased, however, we have come to realize that these parts actually do their job in different ways. In humans, for example, the mathematical and analytical reasoning that we do takes place mostly in the left lobe of the cerebrum, where speech is also controlled. The right side of the cerebrum deals with mostly nonverbal responses, as well as creativity.

When the brain sends signals to different parts of the body in order to control bodily functions or coordinate responses to stimuli, the signals are sent down the medulla oblongata and into the **spinal cord**. The spinal cord runs inside the vertebral column and sends signals to and from the brain. The spinal cord has several nerves (**spinal nerves** and **appendage nerves**) running out from it, so that the signals can be sent to the place that the brain intended for them to go. The nerves serve a dual purpose. They receive signals from the brain and interpret those signals so as to make the body do what the brain wants it to do, but they also transmit signals back to the brain. These signals come from the many receptors that are located throughout the body, sensing any changes that occur in the environment.

Think about what we have here. Receptors take in various stimuli from the environment. The receptors in the eyes take in light; the receptors in the nose take in gaseous chemicals; the receptors in the skin sense motion and touch, etc. These receptors then translate those stimuli into electrical signals that are sent to the brain. The brain processes those signals, figures out what they mean, and decides on a course of action. That course of action is then translated back into electrical signals, which are sent to a *specific set* of nerves in the body. If the brain wants the right arm to move, it sends those signals to nerves in the right arm. The signals go only to the nerves for which the instructions are meant. Those nerves then react to the signal and make the body perform according to the dictates of the brain.

Isn't that amazing? Vertebrates have a nervous system that responds to stimuli better, faster, and more efficiently than any "command and control center" that human science can devise. This kind of processing occurs in *every vertebrate in Creation*! Now, add to that the fact that humans actually have the ability to think originally, reason deductively, feel emotion, and pass knowledge from generation to generation, and you will soon appreciate how incredibly complex this all is. Clearly something more structured and efficient than the most complicated

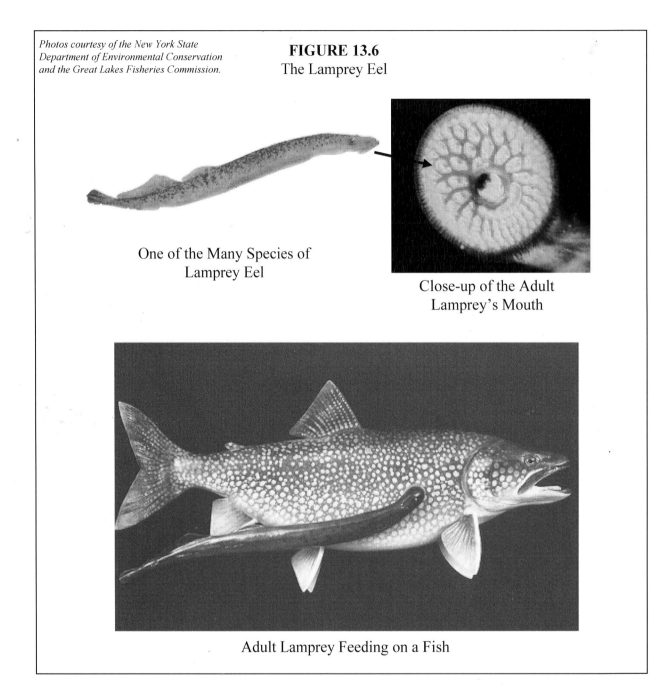

Photos courtesy of the New York State Department of Environmental Conservation and the Great Lakes Fisheries Commission.

FIGURE 13.6
The Lamprey Eel

One of the Many Species of
Lamprey Eel

Close-up of the Adult
Lamprey's Mouth

Adult Lamprey Feeding on a Fish

Some lampreys live in fresh water, and the rest are anadromous. When lamprey eggs hatch, the young are in larva form and are called **ammocoetes** (am oh coh' et eez). While in larva stage, the young behave and look very different from their adult form. They feed by producing strands of mucus which trap food particles that float in the water. They live as larvae for as many as seven years until metamorphosis takes place and they develop into an adult.

As an adult, the lamprey has a round, sucker-like jawless mouth filled with horny teeth and a raspy tongue. A ring of cartilage supports the mouth, in place of a jaw. It feeds as a parasite, mostly preying on other vertebrates. It does so by attaching to the prey with its mouth. A very complex pumping mechanism in the mouth keeps the lamprey attached by suction, while it scrapes the skin of the prey with its tongue. It then sucks the blood that comes out of the skin as the result of this scraping. The lamprey will continue to feed like this until it is not getting enough blood, and it will then move on to another prey. The unfortunate creature to which it was attached is severely weakened by the loss of blood and usually dies.

The skeleton of the lamprey is made of cartilage. It never forms true bones. The skull is made of overlapping plates of cartilage which encase and protect the brain. The brain has a very small cerebellum but large optic lobes. This illustrates a common phenomenon among vertebrates. Many vertebrates tend to have one set of brain lobes larger than that of the "typical" vertebrate because it is designed to "specialize" in a particular function. Lampreys, for example, need good eyesight to find prey. As a result, the optic lobes are large, so that the lamprey can process visual information very effectively. Humans, for example, have a large cerebrum, because God has designed us to think. Since the cerebrum is where the processing of information occurs, we need large cerebrums.

Although lampreys are considered a delicacy in some parts of Europe, they are generally considered a nuisance. When allowed to breed in a region, they can devastate the local fish population. Since 1835, for example, lampreys have been making their way into the Great Lakes through manmade canals. By the 1940s, their population had grown to the point where they were devastating the population of fish that are the lifeblood of the commercial fisherman. As a result, the U.S. and Canadian governments invested a lot of effort in controlling the lamprey population. They used nets to catch adult lamprey and "poisoned" the water with a chemical that killed lamprey larva but was harmless to the rest of the fish population.

By the late 1980s, the lamprey population was under control and the fish populations began to grow again in the Great Lakes. Although the U.S. and Canadian governments claim credit for the lowering of the lamprey population in the Great Lakes, there are many biologists who think, ironically enough, that the pollution problems in the Great Lakes are what actually brought the lamprey population under control. It turns out that lamprey larvae are much more sensitive to pollution than other types of fish, due to their feeding method. As a result, there is a lot of evidence to suggest that the manmade pollution counteracted the effect of the manmade canals and actually saved the commercial fish populations in the Great Lakes!

ON YOUR OWN

13.12 Are there any lampreys that reproduce in marine environments?

13.13 A student says, "lampreys are parasites." How should the student's incorrect statement be modified to make it correct?

13.14 If a vertebrate relies on its sense of smell for survival, what brain lobes will be the largest, compared to those of other vertebrates?

Class Chondrichthyes

We now turn our attention to one of the more frightening classes in Subphylum Vertebrata, class Chondrichthyes (kahn drik' thih eez). In Greek, the word "khondros" means cartilage and "ichthyes" means fish. Thus, the creatures in this class are often referred to as "cartilaginous fish." This, of course, gives away their most important feature: their endoskeleton is made of cartilage, like that of the lamprey. Unlike lampreys, however, members of this class have a very elaborate endoskeleton. This elaborate skeleton stays flexible because it is not reinforced with the calcium salts that bone tissue is reinforced with. Thus, we sometimes say that the skeleton of these creatures is "not calcified." The major representatives of class Chondrichthyes are pictured in Figure 13.7.

FIGURE 13.7
Members of Class Chondrichthyes

a. Blue Shark
Photo by Cyber Sea Photography

b. Sand Tiger Shark
Photo from the MasterClips collection

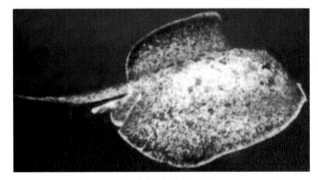

c. Marbled Sting Ray
Photo from the MasterClips collection

d. Gray Skate
Photo by Dr. Jay L. Wile

Now you see why we say this is one of the more frightening classes of vertebrates. Class Chondrichthyes is home to sharks, rays, and skates.

Sharks

Most sharks have been specifically designed to be deadly predators. They have a sleek, torpedo-like shape that allows them to move easily and quickly through the water. Although their skin looks smooth, it is actually covered with tiny scales that make it rather rough. The major external features of the shark are illustrated in Figure 13.8.

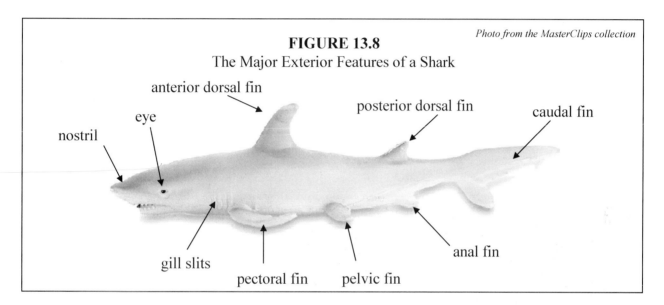

FIGURE 13.8
The Major Exterior Features of a Shark

Photo from the MasterClips collection

anterior dorsal fin

eye

nostril

posterior dorsal fin

caudal fin

anal fin

gill slits

pectoral fin

pelvic fin

The shark has 6 basic types of fins. The **pectoral fin,** near the head, and the **caudal fin,** at the end of the tail, are the fins most used for swimming. By swishing the caudal fin back and forth, the shark can propel itself through the water at great speeds. The **anterior dorsal fin** and the **posterior dorsal fin** are used mostly to stabilize the shark and keep it upright in the water. The **anal fin** performs essentially the same task. The **pelvic fin** is also used for stabilization, but in males, it plays a role in reproduction.

The shark's mouth is typically lined with rows of razor-sharp teeth. Although their endoskeletons are not calcified, their teeth usually are. This makes them strong enough to bite through the skin, flesh, and bones of their prey. If a shark loses a tooth, one of the teeth from a row further back in the mouth advances to replace it, and another grows in place of that one. Thus, the back rows of shark teeth are simply "backup teeth." This allows the shark to feed constantly, regardless of minor injuries.

The shark breathes through gills which are rich with capillaries. As water passes over the gills, the blood in the capillaries rids itself of carbon dioxide, which diffuses through the capillary walls and into the water. At the same time, oxygen that has been dissolved in the water diffuses through the capillary walls and is absorbed by the hemoglobin in the blood. It was once thought the sharks had to continually move in order to supply their gills with fresh water. We now know that this idea was a mistake. Sharks have an independent means of pumping water through their gills, so they can breathe when they are motionless in the water.

Although sharks do have eyes whose receptors are connected to the optic lobes of the brain, their eyesight is not very keen. The shark also has nostrils which lead to receptors that communicate with the olfactory lobes of the brain. This allows the shark to have a sense of smell, which is keener than its eyesight. This sense of smell helps lead them to prey. When a shark smells blood, it often goes into a feeding frenzy, madly eating anything it can sink its teeth into.

Even though the shark does use both its weak eyesight and its sense of smell to hunt prey, neither one of those senses alone would lead it to enough prey to sustain its voracious appetite. The shark's eyesight is weak, and even though its sense of smell is better, smells do not travel well under water. As a result, these senses do not provide the shark with enough range to effectively hunt for prey. How, then, does the shark find enough prey to satisfy its seemingly unquenchable appetite? The shark has two highly-developed means of searching for prey. A shark has a very sensitive vibration detector and an even more sensitive *electrical field sensor*.

Most sharks have a **lateral line**, a canal that runs the length of the shark's body. This canal is full of receptors that can detect very small vibrations which occur underwater. These receptors indicate to the shark's brain where the vibrations come from, and the shark will investigate them as a possible source of food.

The most sensitive means that the shark has for detecting prey, however, is its three-dimensional electrical field sensor. You see, the muscles of vertebrates are controlled by electrochemical reactions that involve the transfer of electrical charge from one molecule to another. Each time a vertebrate moves a muscle, these electrochemical reactions send out a tiny electrical signal. The shark has the ability to sense those electrical signals and filter out the ones in which it is not interested. It can then wait to sense the electrical signals that come from the fish which it is interested in eating. When it detects such a signal, its electrical field sensor determines the *precise location* of the prey, and the shark can catch the unfortunate creature *without either seeing or smelling it*.

This incredible phenomenon was first discovered when marine biologists observed sharks feasting on flounders. Flounders are flat fish that lie at the bottom of the ocean, often covered in sand. Because of their coloring, they blend into the sand, and it is very hard to see them, even if you are looking right at them. Well, biologists would observe sharks that were just swimming along, and then suddenly they would dive down and bite the sand, each time unerringly coming up with a flounder! Even when the scientists could not see the flounder resting on the bottom of the ocean, the shark knew it was there and would deftly grab it.

After extensive research, biologists determined that the shark was homing in on the electrical signals that the flounder sent out whenever it made even the slightest move! Although the ability to detect electrical signals and use them to home in on prey is amazing enough, the marine biologists were really amazed at the electrical field sensitivity of the shark. It turns out that the shark's ability to detect electrical fields is so sensitive that it is comparable to someone detecting the *precise location of a transistor radio battery 1,184 miles away*!

Think about that for a moment. The shark has a *natural* electrical field sensor that is more sensitive (by far) than *anything* that human science can develop. Think what the navy could do with such an electrical field sensor! Our best scientists, however, cannot produce anything close to what the shark has *always had*. The shark's method of finding prey is clearly one of the great testaments to the fact that this world was *designed*!

Although sharks are best known as terrifying predators, there are some species of shark that are quite gentle. The whale shark, for example, is the largest fish in Creation. Unlike other sharks, it swims slowly and quietly in the ocean, filtering out the plankton in the water as its only food source.

When sharks reproduce, fertilization is internal. The male transfers sperm to the female with a **clasper**, which is a specially designed portion of the male's pelvic fins. Most sharks are oviparous, releasing the fertilized egg in a capsule that attaches to stationary objects under water. A few sharks, however, are viviparous. This is very unusual in any vertebrates except those of class Mammalia.

An interesting symbiotic relationship exists between a fish called the **remora** (ruh more' uh) and many species of shark. Although most fish are wary of sharks, the remora comes right up to a shark and attaches to its body with a sucker organ that is on the top of its head. The remora then begins to eat parasites off of the shark's scales. This, of course, provides an enormous health benefit to the shark and, at the same time, provides a food source for the remora. When you see a group of sharks, you will almost always see remoras attached to them.

Rays and Skates

Rays and skates (Figures 13.7c and 13.7d) are flat and thin, not torpedo-shaped like a shark. They are often called "birds of the sea," because their pectoral fins look like huge wings that propel them through the water. Their tails are slender, whip-like structures that they typically use like a rudder.

The difference between rays and skates might be a bit difficult to see from Figure 13.7. Rays are designed to lie at the bottom of the ocean. Their eyes are on top of their bodies in order to look for danger. Their teeth are blunt, ideal for crushing the shells of mollusks and the exoskeletons of crustaceans that inhabit the ocean floor. Some rays protect themselves by changing color and blending in with the sand that they inhabit. Others, such as the stingray, have sharp spines that run along their tails. When the stingray feels threatened, it can lash out at the potential predator with its tail, inflicting painful and slow-healing wounds.

Skates, on the other hand, are designed to swim near the bottom of the ocean. Unlike rays, they do not typically dwell in the sand. Sometimes they rest on the sand at the bottom of the ocean, but they typically swim a few feet above the ocean bottom, looking for prey. Their eyes are typically near the front of their head, not on top of their body.

ON YOUR OWN

13.15 If you cut the anterior and posterior dorsal fins off of a shark, what will happen when it tries to swim?

13.16 When a creature dies, under the right conditions, its remains can be preserved as fossils. Typically, the harder the remains, the more likely they are to fossilize. Thus, if a clam dies, the inside body parts are rarely fossilized, but the shells often are. What parts of the shark would be most likely to fossilize?

13.17 Underwater photographers have batteries in their cameras. When the batteries go dead, they often throw them into the water, which is not a good thing to do, because those batteries pollute the water with their chemicals. Some shark photographers have noticed that the sharks sometimes go after those batteries, as if they are prey. Why?

13.18 You see something flat on the ocean floor. It has large pectoral fins like a ray or a skate. You watch it for a while as it moves along the ocean floor, staying in direct contact with the sand. Is it most likely a ray or a skate?

Class Osteichthyes

What most people think of as fish can be found in class Osteichthyes (ahs tee ik' thih eez). The Greek word "osteon" means bone while, as we already learned, "ichthyes" means fish. Thus, the members of this class are often called the "bony fish." Unlike the creatures that we studied in the last two sections, members of this class have endoskeletons that are calcified. Thus, they have "true" bones. Now remember, just because a creature has bones, this does not mean that the endoskeleton is void of regular cartilage. Vertebrates with truly bony skeletons use cartilage as a connective tissue to keep the bones together.

In fact, most members of this class have only certain regions of their endoskeleton which are calcified. All bony fish have calcified vertebral columns and skulls, but most of them have cartilaginous ribs. The pectoral and pelvic girdles, in the fish that have them, are also usually composed of non-calcified cartilage. Thus, even a "bony fish" is not as bony as you might think! The major exterior features of a "typical" bony fish are shown in Figure 13.9.

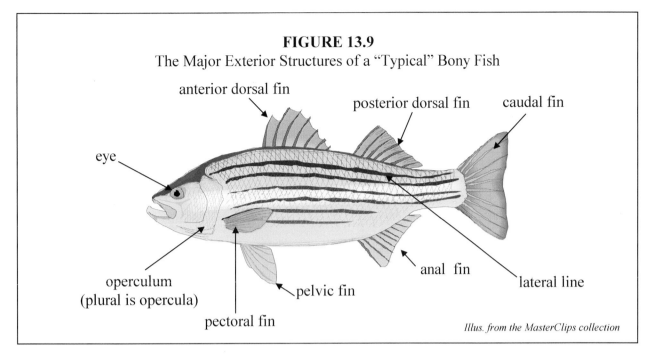

FIGURE 13.9
The Major Exterior Structures of a "Typical" Bony Fish

anterior dorsal fin

posterior dorsal fin

caudal fin

eye

operculum
(plural is opercula)

pectoral fin

pelvic fin

anal fin

lateral line

Illus. from the MasterClips collection

Notice that, like the shark, a bony fish has anterior and posterior dorsal fins, a caudal fin, pelvic fins, and pectoral fins. Notice also that the lateral line, which senses vibrations in the water, is also present in the typical member of class Osteichthyes.

Unlike the shark, the typical bony fish has **opercula** (oh per' kyou lah), which cover the gills. These structures get their name from the Latin word "opercula," which means "cover." In order to breathe, the fish opens its mouth and draws water into it. It then closes its mouth and opens its opercula. It then forces the water to run out of its opercula. In order to do this, the water must first travel through the gills, where the oxygen dissolved in the water diffuses into the capillaries of the gills. At the same time, of course, carbon dioxide diffuses out of the capillaries and into the water. This process is so efficient in bony fish that more than 80% of the oxygen dissolved in the water can be absorbed by the blood in just one pass through the gills!

Since the typical bony fish is flatter than the shark, it needs more help in balancing. This is why the dorsal fins are usually a lot longer than those of the shark. In addition to being longer than the shark's, most bony fish have sharp spines in their anterior dorsal fin. This acts as a defense against predators, because the spines can cause deep jagged wounds. If you have ever grabbed a fish the wrong way while fishing, you have probably experienced the defensive capabilities of the bony fish's anterior dorsal fin!

The body covering of most fish consists of overlapping scales. Special glands beneath these scales secrete **mucus**, a slimy substance that gives fish the "slimy" feel that they have when you touch them. The mucus waterproofs the scales, protects the fish from parasites, and makes the fish more mobile in the water. Studies indicate that this slimy mucus reduces water drag by more than 2/3, allowing the fish to swim faster than it would normally be able to swim.

The members of class Osteichthyes have an interesting internal structure that is ideal for an introductory lesson in anatomy. Their internal structure is complex enough for us to point out some features common to all of the other members of subphylum Vertebrata which we will study. At the same time, however, their internal structure is not so complex that it becomes overwhelming, as is the case with many vertebrates. The major internal features of a bony fish are illustrated in Figure 13.10.

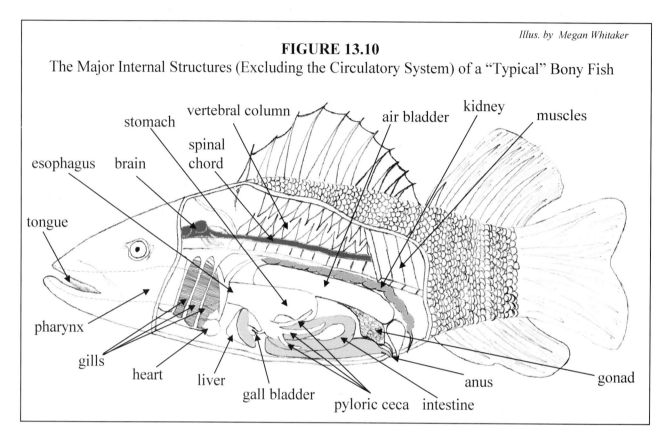

FIGURE 13.10

The Major Internal Structures (Excluding the Circulatory System) of a "Typical" Bony Fish

Illus. by Megan Whitaker

When a fish feeds, the food travels through the mouth and across the **tongue**, which is lined with **taste buds**. These taste buds are receptors that send signals to the brain, providing the fish with a sense of taste. The food then passes through a muscular pharynx (throat) and into the esophagus, which leads to the stomach. The food is broken down and stored in the stomach until it needs to be digested. Then, it is sent to the intestine, where it is digested. Any undigested remains leave the fish through the anus.

The fish has two main organs that aid the intestine in digestion. The **pyloric ceca** secrete some digestive enzymes into the intestine. At the same time, they also secrete chemicals into the stomach that aid in the breakdown of food. The **liver**, which is connected to the small intestine, secretes **bile**.

<u>Bile</u> - A mixture of salts and phospholipids that aids in the breakdown of fat

Bile tends to speed up the digestion of fats in the intestine. Without bile, the fish would release plenty of digestible fats through the anus, because there would not be enough time to digest them before they traveled all the way through the intestine. When food is not flowing through the intestine, a valve shuts the liver off from the intestine, and the bile is concentrated in the **gall bladder**.

Most bony fish have an **air bladder**. This organ helps the fish to stay afloat in the water. You see, bony fish are heavier than water. Without an air bladder, most of them would simply sink to the bottom. However, the fish can direct gases from the blood and digestive system to diffuse into the air bladder. This increases the buoyancy of the fish, allowing it to float. If the fish wants to rise to a shallower depth in the water, it simply increases the amount of gases in its air bladder. This increases its buoyancy, allowing it to rise. Alternatively, if the fish wants to move to a lower depth, it simply releases gases from its air bladder, causing the fish to sink.

The fish's brain is composed of the lobes that exist in a typical vertebrate brain. The olfactory lobes are large for such a small brain because the fish's sense of smell is probably its keenest sense. Fish are able to detect incredibly small amounts of substances by their smell. This is the main way that predatory fish (other than sharks) seek out prey. Even though fish have good eyes and large optic lobes in their brains, the sense of sight is weak underwater, mostly due to the fact that light does not penetrate water effectively.

As with all vertebrates, the sexes are separate in fish. The gonad of a male fish is its testes while the gonad of a female fish is its ovaries. Most bony fish are oviparous. The female lays eggs. The male then covers them with **milt**, a milky-white substance that contains its sperm. Fertilization, therefore, is external in most bony fish. The eggs are then left to develop and hatch. When the female lays its eggs, we usually say that she is **spawning**.

The circulatory system of a fish is also instructive. Figure 13.11 is a sketch of the basic anatomy of a fish's circulatory system .

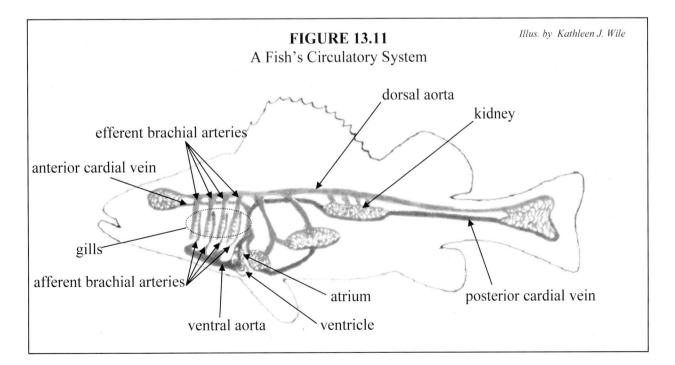

FIGURE 13.11
A Fish's Circulatory System

Illus. by Kathleen J. Wile

Notice that we have colored some of the blood vessels blue and the rest red. This is the typical convention in biology. Red represents blood with oxygen, blue represents blood without oxygen. To be accurate, the colors should really be bright red for blood with oxygen and dark red for blood without oxygen. It would be hard to tell those two colors apart, however, so we use red and blue instead. Also realize that the blood vessels drawn in Figure 13.11 are not all of the blood vessels in a fish! These are just the major blood vessels along with the capillaries that service the major structures in the body. In order to deliver nutrients and necessary gases to every cell in the body, there are obviously many capillaries that we just do not show.

The blood flow begins with a **two-chambered heart**. In most of the organisms that we studied in previous modules, the heart was a single muscle with a single cavity that held blood before it was pumped. As the circulatory system of an animal gets more complex, however, its heart is designed to accomodate these features. The next step in complexity is the two-chambered heart of the fish. The two chambers are called the **atrium** (ay' tree uhm) and the **ventricle** (ven' trih kul).

Atrium - A heart chamber that receives blood

Ventricle - A heart chamber from which blood is pumped out

As its definition implies, the atrium receives oxygen-poor blood from the veins. The blood is stored there until the ventricle is ready to receive it. At that time, the atrium empties into the ventricle, at which point the ventricle pumps the blood back out again. While that pumping action is going on, the atrium fills again, waiting for the next chance to empty into the ventricle.

The blood leaves the heart through the large blood vessel called the **ventral aorta** (ay or' tah), which branches into a series of smaller vessels called **afferent** (ah' fuh rent) **brachial** (bray' kee uhl) **arteries**. These arteries supply blood to the capillaries in the gills, where the blood releases the carbon dioxide that it carries and accepts oxygen from the water that pours over the gills. Now the blood is oxygen-rich, and it leaves the gills through **efferent** (ef' uh rent) **brachial arteries**. These arteries dump blood into the **dorsal aorta**. Some of the blood goes to the brain and other cells in the head, supplying them with nutrients and oxygen. At that point, this blood is oxygen-poor and is carried back to the heart in the **anterior cardial** (car' dee uhl) **vein**.

If the blood goes the other way in the dorsal aorta, it is sent to various organs and cells in the rest of the body. Blood that passes through the capillaries of the intestine picks up nutrients as well as waste while it drops off oxygen to the intestine's cells. The blood will carry those nutrients to the rest of the cells in the body. The blood that passes through the **kidney** is filtered of its waste materials. Once the blood has done its job, the **posterior cardial vein** carries it back to the heart, where the whole process starts again.

Fish are **ectothermic** (ek toh thur' mik), which means they are "cold-blooded."

Ectothermic - Lacking an internal mechanism for regulating body heat

Since they are cold-blooded, fish are very sensitive to the temperature of the water in which they live. If the water temperature changes too much, the fish must either move or die.

It is important for you to know the organs and circulatory system of the fish, because they are similar to those of humans. Although the human organs and systems are more complex, those of the fish do essentially the same job. The intestine, liver, kidneys, mouth, tongue, and esophagus perform basically the same functions in humans as they do in fish. Although humans have lungs while fish have gills, and although the human heart has four chambers instead of two, the fish is still a reasonable model of what goes on in the circulatory system of a human. At the same time, however, the fish's systems and organs are not as complex. Thus, they are a reasonably easy set of systems and organs to learn and, at the same time, provide some insight into the systems and organs of your own body. Thus, you will be expected to know the ins and outs of Figures 13.10 and 13.11, as well as how the systems and organs illustrated there work.

The Diversity of Class Osteichthyes

The diversity observed in this class is dizzying. There are an incredible number of different species of fish in both freshwater environments and marine ecosystems. We have studied some of them already in other modules. The goby, clownfish, Oriental sweetlips, and blue-streak wrasse all participate in amazing symbiotic relationships. There are, of course, thousands of other fish to study - far too many to discuss in an entire course, not to mention a single module. Nevertheless, Figure 13.12 shows a few of these diverse species.

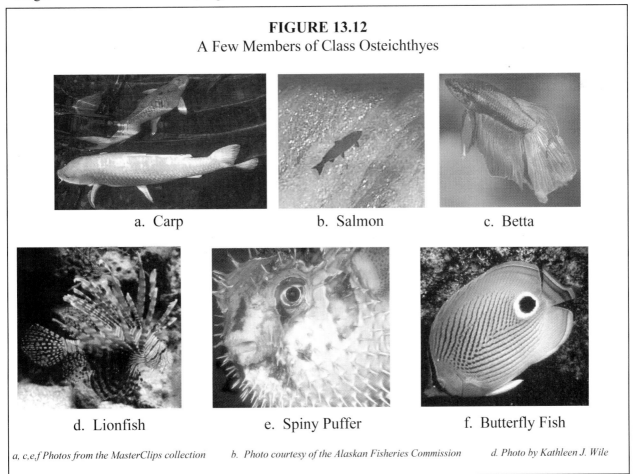

FIGURE 13.12
A Few Members of Class Osteichthyes

a. Carp   b. Salmon   c. Betta

d. Lionfish   e. Spiny Puffer   f. Butterfly Fish

a, c,e,f Photos from the MasterClips collection   b. Photo courtesy of the Alaskan Fisheries Commission   d. Photo by Kathleen J. Wile

The **carp** (Figure 13.12a) is a freshwater fish typically referred to as a "trashfish." This is because it typically feeds off of the bottom of the lake or river in which it lives, giving the illusion that it is eating trash and decaying matter. It is also called a trashfish because it can live in water that is too polluted to support most other species of fish. Finally, carp are so bony that it is very hard to eat them. Thus, they are not worth much. Some species, however, are used as ornamental fish, because they often come in interesting colors. Many ornamental fish ponds are filled with carp. Carp are prolific and breed rapidly. They also tend to "take over" the waters in which they live, driving other fish away. Typically, they do this by stirring up mud and uprooting vegetation.

The **salmon** (Figure 13.12b) has a lifestyle much like the lamprey eel. There are many different species of salmon, but the most popular is the Atlantic salmon. It lives its adult life in the cold waters of the Atlantic ocean. In late spring or early summer, it travels back to the cold, swiftly flowing, freshwater stream in which it was born. It goes back to that stream no matter what obstacles lie in its way. The salmon pictured in the figure, for example, is jumping a 7 foot waterfall to get back to its stream. As the female salmon spawns, it lays as many as 20,000 eggs. The males fertilize them, and then the salmon float back down the stream into salt water. Some salmon (the Pacific salmon, for example) spawn only once and then die afterwards. When the eggs hatch, the young stay in freshwater for two years and then head to the ocean. Salmon are very popular food fish. The adults are heavily fished, which has severely reduced the salmon populations in some regions.

In Figure 13.12c, we see the **betta**. This small fish lives in the warm freshwaters of Southeast Asia. Because they are so beautiful, bettas are cultivated there as aquarium fish. They are sometimes called Siamese fighting fish because the males are so territorial that they will attack any other bettas that they see. They will even attack their own reflection in a mirror! Surprisingly enough, bettas are usually tolerant of other species of fish. Many aquarium owners have one (and only one) betta in their aquariums, peacefully coexisting with many other species of fish.

Moving to marine environments, we see the lionfish in Figure 13.12d. Its bright colors come from specialized cells called **chromatophores** (kro mat' uh fores). These chromatophores produce pigments that give the fish its colors. Sometimes the pigment is of one color. Other times, several overlapping cells produce different colors of pigments, and the coloring of the fish is the result of those colors mixing. The lionfish's colors are meant as a warning. Their enlarged pectoral fins are tipped with poisonous glands. Fish seem to associate the colors of the lionfish with danger and steer clear. Humans should steer clear as well, because the poison of the lionfish can be deadly.

The **spiny puffer** (Figure 13.12e) is, most of the time, a small, harmless looking fish. It uses its beak-like mouth for breaking open hard corals or mollusks and eating the body inside. It only appears as it does in the figure when it feels threatened. When this happens, it takes in large quantities of water and "puffs" itself up to many times its normal size. The spines then stand out, giving the spiny puffer the appearance of an underwater porcupine. It is therefore sometimes called the "porcupine fish." Although considered very tasty , especially in Oriental cultures, the spiny puffer's flesh can be deadly if not prepared properly by an expert chef.

Finally, the **butterfly fish** (Figure 13.12f) is a general name for several brightly-colored tropical marine fish. These fish have an interesting defense mechanism. The chromatophores in their posterior region produce a very dark pigment. They are arranged in a large disk, giving the visual impression of a large eye. This large eye is mistaken by other fish as being indicative of a large predator. As a result, most potential predators stay away.

ON YOUR OWN

13.19 A student sees the ribs of a fish and notices that they are made of cartilage. Based on this, the student says that the fish cannot be a member of class Osteichthyes. Is the student correct? Why or why not?

13.20 A fish can't seem to digest fats. What organ is not functioning properly?

13.21 Most arteries carry oxygen-rich blood. Which artery in the fish carries oxygen-poor blood?

13.22 Why don't remoras have symbiotic relationships with bony fish?

If you purchased the dissection kit, then you will want to perform the following optional dissection experiment.

Optional Dissection Experiment #3
The Perch

Be careful! Dissection tools are SHARP!

Supplies :

- Dissecting tools and tray that came with your dissection kit
- Perch specimen
- Magnifying glass
- Laboratory notebook
- Water
- A small bowl

Object: To become more familiar with the anatomy of the perch through dissection

Procedure:

A. Examine your specimen. Identify all of the structures indicated in Figure 13.9. List in your laboratory notebook any structures in that figure which you could not identify.

B. Open the specimen's mouth and examine the teeth. Write a description of them in your laboratory notebook.

C. Examine the tongue. Where is it attached to the mouth? Record this in your laboratory notebook.

D. Examine the anterior dorsal fin. Fins can be supported by one of two structures: rays or spines. The way to tell the difference is to put your finger on the tip of the supporting structure and push lightly. If the structure bends, it is a ray. If it does not bend, it is a spine. Raise the anterior dorsal fin so that it looks like the anterior dorsal fin in Figure 13.9. Now put your finger on the tip of one of the supporting structures and push gently. Don't push too hard, or you will get stuck! Is the anterior dorsal fin supported by rays or spines?

E. Repeat part D for all other fins, and record in your notebook which are supported by spines and which are supported by rays.

F. Pull a scale off of the specimen and observe it under the magnifying glass. Draw what you see. Can you make out the rings? These are growth rings, which show you how the scales of the fish have grown as the fish grew.

G. Raise the right operculum, and use your probe to count the gills. Now do the same for the left operculum. Record the number of gills on each side in your laboratory notebook.

H. Using your scissors, cut the left operculum away and remove one set of gills. Place the gills in the bowl and cover them with water. Note the structure. You should be able to see a strong **arch** in each gill. The arch should have a comb-like structure on one side and feathery extensions on the other. The "teeth" of that comb-like structure are called the **rakers** while the feathery extensions are called the **filaments**.

I. Draw a gill in your laboratory notebook, labeling the arch, rakers, and filaments.

J. Now you are ready to look at the internal structures of the specimen. Hold the specimen ventral side up with the head pointing away from you. Use your scalpel to make a cut from just anterior of the anus all the way to the operculum.

K. Now turn the specimen so that its left side is facing you and so that the head is pointing left and the tail is pointing right. Now make a new cut from the point at which you left off up towards the dorsal fins. Continue until your cut is just above the level of the fish's eye.

L. Make a similar cut from the anus straight towards the dorsal fins until that cut reaches essentially the same level.

M. Lift up the body wall and completely remove this flap of body covering by making a final cut with your scalpel that runs from the cut you made in (K) to the cut you made in (L). The outline of your cut should look like the white line in the drawing below:

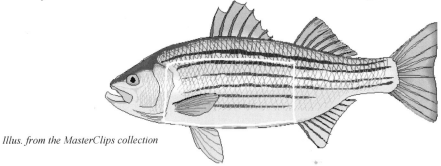

Illus. from the MasterClips collection

N. Once you have removed the body wall, you should be able to see the internal organs as illustrated in Figure 13.10. Draw your specimen.

O. Using Figure 13.10 as a guide, try to locate the following organs:

> **Liver** - It should be beige and it should lie in the anterior region of the fish.
> **Gall bladder** - If you gently raise the lobes of the liver, you should see the gall bladder, which looks like a deflated balloon.
> **Stomach** - This should be dorsal and posterior to the liver.
> **Esophagus** - This runs from the stomach to the gills.
> **Intestine** - This should be posterior to the liver and should run to the **anus**.

P. Remove the organs you just identified, trying not to disturb anything else. In order to keep from disturbing other organs, feel free to cut the organs you identified as much as you need to. If you want, cut open the stomach once you have removed it and see if you recognize the perch's last meal!

Q. With the digestive organs gone, you should be able to see the **gonad**. For females, the ovary will be orangish if it contain eggs. If the female was not ready to lay eggs, it should be beige and look like a drumstick. For the male, the testes will be small and cream-colored. Determine the sex of your specimen by whether you find an ovary or testes. Record the sex.

R. You should also be able to see red strands which are just ventral of the vertebral column. These make up the **kidney**.

S. You should also be able to see the **air bladder**, as long as you did not destroy it when you removed the body wall.

T. Now find the **heart**. Notice that it has two chambers. The upper one is the atrium and the lower one is the ventricle.

U. It is now time to find the fish's brain. Hold the fish with its dorsal side up and position its head so that it points away from you. Using your scalpel, cut the skin away from the skull.

V. Once you expose the skull, begin scraping it with your scissors to wear away the bone.

W. As the bone gets thinner, start picking it away with your forceps instead of scraping it. If you are careful, you can expose the brain in this manner.

X. Once the brain has been exposed, look at Figure 13.5 and see if you can identify the regions of the brain as pointed out in that figure.

> **Olfactory lobes** - These should be two small bulges in the front of the brain.
> **Cerebrum** - Made of up the two lobes behind the olfactory lobes.
> **Optic lobes** - These will be the largest and are just behind the cerebrum.
> **Cerebellum** - A single lobe behind the optic lobes.
> **Medulla oblongata** - Just underneath the cerebellum.

Y. Draw the brain as you see it in your specimen, labeling all parts that you can identify. List those parts you could not identify.

Z. Clean up everything. Wash and dry your tools and tray. Put everything away.

Class Amphibia

In Greek, the term "amphi" means "both sides" while "bio" means life. Thus, "amphibia" (am fib' ee uh) means "dual life." This is an excellent description of the organisms in class

Amphibia. Most amphibians begin their lifecycle in the water. The young hatch from eggs that are allowed to develop outside of the mother. They typically begin life in the larva stage. These larvae have gills and breathe in the water. When metamorphosis occurs, the gills degenerate and an air-breathing respiration system develops. Fins disappear and legs appear. The resulting adult is an air-breathing, land-dwelling creature that looks nothing like the larva. The adult female lays eggs which are externally fertilized by the male in the water, and the whole process begins again.

Although this metamorphosis lifecycle is common to all amphibians, it takes on a variety of different forms in this class. The larva of a salamander actually looks a lot like the adult, but it has gills for respiration whereas the adult does not. The larva of a frog, on the other hand looks nothing like the adult. This larva, often called the **tadpole**, looks like some sort of bulbous eel. It has an oval body which tapers down to a tail. When it matures into an adult, the tail disappears along with the gills. Limbs and lungs take their place.

Amphibians have several characteristics that separate them from the rest of Subphylum Vertebrata:

- Their endoskeleton is made mostly of bone.

- Their skin is smooth with many capillaries and pigments. Amphibians do not have scales.

- They usually have two pairs of limbs with webbed feet.

- They have as many as four organs of respiration.

- They have a three-chambered heart.

- They are oviparous with external fertilization.

These characteristics make amphibians a very interesting class of animals to study!

Perhaps the most interesting aspects of amphibians are their respiratory and circulatory systems. As was mentioned above, amphibians have as many as four respiratory organs. First, all amphibians have gills during their larva stage. When the gills degenerate, they are often replaced by **lungs**, which allow the exchange of gases between the blood and the atmosphere to occur internally. Lungs, however, are generally used only as a backup. Like the earthworm, the main organ of respiration for amphibians is their skin, which is rich in capillaries. For some amphibians, more than 90% of their respiration is accomplished through their skin. Finally, some amphibians (like the frog) use their mouth as a respiratory organ. Like the skin, the lining of a frog's mouth is rich with capillaries. A large amount of gas exchange can take place there.

Because amphibians have so many means of respiration at their disposal in their adult form, they usually are comfortable in or out of water. Most amphibians can go for very long periods of time without filling their lungs. As a result, they can breathe through their skin while they are under water. This allows them to stay submerged for extensive periods of time. Thus, even in their adult form, amphibians can spend a lot of time in the water, even though they are technically land-dwellers.

Like fish, amphibians are ectothermic. Thus, the temperature of their surroundings affects them greatly. When the temperature drops, as it does in the winter, a typical amphibian will burrow into the mud. As the mud gets colder, so do the internal organs and blood in the amphibian. This slows down all of the chemical reactions in the amphibian's body, putting it in a deep state of low activity which we call **hibernation**.

<u>Hibernation</u> - A state of extremely low metabolism

Since the amphibian's life processes are dramatically slowed during hibernation, it does not need to eat. It continues to breathe through its skin, and can simply "wait out" the cold weather. When warm weather returns, the chemical reactions in the body speed up, and the amphibian resumes its normal life.

The circulatory system of a frog is similar to that of a fish, but there are some major differences. Predominant among these differences is the amphibian's heart, which is composed of three chambers, not two. A typical amphibian heart is shown in Figure 13.13.

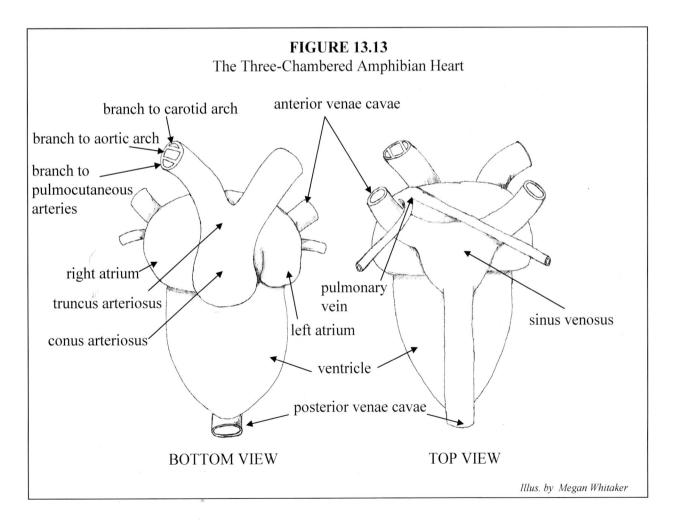

FIGURE 13.13
The Three-Chambered Amphibian Heart

BOTTOM VIEW   TOP VIEW

Illus. by Megan Whitaker

A three-chambered heart has a single ventricle and two separate atria (plural of atrium), the **left atrium** and the **right atrium**. The left atrium of the heart is filled with oxygenated blood that has come from the lungs, mouth, or skin. This oxygenated blood is dumped into the left atrium by a large vein called the **pulmonary vein**. Three large veins called the **venae** (vee' nuh) **cavae** (kav ay') fill the **sinus** (sigh' nus) **venosus** (veh no' sus) with deoxygenated blood that has come from the various internal organs of the amphibian. The sinus venosus then empties the deoxygenated blood into the right atrium.

Once filled, both atria dump their contents into the ventricle. This mixes the oxygenated blood with the deoxygenated blood. The ventricle then pumps this mixture out into a single artery, the **conus** (cah' nus) **arteriosus** (are tear' ee oh sis), which then splits into two smaller arteries called the left and right **truncus** (truhn' kus) **arteriosus**. Each of these arteries splits into three smaller arteries. The first, the **carotid** (kuh raht' id) **arch**, sends blood to the head. Since there are two truncus arteriosus, there are two carotid arches. Each of them provides blood to its side of the head. The second pair of arteries are the **aortic arches** which fuse into a single dorsal aorta, like that of a fish. Finally, the last pair of arteries are called the **pulmocutaneous** (pool mug kyou tay' nee us) **arteries**. These arteries take blood to the lungs, skin, and mouth, to get re-oxygenated.

Specific Creatures in Class Amphibia

The majority of amphibians can be lumped into two orders: Caudata (caw dah' tuh) and Anura (uh nur' uh). Members of each of these orders are shown in Figure 13.14.

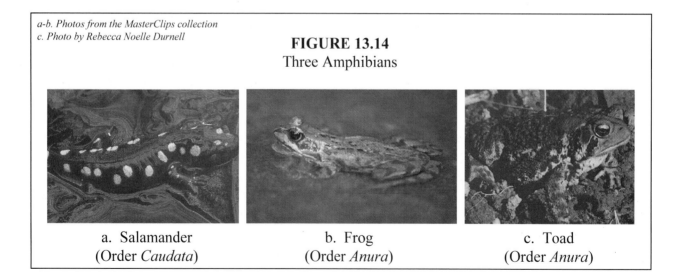

a-b. Photos from the MasterClips collection
c. Photo by Rebecca Noelle Durnell

FIGURE 13.14
Three Amphibians

a. Salamander
(Order *Caudata*)

b. Frog
(Order *Anura*)

c. Toad
(Order *Anura*)

Salamanders, like the one pictured in Figure 13.14a, are often mistaken for lizards. Unlike lizards, however, salamanders have the webbed feet of an amphibian and do most of their respiration through their skin. As a result, they cannot be grouped with the reptiles. Instead, they are in Order Caudata of class Amphibia.

Most salamanders are small and have little color. There are, of course, exceptions to this general rule. The giant salamander of Japan can reach lengths of up to 5 feet, and the tiger salamander gets its name from the bright markings on its skin. Most salamanders have no lungs. As much as 95% of their respiration occurs through their skin. As a result, salamanders can spend a great deal of time under water.

Frogs, like the one pictured in Figure 13.14b, and toads (Figure 13.14c) belong to order Anura. What's the difference between frogs and toads? Well, frogs have smooth, shiny skin that dries easily. Since frogs must keep their skin wet, they spend a great deal of time in the water. Toads, on the other hand, have dry, warty skin. They return to the water only to reproduce. Despite many ideas to the contrary, toads *do not* cause warts. They got that reputation simply because of the appearance of their skin.

ON YOUR OWN

13.23 An amphibian is breathing with gills. Is it in its larva or adult stage?

13.24 A frog that has overexerted itself leaves its mouth wide open. Why?

13.25 You find a creature from order Anura. If you find it far from any body of water, what (most likely) is it?

Optional Dissection Experiment #4
The Frog

Be careful! Dissection tools are SHARP!

If you purchased the dissection kit, you will want to dissect the frog at this time. In order to give you some experience in doing things more like a real scientist would, we will not give you any directions in how to perform this dissection. After all, a real scientist does not have directions at his or her disposal! In order to perform experiments, real scientists do literature research to find experimental protocols which they can use. That's what we ask you to do. Go to your local library or look on the INTERNET for a protocol that will tell you how to dissect the frog. There are many, many resources which give you such instructions. Use a protocol you find and try to learn as much as you can about the frog's anatomy. Compare that to how much you learned in the other dissections you did.

Summing Up

One of the most important things that experimental biologists can do is to make **field studies**. In these studies, biologists examine a part of nature, looking at the organisms that inhabit that region and the relationships those organisms have with one another. You have already made one field study. In Module #2, you went to a pond and observed the life that existed near the edge of that pond. You then took samples of the pond water home for further investigation.

We now want you to perform your second field study. Since you have learned a great deal more about Creation since your first field study, this one should be even more rewarding. Of course, depending on your homeschooling schedule, the time of year might not be ideal for a field study, so you should feel free to put this experiment off if the weather is not suitable. Wait until spring, find a nice, warm sunny day, and then do the following experiment. If you performed the dissection labs in this module, you can choose to skip this experiment if you like.

EXPERIMENT 13.1
Field Study II

Supplies:

- Lab notebook and pencil
- Colored pencils
- Magnifying glass (if available)
- Field Guide (This will help you identify the organisms that you see. Most libraries have field guides. Try to find one on plants and one on animals.)

Object: To observe how God's wonderful systems work together to sustain life in the world around us.

Procedure:
    A. Locate an area that is known for its wildlife like a creek, a pond, a woods or a lake area. Suggested field trip areas would be a state or city park or wildlife preserve near your home. (Take the whole family and make it a day outing.)

    B. On arriving in the area, find a relatively isolated area and sit down with your notebook and pencils. Remain in the area 1 to 2 hours.

    C. Note every living thing, plant and animal, as you sit quietly.

    D. Try to identify each specimen with your field guides.

    E. Make up a chart in your notebook including the following:
        Date
        Time
        Weather Conditions

Plants: (Identify if possible and note the number seen)
     1. Trees
     2. Shrubs
     3. Other Plants

Animals (Identify if possible and note the number seen)
     1. Insects
     2. Birds
     3. Fish
     4. Amphibians
     5. Other animals

Try to draw a few of the organisms that you see. The drawings need not be artistic. When biologists draw organisms, it forces them to look at the organism more analytically than usual. This allows them to notice details that they otherwise would not notice.

F. Write a summary approximately one page long and tell what you observed and how different specimens interacted. This will serve as the summary that you normally write at the end of an experiment.

Even though we have not finished up phylum Chordata (we still have three classes to go), we will actually end our discussion of the vertebrates for now. Never fear, however. We will pick up our discussion again in Module #16 where we will study reptiles, birds, and mammals. For this particular module, you should be most familiar with the bony fish. You need to have the organs and systems illustrated in Figures 13.10 and 13.11 committed to memory, as well as a strong knowledge of how these systems work. For the rest of the classes discussed here, you need to have a more general knowledge of the facts presented. As always, use the study guide and "on your own" problems to indicate what you need to concentrate on for the test.

ANSWERS TO THE OWN YOUR OWN PROBLEMS

13.1 The sea squirt is not classified with sponges because, other than its feeding habits and the fact that it stays immobile as an adult, it has nothing else in common with them. For example, sponges do not go through metamorphosis, they do not support themselves with leathery coverings; they do not have one specific entrance and another specific exit for water; and they do not have the internal organs that a sea squirt has.

13.2 The tunic, or leathery covering, in the adult supports the organism. That's what the notochord does in the larva.

13.3 The oral cirri were not functioning properly. Usually, large debris does not clog up the slits of the pharynx because the oral cirri filter it out of the water that the lancelet is taking in.

13.4 Blood cells are most affected by this procedure. After all, the bone marrow makes blood cells. If you replace a person's bone marrow, new blood cells will be created.

13.5 A shark's skeleton is made of cartilage. Since the shark is in Subphylum Vertebrata, its endoskeleton must be made of either bone or cartilage. Since the human skeleton is made of bone, and since cartilage is more flexible than bone, the shark's skeleton must be made of cartilage.

13.6 The blood came from an artery. In general, arteries carry blood away from the heart and, most of the time, the blood is oxygenated and is therefore bright red. Veins carry deoxygenated blood back to the heart, and deoxygenated blood is dark red.

13.7 It is in a vein. Arteries carry blood away from the heart, while veins bring it back.

13.8 The optic lobes are not working. In order to see, the eyes must detect light and send signals to the brain, which it interprets as an image. If the optic lobes are not working, the second half of the process fails.

13.9 The brain sends signals down the spinal cord to nerves that branch out to the limbs. That's how a person moves his or her limbs. If the spinal cord is cut, the signals cannot travel any farther than the cut. Thus, any limbs below that cut will never receive the signals sent to them.

13.10 The placenta is not working. In viviparous animals, the young develop in the mother and are nourished through the placenta. If the placenta does not work, the young will starve.

13.11 This reproduction involves external fertilization and oviparous development.

13.12 No. Remember, lampreys either live their entire lives in freshwater or migrate back to freshwater to reproduce. Thus, no reproduction occurs in saltwalter.

13.13 The phrase should really be "lampreys are parasites when they are adults." In the larva stage, the lamprey feeds on plankton and other small organisms.

13.14 The olfactory lobes will be largest. Vertebrates tend to have large brain lobes that control the most important processes. Since olfactory lobes control the sense of smell, they will be proportionally larger in any creature that uses the sense of smell as a primary means of survival.

13.15 The shark will not be able to stay upright in the water. The dorsal fins allow the shark to balance itself in the water. Without them, the shark would tumble around, not being able to stay upright.

13.16 The hardest parts of a shark are its teeth. Thus, its teeth are most likely to be fossilized. Shark teeth are, in fact, the most abundant shark fossil that can be found.

13.17 If the battery is not quite dead, but it is instead very weak, it still emits an electrical signal. If the electrical signal just happens to be like that of the prey a shark is hunting, the shark can mistake the electrical signal from the battery for the electrical signal of its intended prey.

13.18 It is a ray. Skates tend to glide in the water above the ocean floor, rays stay in direct contact with the ocean floor.

13.19 The student is not correct. Just because a part of the skeleton is cartilage, it does not preclude membership in class Osteichthyes. Many bony fish have only their skull and vertebral column made of bone. The rest of the skeleton is still cartilage.

13.20 The liver is not functioning properly. The liver produces bile which aids in the digestion of fats. Without bile, the fish cannot digest fats very well. You could also say gall bladder here as well, since it concentrates the bile, making it more effective.

13.21 The ventral aorta is an artery, because it carries blood away from the heart. However, the blood has not reached the gills yet, so it is still oxygen poor. Also, afferent brachial arteries hold oxygen-poor blood until they reach the gills. Thus, either the ventral aorta or the afferent brachial arteries are acceptable answers.

13.22 Bony fish don't need remoras because they are protected from parasites by the mucus that covers the scales.

13.23 The amphibian must be in the larva stage. Amphibians only use gills in their larva stage.

13.24 The frog can breathe through the lining of its mouth. By keeping the mouth open, the frog is breathing extra hard.

13.25 It is most likely a toad. Only frogs and toads are in Order Anura. Frogs, however, stay very close to water to keep their skin moist.

STUDY GUIDE FOR MODULE #13

*1. Define the following terms:

a.	Vertebrae	n.	Cerebellum
b.	Notochord	o.	Medulla oblongata
c.	Endoskeleton	p.	Internal fertilization
d.	Bone marrow	q.	External fertilization
e.	Axial skeleton	r.	Oviparous development
f.	Appendicular skeleton	s.	Ovoviviparous development
g.	Closed circulatory system	t.	Viviparous development
h.	Arteries	u.	Anadromous
i.	Capillaries	v.	Bile
j.	Veins	w.	Atrium
k.	Olfactory lobes	x.	Ventricle
l.	Cerebrum	y.	Ectothermic
m.	Optic lobes	z.	Hibernation

2. Assign the following creatures to one of these classifications: subphylum Urochordata, subphylum Cephalochordata, class Agnatha, class Chondrichthyes, class Osteichthyes, class Amphibia

    a. Frog   b. Shark   c. Lancelet   d. Carp   e. Sea squirt   f. Lamprey eel

3. What do sea squirts, lampreys, and amphibians have in common?

4. What is the difference between cartilage and bone?

5. You see a blood vessel from a creature. You have no idea what creature and you have no idea where it came from. You do notice, however, that the blood vessel wall is very thin. What kind of blood vessel is this?

6. What do red blood cells do?

7. What protein gives red blood cells their color?

8. Frogs and toads are quite uncoordinated. They move their muscles in a very jerky manner. Which brain lobe is small in amphibians?

9. An owl has very sensitive vision. Which brain lobes are larger in the owl compared to the "average" vertebrate?

10. A creature reproduces when the female receives sperm from the male and then lays an egg which hatches. Is fertilization internal or external? What kind of development is this?

✳11. Which has the most inflexible skeleton, a ray, a lamprey, or a salmon?

✳12. What do Atlantic salmon and many lampreys have in common?

✳13. What is the shark's most sensitive means of finding prey?

✳14. What function does the lateral line perform in sharks and bony fish?

✳15. What function do the dorsal fins perform in both sharks and bony fish? What function does the anterior dorsal fin play only in bony fish?

✳16. What are the behavior differences between rays and skates?

✳17. Identify the structures in this figure:

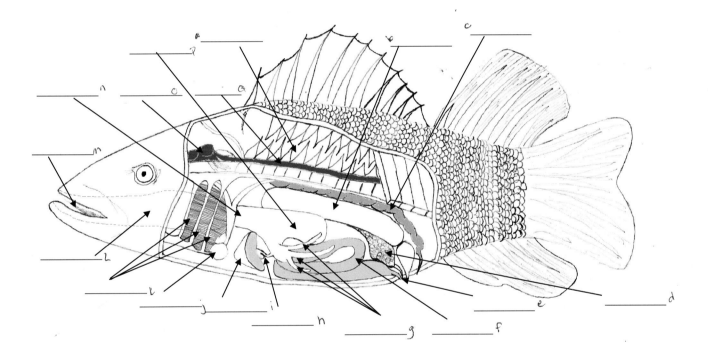

✳18. Describe the basic function of each organ in problem #17.

19. Identify the structures in this figure:

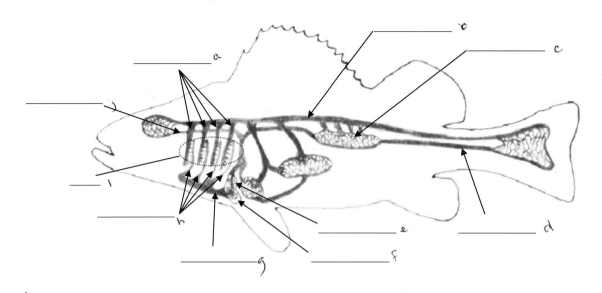

*20. Of the structures listed in problem 19, which are veins, which are arteries, and which are neither?

21. List the six common characteristics of amphibians.

22. What is the difference between a toad and a frog?

23. For most amphibians, what is the major respiratory organ?

Module #14: Kingdom Plantae: Anatomy and Classification

Introduction

We are now going to take a break from studying animals and begin a study of plants. Kingdom Plantae (plan' tay) is large and diverse, so we will spend two modules studying it. In this module, we will look at basic plant anatomy and classification. In the next module, we will turn our eyes to the life processes of plants. After you get through these two modules, you should have a much deeper appreciation for the plants that God has placed in His Creation.

The study of plants is usually referred to as **botany**, and biologists that study plants are called **botanists**.

Botany - The study of plants

Since most people know that plants perform photosynthesis, they assume that plants are our main source of oxygen. It turns out that this idea is a common misconception among science students, and even some scientists. Although plants do contribute to our planet's oxygen supply, they are not the principal producers of oxygen. The majority of oxygen comes from phytoplankton that live in the oceans, lakes, and streams of the planet.

Basic Plant Anatomy

There are many, many different ways to look at and analyze plants. One very useful way to examine plants is to determine whether or not they are **woody** or **herbaceous** (her bay' shus). If a plant has woody parts, such as trunks and/or stems, it typically grows year after year. We call these plants **perennials**.

Perennial plants - Plants that grow year after year

Plants that do not have woody parts are called herbaceous, and they typically live for only one year. These plants are called **annuals**.

Annual plants - Plants that live for only one year

A few types of herbaceous plants are **biennial**. These plants live for two years. Typically, they store food during the first season of growth and then flower the second season.

Biennial plants - Plants that live for two years

When you begin to study a given plant, you will notice that, much like an animal, it has certain organs and tissues. In general, plant organs can be categorized as either **vegetative organs** or **reproductive organs**.

Vegetative organs - The stems, roots, and leaves of a plant

Reproductive plant organs - The flowers, fruits, and seeds of a plant

Many of the things that we eat are either vegetative organs or reproductive organs from plants. Have you ever wondered whether a food item (like a tomato) is a fruit or a vegetable? Well, from a biological point of view, the answer is rather simple. If a food item is a reproductive plant organ, then it is a fruit. If it is a root, leaf, or stem, then it is a vegetable. Thus, a tomato (which contains seeds) is definitely a fruit, because it is a reproductive organ. We know it is a reproductive organ because it encases seeds. Something that comes from a root (like a carrot), however, is definitely a vegetable. Many of the foods that we call "vegetables" are, in fact, seeds. Peas and corn, for example, are technically seeds, not vegetables.

Plants have three basic kinds of tissues: **meristematic** (mehr uh stem ah' tik) **tissue**, **vascular** (vas' kyou luhr) **tissue**, and **structural tissue**. Meristematic tissue contains cells that are undifferentiated.

Undifferentiated cells - Cells that have not specialized in any particular function

Since these cells are undifferentiated, they can develop into any tissue that the plant needs. The cells capable of mitosis are a part of the meristematic tissue as well. Since the growing parts of a plant must be supplied with cells that are ready to turn into whatever specialized tissue is needed for growth, you can find meristematic tissues in any part of a plant that is growing.

Vascular tissue is used to carry water and dissolved material throughout the plant. In a way, these tissues are much like the blood vessels in animals. Whereas blood vessels transport blood that carries nutrients throughout the animal, vascular tissues transport water that has nutrients dissolved in it throughout the plant. Vascular tissue is generally categorized as either **xylem** (zy' lum) or **phloem** (floh' ehm).

Xylem - A vascular tissue that carries substances upward in a plant

Phloem - Vascular tissue that carries substances downward in a plant

The cells which make up the xylem have thicker walls than that which make up the phloem. In addition, xylem cells die when they mature, leaving long "tubes" in the plant. Phloem cells continue to live after they mature.

The last kind of tissue in a plant is the structural tissue. Any cells that participate in photosynthesis are considered part of the structural tissue. In addition, cells that store food, strengthen, or protect the plant are a part of the structural tissue.

ON YOUR OWN

14.1 A gardener plants a group of flowers that grow beautifully over the course of a year. In order to have the same flowers each year, however, he must replant them. Are these most likely woody plants or herbaceous plants?

14.2 A carrot is actually the root of a carrot plant. What kind of organ is it?

14.3 A section of plant has stopped growing. What kind of tissue should be absent in that section?

The Macroscopic Structure of a Leaf

In the next module, we will concentrate on the tissues of a plant as well as its reproductive organs. In this module, we want to spend some time on the vegetative organs in a plant. We begin with the **leaf**. Leaves come in all shapes and sizes. The major parts of a leaf are shown in Figure 14.1.

FIGURE 14.1
The Major Parts of a Leaf

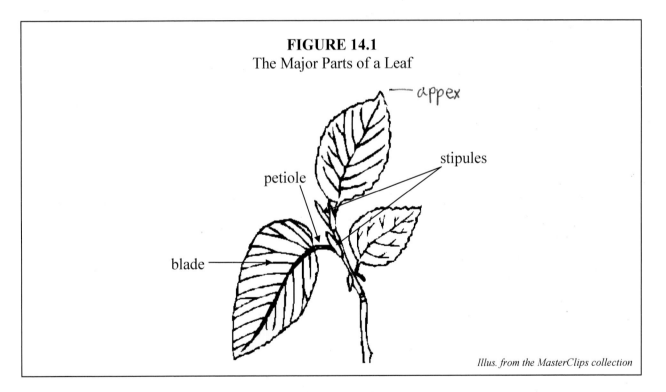

Illus. from the MasterClips collection

The primary portion of the leaf is called the **blade**. The blade is attached to the stem with a small stalk called the **petiole** (peh' tee ohl). At the base of the petiole, most plants have **stipules** that grow. These small stalk-like or leaf-like growths are usually the structure that covered the leaf when it first began to grow.

The most important job that the leaf performs is photosynthesis, so that the plant can obtain food. As a result, leaves are usually arranged in such a way as to expose the largest amount of surface area to the sun. The arrangement of leaves on the stem of a plant is called the **leaf mosaic**.

Leaf mosaic - The arrangement of leaves on the stem of a plant

There are many different leaf mosaics in Creation, but we will concentrate on the top three types, illustrated in Figure 14.2.

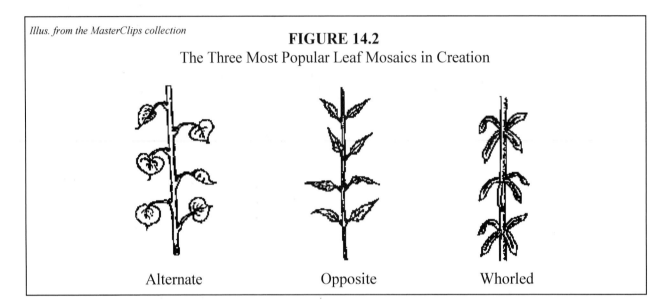

Illus. from the MasterClips collection

FIGURE 14.2
The Three Most Popular Leaf Mosaics in Creation

Alternate   Opposite   Whorled

Whereas the alternate leaf mosaic is probably the most common among plants, tropical plants tend to favor the whorled mosaic.

Looking at the leaf in detail, there are three characteristics that botanists can use to classify the plant from which it comes: **shape, margin**, and **venation** (ven ay' shun). Although the shape of a leaf is rather straightforward, the terms margin and venation are probably rather new to you, at least in their relationship to botany.

Leaf margin - The characteristics of the leaf edge

Figure 14.3 illustrates the major leaf shapes and margins in Creation by showing you several different kinds of leaves.

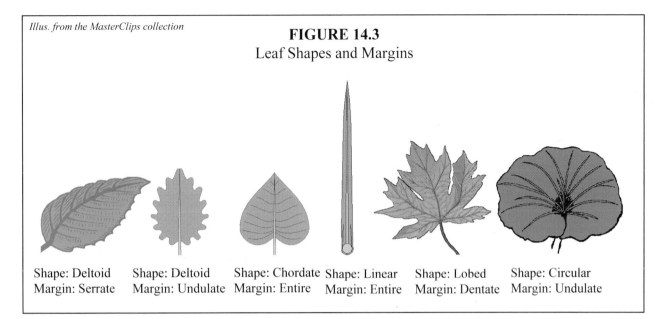

FIGURE 14.3
Leaf Shapes and Margins

Shape: Deltoid
Margin: Serrate

Shape: Deltoid
Margin: Undulate

Shape: Chordate
Margin: Entire

Shape: Linear
Margin: Entire

Shape: Lobed
Margin: Dentate

Shape: Circular
Margin: Undulate

When a leaf has small, sharp serrations around its edge, we say that the margin is **serrate**. If those serrations are larger and smoothly sloping, we say that the margin is **undulate**. When the edge is smooth, we say that the margin is **entire**, and when there are large, sharp serrations, we say that the margin is **dentate**. Notice how the margin can affect the shape. The shapes of the first two leaves might look different to you. That's because of the margin. They both have a deltoid shape, but the margin makes them look different.

The other characteristic used to classify leaves is leaf venation. If you look closely at a leaf, you will find that there are veins which run through it. These veins (comprised of xylem and phloem) form different patterns in different leaves. The patterns can be categorized into two general groups: **parallel venation** and **netted venation**. Netted venation can be further divided into two classes: **pinnate** (pin' ate) and **palmate** (pal' mate). These are illustrated in Figure 14.4.

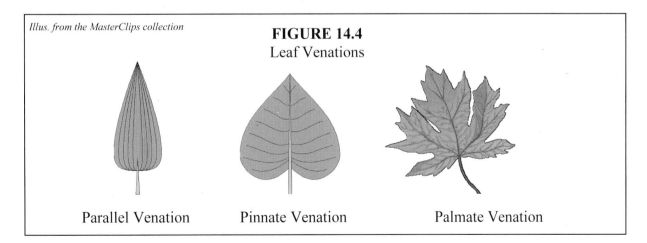

FIGURE 14.4
Leaf Venations

Parallel Venation

Pinnate Venation

Palmate Venation

If a leaf has parallel venation, the veins run up and down the leaf in a roughly parallel fashion. If this is not the case, then the venation is netted. If the veins branch off of one main vein, the netting is called pinnate venation. If, instead, there are several main veins that each have branches, the netting is called palmate venation.

It turns out that the venation of a leaf can tell us something about the type of plant it comes from. In Module #1, you were presented with a biological key that allowed you to determine what class a plant belonged to based on the venation of its leaf. If the venation is parallel, the plant belongs to class Monocotyledonae and is typically called a **monocot**. On the other hand, if the venation is netted, the plant belongs to class Dicotyledonae and is typically called a **dicot**. We will learn later that the distinction between monocots and dicots actually depends on the structure of seed that is produced. However, since all monocots have parallel venation and all dicots have netted venation, the kind of venation you see on a leaf can tell you what kind of seed the plant produces!

Since the venation (and other physical characteristics) of the leaf can tell you so much about a plant, it is important for you to be able to properly identify the shape, margin, and venation of a leaf. Thus, try your hand at using these terms in classifying the leaves shown in "on your own" problem 14.4.

ON YOUR OWN

14.4 Determine the shape, margin, and venation of the following leaves.

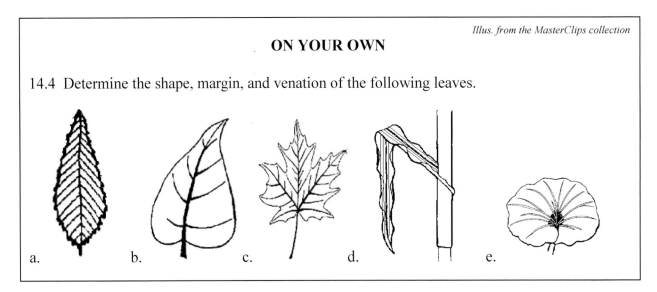

a.   b.   c.   d.   e.

To get more experience identifying leaves, perform the following experiment. Depending on your school schedule, you may have to put this one off for a while. You should perform this experiment in the spring, once new leaves have grown on the trees.

EXPERIMENT 14.1
Leaf Collection and Identification

<u>Supplies:</u>

- Leaf press or old magazines
- Laboratory notebook
- Tree identification book (from library)

Object: To become familiar with the various trees in your area.
           To compare the similarities and differences of these trees.

Procedure:

A. Begin collecting leaf specimens from trees in your area. You should have around 20 different types of leaves. Make good notes about each tree from which you collect a leaf.

B. Allow sufficient time for the leaves to dry. Leaves that are still damp will mold later.

C. Press your leaves in a leaf press or between the pages of old magazines.

D. If you are using magazines, make sure you add weight to the stack of magazines to help press the leaves.

E. For those in a hurry, leaves can be pressed between two layers of wax paper with a low temperature iron. Trim the paper a short distance from the leaf before mounting.

F. Attach the leaves carefully to the pages of your laboratory notebook. Each page should contain one leaf and the following information:

- Leaf mosaic: alternate, opposite, or whorled
- Shape of leaf: deltoid, linear, chordate, lobed or circular
- Venation of leaf: parallel, pinnate, palmate
- Leaf margins: entire, undulate, serrate or dentate
- Bark:
  ⇒ color
  ⇒ rough or somewhat smooth
  ⇒ other outstanding characteristics
- Any evidence of fruits (including nuts)?
- Identification of tree (if possible)

The Microscopic Structure of a Leaf

If we take a closer look at the leaf with a microscope, we can actually see that it is made up of a rather complex microscopic structure. Figure 14.5 shows a magnified image of the cross section of a leaf.

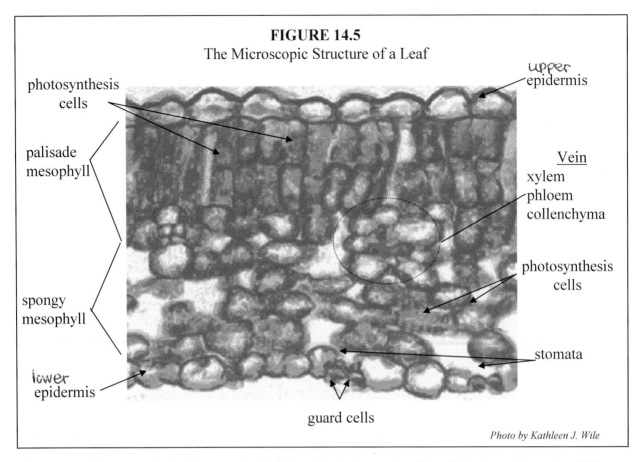

FIGURE 14.5
The Microscopic Structure of a Leaf

photosynthesis cells

palisade mesophyll

spongy mesophyll

lower epidermis

upper epidermis

Vein
xylem
phloem
collenchyma

photosynthesis cells

stomata

guard cells

Photo by Kathleen J. Wile

The top and bottom of a leaf are covered with a single layer of cells called the **epidermis**. This epidermis protects the inner parts of the leaf. Sometimes, the epidermis secretes a waxy substance called a **cuticle** (kyou' tuh cuhl). Leaves that have cuticles are often shiny in appearance. If the epidermis of a leaf does not secrete a cuticle, it often grows hairs. These hairs give the leaf a velvety appearance. The African violet, for example, has velvety-looking leaves that get their appearance from the epidermal hairs. Many plants use these hairs as a defense mechanism. Stinging nettles, for example, have hairs that stick into your skin and inject an irritant. This is an effective means of keeping people and animals from walking through them.

On the underside of most leaves, there are tiny holes called **stomata** (stoh mah' tah). These stomata allow for the exchange of gases with the atmosphere, which is absolutely necessary for the survival of the plant. Each stoma (singular of stomata) is flanked by two cells called **guard cells**. These amazing little cells open and close the stoma. They contain chlorophyll and do photosynthesis to produce specialized sugars. These sugars cause the water pressure inside the cell to change, swelling or shrinking the cell.

When water is in great abundance and photosynthesis is taking place, the guard cells open the stomata. This allows the leaf to take in carbon dioxide for photosynthesis and release oxygen. When water is scarce, however, the leaf cannot afford to have the stomata open, because water tends to evaporate from the leaf through the stomata. Since photosynthesis cannot take place when water is scarce, there is no real reason for the stomata to be open anyway. Thus, under these conditions, the guard cells actually close the stomata. This keeps the leaf from losing too much water. When water becomes abundant again, the guard cells open up, allowing photosynthesis to begin again.

Under the epidermis on both sides of the leaf are **parenchyma** (pair en' kye muh) **tissues**. These tissues are comprised of cells that do the photosynthesis. The parenchyma tissue is comprised of two layers, the **palisade** (pal ih sade') **mesophyll** (mez' uh fil) and the **spongy mesophyll**. The palisade mesophyll is made of oval-shaped cells that are rich in chloroplasts. In this layer of the parenchyma tissue, the cells are packed tightly, so as to maximize the number of cells in the tissue. At first, you might think that this reduces the amount of sunlight that the cells can absorb, because they all block each other's light. The Designer of these cells, however, has worked out an ingenious way around this problem.

The chloroplasts in the cells are constantly moving due to cytoplasmic streaming. This cytoplasmic streaming directs the chloroplasts to the top of the cell. Once at the top of the cell, the chloroplast absorbs all of the sunlight that it can handle and then is moved to make room for the next. In the end, then, the chloroplasts absorb the energy they need when they reach the top of the cell. That way, the cells can be packed tightly, to have as many food-producing cells as possible in the leaf.

Now wait a minute, you might be thinking. In order to perform photosynthesis, the food-producing cells need to absorb carbon dioxide. If the cells are packed together very tightly, there will not be room for the carbon dioxide to get into the cells. Once again, the leaf's Designer has worked a way around this problem. On the bottom of the leaf, the parenchyma tissue is made up of the spongy mesophyll. This tissue has plenty of room for air, because the food-producing cells are packed very loosely. Thus, the stomata open into the spongy mesophyll, so that air can come in and fill the space there. The cells in the spongy mesophyll are arranged so that the air which occupies the space provided for it touches *every food-producing cell in the leaf*! That way, each cell can absorb carbon dioxide for photosynthesis and can release the oxygen that photosynthesis produces.

The veins that you see running through a leaf are made up of three tissues. These tissues come in two types: the vascular tissue, comprised of xylem and phloem, and another tissue called the **collenchyma** (kuh len' kye muh). The collenchyma is made up of thick-walled cells that support the vein. Towards the end of the leaf, the veins get so small that there is no collenchyma tissue any more.

Think about what you have just read. Many engineers have looked at the microscopic structure of a leaf, and they all come to the same conclusion. The leaf is designed in the most efficient way so as to maximize its food-producing capabilities. It packs cells so that the total number of food-producing cells is as large as it can possibly be without hampering the availability of the chemicals and sunlight necessary for photosynthesis. At the same time, the leaf has stomata, which open when the leaf can use the gases in the air and close when the leaf cannot use them! That way, the leaf is exposed to the outside only when it benefits the plant. If exposure to the air does not benefit the plant, then the plant shuts down! This is an incredibly efficient, well-designed food producing machine!

ON YOUR OWN

14.5 A leaf cannot get the carbon dioxide that it needs for photosynthesis. What, most likely, is wrong?

14.6 Why can't the parenchyma be made of two layers of palisade mesophyll?

<u>Leaf Color</u>

Most leaves are green. Why? Well, the chlorophyll in their chloroplasts gives them their color. The more chlorophyll-containing cells there are in a section of leaf, the darker green that section is. Many leaves, for example, have a much deeper green color on the top of the leaf than on the underside. This is because the top of the leaf has the palisade mesophyll, where the chlorophyll-containing cells are packed together tightly. The underside of the leaf, however, contains the spongy mesophyll, where the chlorophyll-containing cells are not tightly-packed. As a result, the underside of the leaf is usually a lighter shade of green than is the top side of the leaf.

Of course, not all leaves are green, are they? Many plants have leaves that are not green at all or are only partially green. In addition, many leaves lose their green color in the fall. Consider, for example, the leaves shown in Figure 14.6.

FIGURE 14.6
Leaves That Are Not Green or Are Only Partially Green

a. Caladium leaf

Photo from the Expert 3000 collection

b. Dogwood leaves

Photo from the MasterClips collection

c. Autumn leaves

Photo from the MasterClips collection

In Figure 14.6a, we show the leaf of a caladium plant. Although there is clearly a lot of green on the leaf, there is also some red. In Figure 14.6b, we show some leaves from a dogwood tree. Many people mistakenly call the white (or pink, depending on the species of dogwood) blossom that you see on a dogwood tree a flower. It is not a flower, however. The flower of a dogwood tree is actually the small green structure in the middle of the white leaves. Thus, the white blossoms that you see there are leaves, not flowers. The dogwood tree, then has both green and white (or pink) leaves. Finally, in Figure 14.6c, we see a tree in autumn. The leaves are no longer green at all, but a mixture of red, purple, and yellow.

Where do all of these colors come from? Well, as we said before, the green is due to chlorophyll. Remember, in Module #6, we called chlorophyll a "pigment." It gives the leaf color, although its real job in the leaf is to absorb sunlight for photosynthesis. Plants contain pigments other than chlorophyll, however. In Module #6, we learned that cells contain **plastids**. These plastids hold pigments, starches, and oils. Well, the plastids of some plants contain a group of pigments called **carotenoids** (kehr uh ten' oyds). These carotenoids typically have yellow or orange hues. The orange color of pumpkins and carrots, for example, comes from one type of carotenoid. The yellow in yellow peas or corn comes from another type of carotenoid. Well, it turns out that most leaves contain carotenoids as well. In most leaves, the green color of the chlorophyll overwhelms the oranges and yellows of the carotenoids. In other plants, however, the amount of carotenoids is greater, and the leaf picks up some color from them.

In the vascular tissue of the leaf, there is another pigment called **anthocyanin** (an tho sye' uh nin). This interesting pigment has different colors, depending on the pH of the leaf tissue. To give you an idea of what this means, perform Experiment 14.2.

EXPERIMENT 14.2
How Anthocyanin and pH Help Determine Leaf Color

Supplies

- Red (some people call it purple) cabbage (just a few leaves)
- Stove
- Stirring spoon
- Boiling pan
- White vinegar (It must be clear. Apple cider vinegar will not work.)
- Clear Windex® (or any clear, ammonia-containing cleaning liquid)
- 2 medicine droppers
- 2 small cups or glasses
- 1 small glass (It must be see-through!)
- A white sheet of paper (preferably without lines)
- Measuring cups (1 cup and 1/4 cup)

Object: To see how anthocyanin changes color with differing pH levels

Procedure:

1. Put one cup of water and a few leaves of red cabbage into a boiling pan.

2. Turn on the heat and bring the water to a boil. Allow the water to boil vigorously for 5 minutes, stirring continuously.

3. Allow the water to cool for a few minutes.

4. While the water is cooling, pour some clear Windex into one of the small glasses or cups and some white vinegar into the other. Place a medicine dropper in each.

5. Once the boiled water has cooled a little, pour it into the measuring cup until you have 1/4 cup of the boiled water. Be sure to remove any leaves that got into the cup.

6. Place the small glass on the white sheet of paper and pour the 1/4 cup of boiled water into it. Observe the color. Record that color in your notebook.

7. Depending on the type of water you have in your home, the boiled water may be one of a few colors. Add 5 drops of clear Windex and swirl the glass so that the Windex mixes in. Is it green yet? If it is, stop adding Windex. If not, add the Windex 5 drops at a time, swirling each time after adding the drops, until you get a green color. Even if your water was green as soon as you poured it into the glass, you still need to add those first 5 drops of Windex. Record how many drops of Windex you added, and that the color is now green.

8. Now add 2 drops of vinegar. Swirl the glass so that the vinegar mixes in. Note the color. Record the fact that you added two drops of vinegar and record the color.

9. Continue to add vinegar 2 drops at a time, swirling each time. Record the total number of drops of vinegar added and the color after adding those drops. Continue to do this until the solution has reached a pink hue and it has not noticeably changed from that shade of pink for 3 additions of vinegar.

10. How many different colors did you see? You should have seen 4 distinct colors: green, blue, purple, and pinkish. If you want, you can reverse what you did by now adding clear Windex. You will have to add the Windex 5 or more drops at a time rather than 2 drops at a time, and the colors will be weaker because the solution is becoming less concentrated the more liquid you add. Nevertheless, with enough Windex, you should be able to get back to the green color that you started with.

Conclusion: What happened in this experiment? Well, red cabbage contains anthocyanin. When you boiled the red cabbage leaves in water, you extracted the anthocyanin in the leaves and dissolved it in the water. Clear Windex is a base. As we learned in Module #5, bases have a pH greater than seven. In fact, the pH of Windex is about 10. When you initially started, you gave the boiled water (which contained anthocyanin) a pH of more than 9. At that pH, anthocyanin is green; thus, the solution was green. Vinegar is an acid. As you began adding acid, you started lowering the pH. When the pH dropped below 9, the anthocyanin turned blue. As you continued to add more vinegar, the pH continued to drop. When it dropped below 7, the anthocyanin turned purple. Pure vinegar has a pH of about 4. As you added more and more vinegar, you got the pH closer and closer to 4. Once it got near 4, the anthocyanin turned pink. We see from this experiment, then, that a single pigment (anthocyanin) can give leaves different colors depending on the pH that the leaf has.

So now you know why some leaves are never completely green. Depending on the amount of carotenoids in the leaf, it might have an orange or yellow hue. If this orange or yellow hue mixes with the green of chlorophyll, then another color might develop. Finally, depending on the pH of the leaf, anthocyanin will provide a blue, purple, or pink color to the leaf.

What we haven't explained yet, however, is why leaves that are green in the spring and summer turn different colors in the fall. The explanation for that is partly rooted in what we have discussed, but there is another part to the explanation as well. Before we launch into that explanation, however, we need to make one thing clear. As you are well aware, not all leaves turn colors in the autumn. The needles on evergreen trees (those needles are the leaves for the tree), for example, never turn colors in the autumn. Also, the leaves on tropical trees do not change color in the autumn. Only certain trees have leaves that change color in the autumn. They are the trees that lose all of their leaves before winter. We call those trees **deciduous** (duh sid' you us) trees.

Deciduous plant - A plant that loses its leaves before winter

Deciduous plants lose their leaves in order to conserve water throughout the winter.

At the base of each petiole in deciduous trees, there is a thin layer of tissue called the **abscission** (ab cih' shun) **layer**. These cells perform a very specialized task. When the days start getting shorter, the cells in the abscission layer begin to block off the xylem and phloem running through the petiole of the leaf. Once they have succeeded in blocking off the xylem and phloem, they begin to die. As the cells in the abscission layer die, the layer itself begins cracking. The cracks eventually become so severe that the leaf falls off of the tree under its own weight. Thus, the cells in the abscission layer have only one job: wait for the days to turn shorter, block off the leaf's supply of water and nutrients, and then die. If you have a deciduous indoor plant, you will notice that it loses its leaves in the autumn as well. This is because the abscission layer is not affected by temperature. It begins doing its job when the time it is exposed to the sun begins decreasing.

Once the abscission layer has blocked off the water and nutrient supply for the leaf, the leaf can no longer produce chlorophyll. As a result, the green color slowly fades away. This reveals whatever pigments were covered by the vibrant green of the chlorophyll. At the same time, when the leaf stops doing photosynthesis because it has no chlorophyll, other chemical reactions begin to occur. Under the right conditions (sunny and cool weather), these chemical reactions will produce anthocyanin and change the pH in the leaves. This, in turn, changes the color of the anthocyanin in the leaf. In areas of the world in which there are sunny, cool autumns, you will see brilliant reds, yellows, and purples in the leaves. This is mostly due to the anthocyanin in the leaf and the pH that the leaf has as a result of the chemical reactions which are occurring in it. In parts of the world where the autumn is cloudy and warm, the leaves do not have these brilliant colors, because anthocyanin has not been produced.

Some leaves do not produce pigments other than chlorophyll. As a result, when they die, they turn brown. This brown color is the result of a chemical called **tannic** (tan' ik) **acid**.

Tannic acid is one product of the breakdown of plant cell contents. With no pigments to mask the color of tannic acid, the leaf is brown. If the leaf contains a high concentration of tannic acid, the leaf can be boiled, and the tannic acid will dissolve into the water, making tea. Since all dead leaves have tannic acid in them, you could make tea out of any leaf. However, when you boil leaves, chemicals other than tannic acid dissolve in the water as well. If human taste buds respond pleasantly to those chemicals, then a pleasant-tasting tea is made. If not, the tea can be quite revolting! Thus, the next time you drink tea, just remember that you are actually drinking the remains of dead leaves!

ON YOUR OWN

14.7 There are a few leaves in which the stomata and the spongy mesophyll are on the top side of the leaf while the palisade mesophyll is on the bottom. In these leaves, which side will have the darker green color?

14.8 If a leaf isn't green, does that mean there is no chlorophyll in it?

14.9 If a green leaf has no abscission layer, what color will the leaf be in the winter?

<u>Roots</u>

The roots of a plant perform three very important functions. First, they absorb water and nutrients from the plant's surroundings and transport them to where they are needed. Second, they anchor the plant. Finally, they are often used as a place for food storage. Many times, we exploit that last function for our own good. When you eat a carrot, you are actually eating the root of the carrot plant. The substances in that carrot which provide you with nutrition are actually food that the plant had stored for later use.

Most roots are below ground, in the soil. This is not true for all plants, however. The roots of an epiphytic (ep uh fie' tik) orchid, for example, wind around the branches of tropical trees. They absorb the materials that collect in the cracks of these branches. Other, parasitic plants, such as the mistletoe, for example, sink their roots into the host. These roots steal the nutrients that the host had absorbed for itself. Other plants, such as ivy, have roots that hold them to rough surfaces like brick walls or rough tree bark. Even though we will discuss roots as if they all exist under the soil, it is important to note that there are exceptions, such as these.

There are basically two kinds of root systems in plants: **fibrous root systems** and **taproot systems**. When a seed begins to sprout, the first root that comes out is called the "primary root." If the primary root continues to grow and stays the main root, then the plant has a taproot system. Carrot plants have a taproot system. The carrot is the primary root, and it continues to grow as the plant's main root. Aside from a few, tiny branches, the carrot is the predominant root in the plant. If, on the other hand, the primary root begins branching and branching until the root system looks like an underground "bush," then the plant has a fibrous root system.

The vast majority of plants need significantly more surface area in their root system than in their leaves. Thus, the root systems of most plants are significantly larger than the part of the plant that exists above the soil. Corn, for example, usually reaches a height of 8-10 feet. The fibrous root system, however, will have so many roots that, if you attached them end-to-end, they would stretch more than 150 feet! This fact is especially true of trees. The roots of a tree, if attached end-to-end, will generally be 5 to 10 times longer than the tree is tall.

A magnified image of a **longitudinal cross section** of a young, primary root is shown in Figure 14.7 In a longitudinal cross section, we take a slice of the root along its length. Then, we look at it as if we are looking at the side of the root.

FIGURE 14.7
Magnified Image of a Longitudinal Cross Section of a Primary Root

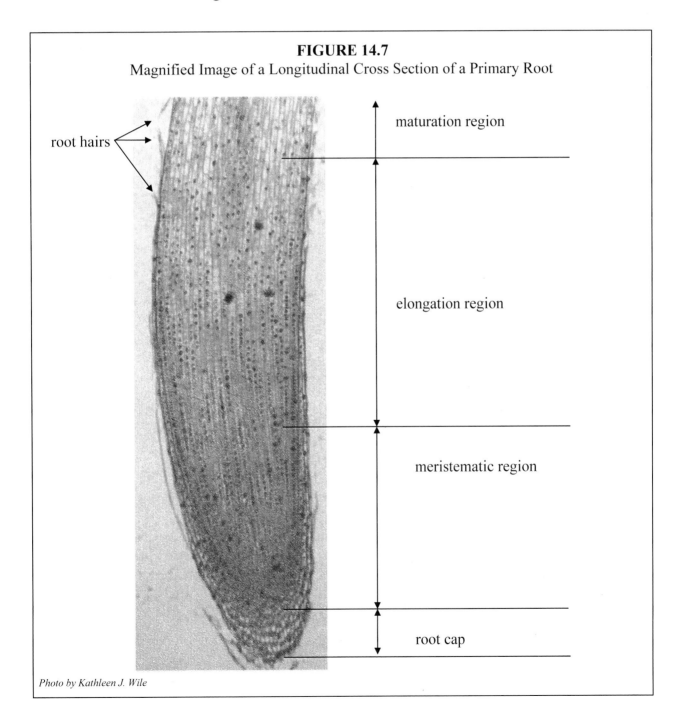

root hairs

maturation region

elongation region

meristematic region

root cap

Photo by Kathleen J. Wile

Notice that the root is split into four basic regions: the **root cap**, the **meristematic region,** the **elongation region**, and the **maturation region**. The root cap is composed of dead, thick-walled cells. These cells protect the root as it shoves its way down into the soil. Just above the root cap is the meristematic region. In this part of the root, undifferentiated cells carry on mitosis. This is where most of the growth of the root takes place. After all, in order to grow, a root needs more cells. It makes sense, then, that the majority of growth will take place where mitosis occurs. In the elongation region of the root, cells are beginning to differentiate into root cells. They stretch out, forming large vacuoles inside the cells. Since the cells are stretching out, some growth takes place in this portion of the root. Finally, in the maturation region, cells are becoming fully differentiated. Often, **root hairs** are produced. These hairs increase the surface area of the root, allowing it to absorb more water and nutrients from the soil.

Although a longitudinal cross section of the root, as shown in Figure 14.7 is instructive, another interesting way to look at a root is by taking a **lateral cross section** of the maturation region. In a lateral cross section, we slice a thin layer of the root from side-to-side so that the center of the root is the center of the slice. We then look at it as if we are looking straight down the root. This tells us what the inside of a root is made of. Figure 14.8 shows just such a cross section.

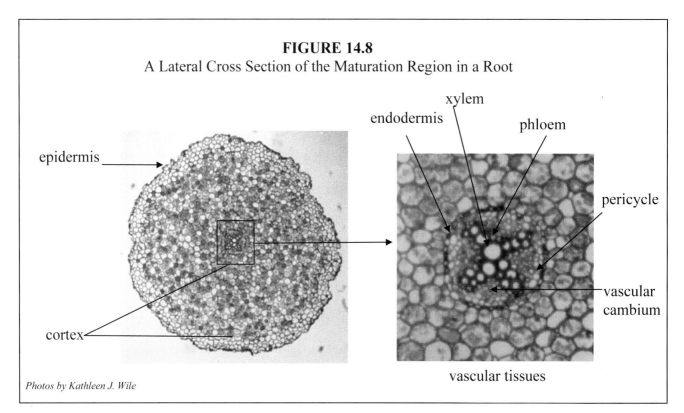

FIGURE 14.8
A Lateral Cross Section of the Maturation Region in a Root

epidermis

endodermis

xylem

phloem

pericycle

vascular cambium

cortex

vascular tissues

Photos by Kathleen J. Wile

As you can see from the figure, the root is protected by an epidermis. The cells inside the epidermis are called the **cortex**. This is where substances are stored for later use. Towards the center of the root, there is another one-cell-thick layer called the **endodermis**. These cells surround the **vascular chamber**, so that any substances which go to the xylem and phloem must first pass through them. This is the way the root controls what substances enter the plant. If the endodermis does not let a substance through, there is no way it can reach any other part of the plant. Thus, the endodermis guards the xylem and phloem, keeping out unwanted substances.

Inside the vascular chamber are the xylem and phloem, which transport nutrients and water throughout the plant. In the root of a dicot (the plants that have leaves with netted venation), the xylem usually form an X-shape. The phloem then fit in between the arms of the "X." Between the xylem and the phloem lies the **vascular cambium** (cam' bee uhm). This tissue can become either xylem or phloem, depending on what the root needs. In between the endodermis and the vascular tissues is the **pericycle**. These cells are undifferentiated. If the root sends out a branch, the branch will form in this tissue and then "break out" of the endodermis, eventually traveling through the cortex and breaking out of the epidermis, thus forming a new root branch.

ON YOUR OWN

14.10 A twelve-foot high plant has a root system that travels to a soil depth of only three feet. Does this plant have a taproot system or a fibrous root system?

14.11 A group of leaf cells that are about to undergo mitosis are examined and compared to root cells. In which region of the root will the root cells most resemble these leaf cells?

14.12 If a root contains little or no cortex tissue, what function can you conclude that the root does **not** perform?

Stems

Unlike roots, stems are about as diverse as the plant kingdom itself. Stems can be erect (as they are in trees and flowers), climbing (as the are in creeping vines), prostrate (as they are in watermelons and cucumbers), or even subterranean (as in the potato). They can be either woody or herbaceous. Regardless of the variety, however, stems perform three basic functions in a plant. First, they support and manufacture the plant's leaves. Second, they conduct water and nutrients to and from the leaves. Finally, they carry on photosynthesis. In some plants, stems only carry on photosynthesis when they are young. In other plants, like the cactus, the stems are actually the primary photosynthesis organs.

Herbaceous Stems

The herbaceous stems of dicots and monocots are slightly different. In Figure 14.9, a lateral cross section of a herbaceous monocot and a lateral cross section of a herbaceous dicot are shown.

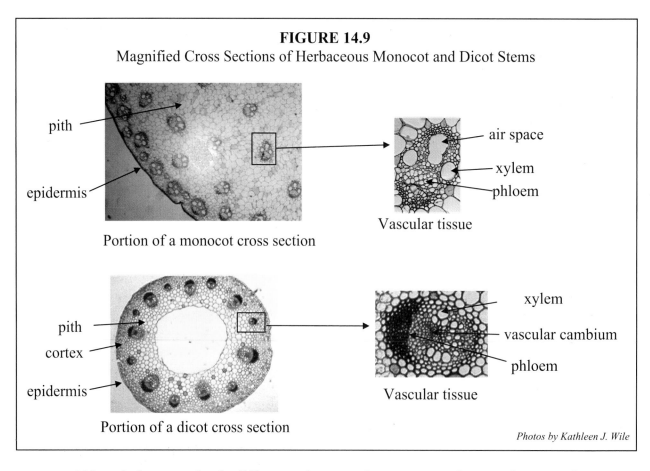

FIGURE 14.9
Magnified Cross Sections of Herbaceous Monocot and Dicot Stems

pith

epidermis

Portion of a monocot cross section

air space

xylem

phloem

Vascular tissue

pith

cortex

epidermis

Portion of a dicot cross section

xylem

vascular cambium

phloem

Vascular tissue

Photos by Kathleen J. Wile

Although there are clearly differences between the two types of stems, there are also some similarities. Both are covered with an epidermis, which protects the stem. They also both contain **fibrovascular** (fie broh vas' kyou ler) **bundles**. These bundles contain the xylem and phloem that transport substances throughout the plant. Throughout each stem is a region of cells called the **pith**, which conduct substances to and from the vascular bundles. In young stems, the pith usually carries on photosynthesis as well. In some stems, the pith is also used to store substances for later use.

The main differences between monocot and dicot stems are the fibrovascular bundles and the ways in which they are arranged. In monocot stems, the bundles are located throughout the stem. In dicot stems, however, they tend to form a ring near the outer part of the stem. In between this ring and the epidermis lies the other difference between monocot and dicot stems. In dicot stems, a region of cortex tissue lies between the epidermis and the fibrovascular bundles. In a monocot stem, there is no cortex tissue.

Notice that the xylem and phloem within the fibrovascular bundles are arranged quite differently. In a monocot stem, the xylem and air spaces are arranged to form what almost looks like a face. The phloem form the "forehead" of the face. In the dicot stem, however, the xylem are smaller and there are several of them. The phloem are bundled together in one region of the fibrovascular bundle.

One other difference between monocot and dicot stems is the presence of a **vascular cambium**. In the fibrovascular bundles of a dicot stem, the vascular cambium can form new xylem or phloem if the stem needs the ability to transport more materials. The amount of new xylem and phloem that can be produced is limited, however. The epidermis of the stem cannot grow once it is mature. Thus, if too many new xylem or phloem are created, the stem will crack through the epidermis, exposing its inner tissues to the environment. This will usually kill the stem.

Woody Stems

Woody stems have much the same features of herbaceous stems, but there are, of course, a few differences. One of the major differences between herbaceous stems and woody stems is the **bark**. The bark of a woody stem is actually composed of two layers: the **inner bark** and the **outer bark**. The inner bark is composed of phloem and cortex tissue. In between the inner bark and the outer bark is a layer of tissue called the **cork cambium**. This layer continually produces **cork cells**. These cork cells die quickly, and are impenetrable to water, gases, and most parasites. They form the **outer bark**.

The formation of bark allows a woody stem to continue to grow, unlike most herbaceous stems. Remember, if a herbaceous stem grows too much, the epidermis cracks, exposing the inner parts of the stem to the environment. In a woody stem, however, the growth causes the outer bark to crack and break, but that's okay, because the cork cambium simply produces new outer bark to protect the stem. Now as far as we are concerned, the trunk of a tree is really just the main stem of the tree. So now you know why the bark on the outside of a tree is cracked and rough. It is the result of the tree "outgrowing" its bark shield and breaking through it.

Not all trees have rough bark, however. The white birch tree, commonly found in Canada but also found in the United States, has thin, white bark that peels off. When this bark peels off, the cork cambium just produces new bark to take its place. This gives the stems of the white birch a very smooth feel. Because the bark is so thin, it is also an excellent fire starter. Boy Scouts and Girl Scouts are taught to find a white birch and peel off its bark in order to get easy-to-light kindling for a fire.

Because there is really no limit to the growth of a woody stem, new xylem and phloem must always be produced inside the stem. During the spring when there is plenty of water, the xylem produced are quite large. As the summer goes on and water becomes scarce, the xylem produced become smaller. When the xylem cells die, they form what we call the "wood" of the tree. Since the xylem produced in the spring are much bigger than those produced in the later part of the growing season, the wood of the tree is lighter when it is produced in the spring and darker when it is produced in the late summer. These alternating areas of light and dark form what we call **tree rings**. If you look at a lateral cross section of a woody stem, you can tell how old it is by counting these rings. Also, the thickness of the rings can tell you a lot about the weather conditions during that year of formation.

In case this little fact has slipped past you, we need to emphasize it. In herbaceous stems, the xylem and phloem are packed together in little bundles. In woody stems, however, this is not the case. The xylem are in the inner part of the stem, what we normally call the "wood." The phloem, however, are a part of the tree's inner bark. Thus, the phloem are always near the outside of the stem, while the xylem are always in the inner part of the stem. This leads us to a discussion of **girdling**.

Girdling - The process of cutting away a ring of inner and outer bark all the way around a tree
trunk

When a tree is girdled, it at first appears that the tree is okay. For quite a while, the leaves of the tree will stay green and the tree will grow. However, it never grows *below* the point at which it is girdled.

Remember, the xylem (which carry nutrients and water up the stem) are in the inner part of a stem (or tree trunk), while the phloem (which carry nutrients and water down the stem) are a part of the inner bark. If you cut away a ring of the inner bark, the water and nutrients can still move up the tree (in the xylem), but they cannot move down the tree past the ring, because all of the phloem have been cut. As a result, the roots can still send water and nutrients up to the leaves, but the leaves cannot send food back down to the roots. Eventually, the roots will starve to death, and that will kill the rest of the tree. This process can take several years, however.

Specialized Stems

As we mentioned before, some stems actually exist underground. They are sometimes mistaken for roots, but they are, nevertheless, stems. The onion plant, for example, produces underground **bulbs**. These bulbs are not roots. They are, instead, a collection of underground leaves that sprout from an underground stem. These leaves do not perform photosynthesis, of course, because there is no light underground. Instead, the leaves are used for storage. Another stem often mistaken for a root is the **tuber**. Potato plants produce tubers, which are underground stems that are used to store excess food. We harvest these tubers, of course, and we call them potatoes.

ON YOUR OWN

14.13 If a stem has no cork, is it woody or herbaceous?

14.14 If a stem has no limits to its growth, is it woody or herbaceous?

14.15 If a stem has xylem and phloem packed together in bundles, is it woody or herbaceous?

To better familiarize yourself with the roots, stems, and leaves of plants, perform the following microscope experiment. If you do not have a microscope, skip the experiment.

EXPERIMENT 14.3
Dicot/Monocot Cross-section of Roots, Stem and Leaf

Supplies:

- *Zea mays* (corn) cross section of stem prepared slide
- *Zea mays* (corn) cross section of root prepared slide
- *Ranunculus* (buttercup) cross section of stem prepared slide
- *Ranunculus* (buttercup) cross section of root prepared slide
- Leaf cross section with vein prepared slide
- Microscope
- Lab notebook
- Colored pencils

Object: To observe the anatomical structure of a dicot and a monocot and compare the two.

Procedure:

A. Set up your microscope and observe the cross section of the leaf. You should observe this on low power first. Then you can increase the magnification to find the optimum magnification with which to view the slide (x100 is usually best). Note that there are a few different kinds of leaves shown. Find one that looks like the cross section presented in Figure 14.5. Draw and label the leaf including the following parts:

1. Upper epidermal cells
2. Palisade mesophyll cells
3. Veins
4. Spongy mesophyll cells
5. Lower epidermal cells
6. Guard cells
7. Stomata

B. Now that you have seen one which looks like that presented in Figure 14.5, look at the others. Try to find the same parts on those cross sections.

C. Observe the slide *Ranunculus* cross section of the root. *Ranunculus* is a dicot. Draw and label what you see including the following parts:

1. Xylem
2. Vascular cambium
3. Phloem
4. Pericycle
5. Endodermis
6. Cortex
7. Epidermis

D. Observe the slide *Zea mays* cross section of the root. *Zea mays* is the genus and species of a common corn plant, which is a monocot. Try to draw and label the following. This might be difficult, since we only showed you a dicot when we discussed roots. Nevertheless, give it a try!

      1. Xylem
      2. Phloem
      3. Endodermis
      4. Cortex
      5. Epidermis

E. Note the differences between the monocot root and the dicot root.

F. Observe the slide *Zea Mays* cross section of the stem. Draw and label the following parts:

      1. Phloem
      2. Xylem
      3. Pith
      4. Fibrovascular bundle

G. Observe the slide *Ranunculus* cross section of the stem. Draw and label what you see.

      1. Phloem
      2. Xylem
      3. Vascular cambium
      4. Cortex
      5. Pith
      6. Fibrovascular bundles

H. Note the differences between the monocot stem and the dicot stem.

I. Clean up and return your equipment to the proper place.

If plant anatomy has caught your interest, you might want to look at a book entitled *The Visual Dictionary of Plants*. This book, which is part of the *Eyewitness Visual Dictionary* series, can most likely be found in your local library.

Classification of Plants

There is a *lot* of controversy about how to classify plants. Thus, the classification scheme that you learn here might be different than other ones that you study later on. In fact, some classification schemes that you learn might even put algae in kingdom Plantae. After all, since algae are photosynthetic, they do have something in common with the plants. In this book, however, we consider plants to be multicellular creatures. Thus, most algae are not really plants. Even the algae that are multicellular do not have the specialized tissues that plants do, so we do not consider them plants. Nevertheless, some biologists do. Depending on the next biology book you read, then, there may be a lot of difference between the classification scheme that we give you and the one that it gives you.

The plant kingdom can be split into two groups: plants with vascular tissue and plants without vascular tissue. All plants without vascular tissue exist in single phylum: phylum Bryophyta (bry oh fie' tuh). Because bryophytes do not have vascular tissue, they cannot grow to be very tall. After all, the taller the plant, the farther nutrients must be transported. Without vascular tissue, it is difficult to transport nutrients very far. There are many, many phyla of vascular plants. We will discuss three of them shortly.

Phylum Bryophyta: The Non-Vascular Plants

The major representatives from phylum Bryophyta are the mosses. Although it is tempting to call any small, green clump a moss, to a biologist, the term **moss** means something quite particular. A moss is composed of many, tightly packed individual plants. These plants are composed of **leafy shoots** and **rhizoids**. The leafy shoots are tiny stems with even tinier, leaf-like structures. These leaflike structures are but one cell thick, and they directly absorb the nutrients that they need from the environment, while they perform photosynthesis. Nutrients can travel through the leafy shoot by falling through the spaces in between the cells of the stem, much like water can travel through a paper towel.

The rhizoids of a moss look like roots, but they are not. Roots can absorb nutrients and then send them to the other parts of the plant. Since mosses have no vascular tissues, the rhizoids cannot send nutrients anywhere. Instead, the rhizoids are simply strands of tissue used to anchor the moss. Since the rhizoids cannot absorb water from the soil and send it anywhere, mosses are dependent on absorbing the water that collects on their leafy shoots. This means that in order to survive, mosses need very moist environments.

Mosses have an interesting lifecycle which biologists often describe as **alternation of generations**.

Alternation of generations - A lifecycle in which sexual reproduction gives rise to asexual reproduction, which in turn gives rise to sexual reproduction

In this lifecycle, the leafy shoot has either male or female reproductive structures. In some rare cases, a single leafy shoot might have both, but usually it has either one or the other. The egg is released into water that has fallen on the leafy shoot and then the sperm are released to swim towards it. Once fertilization occurs, the zygote grows into a stalk that produces spores. While it is growing, the developing zygote actually survives by being a *parasite on the leafy shoots*. In some ways, then, the offspring of a moss actually eats its parents! When the offspring develops, it forms spores which it releases into the wind. When the spores land and the conditions are right, they develop into new leafy shoots.

Do you see what's going on here? The leafy shoot reproduces sexually, by forming gametes. As a result, the leafy shoot is called the **gametophyte** (gam ee' toh fite) generation. The offspring from this sexual reproduction, however, reproduces *asexually* by releasing spores. Thus, the offspring of the sexual reproduction is called the **sporophyte** (spoor' oh fite) generation. When the spores begin to develop, they form leafy shoots again, which are the gametophyte generation.

Do you see why we call this "alternation of generations?" In the first generation, the moss is a sexually-reproducing gametophyte. In the second generation, it is an asexually reproducing sporophyte. In the end, then, the generations alternate between sexual and asexual reproduction! Since the leafy shoot is what we typically see when we examine mosses, we say that the gametophyte is the **dominant generation**.

Dominant generation - In alternation of generations, the generation that occupies the largest portion of the lifecycle

Alternation of generation, although seemingly exotic, happens in phylum Pterophyta (a phylum we will study in a moment) as well.

Although mosses are rather common, they have few uses. About the only kind of moss that is of economic importance to us is the peat moss. These mosses grow floating at the top of a pond. When the moss dies, it sinks to the bottom. Eventually, the dead moss fills up the pond, turning it into a bog. Eventually, plants begin to grow in the peat moss of the bog and, after a while, there is no evidence that a pond was there at all. Farmers use peat moss to make their soil more fertile. It can also be used to pack plants for shipment. Finally, dried peat moss can be burned as fuel.

ON YOUR OWN

14.16 If you see a large plant, can it be from phylum Bryophyta?

14.17 You study a moss that reproduces by making spores. If you study the offspring of this reproduction, how will that offspring reproduce?

Vascular Plants

The vast majority of what you call "plants" are vascular plants. Because they have vascular tissues, these plants can grow to be quite large. Sequoia trees in California for example, can grow more than 250 feet high and have trunks that are as large as 85 feet in circumference! There are many phyla of vascular plants, but we want to concentrate only on the three that contain the vast majority of living species in Creation.

Phylum Pterophyta

The members of phylum Pterophyta (ter uh fye' tuh) are commonly called "ferns." Ferns can be found either on the forest floor, growing on other trees, or with slender trunks that can reach heights of up to 60 feet. Ferns are separated from other vascular plants because they do not produce seeds. Instead, they have an alternation of generations lifecycle. Some examples of ferns are shown in Figure 14.10

FIGURE 14.10
Various Types of Fern

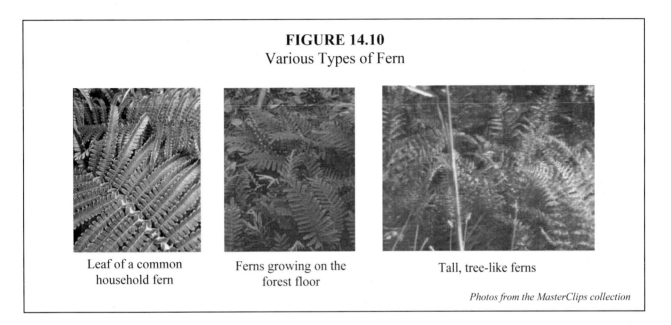

Leaf of a common
household fern

Ferns growing on the
forest floor

Tall, tree-like ferns

Photos from the MasterClips collection

Ferns usually grow from a stem that is either underground or attached to another tree. If the stem is underground, it produces roots. If the stem is attached to another tree, the fern does not act as a parasite. Instead, the stem produces root hairs that absorb the water and nutrients that have collected in the cracks of the tree's outer bark. There are a few species of fern that look like a tree. They have slender trunks that support the stems and leaves. Unlike tree trunks, fern trunks are not a solid woody mass. Instead, they are made up of a network of hard stems. Ferns are typically delicate and often die as a result of even slight environmental changes. Although popular houseplants, many people find it hard to keep them alive, because of their delicate nature.

Ferns are also characterized by their alternation of generations lifecycle. In the fern, however, the sporophyte generation is the dominant generation. If you look closely at a fern leaf, you might see what looks like insect eggs on the underside of the leaf. In fact, these are **sorri** (sore' eye), which produce spores. The spores, when released and conditions are right, will mature into a **prothalus** (pro thal' us), which is a one-cell thick, heart-shaped structure. This is the gametophyte generation of the fern.

ON YOUR OWN

14.18 A biologist studies a plant that has an alternating generations lifecycle. If the dominant generation is the sporophyte generation, is it most likely a fern or a moss?

14.19 A student sees a fern growing on the branch of another tree. The student says that the fern is obviously a parasite. Is the student correct? Why or why not?

Phylum Coniferophyta

The evergreen tree is general name for members of phylum Coniferophyta (con uh fur' oh fye' tuh). These trees, typically called "conifers," reproduce by forming cones. The pinecone, which nearly everyone has seen at one time or another, tells you that pine trees are conifers.

A pine tree actually produces two kinds of cones. The first, which everyone seems to notice, are the large pinecones. These are what biologists call **seed cones**, and they are the female reproductive organ of the pine tree. The seed cones contain the egg cells. On the tips of most pine branches, little knobs develop. Those knobs are actually **pollen cones**, which contain **pollen**.

Pollen - A fine dust that contains the sperm of seed-producing plants

When the pollen cones release their pollen, it is carried by the wind. Some of the pollen will land in seed cones. If the seed cone happens to be on the same tree as the pollen cone was, the result is called **self-fertilization**, which we talked about in Module #8. The seed cone then closes, and the sperm attempt to fertilize the egg. This may take several months. Once fertilization occurs, a seed begins to form. When the seed is ready and conditions are right, the seed cone opens again, releasing the seeds. The seed cones and pollen cones of a conifer are shown in Figure 14.11

FIGURE 14.11
The Reproductive Organs of a Conifer

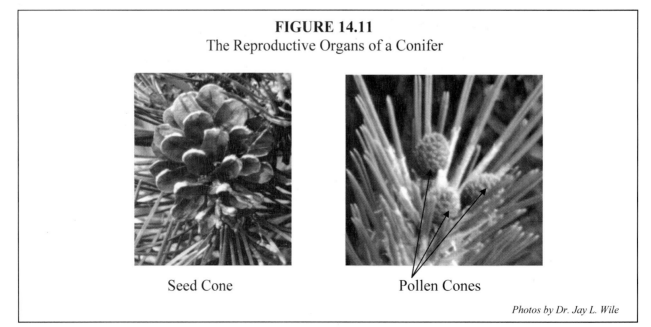

Seed Cone   Pollen Cones

Photos by Dr. Jay L. Wile

Although the pine trees are the best known conifers, there are others. The giant redwood trees and sequoia trees in California are conifers. So are the juniper shrubs and the cypress trees. The slow-growing yew tree, which is typically prized by gardeners as an ornamental tree, is also a conifer.

ON YOUR OWN

14.20 Suppose a conifer self-fertilizes. From a genetic point of view, is this the same as asexual reproduction? Why or why not?

<u>Phylum Anthophyta</u>

Members of this phylum are often called the "flowering plants." This is because anthophytes all have flowers in which their seeds are produced. Although you usually think of flowers as beautiful blossoms, it is important to note that not all flowers are brightly-colored and beautiful. If you pass a corn field towards the end of the corn season, you will see tassels on top of the corn. Those are actually the flowers of the corn plant.

The seeds produced by anthophytes are covered in a structure called the **ovary**. In animals, the ovary is the female gonad. In plants, however, the ovary is the protective coating that covers the seed. It is more commonly called the **fruit** of the plant. Apples, cherries, and tomatoes, then, are fruits. In addition, the pods that peas grow in are fruits, as are the kernels that corn seeds grow in.

Phylum Anthophyta is split into two classes: class Monocotyledonae (mon uh kaht' uh lee' doh nay) and class Dicotyledonae (dye kaht' uh lee' doh nay). These are the monocots and dicots that you read about earlier. Although there are many differences between monocots and dicots (stem structure, root structure, and leaf venation), the fundamental difference between them is the way that the seed is constructed. All seeds produced in phylum Anthophyta have one or two **cotyledons** (cot uh lee' dunz).

<u>Cotyledon</u> - A "seed leaf" which develops as a part of the seed. It provides nutrients to the developing seedling and eventually becomes the first leaf of the plant.

Monocots (such as corn) have a single cotyledon whereas dicots (such as peanuts) have two cotyledons.

This fundamental difference between monocots and dicots leads to a host of other differences. As you already know, the venation of a leaf in a monocot is parallel whereas it is netted in a dicot. The structures of the stem are different in monocots and dicots. Typically, monocots have fibrous root systems whereas dicots have taproot systems. Finally, monocots usually produce flowers in groups of 3 or 6 while dicots produce flowers in groups of 4 or 5.

The ways in which monocots and dicots reproduce are a bit complicated, so we will study them in detail in the next module.

ON YOUR OWN

14.21 Construct a biological key for the classification of plants. Assume that the specimen you are examining is definitely a part of kingdom Plantae. For phyla other than Anthophyta, simply stop at the phylum level. If the plant is a part of phylum Anthophyta, then classify it down to the class level, using the fundamental distinction between monocots and dicots.

ANSWERS TO THE OWN YOUR OWN PROBLEMS

14.1 <u>They are most likely herbaceous plants</u>, because herbaceous plants are annuals.

14.2 <u>It is a vegetative organ</u>. All roots are vegetative organs.

14.3 <u>Meristematic tissue will not be present</u>. Meristematic tissue is the tissue that is undergoing mitosis. If a region of a plant is not undergoing mitosis, no meristematic tissue will be there.

14.4

Letter	Shape	Margin	Venation
a.	Deltoid	Serrate	Pinnate
b.	Chordate	Entire	Pinnate
c.	Lobed	Dentate	Palmate
d.	Linear	Entire (It might look like undulate, but the leaf is just wrinkled around the edges.)	Parallel
e.	Circular	Undulate	Pinnate (This is a tough one. You might think it's parallel, but there is actually a vein in the middle, from which the other veins sprout.)

14.5 <u>The stomata are not opening</u>. This is a problem with the guard cells.

14.6 <u>If both layers were palisade mesophyll, there would be no room for carbon dioxide to get into the leaf or oxygen to get out</u>.

14.7 <u>The underside of the leaf will be darker</u>, because in these leaves, the chlorophyll-containing cells are packed tighter on the underside of the leaf.

14.8 <u>No</u>. The color of the chlorophyll may be masked by other pigments.

14.9 Without an abscission layer, there is nothing to cut off the flow of nutrients. This means the leaf will not fall off of the tree, and will remain <u>green all winter</u>.

14.10 Since almost all plants have roots that, when stacked end-to-end, are much longer than the plant itself, this plant must have a <u>fibrous root system</u>. That's the only way a root system which goes only 3 feet deep will be longer than the plant.

14.11 The cells in the <u>meristematic region</u> of the root will look most like those cells. After all, the meristematic region is where the root cells are undergoing mitosis.

14.12 With little or no cortex tissue, the root will <u>not store substances</u>.

14.13 Without cork, the stem is <u>herbaceous</u>. All woody stems have cork, which makes up the outer bark.

14.14 All herbaceous stems have limits to their growth. This must be a <u>woody stem</u>, because the construction of bark poses no limits to growth.

14.15 <u>The stem is herbaceous</u>. In woody stems, xylem and phloem are separated.

14.16 <u>No</u>. Bryophytes cannot be large due to their lack of vascular tissue.

14.17 Mosses have alternation of generations. Thus, if the generation studied reproduces with spores, <u>the next generation will sexually reproduce</u>.

14.18 <u>It is a fern</u>. The dominant generation of moss is the gametophyte generation, while the dominant generation of ferns is the sporophyte generation.

14.19 <u>The student is wrong. Ferns do grow on other trees, but they do not take nutrients from the trees. Instead, they take nutrients that gather in the cracks of the tree's bark</u>.

14.20 <u>This is quite different from asexual reproduction</u>. Remember, in asexual reproduction, the genetic code is exactly the same between parent and offspring. When sexual reproduction occurs, on the other hand, two haploid cells form a diploid cell. These haploid cells contain only half of the alleles. Thus, if the sperm and egg get the same half, the genotype of the parent will be different than that of the offspring. In Module #8, for example, we saw that when a Tt plant self-fertilizes, it can make a TT plant, a Tt plant, or a tt plant. If the offspring is TT or tt, then it is quite different than its parent!

14.21 Your key may look different. You need to ask these questions, but yours can be in a different order than ours.

1. Vascular Tissue **2**
   No Vascular Tissue... *phylum Bryophyta*

2. Produces seeds **3**
   Does not produce seeds .... *phylum Pterophyta*

3. Seeds not produced in cones ....*phylum Anthophyta*......... **4**
   Seeds produced in cones .... *phylum Coniferophyta*

4. Single cotyledon in seed .... *class Monocotyledonae*
   Two cotyledons in seed .... *class Dicotyledonae*

NOTE: This is by no means a complete biological key for plants. There are phyla that we did not discuss, and, of course, we haven't gone anywhere close to classifying down to the species level!

STUDY GUIDE FOR MODULE #14

⚹1. Define the following terms:

⚹ a. Botany   ⚹ g. Undifferentiated cells   ⚹ m. Girdling

⚹ b. Perennial plants   ⚹ h. Xylem   ⚹ n. Alternation of generations

⚹ c. Annual plants   ⚹ i. Phloem   ⚹ o. Dominant generation

⚹ d. Biennial plants   ⚹ j. Leaf mosaic   ⚹ p. Pollen

⚹ e. Vegetative organs   ⚹ k. Leaf margin   ⚹ q. Cotyledon

⚹ f. Reproductive plant organs ⚹ l. Deciduous plant

⚹ 2. If a portion of a plant is growing, what type of plant tissue will be in that region?

⚹ 3. What do we call the structure that attaches the blade of the leaf to the stem?

⚹ 4. Identify the leaf mosaics in the picture below:

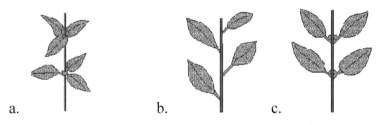

a.   b.   c.

Illus. by Dr. Jay L. Wile

⚹ 5. Determine the shape, margin, and venation of the following leaves:

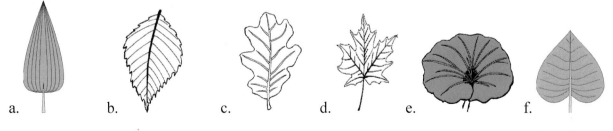

a.   b.   c.   d.   e.   f.

Illus. from the MasterClips collection

⚹ 6. In a leaf, what is the main difference between the palisade mesophyll and the spongy mesophyll?

7. What structure controls when the stomata of a leaf open and close?

8. Why is the bottom of a leaf typically a lighter shade of green than the top of the leaf?

9. Name two pigments that cause leaves to be a color other than green.

10. If a tree has no abscission layer, will it be deciduous?

11. Where is the abscission layer?

12. Name the four regions of a root. Which region contains undifferentiated cells?

13. State which of the following stem cross sections came from a monocot and which came from a dicot:

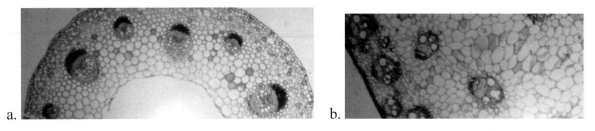

a.   b.

Photos by Kathleen J. Wile

14. What allows woody stems to have no limits to their growth, unlike herbaceous stems?

15. Name two types of plants that have the alternation of generations lifecycle. Which has the sporophyte as the dominate generation? Which has the gametophyte as the dominant generation?

16. Why are plants from phylum Bryophyta relatively small?

17. If a 15 foot plant has a root system that goes 2 feet deep, is it a fibrous or taproot system?

18. What are the male and female sexual organs in a tree from phylum Coniferophyta?

19. What is the fundamental difference between monocots and dicots?

20. Name another difference between monocots and dicots.

Module #15: Kingdom Plantae: Physiology and Reproduction

Introduction

In the previous module, we discussed kingdom Plantae from an anatomy and classification standpoint. We will now conclude our discussion of plants by looking at their **physiology** (fiz ee awl' uh gee) and reproduction.

<u>Physiology</u> - The study of life processes that occur in the daily life of an organism

The definition of physiology tells us that we are going to spend time talking about how a plant functions from day to day. Since reproduction is a part of how a plant functions, it can actually be included in the term "physiology." Reproduction, however, is probably one of the most fascinating aspects of a plant's life, so we will spend a significant amount of time studying it. That's why, in this case, we have separated reproduction from physiology.

Plant Physiology

We already know some of the aspects of a plant's day-to-day life functions. After all, we know that a plant manufactures its own food through photosynthesis. We know that this process involves absorbing carbon dioxide from the atmosphere and releasing oxygen back into the atmosphere. We also know that a plant absorbs water and nutrients through its root system and sends them to the various parts of the plant. There is so much more to the life processes of a plant, however, that we must spend a little more time studying how a plant functions. We will start with something that you probably think you already understand: how a plant depends on water. You might be surprised at the complexity of this subject!

How a Plant Depends on Water

We know all plants need water. However, some plants need a lot more water than others. Many species of flowering plants need to be watered several times each week, whereas other plants (like household ferns), typically need to be watered only once per week. In addition, plants like cacti (the plural of cactus) need to be watered once a month or even less. Each species of plant has been designed by God to fill a particular role in Creation. As a result, their water needs change, depending on what ecosystem they have been designed to occupy.

Regardless of how much water a plant requires, all plants use it for essentially four processes: **photosynthesis, turgor pressure, hydrolysis,** and **transport**. Now, believe it or not, we have discussed all of these processes before. It has been a while, however, so we should spend some time reviewing them. In Module #5, we showed you that to make its own food (glucose - $C_6H_{12}O_6$), a plant uses carbon dioxide and water in a process called photosynthesis. In fact, for each molecule of glucose, the plant needs 6 molecules of water and 6 molecules of carbon dioxide. Obviously, then, without water, the plant would not be able to manufacture its own food and would end up starving to death.

Although we typically think of photosynthesis as the main function of the water in a plant, without the other three processes mentioned above, the plant would die as well. In Module #6, we talked about turgor pressure. In a plant cell, there is a large central vacuole that fills with water by osmosis. As more and more water fills the vacuole, the cell becomes pressurized. This pressure, called turgor pressure, keeps the plant (especially the stems and leaves) stiff. This is why plants, when deprived of water, wilt. Without turgor pressure, plants cannot stand up.

Turgor pressure is also responsible for some of the motion that we observe in plants. For example, some flowers open their petals during the day and close them at night. This is an example of **nastic** (nas' tik) **movement**, and it happens because cells near the base of the flower change their turgor pressure.

Nastic movement - Movement in a plant caused by changes in turgor pressure

Some plants use nastic movement to follow the sun as it moves across the sky each day. The sunflower, for example, will always point towards the sun, no matter where the sun is in the sky. It does this by regulating the turgor pressure in various cells to move the leaves of the sunflower so that they stay facing the sun.

In Module #6, we also talked about hydrolysis. When a cell takes in large molecules, it must first break them down into smaller molecules in order to use them. The cell accomplishes this by adding water to the large molecules. In the presence of specialized enzymes, the addition of water will break down the large molecule. This process is called hydrolysis.

Why does a plant need to break down large molecules? Well, the glucose that it manufactures is a monosaccharide (we discussed this kind of molecule in Module #5). The plant can use that monosaccharide right away for energy, or, if there is plenty of food, it can store the glucose for later use. In order to be efficient, however, the plant will not store glucose as a monosaccharide. The glucose can be stored in a smaller volume if it is first converted into a polysaccharide (also discussed in Module #5) that we call "starch." Thus, the stored food reserves of plants are actually large polysaccharide molecules. When a plant needs to use its stored food supply, it must first break the polysaccharides down into monosaccharides. That's one of the main things a plant does with hydrolysis.

Although a plant uses water for all three of the processes discussed so far, the vast majority of water that a plant takes in is used for transportation. You see, in order to live, plants need a lot more than just carbon dioxide and water. These two substances are all that the plant requires to make its food, but there are a host of other chemicals that a plant must manufacture in order to stay alive. After all, the plant cells must carry on all sorts of biosynthesis, which requires a lot of raw materials. These raw materials are absorbed from the environment and transported throughout the plant. This transportation is accomplished by dissolving the materials in water and allowing the water to carry them up the xylem. Since the motion of water throughout a plant is a very interesting process, we will concentrate on it in the next section.

ON YOUR OWN

15.1 If water is in short supply, a plant wilts but does not die unless the water stays in short supply for a long time. Based on this, which one of the four water-based processes that we discussed can be temporarily neglected?

Water Absorption in Plants

Water is absorbed by a plant through its roots. Contrary to what many people think, a plant cannot absorb water that falls or collects on its leaves. If the water falls off of the leaves and onto the ground, then the roots might absorb it, but that's the only way a plant can absorb the water. Although there are exceptions, most plants have their root systems in the soil, so most roots absorb the water that is in the soil.

The composition of soil is rather complex. There are both organic and inorganic components to the soil. The organic components of the soil include various living creatures such as molds, bacteria, protozoa, worms, and yeasts. In addition, there are organic components in the soil which come from the decaying remains of once-living organisms. The inorganic components of the soil are **gravel**, **sand**, **silt**, and **clay**. The big difference between these inorganic components is the *size of the particles* in that component. Gravel, for example, is composed of relatively large particles with diameters of 1 to 2 millimeters (0.04 to .08 inches). Sand is made up of smaller particles, while silt is made up of even smaller particles, and clay is made up of the smallest particles.

The importance of these components rests in how they form the **pore spaces** of the soil.

Pore spaces - Spaces in the soil which determine how much water and air the soil contains

You see, when a soil is made up of large particles, such as gravel and sand, the soil has large pore spaces. These spaces allow the soil to absorb lots of water in a short time, but the water passes through the soil quickly, so that the soil dries out soon after it gets wet. When a soil is made up of very small particles, however, the pore spaces are small. This means that the soil takes a long time to absorb water, but, once it is wet, it holds the moisture for a long time.

Well, given these facts, the smaller the pore spaces, the more fertile the soil, right? Wrong! Even though the roots of a plant need to absorb water, they need to absorb other things as well. Remember, the roots of a plant are made up of living tissue. Thus, the cells in the roots need to absorb oxygen for respiration. They also need to release carbon dioxide into the environment. Well, the larger the pore spaces, the more oxygen the soil can hold, and the more carbon dioxide the roots can release. Thus, truly fertile soil must have pore spaces large enough to contain plenty of oxygen, but small enough so that the soil can hold moisture for a reasonably long time. The way this can happen is for the soil to be a **loam**.

<u>Loam</u> - A mixture of gravel, sand, silt, and clay

Soil scientists say that the most fertile soil would be a loam consisting of a tiny amount of gravel, 40% silt, 40% sand, and slightly less than 20% clay. This mixture gives the ideal pore spaces for roots to absorb plenty of water while still being able to absorb oxygen and release carbon dioxide.

ON YOUR OWN

15.2 Some people think that putting plant roots underwater will allow the plant to absorb all of the water it needs. If you do this to most plants, however, it will kill them. Why?

Water Transport in Plants

Now that we know what conditions are necessary for roots to absorb water, the next thing we need to know is how that water makes its way to the other parts of the plants. We know from the previous chapter that vascular plants have xylem and phloem which allow water to travel up and down the plant, but that doesn't tell us *how* the water actually moves. After all, in the animals we have studied (with the exception of the earthworm), the creatures' blood vessels did not move the blood through the organism. They provided a "path" for the blood flow, but the organism needed a heart to actually pump the blood through the vessels to the various parts of the organism's body. Well, in a plant, the xylem and phloem provide a path for the water to travel, but what actually makes water move? Plants don't have a "water pump" like a heart. What, then, causes the water to move?

Believe it or not, we aren't completely sure how to answer this question. Although we have a good theory as to why this happens, the theory has not been tested enough to become a scientific law. Thus, the theory provides us with an explanation that is consistent with what we know about plants, but we cannot say for sure that it is the right explanation. This theory, called the **cohesion-tension theory**, says that transpiration causes the water to move up the xylem in a plant.

Remember that we defined transpiration back in Module #10. We said that transpiration is the evaporation of water from the leaves of a plant. We didn't go into much detail then, because it wasn't necessary to at that time. Now it is. In transpiration, water actually evaporates out of the inside of a leaf. It does this because, in order to have carbon dioxide for photosynthesis and oxygen for respiration, the leaves must have their stomata open. When the stomata are open, the water in the leaves is exposed to the air, and it begins to evaporate. Thus, transpiration is really just evaporation, but it is evaporation that occurs inside the leaf. This evaporation is a consequence of the fact that a leaf must have its stomata open in order for the plant to breathe and perform photosynthesis. This is, of course, why a plant's stomata close when it cannot perform photosynthesis. After all, every time the stomata are open, the plant loses water. It doesn't want to lose water; therefore, the stomata will stay open only when the plant absolutely needs them open.

How in the world can transpiration cause water to move up through a plant? Well, water has a very strong tendency towards **cohesion** (coh he' shun).

Cohesion - The phenomenon that occurs when individual molecules are so strongly attracted to each other that they tend to stay together, even when exposed to tension

Now this definition might be a little confusing to you, so we want to explain it by means of an illustration. Have you ever seen a water strider like the one illustrated in Figure 15.1?

FIGURE 15.1
A Water Strider Can Walk On Water Because of Cohesion

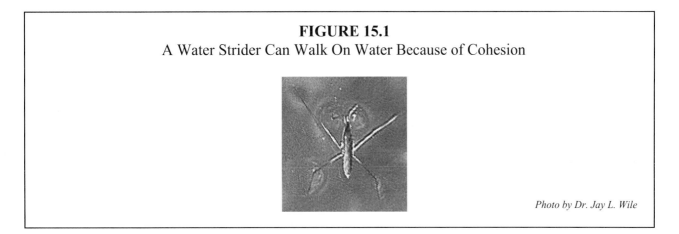

Photo by Dr. Jay L. Wile

Why don't the strider's legs just sink into the water? Well, the strider's weight is pressing down on the surface of the water, evenly distributed by its four legs. Thus, each of the strider's legs is exerting 1/4 of its weight onto the water. If, in response to this weight, the individual water molecules were to move away from each other, then the strider's legs would sink into the water. This doesn't happen, however, because of cohesion. The water molecules are attracted to one another and would therefore like to stay close together. The strider's weight, evenly distributed over four legs, is not large enough to counteract this desire to stay together. Thus, the water molecules do not move away from each other, and the strider's legs do not sink.

Now, of course, cohesion has its limits. Most animals, when they step onto water, exert so much weight on the water that it counteracts the cohesion of the water molecules, and the animals sink. In order to make a difference, then, the strength of cohesion between the water molecules must be greater than whatever force is trying to pull them apart from one another.

How does this explain the transport of water in plants? When water evaporates from the leaves, there is a deficit of water in the leaves. This causes a **tension**, which pulls up on the water molecules just below where the deficit of water exists. These water molecules, in response to the tension, move up to replace the evaporated water molecules. The water molecules near the ones that move up, however, do not want to be separated from them. Because of cohesion, then, these water molecules move up to stay near the others that have moved up. This results in a "chain reaction," with all water molecules moving upwards due to their cohesion with the water molecules directly above them. Like a reverse domino effect, this motion goes all the way down to the root, eventually causing water in the roots to move upwards towards the other parts of the plant.

Although we cannot yet be sure that this is the proper explanation for how water moves upwards in a plant, there is a lot of experimental evidence to back up this theory. As a result, the cohesion-tension theory has become a widely accepted explanation for water transport in plants. Even if the plant isn't vascular, the cohesion-tension theory still works as an explanation for how water moves in the plant. Non-vascular plants have leaves as well, and they lose water through these leaves, which causes other water molecules to move upwards in order to take their place. In these plants, however, there is no clear "path" over which this motion can take place. Instead, the water molecules must weave in and out between the cells in order to follow the other water molecules. This is more difficult, limiting the range over which the water can travel. That's why plants in phylum Bryophyta are smaller than the vascular plants.

Now, of course, a plant needs to transport things besides just water. Minerals and other raw materials for biosynthesis are absorbed by the roots. These must make it up to the remote parts of the plant as well. That's why we said that any materials which the plant needs to transport upwards must be dissolved in water. Since the water flows upwards in accordance with the cohesion-tension theory, any substances dissolved in the water will move up as well.

Although the cohesion tension theory explains how substances move upwards through the xylem in a plant, it does not explain how substances travel down the plant in the phloem. This process, called **translocation**, is quite different.

Translocation - The process by which organic substances move down the phloem of a plant

If you think about it, although water and minerals need to travel up the plant, the only substances that need to move down a plant are the products of photosynthesis and other types of biosynthesis. After all, the roots and stems are comprised of living tissue which need food. The food is produced primarily in the leaves of the plant, so the food must move down from the leaves to the other parts of the plant, including the roots. As we have already learned, some plants store their excess food in their roots. Thus, the roots need food to stay alive and are often the repositories for excess food. As a result, the products of photosynthesis, which are organic chemicals, must move down the phloem in the plant.

How is this accomplished? Do you remember one of the big differences we mentioned between xylem and phloem tissue? The cells that make up the xylem do not live long. This is because the xylem are simply tubes through which the water flows. Thus, the cells that make them up need not be alive, because they do nothing to promote the transport of water. Phloem cells, on the other hand, need to be alive, because they actively expend energy to guide the flow of the organic molecules down the phloem. Interestingly enough, the phloem cells spend most of their energy trying to slow down the transport of organic substances, because if they were left to the force of gravity, they would fall so fast that only the roots would receive nourishment! The details of this process are a bit beyond the scope of this course, so you just need to remember that this is the reason why phloem cells are alive, as opposed to xylem cells which die soon after the xylem forms. Since the phloem cells actively participate in translocation, they must be alive to perform their duties.

ON YOUR OWN

15.3 A botanist has two samples of liquid. Sample A is composed primarily of organic materials, while sample B is composed mostly of water and minerals. Which liquid was extracted from the xylem of a plant and which came from the phloem?

15.4 Typically, a plant opens its stomata during the day and closes them at night. When would you expect transportation of water to occur, in the day or during the night?

Plant Growth

As a plant develops, it grows. As we have mentioned before, this growth occurs in the meristematic tissue, where undifferentiated cells exist and where mitosis takes place. Although you already know the ins and outs of mitosis (discussed in Module #8), you probably don't know what controls mitosis. After all, each cell in the meristematic tissue could, if it wanted, go through mitosis about once every 30 minutes. This would lead to an *enormous* amount of growth in the plant. Under most conditions, the plant could not sustain so much growth, because it could not develop enough of the supporting tissue (such as vascular tissue) to keep these new cells alive. Thus, plants (and animals) must have some means of controlling the rate at which mitosis occurs.

In addition, once a new cell has formed, its development must be regulated so that it develops into the tissue that the plant needs. Therefore, plants also need chemicals which will regulate the development of new cells. Controlling mitosis and regulating plant cell development are accomplished with chemicals called **hormones**.

Hormones - Chemicals that affect the rate of cellular reproduction and the development of cells

In plants, there are five identifiable groups of plant hormones: **auxins** (awx' uhns), **gibberellins** (gib uh rehl' ins), **cytokinins** (sigh toh kie' nins), **abscisic** (ab sih' sik) **acid**, and **ethylene** (eth' uh lean). In addition, botanists suspect there is at least one more kind of plant hormone that, as of yet, we have not isolated. This mystery group of hormones is usually referred to as **florigen** (floor' uh jen).

Auxins were the first group of plant hormones discovered. These hormones regulate the development of cells, altering the amount that they elongate. This primarily affects the length of a plant's stems. The more the stem cells elongate, the longer the stem grows. Auxins are considered the driving force in **phototropism** (foe' toe trohp iz uhm), **gravotropism** (grav' uh trohp iz uhm), and **thigmotropism** (thig' muh trohp iz uhm).

<u>Phototropism</u> - A growth response to light

<u>Gravotropism</u> - A growth response to gravity

<u>Thigmotropism</u> - A growth response to touch

If you have experimented with plants, you are probably familiar with these growth responses.

If you place a houseplant near a window which the is only source of light for the plant, you will notice that over time, the plant will grow so that its leaves are all pointed towards the window. If, then, you turn the plant around so that the leaves point away from the window, the plant will change its growth patterns so that, eventually, the leaves will face the window again. This is the phenomenon of phototropism. Most botanists believe that auxins are destroyed by light. Thus, they end up in highest concentration where the plant is exposed to the least light. This causes the plant tissue in the darker regions to grow quickly, while the plant tissue which is exposed to a lot of light doesn't grow much. The result of this growth imbalance is that the plant stems tend to bend toward the light.

Etiolation (ee tee uh lay' shun) is what might be considered an extreme form of phototropism. If a plant is placed in a rather dark area, none of its auxins are destroyed. As a result, the stems grow far too rapidly, becoming very long and thin. Since they are long and thin, the stems do not stand up well, but tend to crawl along the ground. Because there is little light, there is little need for chlorophyll, so the stems tend to have a very light color and there tend to be few leaves. If the stems eventually grow into an area that does have light, the plant will produce leaves there. In the end, it is almost as if the stems grow rapidly in search of light. If sufficient light is not found in a reasonable time frame, the plant dies.

Gravotropism is illustrated in Figure 15.2.

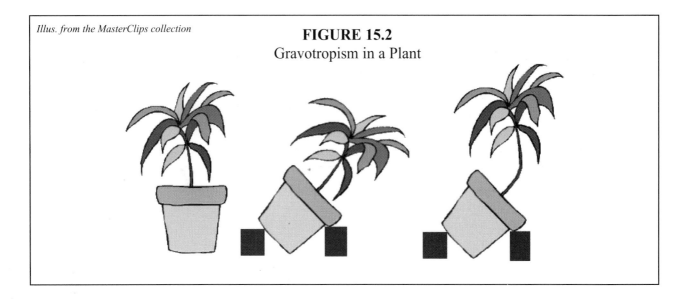

Illus. from the MasterClips collection

FIGURE 15.2
Gravotropism in a Plant

If you tilt a plant on its side and prop it in some way so that it stands diagonally, the plant will bend in a few days so that the leaves are pointed upwards again. This is because the auxins in the plant are affected by gravity. They tend to collect wherever gravity causes them to fall. If a plant is tilted, the underside of the stem will have more auxins than the top side of the stem. This will cause a growth imbalance, with the tissues on the underside growing faster that those on the top side of the stem. This bends the stem so that the plant begins to grow upwards.

Thigmotropism is best illustrated by a vine. If you ever look at a vine, you will see that it tends to wrap around whatever it touches, and the leaves of the vine always point away from the structure upon which it is growing. This is due to thigmotropism. Auxins tend to move away from surfaces which a plant touches. Thus, when a vine touches a structure, the auxins travel to the other side of the vine. That's why stems and leaves grow there. Also, the same growth imbalance which you see in phototropism and gravotropism occurs in this case. Thus, if the vine can, it tends to bend around the surface which it touches. Over time, this causes the vine to wrap around the surface to which it is attached.

The gibberellin family of hormones also promote elongation in stems, but they do other jobs as well. Gibberellins affect mitosis rates, and they also induce seeds, buds, and flowers to grow. Cytokinins also affect mitosis rates, but in addition, they induce leaf cells to elongate, allowing the leaf to expand. Finally, some experiments indicate that cytokinins actually help retard the effect of aging in leaves. Abscisic acid's primary job is to open and close leaf stomata. In addition, it is an "anti-growth" hormone, retarding growth when the plant cannot sustain it. Finally, ethylene promotes the ripening of fruits and controls the abscission layer, causing it to begin its growth so that the leaves of a deciduous plant will fall. Since it promotes fruit ripening, it is often used by farmers to artificially ripen fruits that were picked before they had a chance to ripen on the plant.

As we mentioned above, botanists also believe in the existence of a heretofore undetected hormone called florigen. Why do botanists believe that this hormone exists even though it has not been discovered yet? Well, something controls the budding of leaves and the formation of flowers. Botanists believe that it must be some sort of hormone, but the search for this hormone has so far yielded nothing. Nevertheless, since flowering and budding can be considered "growth," it is assumed that some growth hormone controls these processes.

As we mentioned before, animals have hormones that promote the growth and development of certain tissues as well. For example, some athletes take specialized hormones called steroids which enhance the growth of muscle tissue. This tends to "bulk" the athelete up, giving him or her strength that he or she would not normally have. This is a very dangerous practice, however, because each person's body chemistry is delicately balanced. By artificially adding hormones to their bodies, these players do become stronger, but they also throw off their body chemistry, resulting in brain damage, respiratory problems, and sometimes death!

ON YOUR OWN

15.5 Regardless of how you plant a seed, the seedling always sprouts up through the soil. What growth hormone is most likely responsible for this amazing ability?

15.6 Conifers are not deciduous and do not produce fruits. What hormone is, most likely, not present in conifers?

Insectivorous Plants

The **insectivorous** (in sek tiv' or us) **plants** have a rather interesting physiology that is worth discussing. These plants have leaves that are designed to trap and "digest" insects. A common example of such a plant is shown in Figure 15.3.

FIGURE 15.3
Venus's Flytrap

Photo courtesy of
www.PDImages.com

When an insect lands on the inside (red portion) of a Venus's flytrap's leaf, the leaf constricts, trapping the insect. Digestive juices are then excreted from specialized cells in the leaf's tissue, decomposing the insect.

Contrary to popular belief, insectivorous plants *do not eat* the insects that they trap. This is why we put the term "digest" in quotes above. Instead of using the insects for food, insectivorous plants have photosynthetic cells that produce the plant's food, just like other plants. Why, then, do insectivorous plants trap and "digest" insects? Well, remember that a plant needs more than just food to survive. It needs certain raw materials for biosynthesis as well. Most plants get these raw materials from minerals and other substances in the soil. Insectivorous plants, however, have been designed by God to live in soils that have little or none of these raw materials. Thus, they must get them from another source. Insectivorous plants get their raw materials for biosynthesis from the chemicals that come from digesting the insects they catch.

One of the most important raw materials that a plant needs for biosynthesis is nitrogen. Now there are plenty of nitrogen atoms in the air around the plant. The atmosphere, after all, is 78% nitrogen gas. Unfortunately, plants do not have the ability to extract the nitrogen atoms from this nitrogen gas. You will learn why this is the case in chemistry. As a result, plants must get their nitrogen from some other source. Most plants get their nitrogen from ammonia, a chemical that is abundant in most soils. In soils that contain little ammonia, plants such as the Venus's flytrap flourish. Interestingly enough, if you place a Venus's flytrap in soil that is rich in ammonia and other minerals, it will not produce any insect-trapping leaves.

ON YOUR OWN

15.7 Most fertilizers are rich in ammonia, so that plants grown in fertilized soil will have a plentiful supply of nitrogen. A person buys a Venus's flytrap from the store and decides to take "really good care of it" by fertilizing the soil in which it is planted. What will happen to the Venus's flytrap?

Reproduction in Plants

In kingdom Plantae, virtually all forms of reproduction can be found. Most plants have the ability to reproduce asexually. Often, this is accomplished through a process called **vegetative reproduction**, which we will discuss in a moment. Plants also have the ability to sexually reproduce. As we learned in the previous module, conifers produce seed cones and pollen cones, which are sexual organs. Flowering plants use their flowers in sexual reproduction, as we will discuss a bit later. Finally, as we learned in the previous module, ferns and mosses reproduce through the interesting process of alternation of generations, a process involving both asexual and sexual reproduction. Clearly, kingdom Plantae is quite diverse in its means of reproduction! In this module, we will concentrate on vegetative reproduction and the reproduction of flowering plants.

Vegetative Reproduction

Vegetative reproduction in plants comes in many different forms. Some plants, like the piggyback plant, grow small "plantlets" right on their leaves. If these plantlets reach the soil, they will grow into new, separate plants. Other plants produce underground stems that originate in their roots. These stems eventually grow into a new plant. If, for example, a gardener plants a single mint plant in his or her garden, dozens of mint plants will appear around the original. This happens because as the roots of the mint plant grow outwards, away from the plant, they develop specialized stems that grow into a new plant. Grass also reproduces in this way, as do many forms of weed. The Irish potato produces tubers, another specialized stem that is formed in the roots of the plant. The tuber is what most people think of as the potato itself. These tubers have many buds on them, which are often called "eyes." These eyes can produce new plants. Often, gardeners propagate potatoes by simply cutting up their tubers into several pieces, each with an eye. They then plant these pieces, and the eyes mature into new potato plants.

Other plants can vegetatively reproduce using above-ground stems. Plants such as the strawberry plant can produce **runners**. These above-ground stems are long and spindly, and they produce a small plantlet at the end. When the plantlet grows too heavy for the runner, it touches the ground, and the plantlet begins to form roots. Once the roots take hold, the runner dies, and the result is a new strawberry plant.

The regular stems of certain plants are capable of asexual reproduction as well. In plants such as blackberry plant, if a stem is broken or bent so that it touches the soil, it will often produce roots and begin to grow as a separate plant. If the original bend or break was severe enough to kill the stem, then the stem will die, and the new plant will be separated from the original one. If the bend or break heals, however, the new plant will stay attached to the original one. Sometimes, gardeners and farmers use this method to propagate their plants. They will cut a stem from a plant, often referring to it as a "stem cutting," and then replant the stem. The stem will form roots, just as if a break had occurred, and a new plant will form.

Yet another form of vegetative reproduction comes from the leaves of certain plants. A leaf of the African violet, for example, can be planted in soil and the meristematic tissue in the leaf can actually form roots and eventually mature into a new African violet plant. In principle, any leaf *should* be able to do this. Under normal conditions, however, most leaves wilt and die long before the meristematic tissue has time to form roots. If an expert manipulates the conditions under which a leaf is planted in the soil, however, almost any plant can be reproduced in this way.

Now remember, all of these forms of vegetative reproduction are asexual. This means that the offspring are a genetic copy of the parent. Thus, if an African violet reproduces when a leaf is broken off of the plant and drops into the soil, the new plant will be genetically identical to the original plant. This kind of reproduction can be very useful to commercial gardeners. For example, suppose a gardener spends a lot of money on a rose plant that produces ideal-looking roses. If that plant were to sexually reproduce with itself or another rose plant, the resulting offspring will not be genetically identical to the original rose plant. As a result, the gardener will not be able to guarantee that the roses produced by the new plant will be ideal. However, if the gardener vegetatively reproduces the rose plant, the roses produced by the offspring will be the same, ideal roses produced by the original plant. Thus, vegetative reproduction is often used by commercial gardeners to ensure that their plants are as close to ideal as possible.

One other process that is sometimes considered vegetative reproduction is **grafting**. In grafting, a stem is cut from one plant and attached to another. The stem is called the **scion** (sy' un), and the plant to which it is attached is called the **stock**. Typically, the scion is attached to the stock so as to ensure that the vascular cambiums of the two are as close to one another as possible. This allows the xylem and phloem to merge, making a seamless interface between scion and stock. In order for grafting to work, the stock and scion generally have to be rather similar. Usually, this means that they must come from the same genus. Also, the grafting should take place when both plants are not growing. The point at which the graft takes place is wrapped in tape and wax, to make an airtight, watertight seal.

The neat thing about grafting is the fact that no genetic material is exchanged in the process. Thus, the stock remains the same as before the graft and the scion remains the same as well. Thus, if the stock has stems, those stems will produce leaves, flowers, and/or fruits in accordance to its species, and the scion will produce leaves, flowers, and/or fruits in accordance to its species! Grafting is often used in orchards. Gardeners take stems from a certain tree that produces nice fruit. They will then take those stems to a similar tree that produces fruits which aren't nearly as nice. The stems of this tree are cut away, and the scions from the first tree are then grafted on to the second tree. In a few years, the second tree will produce the same, nice fruits as the first tree does. This allows the caretaker of the orchard to ensure a quality crop of fruits from all of the trees.

The New Testament provides an excellent illustration of grafting. The Apostle Paul says that the Gentiles are like "scions" that have been grafted onto the "tree" of Israel. The Jews that believe in Christ are like the stems from the original "stock," serving the Lord as members of the original chosen race. Gentiles who believe (the grafted stems) are also members of the chosen race. After all, even though they come from a different race, they have been grafted onto the "stock" of Israel. Thus, Jews who believe in Christ and Gentiles who believe in Christ are both members of the chosen people, even though they originally came from different stock!

ON YOUR OWN

15.8 A gardener is experimenting with one of her African violet plants. This plant produces large, deeply-colored flowers. She shows you two offspring from the plant. The first produces small, lightly-colored flowers and the second produces large, deeply-colored flowers. Which offspring was produced with vegetative reproduction? Which one was produced with sexual reproduction?

15.9 A gardener wants to graft a limb from a Macintosh apple tree onto another tree so that he can have more Macintosh apples. Should the gardener graft this limb to his Red Delicious apple tree or to his wild cherry tree?

Sexual Reproduction in Plants

In the previous module, we spent some time discussing the sexual reproduction that occurs in the alternate generations of phyla Bryophyta and Pterophyta. In addition, we discussed the sexual reproduction in phylum Coniferophyta which occurs by seed cones and pollen cones. When we got to phylum Anthophyta, however, we decided to put off a discussion of sexual reproduction because we said it was a little complicated. Well, we are going to have that discussion now.

Members of phylum Anthophyta are often referred to as the "flowering plants." This is, of course, because all members of this phylum produce flowers at some point in their lifecycle. These flowers contain the reproductive organs that allow the flowering plants to reproduce sexually. Although the color, size, odor, and shape of flowers differ greatly from species to species within phylum Anthophyta, there are some general characteristics that all flowers share. These are illustrated in Figure 15.4.

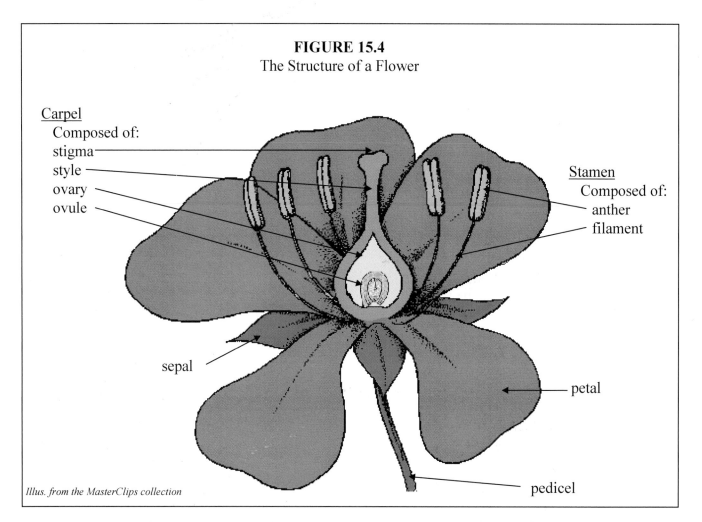

FIGURE 15.4
The Structure of a Flower

Carpel
Composed of:
stigma
style
ovary
ovule

Stamen
Composed of:
anther
filament

sepal

petal

pedicel

Illus. from the MasterClips collection

Notice from the figure that there are really only five basic parts to a flower. The **pedicel** (ped' uh sil) is the stem that holds the flower; the **sepals** (see' puls) are the green leaf-like structures above the pedicel; the **petals** are the leaf-like structures above the sepals; the **stamens** (stay' men) are the male reproductive organs of the flower; and the **carpels** (car' puls) are the female reproductive organs. The carpels used to be known as "pistils," but botanists have changed the name over the past few years for international consistency. Look at how these five structures are arranged in the flower. If you were to "fold up" the flower, the reproductive organs would be enclosed in the petals, which would be enclosed in the sepals. This is the way the flower is designed, in order to protect its sexual organs before the flower opens.

If you look again at Figure 15.4, you will see that the carpels and stamens are actually comprised of other structures. The stamens have long **filaments** that serve as "stalks" which support the **anther** (an' thur). The anther contains **pollen grains**, which we will discuss more in a moment. These pollen grains contain haploid cells that function as the sperm of the plant. The carpels are comprised of the **stigma, style, ovary**, and **ovule** (oh' vyoul). The stigma is covered in a sticky substance designed to catch pollen grains; the style is an extension of the ovary designed to hold the stigma up where it can be exposed to pollen grains; the ovary protects the ovule, where the **embryo sac** develops. The embryo sac functions as the egg of the plant, as well as the beginnings of the plant's seed. We will talk more about that in a moment.

It is important to note that while these five structures make up the "typical" flower, not all flowers have all of them. As you should be aware of by now, there are exceptions to nearly every rule in biology. The structure of a flower is no different. Some flowers do not have both stamens and carpels. Flowers that have both reproductive organs are called **perfect flowers** while those that have either stamens *or* carpels are called **imperfect flowers**.

Perfect flowers - Flowers with both stamens and carpels

Imperfect flowers - Flowers with either stamens or carpels, but not both

Remember in Module #8 when you first learned that plants can sexually reproduce with themselves? In Module #14, it became apparent how conifers can do so, and now it should be clear how flowering plants can do so as well. After all, if the pollen grains (sperm) produced by the stamen of a flower fertilize the embryo sac (egg) in the ovary of a flower on the same plant, that plant will have sexually reproduced with itself! Remember, however, that when a plant sexually reproduces with itself, the offspring is not necessarily a genetic copy of the parent.

While the function of the carpels and stamens is rather obvious, what are the functions of the other 3 parts of the flower? Well, the pedicel holds the flower up. It actually has a swelling on its tip (not shown in the figure) called the **receptacle.** This receptacle holds the flower bud as it develops and supports the flower. The sepals enclose the bud of the flower, protecting it while it develops. When the flower is properly developed, the sepals peel back, revealing the petals. The petals also peel back, revealing the stamens and carpels.

What function do the petals perform? Are they just an extra layer of protection for the carpels and stamens as they form? Absolutely not. The petals contain carotenoids and other pigments that often result in beautiful colors. In addition, the epidermis of the petals often contains oils that let off a fragrant smell. Why the bright colors and fragrant smells? Well, as you will learn in more detail later, in order for the pollen grains of one flower to reach the ovaries of another, the pollen grains must be carried from flower to flower. This is often accomplished by wind, but it is also common for insects and birds to carry pollen from flower to flower. The bright colors and fragrant smell of flowers are designed to attract these animals so that they can transport pollen. Thus, even though we enjoy them, the petals of flowers are not just pleasing to the eyes and nose; they are designed to be very functional!

Although Figure 15.4 shows you the basic anatomy of a flower, please realize that there are many variations on this basic structure. As we have already mentioned, imperfect flowers have only one of the two reproductive organs. In addition, some flowers have several carpels (instead of just one as is shown in the figure). A single carpel can also have several ovules as well. Finally, some flowers are actually **composite flowers**, which are made up of several individual flowers. The sunflower, for example, looks like a single flower from far away. If you examine it closely, however, you will see that it is actually several tiny flowers that form in a single receptacle.

ON YOUR OWN

15.10 Can a plant which produces imperfect flowers sexually reproduce with itself? Why or why not?

15.11 Suppose a flower has a carpel with a very short style. Which would you consider to be the best means of transferring pollen grains to the carpel: the wind or an animal?

Before going on to see how flowers are used in the reproductive process of plants, perform the following experiment, which will give you more experience with the anatomy of a flower. Even though this experiment calls for a microscope, parts A-E can be done without one, so you should perform those steps even if you do not have a microscope.

EXPERIMENT 15.1
Flower Anatomy

Supplies:

- Sharp scissors
- Sharp blade
- Slides and coverslips
- Water
- Dropper
- Microscope
- Lab notebook and colored pencils
- A variety of flowers (Most flower shops will save old flowers for you, if you contact them ahead of time and tell them what you are using them for.)

Object: To observe various types of flowers and compare how they are alike and how they are different.

Procedure:

A. Obtain flowers. They do not need to be fresh, but should be a good variety. An example of a good variety would be: a rose, a carnation, a daisy, a lily and a tulip.

B. Begin dissecting the flower by carefully pulling off the petals and sepals on one side, making sure you **do not disturb** the other parts.

C. Use your knife to cut the carpel vertically through the middle. This will expose the ovary and ovules, if they are developed enough. Try to do this without breaking off the carpel, so that you see a good cross section of the flower intact.

D. Make a drawing of each flower and label the parts.

E. The following is a list of things to look for and **label** in your drawings:

    1. **Pedicel** - the stalk which supports the flower
    2. **Receptacle** - "bulge" at the tip of the pedicel
    3. **Sepals** - often green, at the base of the flower
    4. **Petals** - all of the petals together are collectively called the **corolla**
    5. **Stamens** - male reproductive parts
        a. **filament** - stalk
        b. **anther** - forms and holds pollen
        c. **pollen grains** - contains the sperm nuclei in plants
    6. **Carpel** - female reproductive parts
        a. **stigma** - top of carpel, receives pollen
        b. **style** - supports stigma
        c. **ovary** - contains ovules
        d. **ovules** - holds egg, develops seed

F. Remember flowers like the daisy and sunflower are **composite flowers** which means they are made up of hundreds of complete individual flowers.

G. Scrape some pollen from each flower. Place the dust on a slide and add a drop of water. Cover the wet sample with a coverslip. Observe with the microscope. Note the varying shapes and colors of the pollen. Draw each type of pollen.

H. Investigate with the microscope any other part of the flower that you wish to see better.

I. Clean up and put away all equipment.

The Reproductive Process in Plants: Part 1 - Forming Pollen and Embryo Sacs

Now we already know that the stamens produce the pollen grains (sperm) of a flowering plant while the embryo sacs (eggs) are produced in the carpel. How does this happen? Well, we need to look at the formation of each of these gametes individually, because God has designed the process of pollen grain formation quite differently than He has designed the process of egg formation. We will start with pollen grain formation, which is illustrated in Figure 15.5.

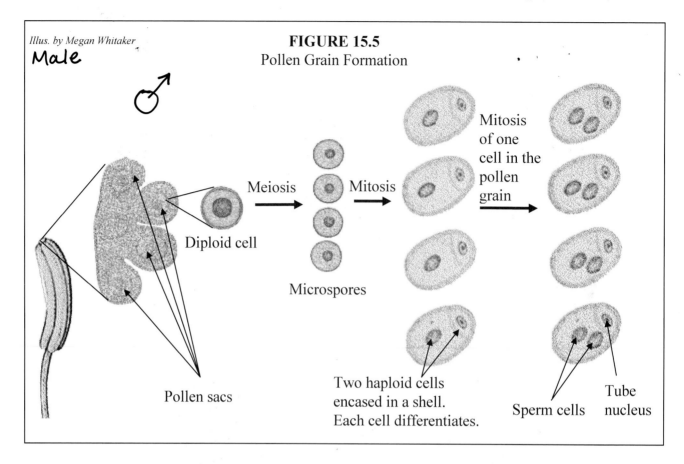

Illus. by Megan Whitaker

Male

♂

FIGURE 15.5

Pollen Grain Formation

Diploid cell

Meiosis → Mitosis → Mitosis of one cell in the pollen grain →

Microspores

Pollen sacs

Two haploid cells encased in a shell. Each cell differentiates.

Sperm cells

Tube nucleus

Each anther in a flower's stamen is divided into four regions called **pollen sacs**. In these sacs, diploid cells undergo meiosis in order to form haploid cells. Unlike the process of sperm formation in animals, however, this is just the beginning of the formation of pollen grains. Once meiosis has occurred, there are 4 haploid cells, which are often referred to as **microspores**. Each microspore becomes encased in a thick wall that is designed to protect it from unfavorable conditions. Once the microspore is encased, it undergoes mitosis. This results in two haploid cells inside the casing. In some plants, mitosis occurs more than once inside the casing, making even more haploid cells.

Once there are at least two cells in the casing, one of the cells differentiates and becomes a **tube nucleus**. The other cell divides one more time by mitosis, and the products of that mitosis differentiate into sperm. Remember, the tube nucleus and the sperm are still encased in a tough cell wall, so they are protected from unfavorable conditions. These cells, encased in their protective coating, make up a pollen grain. Each pollen sac in a mature anther contains hundreds or even thousands of pollen grains.

In the carpel, a completely different process, illustrated in Figure 15.6, is responsible for the formation of egg cells.

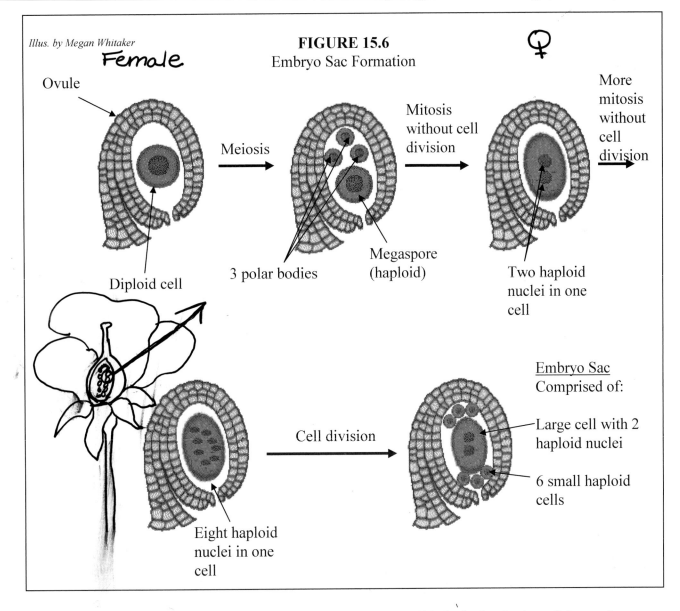

Illus. by Megan Whitaker

FIGURE 15.6
Embryo Sac Formation

Female ♀

On the inside wall of the ovary, a swelling begins to grow. This is the beginning of the ovule. As the ovule grows, a single diploid cell undergoes meiosis, resulting in 4 haploid cells. As is the case in animal female meiosis, three of these haploid cells (polar bodies) are useless and eventually disintegrate. The remaining haploid cell, called a **megaspore**, then undergoes a very interesting form of mitosis. In this form of mitosis, the DNA is copied, and a new nucleus is made, but the cell never divides. In then end, then, this form of mitosis produces a single cell with two nuclei. These nuclei continue mitosis, making a total of 8 nuclei in a single cell.

Once the megaspore has 8 haploid nuclei, three of the nuclei travel to one side of the cell, two stay in the middle, and the remaining three move to the other side of the cell. Once the nuclei are in their proper place, cell walls and membranes form around the six individual nuclei on each side of the cell. This results in a large, central cell with two nuclei and six smaller cells, three of which are on one side of the large cell and three of which are on the other. This arrangement of seven cells is called an **embryo sac**.

ON YOUR OWN

15.12 If a pollen sac contains 1,000 diploid cells before the formation of pollen grains, how many pollen grains will there be in the pollen sac once pollen formation is complete?

15.13 After the original cell of an ovule undergoes meiosis and forms one viable megaspore, how many times does mitosis occur in order to get a total of 8 nuclei in the megaspore?

<u>The Reproductive Process in Plants: Part 2 - Pollination</u>

By early spring, most flowers have completed forming pollen grains and embryo sacs and are ready to begin the process of **pollination**.

<u>Pollination</u> - The transfer of pollen grains from the anther to the carpel in flowering plants

If you have hay fever, you are already acutely aware of the process of pollination. When the anthers in a flower release their pollen grains, the pollen grains often travel on the wind in order to reach the carpels of another flower. Many people are allergic to these pollen grains. Thus, when they breathe the pollen-laden air in the spring, they have an allergic reaction to the pollen grains, causing stuffed-up noses, coughing, congestion, and respiratory difficulties. Although many pollen grains travel on the wind in order to reach the carpels of other flowers, it is important to note that God has designed several other means of transport in order to ensure that the pollen grains of one plant make it to the carpels of another.

You see, the plants that produce flowers send some of the glucose produced by photosynthesis up to the flowers. The flowers store this glucose in a sticky-sweet solution called **nectar**. This nectar attracts birds, bees, moths, and butterflies that eat the nectar for food. As these creatures eat the nectar in the flower, the pollen in the anther's pollen sacs are released onto them. As they travel from flower to flower, then, they will brush up against the carpels of the flowers, and the pollen on them will brush off onto the sticky surface of the carpel. This completes the pollination process. Each flower that these creatures visit receives pollen from the other flowers that they have visited and donates pollen for the next flowers that will be visited. This results in an incredibly efficient means of making sure that the pollen of a flower reaches the carpels of many, many other flowers.

It actually turns out that the process of pollination is even more efficient than we have described up to this point. As we mentioned before, the bright colors and fragrant smells of flowers have been designed to attract creatures for help in pollination. What we didn't mention before, however, was that *specific flowers are designed to attract specific creatures!* For example, birds are typically attracted to certain types of red flowers. These flowers store their nectar in long tubes. Birds with long, thin beaks are designed to fly to the flower and suck up the nectar through their beaks. Insects are not drawn to these flowers, however, because there is so much nectar in the flower that most insects can easily drown in the substance that they would normally use for food. Thus, God has designed insects to ignore these red flowers, but He has designed birds to be attracted by them. Interestingly enough, birds have a poor sense of smell. As a result, flowers that have been designed to be pollinated by birds usually have no fragrant smell, because the smell would serve no purpose. Now, of course, not all red flowers have no smell. Some red flowers (roses, for example) do not depend on birds for pollination; thus, they have a fragrance. In general, however, red flowers are much less likely to be fragrant than flowers that are not red, because they are more likely to depend on birds for pollination.

Flowers that attract bees for pollination, on the other hand, usually produce fragrant, sweet odors. These odors attract the bees to the flower. As the bee feeds on the nectar of the flowers, the anthers release pollen onto their bodies. Before the bee leaves the flower, it grooms itself, placing all of the pollen that has been scattered over its body into specialized structures (also called pollen sacs) formed by the hairs on its legs. It does this because as the bee crawls over the flower, its legs tend to brush up against the carpel much more often than any other part of its body. Thus, as the bee forages for food, it actually takes the time to rearrange the pollen that has been scattered all over its body so that it ends up in the best place for pollination of the next flower!

Beetles are also attracted by smell, but the smell is different than that which attracts bees. Beetles feed on dung and decaying matter. Thus, the flowers that God has designed to be pollinated by beetles actually have the foul odor of organic wastes and decaying organisms! These flowers are generally produced by short plants or plants that produces runners which run along the surface of the soil. As beetles crawl around them, they smell the odor of the flower and think it is food. Thus, they crawl all over the flower looking for dung or decaying matter, and all they get is a body covered by pollen. When they head to the next flower, some of that pollen ends up sticking to the carpel, completing pollination.

Moths and butterflies also aid in flower pollination. Butterflies typically forage for food in the daytime. They are attracted to flowers that have fragrant smells but also a wide, relatively horizontal surface upon which they can land. Daisies are one such flower. Moths, on the other hand, tend to forage at night. They tend to be attracted to white or otherwise pale colors, because these colors show up better at night. Fragrant odors also attract moths. Once again, when these insects land on a flower, pollen is scattered on their bodies. As they travel from flower to flower, they transfer this pollen to the carpels of the flowers upon which they land.

Why has God gone to all of the trouble to design flowers so that birds, bees, beetles, butterflies, and moths can transport pollen from one plant to another? If a plant can reproduce with itself by simply transferring the pollen grains of one flower to the carpel of another flower on the same plant, why did God bother to design flowers so as to ensure that pollen from one plant can get to the carpel of another plant? Well, remember what we learned in Module #9. God has designed His Creation to be able to change and adapt in order to survive in the changing environment. One way He has seen fit to do this is to design a genetic code that is incredibly flexible. In order to retain this flexibility, though, the genetic codes must be continually mixed between individuals. If a plant simply reproduced with itself, then only a small number of phenotypes are possible in the offspring. If a radical change in environment took place, the offspring of such plants might not be able to adapt to it. If, however, several different individual plants reproduce with each other, then the genetic codes of many individuals are involved. This means there are a LOT more possibilities of phenotypes, and this ensures that the plants can adapt to changes in the environment.

ON YOUR OWN

15.14 Suppose you were blindfolded and asked to smell several flowers. If you encountered a flower with little or no smell, what would be a good guess as to its predominant color?

15.15 Suppose you were on an island with plenty of plant life but essentially no insects or birds. Would you expect to see many flowering plants? Why or why not?

The Reproductive Process in Plants: Part 3 - Fertilization

When pollination occurs by any one of the several means God has designed, the pollen grains stick to the sticky substance at the top of the stigma. Although the "sperm" of an anther has been transferred to the ovary of the flower, fertilization has not yet taken place. After all, the sperm cells have to make it to the egg cell in order to really have fertilization. Well, in order to get to the egg cells, these sperm cells are going to have to travel down the carpel to that ovule. How does this happen? The process is illustrated in Figure 15.7.

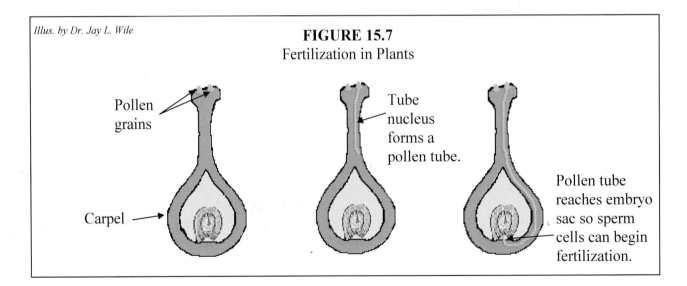

Illus. by Dr. Jay L. Wile

FIGURE 15.7
Fertilization in Plants

Pollen grains

Carpel

Tube nucleus forms a pollen tube.

Pollen tube reaches embryo sac so sperm cells can begin fertilization.

Remember the tube nucleus that formed in the pollen grain? It is separate from the sperm cells encased in the same grain, because its job is quite different than theirs. When pollen lands on the stigma, the tube nucleus begins to undergo mitosis and differentiation. This results in a tube, which is called the **pollen tube**. The pollen tube grows through the tissue of the ovary, eventually finding its way to the embryo sac. The sperm nuclei travel from the top of the carpel to the embryo sac through this tube that has grown from the tube nucleus.

Once the two sperm nuclei have reached the embryo sac, a rather unusual thing happens. One of the sperm cells fuses with one of the six small cells located on either side of the large, double-nucleus cell that makes up the majority of the embryo sac. When this fertilization takes place, a zygote is formed and the other 5 small cells disintegrate. What happens to the other sperm cell? It travels to the large, double-nucleus cell in the center of the ovule and fuses with its two nuclei. Remember, since each of the two nuclei in the large cell is haploid, and since the sperm cell that fuses with them is also haploid, the cell that results, called the **endosperm**, contains extra genetic information. This extra genetic information causes the endosperm to develop differently than a normal zygote. Whereas the zygote begins to develop into the seedling of the plant, the endosperm begins to develop into a food source for that zygote.

Since two sperm nuclei fuse with two different cells in the ovule, we say that plants go through **double fertilization**.

Double fertilization - A fertilization process that requires two sperm to fuse with two eggs

Double fertilization is unique to kingdom Plantae.

ON YOUR OWN

15.16 In Module #8, we used the letter "n" to represent haploid cells, and "2n" to represent diploid cells. Using this notation, how would you represent the zygote formed during double fertilization? How would you represent the endosperm?

15.17 The process of a seedling sprouting from a seed is often called "germination." One of the steps in the fertilization process described above is often called germination as well. What part of the fertilization process discussed above could be called germination?

Seeds, Fruits, and Early Plant Development

Once fertilization takes place, the endosperm and the zygote begin to develop by undergoing mitosis. The endosperm develops into the food source for the developing zygote. It does this by absorbing food that the plant continues to send it and holding it until the embryo absorbs it. Early on, the zygote forms a stalk of cells that anchor to the endosperm and begin to absorb nutrients from it. This powers the mitosis that begins to form the embryo of the plant. The next structure that forms in the plant is the **cotyledon**. We mentioned this structure in the previous module. It is the fundamental characteristic that separates monocots from dicots. Monocots produce one cotyledon in the embryo, and dicots produce two. The cotyledons either absorb the remaining endosperm and then provide nutrition to the embryo, or they produce enzymes that help the embryo to absorb more nutrients from the endosperm.

Once the embryo has fully formed its cotyledon or cotyledons, it is considered mature. At this point, it needs no more nutrients, because it stops its development. Since its need for nutrients has ceased, the endosperm "fills up" with as much food as it can, and then the connection between the ovule and the plant is severed. After this connection is severed, the ovule begins to develop a protective coating. Once the protective coating has been formed, the ovule is called a **seed**.

Although the seed has no connection to the plant, it is still inside the ovary. This ovary begins to swell and mature, becoming the **fruit** of the plant. Before we discuss the different kinds of fruits that result from the reproduction of flowering plants, it is important to make sure you understand the difference between a fruit and a seed.

Seed - An ovule with a protective coating, encasing a mature plant embryo and a nutrient source

Fruit - A mature ovary which contains a seed or seeds

So you see that a fruit contains a seed, but the fruit does not grow into a plant. The seed which is contained in the fruit is what eventually grows into a plant.

Plants in phylum Anthophyta produce a wide range of fruits. There are many ways that we can classify them. First of all, if a fruit forms from a single ovary, it is considered a **simple fruit**, whereas fruits that form from many ovaries are called **compound fruits**. Compound fruits can be further classified into either **aggregate fruits** or **multiple fruits**. If the fruit is formed by several ovaries from the same flower, it is an aggregate fruit. If the collection of ovaries come from different flowers, then it is a multiple fruit. Raspberries and strawberries, for example, are aggregate fruits. The multiple ovaries are all a part of the same flower on a raspberry bush or strawberry plant. You can tell this because the individual "bulges" on these fruits ripen at different times. This tells you that each bulge is an individual ovary. Pineapples, figs, and mulberries, however, form when ovaries from several different flowers fuse together to form a multiple fruit. When this happens, the ovaries are no longer individual, and the fruit still has bulges, but it ripens as a whole.

Simple fruits can be further classified into many, many different categories. To begin with, a simple fruit can be a **fleshy fruit** or a **dry fruit**. Fleshy fruits such as apples, tomatoes, cucumbers, and cantaloupes have fleshy tissue that forms between the seeds and the ovary covering. Dry fruits, on the other hand, have no fleshy tissue between the seeds and the ovary's covering. Nuts, grains, and pods are examples of dry fruits.

Fleshy fruits are further classified as **pomes** (pohms), **drupes** (droops), **berries**, or **modified berries**. Pomes are formed when the fleshy part of the fruit is not formed from the ovary, but instead comes from the tip of the pedicel. The ovary just forms a leathery covering for the fleshy tissue. Apples and pears are examples of pomes. If the ovary itself develops the fleshy tissue of the fruit, there are two possible structures for the fruit. The ovary can form a hard covering around the seed and then enclose that covering in fleshy tissue. These fruits are classified as drupes. Olives and peaches are examples of drupes. Typically, the seed encased in

the covering is called the "pit" of the drupe and the fleshy tissue is mistakenly called the "fruit."
If the ovary doesn't form a hard covering around the seed but instead simply encloses the seed or
seeds in fleshy tissue, it is some form of berry. If the ovary has a thin covering around this fleshy
tissue, it is just called a berry, whereas if the ovary has a thick, tough covering, the fruit is called
a modified berry. Whereas tomatoes and cherries are examples of berries, cucumbers and
oranges are modified berries.

Dry fruits can also be classified into several different groups as well. If the ovary has a
single chamber containing many seeds, it is called a **pod**. Beans and peanuts are examples of
pods. If the ovary has many chambers, each of which contains many seeds, we call the fruit a
capsule. Poppy seeds form in capsules. If the fruit forms a long, thin wing from the ovary wall,
it is called a **samara** (suh mar' uh). Maple tree fruits are examples of samaras. If the ovary
forms a hard, woody covering around a single seed, it is called a **nut**. Walnuts and acorns are
examples of nuts. It is important to note that although a peanut contains "nut" in its name, it is
not a nut. It is a pod, because its single chamber contains 2 seeds. If the ovary wall is actually
connected to the seed, the fruit is called a **grain**, while an ovary that is separated from the seed
forms an **achene** (uh keen'). Corn is a grain, while the fruit of a sunflower is an achene.

To give you some more experience with the classification of fruits, perform the following
experiment.

EXPERIMENT 15.2
Fruits

Supplies:

- Sharp knife
- Lab notebook
- A variety of different fruit (suggested fruit: apple, plum, orange, tomato, walnut, sunflower
  seed, maple seed, pea in pod, strawberry and raspberry)

Object: To observe the various types of fruit and compare their many differences.

Procedure:

A. Dissect and investigate each fruit and place it in a class. Here is an abbreviated list of each
fruit class and its description:

1. **Aggregate fruit**: Several separate ovaries of a single flower which ripen individually

2. **Multiple fruit:** Several ovaries from separate flowers which ripen fused together

3. **Pod**: Single chambered ovary with more than one seed, opens along side when ripe

4. **Capsule**: Multiple chambered ovary, each with many seeds, opens when ripe

5. **Samara**: Fruit with thin wings

6. **Nut:** Thick, hard, woody ovary wall enclosing a single seed

7. **Grain:** Ovary wall fastened to a single seed

8. **Pome:** Fleshy portion develops from the receptacle, ovary forms leather core with seeds inside

9. **Drupe**: Ovary has outer layer fleshy, inner layer forms hard, woody stone or pit, usually only one seed

10. **Berry:** Thin-skinned fruit with ovary divided in sections that contain the seeds

11. **Modified berry**: Like a berry, but with a thick, tough skin

Now don't get wrapped up in memorizing all of the different classifications of fruits. We wanted to discuss them so that you would get an idea of the diversity of fruits that exist in phylum Anthophyta, but we don't want you to waste time memorizing them. It is more important for you to understand how fruits aid in the dispersal of seeds.

Fruits have been designed to move the seeds of a plant away from the parent. After all, if a seed ends up on the ground near the parent plant, then the plant produced by the seed will compete with the parent for nutrients and resources. If, instead, the seed can end up on the ground some distance away from the parent plant, the developing plant will not compete with its parent. Fruits disperse the seeds far from the parent plant in many ways. Samara, for example, will float on the wind because of their wings. This allows them to drop on the ground far from the parent plant. Nuts and pods are often carried away from the plant by animals. When the animals open the nut or pod to eat the seeds, some seeds spill on the ground and are lost by the animal. Compound fruits, as well as many fleshy fruits, are eaten by animals. The animals usually carry the fruits away from the plant before eating them. When they eat the fruit and discard the seeds, they have succeeded in moving the seed away from the plant. Other fruits are eaten, seeds and all, by animals. The seeds have hard enough coverings to prevent them from being digested. Thus, when the animal excretes its undigested food, the seeds come out as well, ready to begin development.

Now remember, the growth of the embryo stops when the seed is mature. When the seed is dispersed, however, the embryo can "turn on" its development again. Typically, this happens when the seed embryo begins to absorb a significant amount of water. There are other factors, however. Often, seeds need the right temperature or amount of oxygen in the soil before the embryo will restart its development. Once these factors are in place and the embryo starts growing again, it breaks through the seed wall. Gravotropism allows the plant to grow its roots down into the soil, and allows the seedling to grow up, breaking through the soil surface. Typically, this process is referred to as **germination**.

The seed has stored a large amount of nutrients in its endosperm, and these nutrients sustain the germination process. Once the seedling has broken through the soil surface, however, it can begin photosynthesis to supplement and eventually replace the endosperm as the nutrient source. The cotyledon or cotyledons are the first leaves formed by the plant, and they begin the photosynthesis. In dicots, the two cotyledons only survive for a few days and eventually die and fall off as the plant produces more leaves.

⟍ ON YOUR OWN

15.18 If you plant a seed too deep, the seedling will break through the seed and begin to grow. It will die before it reaches the surface, however. What does it die from?

ANSWERS TO THE ON YOUR OWN PROBLEMS

15.1 <u>Turgor pressure can be temporarily neglected</u>. Since turgor pressure keeps a plant from wilting, as soon as the plant starts wilting, you know that it is not maintaining turgor pressure. Since a plant can wilt without dying, then we know that a lack of turgor pressure is not necessarily fatal.

15.2 Plant roots also need oxygen for respiration. Underwater, the plants cannot absorb oxygen. Thus, <u>putting plant roots under water deprives them of oxygen</u>.

15.3 <u>Sample B came from the xylem and sample A came from phloem</u>. Xylem transfer water and minerals up the plant. Phloem transport organic materials (such as the products of photosynthesis) from the leaves down the plant.

15.4 Water transport takes place due to evaporation of water through the leaves. When the stomata are closed, evaporation does not take place. Thus, <u>water transport takes place during the day</u>.

15.5 Auxins are responsible for gravotropism. Gravotropism allows a plant to know which way is up and which way is down. Thus, <u>auxins are most likely responsible for a seedling knowing which way to grow</u>.

15.6 Ethylene promotes the formation of the abscission layer and the ripening of fruit. Thus, <u>conifers do not contain ethylene</u>.

15.7 A Venus's flytrap makes its leaves and traps insects to get nitrogen and other raw materials for biosynthesis. If the Venus's flytrap can get them from the soil, then it need not produce those specialized leaves or trap insects. Thus, <u>the insect-trapping leaves will stop working and the plant will not produce more</u>.

15.8 <u>The second was the result of vegetative reproduction, while the first was the result of sexual reproduction</u>. The offspring of vegetative reproduction are genetically identical to the parent plant. Thus, the flowers of the offspring should look like the parent plant.

15.9 Grafts work when the scion and stock are close to the same species. Thus, <u>the gardener should graft to the Red Delicious apple tree</u>.

15.10 <u>Yes, because one of its flowers can produce a stamen and another can produce a carpel. If pollen goes from the stamen of one flower to the carpel of another on the same plant, the plant has sexually reproduced with itself</u>.

15.11 If the style is short, the carpel will not be very well exposed to the wind. Thus, <u>animals will be the best means of pollen transfer</u>.

15.12 When a diploid cell undergoes meiosis, four haploid cells are formed. In a plant, each of those haploid cells forms a pollen grain. Thus, there should be <u>4,000 pollen grains</u>.

15.13 When the first mitosis occurs, there will be two nuclei. When each of those undergo mitosis, there will be four. When each of those undergoes mitosis, there will be eight. Thus, <u>the megaspore undergoes mitosis 3 times</u>.

15.14 Red flowers tend to attract birds. Since birds have a lousy sense of smell, they usually do not have a fragrance. Thus, <u>if a flower has no smell, it more likely to be red than any other color</u>.

15.15 <u>You would see little or no flowering plants</u>. Birds and insects are vital to the pollination of flowering plants. Without them, the reproductive process of flowering plants is so inhibited that flowering plants will not flourish.

15.16 In this notation, "n" represents half of the genetic code. Thus, a haploid cell is "n" and a diploid cell is "2n." Since the zygote is the product of fusion between two haploid cells, <u>the zygote would be labeled as "2n."</u> The endosperm is the product of the fusion of 3 haploid nuclei, so <u>the endosperm would be labeled as "3n."</u>

15.17 When the pollen tube begins to form, it grows, much like a plant grows. Thus, <u>the formation of the pollen tube is often called germination</u>.

15.18 <u>The plant dies from starvation</u>. Remember, the endosperm has some nutrients to allow the embryo to start growing. If the cotyledon or cotyledons of the embryo do not get to the surface and begin photosynthesis before the endosperm runs out of nutrients, the plant will have no more food and will die.

STUDY GUIDE FOR MODULE #15

1. Define the following terms:

a. Physiology
b. Nastic movement
c. Pore spaces
d. Loam
e. Cohesion
f. Translocation
g. Hormones
h. Phototropism

i. Gravotropism
j. Thigmotropism
k. Perfect flowers
l. Imperfect flowers
m. Pollination
n. Seed
o. Fruit

2. Name the four processes for which plants require water. Which of these processes can be neglected for a short amount of time?

3. A biologist studies two plants. The leaves of the first plant follow the sun as it travels across the sky each day. The second plant does not follow the sun each day. However, if the plant is placed so that one of its sides is in the shade and the other is in the sunlight, the plant will eventually grow so that all of its leaves point towards the sunlight. Which plant is using nastic movement and which is using phototropism?

4. Briefly describe the cohesion-tension theory of water transport in plants.

5. Do xylem cells need to be alive in order for xylem to do their job? Why or why not?

6. Do phloem cells need to be alive in order for phloem to do their job? Why or why not?

7. What substances do xylem contain? What substances do phloem contain?

8. Do insectivorous plants really eat insects? Why or why not?

9. From a genetic point of view, what is the difference between vegetative reproduction and sexual reproduction in plants?

10. A gardener says that one limb of his crabapple tree now produces normal-sized apples. What must the gardener have done to make this happen?

(STUDY GUIDE CONTINUES ON NEXT PAGE)

11. Identify the structures in the figure below:

_____

Composed of:

_____

_____

_____

_____ Composed of:

_____

_____

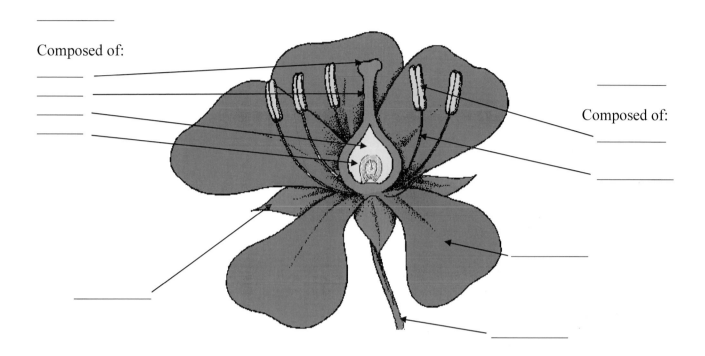

_____

_____

12. What is the male reproductive organ of a flower? What is the female reproductive organ?

13. Typically, how many cells are in a pollen grain? How many of those are sperm cells?

14. Typically, how many cells are in an embryo sac? How many of them get fertilized?

15. Where does the endosperm come from? What is its purpose?

16. The cotyledon or cotyledons help provide food for the plant both before and after germination. How do cotyledons accomplish each task?

17. What is the purpose of a fruit?

18. Name at least three ways in which pollen is transferred from the stamens of one flower to the carpels of another.

MODULE #16: Reptiles, Birds, and Mammals

Introduction

In the previous two modules, we took a break from discussing the animal kingdom and concentrated instead on the plant kingdom. Well, in this, the final (yeah!) module of the course, we are going to revisit the animal kingdom, concentrating on reptiles, birds, and mammals. Those who believe in macroevolution call these the "higher" animals, because they are under the mistaken impression that reptiles, birds, and mammals evolved from the other animals that we have already studied. Of course, you know that the whole idea of macroevolution is scientifically absurd, but nevertheless, you will probably hear the term "higher animal" in the future, so it is important for you to know where it comes from.

Class Reptilia

Class Reptilia contains the turtles, snakes, lizards, alligators, and crocodiles of Creation. In addition, biologists put dinosaurs in this class. The members of class Reptilia have several characteristics in common:

- They are covered with tough, dry scales.

- They breathe with lungs.

- They have a three-chambered heart with a ventricle which is partially divided.

- They produce **amniotic** (am nee ah' tik) **eggs** covered with a leathery shell. Most are oviparous, some are ovoviviparous.

- If they have legs, the legs are paired.

These characteristics serve to distinguish reptiles from other creatures in Creation, so they are worth discussing.

First, reptiles are covered with tough, dry scales. Many people call snakes "slimy." However, since they belong to class Reptilia, they are not slimy at all. If you touch any reptile (including a snake), its skin feels cool, dry, and leathery. This is due to its scales. Unlike fish, the scales on a reptile are not covered in mucus. Instead, God has designed reptile scales to prevent water loss. Many reptiles live in regions where water is scarce. Thus, it is important that they do not lose water through evaporation. The scales make sure that this does not happen. Scales serve another purpose as well. Reptiles are ectothermic, which means they have no internal mechanisms to keep themselves warm. This is why reptiles like to lay on warm rocks and sun themselves. This allows them to warm up their bodies. Once they are warm, their scales help insulate them, keeping the warmth in. Although reptiles love to sun themselves, it is possible for a reptile to get too much sun. If this happens, the reptile will actually get too hot and die. Thus, during the hottest part of the day, you rarely see reptiles sunning themselves. During that time, they are actually hiding from the sun in a cool, shady spot.

The scales on a reptile are not living tissue. As a result, they cannot grow with the creature. Much like the arthropods, reptiles must periodically shed their old skin in order to continue growing. In order to accomplish this, a new, larger set of scales begins growing before the old scales are shed. That way, the reptile is always covered with scales.

The next characteristic of reptiles is that they breathe with lungs. Remember, most adult amphibians have lungs, but their lungs are only one of three possible means of respiration. For reptiles, lungs are the only means of respiration. As a result, a reptile's lungs are more efficient and have more capacity than the lungs of a similarly-sized amphibian. Surprisingly enough, even reptiles that live in the water (like water snakes and sea turtles) still breathe through lungs. Thus, it is possible for such reptiles to drown, despite the fact that they make the water their home!

Fish have two-chambered hearts, amphibians have three-chambered hearts, and reptiles have three-chambered hearts whose ventricles are partially divided. If you remember the amphibian heart you studied in Module #13, you will recall that it was comprised of two atria and one ventricle. This resulted in three chambers. Well, the reptilian heart has a ventricle that is partially-divided. As a result, we can think of reptiles as having an "almost four-chambered" heart.

The next characteristic of reptiles may use a term with which you are not familiar. Amniotic eggs are different than the eggs of fish and amphibians.

Amniotic egg - An egg in which the embryo is protected by a membrane called an amnion. In addition, the egg is covered in a hard or leathery covering.

Remember, fish and amphibians lay their eggs underwater. These eggs are jelly-like, with only a thin membrane covering them. This is okay, though, because they are always bathed in water. If reptile eggs were like that, they would dry out right away when they are laid on the ground. Thus, reptiles lay amniotic eggs. Figure 16.1 is an illustration of a typical amniotic egg.

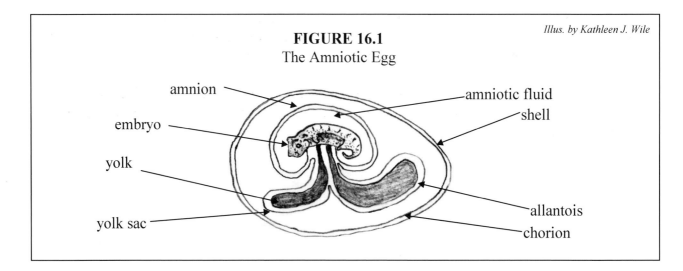

FIGURE 16.1
The Amniotic Egg

Illus. by Kathleen J. Wile

In reptiles, fertilization occurs inside the female. Once this happens, the developing zygote is encased in a protective **shell** and passes out of the female's body. The shell is a remarkably engineered structure, which is just porous enough to exchange gases with the environment but not porous enough for water to evaporate out of the egg! As the egg develops, four membranes form inside. The **amnion** (am nee ahn') grows around the embryo, forming a fluid-filled sac in which the embryo floats. The next membrane is called the **yolk sac**. An amniotic egg has a clump of nutrients that is used to sustain the embryo during development. This clump of nutrients is called the **yolk**, and the yolk sac encloses it. As the yolk is used up by the developing embryo, it slowly disappears, and the yolk sac is drawn into the embryo until it disappears as well. The next membrane, the **chorion** (kor ee ahn'), is a membrane that lines the inner wall of the shell, enveloping all of the contents of the inside of the egg. Finally, the **allantois** (uh lan' toh iss) is a sac of blood vessels that allows for the respiration and excretion of the embryo.

As we already mentioned, the shell of an egg is a marvelously engineered device, but that is only the beginning of the careful design apparent in an egg. Think about it. The egg is a self-contained unit. It provides a home, protection, and nourishment for the developing embryo. Since the egg is porous, the embryo can breathe. Of course, if it were too porous, water would escape, but the egg shell is not that porous. However, whenever a surface is porous enough to let in air, it is also porous enough to let in bacteria and other pathogens. This presents a problem, doesn't it? The embryo must breathe. At the same time, however, if oxygen can get into the egg, so can pathogens. Why don't bacteria and other pathogens infect eggs? Well, the bacteria and pathogens do get *into* the egg, but they never make it to the tissues of the embryo or its food. Why?

Well, when you break open an egg, you see a yellow part and a white part, called the **egg white**. The yellow part is the yolk. If the egg had an embryo in it, the embryo would feed off of the yolk. The egg white is what God has put in the egg to protect the embryo from pathogens. When bacteria and other pathogens enter the shell and pass through the chorion, they immediately encounter the egg white, which is full of chemicals which are toxic to pathogens! Thus, the egg white is the defense mechanism for an egg, overcoming the problem of letting pathogens in with the air that is essential to the embryo's development. Isn't that amazing? If the egg never hatches for some reason, then eventually, these protective chemicals in the egg white will run out. At that point, the bacteria and other pathogens are free to eat the yolk. Once this happens, the egg begins to smell, and we say that the egg is "rotten."

The last common characteristic of reptiles is that of paired legs. Now, of course, not all reptiles have legs. Snakes, for instance, do not have legs, yet they are still considered reptiles. However, if a reptile has legs, those legs are paired.

ON YOUR OWN

16.1 You are blindfolded and asked to feel the skin of two creatures. One is a fish and the other is a snake. If the first creature feels dry and leathery and the second feels slimy and rough, which is the fish and which is the snake?

16.2 If an amniotic egg develops without an allantois, will the embryo live? Why or why not?

16.3 Through X-ray technology, you observe the inside of an egg as the embryo develops. At first, the yolk sac and allantois are the same size, so you cannot tell them apart. After a few days, the one on the left is smaller than it was, and the one on the right is the same size as before. Which one is the allantois?

16.4 You have a tadpole and a baby water snake. Can either of them drown? If so, which one?

Classification of Reptiles

As is the case with all classes of vertebrates, there is plenty of diversity in class Reptilia. Figure 16.2 shows a few of the creatures contained in this class.

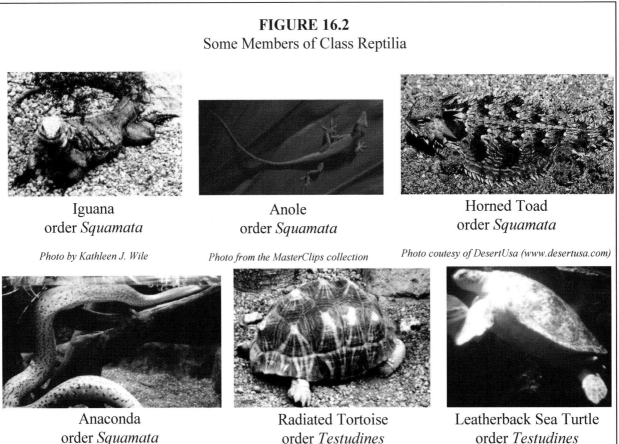

FIGURE 16.2
Some Members of Class Reptilia

Iguana
order *Squamata*

Photo by Kathleen J. Wile

Anole
order *Squamata*

Photo from the MasterClips collection

Horned Toad
order *Squamata*

Photo coutesy of DesertUsa (www.desertusa.com)

Anaconda
order *Squamata*

Photo by Kathleen J. Wile

Radiated Tortoise
order *Testudines*

Photo by Kathleen J. Wile

Leatherback Sea Turtle
order *Testudines*

Photo by Dr. Jay L. Wile

There are many orders within class Reptilia, but many of these orders no longer have any members which are currently living. The majority of reptiles living today belong to one of three orders: Testudines (test uh deen' eez), Squamata (squah mah' tuh), and Crocodilia (crok uh dil' ee uh). Order Testudines is home to the turtles in Creation, while snakes and lizards occupy order Squamata. Alligators and crocodiles make up order Crocodilia. The majority of the other orders are used to classify dinosaurs, which are assumed to be extinct today. There is one order, however, that contains a single species which is still living today.

Order Rhynchocephalia

Order Rhynchocephalia (ring koh suh fale' ee uh) contains many now extinct reptiles (including some dinosaurs) but it also contains one living reptile, called the **tuatara** (too uh tar' uh). Its name means "spine bearer," which points out one of its distinguishing characteristics; a tuatara has spines along its back. Tuataras typically grow to a maximum length of 30 inches and usually eat insects. They are largely nocturnal (active at night), but they spend a good part of the day basking in the sun near the entrance to their home, a burrow in the sand. Tuataras have a very slow reproductive process. They have a life span of 60 years but do not start reproducing until they are 20 years old. Once a tuatara lays an egg, the embryo must develop unattended for more than a year before hatching!

The most distinguishing feature of the tuatara is a third eye, which sits on the top of its head. Although this eye, called a **parietal** (puh rie' uh tul) **eye**, does have light-sensitive cells (in tissue called a retina), it is covered with scales, so there is no way that it can send a real image to the brain. The eye is not useless, however, because it is connected to the brain with a nerve. Many biologists think that this eye functions as a light sensor which helps the tuataras determine how long to bask in the sun before it retreats into its burrow.

Tuataras live on a few small islands near New Zealand. Their population is dwindling, however, because when the Europeans landed on these islands, they brought along rats in their ships. These rats found their way off of the ships and onto the islands, throwing the ecosystem out of balance. The rats began eating the tuatara eggs, which previously had no real predators. Since the eggs lay unattended for more than a year, they are relatively easy pickings for the rats. As a result, not nearly as many tuataras are born now as previously. To correct for this, environmental organizations are moving tuatara to islands that are rat free, hoping that they will thrive there.

ON YOUR OWN

16.5 We assume that the tuatara uses its parietal eye to limit the time that it basks in the sun. Why must it limit its time in the sun?

<u>Order Squamata</u>

Snakes and lizards make up order Squamata. Although they belong to the same order, there are several differences between them:

- Lizards have two pairs of limbs, while snakes have none.

- Lizards have ears and can hear, while snakes are deaf.

- Lizards have the same type of scales all over their bodies, while snakes have specialized scales on their bellies for locomotion.

- Most lizards have eyelids and can therefore close their eyes. Snakes' eyes are always open because they have no eyelids.

These differences allow us to distinguish between lizards and snakes.

Snakes

Most people are afraid of snakes on one level or another. This is probably due to the fact that they have the reputation of biting and poisoning people. Although this is the popular view of snakes, most of them are quite harmless and many are beneficial. After all, snakes feed on rodents and insects that plague humans. Without these useful reptiles, the populations of these pests would quickly rise out of control!

As pointed out above, snakes do not have legs. Instead, they have specialized scales on their bellies. These scales, called **scutes** (scoots), have the ability to grasp the ground upon which the snake is crawling. When the snake wants to move, it forces the scutes to move forward on its skin. The scutes then grasp the ground and the snake can pull itself forward with them. This isn't the only type of locomotion that the snake can use, however. The snake can also wind itself into a tight "S"-shape and then extend itself forward. Also, if there are plenty of twigs, stones, and the like, the snake can simply wind its way around them, pushing against the twigs and stones to move itself forward. Many snakes also live in the water and can swim rather efficiently.

Snakes are carnivorous, feeding on live prey such as rodents, insects, lizards, eggs, and other snakes. Although their eyes are always open (as we pointed out above), they are not especially sharp-sighted. In addition, snakes are also deaf. They have no ears, so they simply cannot hear. They can, however, sense vibrations that travel through the ground. Snakes are drawn to regular ground vibrations, as they represent a possible source of prey.

Since snakes cannot hear and have poor eyesight, they use other senses to hunt. Snakes have a keen sense of smell and typically rely heavily on it for hunting. They have two nostrils on the front of their heads which lead to the olfactory lobes in the brain. To augment this sense of smell, however, most snakes have sensory pits called **Jacobson's organs**. These sensory pits are in the snake's mouth. When a snake sticks out its tongue, it collects chemicals suspended in the air. It then pulls its tongue back into its mouth, transferring the collected chemicals to the Jacobson's organs, which send nerve signals to the brain. The brain correlates these signals with those coming from the nostrils, and the result is probably the keenest sense of smell in Creation. Between the nostrils and the Jacobson's organs, snakes can pick up even the faintest trace of an odor!

Once prey has been detected, the snake captures the unfortunate creature. The simplest means by which this happens is for the snake to simply grab onto the creature with its mouth and begin swallowing it. Some snakes wind themselves around their prey and then squeeze them to death. Other snakes use their bite to inject a poison into their prey. This poison, manufactured in poison glands behind the snake's fangs, can be powerful enough to cause death in humans.

Some snakes use more than one of these methods for capturing prey. The **garter snake**, for example, preys on many creatures, including earthworms and toads. When it eats earthworms, it simply grabs onto the worm with its mouth and begins to swallow it. When it eats a toad, however, it will first constrict the toad, then begin to swallow it. This kind of behavior is generally the exception rather than the rule. Most snakes use only one method to catch their prey. Garter snakes are exceptions to the rule in more than one way. Not only do they use multiple techniques to capture prey, but they are also ovoviviparous, which is rare in class Reptilia.

If a snake uses the constriction method of capturing prey exclusively, it is often called a "constrictor." Two of the more famous constrictors are the **anaconda**, which is found in South America, and the **python**, which lives in tropical and subtropical climates. These snakes can grow very long (up to 40 feet) and have been known on rare occasions to constrict and eat humans. A less intimidating form of constrictor, the **boa constrictor**, usually reaches lengths of only 10 feet or so. These snakes, predominantly found in Central and South America, are often kept as pets by reptile enthusiasts.

Although poisonous snakes are probably the reason that many people are so frightened of this order of reptiles, they actually make up a small percentage of the total number of snakes in Creation. These snakes produce a venom in poison glands which are located behind their teeth. Typically, poisonous snakes produce either **neurotoxins** (noor oh tahk' sinz) or **hemotoxins** (hee muh tahk' sinz).

<u>Neurotoxin</u> - A poison that attacks the nervous system, causing blindness, paralysis, or suffocation

<u>Hemotoxin</u> - A poison that attacks the red blood cells and blood vessels, destroying circulation

Neurotoxins are fast-acting, while hemotoxins take longer to do damage. Hemotoxins are by far the more deadly of the two, however.

There are basically two types of poisonous snakes. The first kind have short, fixed fangs that deliver the poison while the snake chews on its prey. Those in this category predominately produce neurotoxins as their poison. The second type of poisonous snake has long fangs which fold away into pockets of the mouth when they are not in use. These snakes typically produce hemotoxins as their poison, and they usually inject a large amount of the poison through their fangs in a single strike.

Examples of the short-fanged poisonous snakes are the **sea snake, coral snake,** and **cobra**. As we mentioned before, even though sea snakes live in the sea, they must breathe air through lungs. Thus, sea snakes must raise their heads out of the water periodically to get gulps of air. Some species of sea snake have such large lungs that they can stay underwater for more than an hour with just one gulp of air. Thus, to the short-term observer, sea snakes may look like they do not need to breathe air. If you watch them long enough, however, you will see that they do. Although generally unaggressive, the sea snake has perhaps the most powerful poison of any of the short-fanged snakes.

The coral snake lives in many sub-tropical areas, including the southern part of the United States. It is easy to distinguish, being marked with bright red, yellow, and orange bands. Like most snakes, it is not really aggressive. In fact, if you are outfitted properly, the coral snake isn't even dangerous. You see, the fangs of a coral snake are so short that they cannot penetrate thick clothing. Thus, people who hike in the southern parts of the United States need only to wear shoes with thick socks and thick pants to protect themselves from these snakes.

Cobras typically live in parts of Asia. These snakes are responsible for the majority (almost 3/4) of the annual deaths due to snakes. Cobras are best known for their "hoods," an expandable region of skin behind their heads. When a cobra is ready to strike, it expands this hood as a warning. Some cobras, like the **black-necked cobra** of Africa, can spit their venom. They aim for their opponent's eyes, hoping to blind the creature.

The long-fanged variety of poisonous snakes are commonly called **vipers**. Vipers are common in Africa, Europe, and Asia. In the Americas, a special kind of viper called the **pit viper** exists. These vipers have a special means of detecting prey. In addition to their nostrils, pit vipers have two **heat-sensing pits**. These pits, located between the nostrils and the eyes, allow the snake to sense warm-blooded creatures, even if there is no light with which to see them and no way to detect their smell. Most pit vipers like to hunt in total darkness, because their heat-sensing pits give them an advantage in the dark. Common examples of pit vipers in the United States are **water moccasins** (which live in southern swamps and lakes), **copperheads** (which live in the woods on the East Coast), **rattlesnakes** (which are common throughout the United States), and the **western diamondback** (which is found from Texas to eastern California).

Regardless of how the prey is captured, snakes swallow their prey whole, even if the creature is significantly larger than the snake. This can happen because the jaw of the snake is not attached to its skull. Instead, it is attached to the **quadrate bone**, which acts as a hinge. This arrangement, along with its elastic skin, allows the snake to open its mouth very wide so that it can swallow creatures that are much, much larger than it is. An African python, for example, was discovered with a 130-pound impala (an African antelope) inside it!

Lizards

Unlike snakes, lizards have four legs, can close their eyes, can hear, and have scales on their bellies that are the same as the scales on the rest of their bodies. Most lizards are small. The **gecko**, for example, is a small climbing lizard. It rarely reaches lengths greater than nine inches. This interesting lizard is an excellent climber, however. It has sticky pads on its feet that allow it to climb on vertical surfaces. It can even climb upside down on ceilings!

Like the gecko, the **chameleon** is a tree-dwelling lizard that can change color in response to several different stimuli. It can use this color change ability to blend into its surroundings or to communicate with other chameleons. This fascinating lizard has a long tongue that it shoots out to catch insects. The tongue of a chameleon can be almost as long as the chameleon itself! If you bought a chameleon in a pet store, it is most likely not really a chameleon. It is most likely an **anole** (uh noh' lee), which is often called a "false chameleon." This lizard can change its colors, but not nearly with the range that a true chameleon can. In addition, the stimuli that cause the anole to change its colors are not nearly as diverse as those that cause a true chameleon to change its colors.

The **horned toad** is a common lizard in the western United States. Like the tuatara, it has spines on its head and body. When it is afraid, this lizard swells up, making its spines very obvious. In addition, it actually *squirts blood* through its eyes at its attacker! These actions are designed to frighten its would-be attacker, because as fearsome as the horned toad looks, it does not fight well at all. Thus, its best bet is to frighten away its attacker. In fact, many people keep horned toads as pets, an indication of how harmless the lizard really is!

Some lizards are not so small or harmless, however. The **Gila** (he' luh) **monster** is the largest lizard in the United States. It grows to a length of almost 2 feet. Although this is not nearly as large as some of the other reptiles, it is large for a lizard. In addition, the Gila monster produces poison in venom glands that reside in its lower jaws.

<u>Order Testudines</u>

Turtles and tortoises are found in order Testudines. The term "turtle" refers to the members of this order which live in water, while the term "tortoise" is used to refer to those that live on land. Members of this order are distinguished by the large shells that they carry on their backs. Most members of this order can pull all of their appendages (legs, neck, and head) into the shell and close off the openings with little "trap doors." This allows them to become veritable fortresses, almost impervious to attack.

Most members of this order have short, thick legs that end in five-clawed toes, although some turtles have flippers instead. These animals do not have teeth. Instead, their jaws form a tough beak that can be used to snap their food into bits. The nostrils of a turtle are placed very high on the snout, allowing them to breathe air while keeping the vast majority of their body submerged underwater. Because turtles live predominantly in the water but can come onto the shore periodically, they are often confused with amphibians. Turtles are not amphibians, however. They have all of the characteristics of reptiles.

Most turtles and tortoises are rather small. Some can reach great sizes, however. The **giant tortoise**, which lives on the Galapagos Islands, can weigh as much as 350 pounds! The largest member of order Testudines is the **leatherback sea turtle**. These turtles live in the sea, having lightweight, streamlined shells and flippers instead of claws. They typically come to shore only to lay their eggs. When the eggs hatch, the hatchlings make a mad dash for the sea, as predators such as seagulls and crabs try to catch them. Adult leatherback sea turtles have been known to weigh as much as 1500 pounds!

Order Crocodilia

Alligators and crocodiles make up order Crocodilia. What's the difference between an alligator and a crocodile? Well, alligators have thick, blunt snouts. When they close their mouths, their teeth all fit inside. Crocodiles, on the other hand, have thinner, more pointed snouts. In addition, not all of their teeth fit into their mouth. Thus, when a crocodile's mouth is closed, you can still see teeth pointing out of the sides of the mouth.

Aside from the leatherback sea turtles, members of this order are considered to be the largest reptiles in Creation. It turns out that the size of a reptile is really only limited by its lifespan. Thus, in general, the longer a reptile lives, the larger it grows. This cannot be said of most other creatures. Most organisms have some fundamental size limitation that does not allow them to grow past a certain range. This is just not the case for reptiles. Crocodiles and alligators can have very long life spans, which accounts for their large size.

Crocodiles and alligators are stealthy predators. Their eyes and nostrils are placed on the top of their heads, so they can glide through the waters very quietly with only the topmost portion of their heads breaking the surface of the water. One of their favorite means of catching prey is by floating in the water with just their eyes and nose above the surface. They float so still that most creatures mistake them for logs. When a creature gets close enough, however, it soon finds out its mistake! By then, it is too late.

Like most reptiles, members of this order are oviparous. They tend to lay their eggs and cover them with a mound of rotting vegetation. The heat generated by the decomposing organisms eating the rotting vegetation warms the eggs, allowing them to be unattended. Although the mother does not attend the eggs, she does visit them from time to time, because after about 2 weeks the eggs hatch but the young are buried in vegetation. They begin making peeping noises which tell the mother to dig them out and carry them to the water.

ON YOUR OWN

16.6 A reptile has no eyelids and is deaf. Is it a lizard or a snake?

16.7 Why do biologists say that human beings are responsible for the tuatara's dwindling population? Humans do not hunt or kill tuataras.

16.8 If a snake has a heat-sensing pit, is it poisonous or non-poisonous?

16.9 Why do some biologists say that snakes smell with their tongues?

16.10 You see a member of order Testudines that has flippers. Is it a turtle or a tortoise?

16.11 You see a member of order Crocodilia with its mouth closed. Even though its mouth is closed, you still see teeth. Is it a crocodile or an alligator?

Dinosaurs

We could hardly leave a discussion of reptiles without mentioning dinosaurs. The term "dinosaur" actually means "terrible lizard," so we can conclude that dinosaurs are considered part of class Reptilia. Indeed, many of the orders in class Reptilia exist simply to classify dinosaurs. Unfortunately, discussing dinosaurs is a very difficult task because we have never seen them. The only data that we have regarding the dinosaurs are their fossilized remains. This is a serious problem. You see, the most likely part of a vertebrate to fossilize is the skeleton. Fleshy tissue, organs, scales, hair, etc. rarely, if ever, fossilize. Thus, what we know of the dinosaurs is rather limited.

There is also another problem we face when trying to understand dinosaurs. Of the fossilized remains that we do have, there is precious little. You see, 95% of all fossils that we recover are of mollusks. There are, quite literally, fossilized mollusks in every region of the earth. Of the 5% remaining, 95% of those are marine vertebrates such as fish. Of the remaining 5% of those, 95% are insects. The remaining 5% are made up of reptiles, plants, and mammals. Think about that for a moment. If you read most textbooks on fossils, you will think that there are, literally, thousands of fossils of dinosaurs. There aren't. Only five percent of five percent of five percent, or 0.0125%, of the fossils that we have discovered are of reptiles, plants, and mammals. Thus, dinosaurs fossils are a tiny, tiny fraction of the fossils that we have.

Despite the fact that there is precious little data, paleontologists (biologists who study fossils) have been able to learn something about some of the dinosaurs that once existed on earth. Figure 16.3 shows some artists' renderings of what we think some dinosaurs might have looked like. Remember, since all we have is a few of the creatures' bones, these drawings are, at best, educated guesses.

FIGURE 16.3
Types of Dinosaurs

Apatosaurus: These were probably the largest of the dinosaurs. Their bodies were more than 60 feet long, and they could have weighed more than 30 tons. Biologists think that these dinosaurs were swamp-dwelling herbivores.

Stegosaurus: These dinosaurs reached a height of 13 feet or so and probably weighed as much as 10 tons. They had tiny brains, armored plates, and a spiked tail which they used for defense. Biologists think that these dinosaurs were herbivores.

Tyrannosaurus: Often called the "king of the dinosaurs," these creatures reached heights of 20 feet. With teeth that were more than 6 inches long and hind claws as long as 8 inches, biologists assume that these dinosaurs were fierce predators.

Plesiosaurus: An aquatic dinosaur that was probably a carnivore. In 1977, a Japanese fishing boat pulled in what seemed to be a rotting carcass of such a creature. After analyzing the pictures and samples of the catch, Japanese scientists declared that it was the remains of a plesiosaurus. American scientists disputed the conclusion, however, because it would destroy the hypothesis of macroevolution.

Pteranodon: A flying reptile that had a wingspan of as much as 30 feet. The skeletal remains indicate that these creatures were probably carnivores, and they were probably very poor fliers. Most likely, they glided more than they flew. Technically this is not a dinosaur, because it does not have the hips or skull of a dinosaur. Since it is such a large reptile, however, it is usually grouped with the dinosaurs.

Illus. from the Arts & Letters Express collection

Although these drawings might make it look like we know a lot about dinosaurs, we really know very little. Were they ectothermic like other reptiles? We assume so, but we really have no evidence. Some of the latest theories in paleontology are that at least some dinosaurs were warm-blooded, like the mammals. Was the tyrannosaurus really a ferocious predator? We assume he was because of the structure of his teeth, but many paleontologists have noted that while his teeth are really quite sharp, they have very shallow roots and could be dislodged from the skull rather easily. This might be evidence that he was more of scavenger. The point is, we really know very little about the dinosaurs.

Despite the fact the we have very little data regarding dinosaurs, paleontologists have jumped to some rather wild conclusions about them. For example, most paleontologists believe that dinosaurs lived long before human beings. Why? Well, there are really two reasons. First, most paleontologists are committed to macroevolution. Thus, they have to believe that the earth is ancient and that dinosaurs were a step in macroevolution that happened millions and millions of years ago. Second, they think that since we have never found dinosaur fossils and human fossils together, then dinosaurs and humans must have existed at different times. Actually, there are a few fossil finds that seem to consist of dinosaur and human fossils together, but paleontologists generally reject them out of hand.

There are several problems with such an idea, however. First, there is no reason to believe in the idea that the earth is really ancient. If you are committed to the hypothesis of macroevolution, then you must *assume* it is. However, we have already learned that macroevolution is not supported by scientific data. Thus, why should we believe in it? Second, archaeologists have found pictures drawn on cave walls by human beings. These pictures look like tyrannosaurus, brachiosaurus, and plesiosaurus. How in the world could people who lived in caves draw pictures that accurately portray these creatures unless they saw them? Finally, based on the fact that there are so *few* fossils of dinosaurs and even *fewer* fossils of human beings, there is no reason to expect that you would find them together. It is so rare to find them by themselves that it is virtually impossible to find them together.

In the end, then, there is really no reason to think that dinosaurs and man did not live together at one time. In fact, the Bible supports the idea that they did. In Job 40:15-24, God describes to Job a creature (behemoth) that sounds like brachiosaurus. Also, in Job 41:1-8, He describes a creature (leviathan) that sounds like plesiosaurus. In each case, God seems to imply that Job should *recognize* these creatures. If this is true, then Job must have lived with them. Most creation scientists believe that dinosaurs and humans did live together on earth before Noah's flood. They were taken on the ark (there was plenty of room for young dinosaurs), but they did not adapt well to the changes in climate that took place after the flood. Thus, they slowly died out and are, most likely, extinct today.

Class Aves

Birds have always been a fascinating subject for scientists to study. Their ability to fly inspired scientists and engineers to build airplanes so that human beings could fly as well. In some ways, then, these members of class Aves are probably the most closely studied creatures in Creation. To be a member of this well-studied class, a creature must possess the following characteristics:

- The creature must be endothermic.

- The creature's heart must have four chambers.

- The creature's mouth must end in a toothless bill.

- The creature must be oviparous, laying an amniotic egg which is covered in a lime-containing shell.

- The creature's body must be covered with feathers.

- The creature's skeleton must be composed of porous, lightweight bones.

Notice that the ability to fly is *not* on this list. That's because several birds, such as penguins and ostriches, do not fly. They are, nevertheless, members of class Aves.

The first characteristic of birds refers to the circulatory system. Birds are **endothermic**.

Endothermic - A creature is endothermic if it has an internal mechanism by which it can regulate its own body temperature, keeping it constant.

Unlike reptiles and fish, then, birds have internal mechanisms that keep their bodies warm. They need not lie out in the sun or use some other means to regulate their bodies' temperature. Instead, their bodies have mechanisms by which they convert food into energy and use that energy to keep their internal organs and tissues warm. In the same way, they have mechanisms that can cool their internal organs and tissues if they get too hot. When a creature has such mechanisms, we say that the creature is "warm-blooded."

The second characteristic of birds also refers to the circulatory system. Birds have four-chambered hearts. Remember, amphibians have three-chambered hearts, with a single ventricle and two atria. In reptiles, the ventricle has a partial separation; thus, the reptile heart is "almost four-chambered." Well, birds have a true four-chambered heart, with two atria *and* two ventricles. This allows a complete separation between oxygenated and de-oxygenated blood. In a bird's circulatory system, the right atrium receives de-oxygenated blood and dumps it into the right ventricle. The right ventricle then pumps the de-oxygenated blood into the lungs. Once the blood is fully-oxygenated, it travels from the lungs to the left atrium. The left atrium then dumps the oxygenated blood into the left ventricle, which then pumps the blood to the various tissues of the body.

The next characteristic of birds relates to the mouth. A bird's mouth ends in a toothless bill. The structure of a bird's bill varies depending on the way in which it gathers its food. For example, predatory birds such as hawks and eagles have sharp, hooked bills that are ideally designed for tearing meat. Flamingoes, on the other hand, have long bills with sieves on the end, because they feed on tiny organisms that live in or near the mud of rivers and streams. The sieves on flamingo bills serve to filter the mud away from the food that they are eating. Finally, birds such as hummingbirds that feed off the nectar of flowers have long, slender bills ideal for seeking the sticky-sweet substance inside a flower.

Another characteristic of birds involves their method of reproduction. Like reptiles, fertilization is internal in birds, and birds are oviparous, laying amniotic eggs. Unlike reptiles, however, the covering of a bird's egg is not leathery. Instead, it is much harder, because it contains lime, a calcium-containing compound. Remember when we discussed bones, we made the point that bone tissue is hard because it contains calcium. Much like bone tissue, then, the covering of an egg is hard because it is made up of a calcium-containing compound. To get an idea of how a bird embryo develops in its egg, perform the following experiment.

EXPERIMENT 16.1
Bird Embryology

Supplies:

- Micro slide of **The Chick Embryo**
- Magnifying glass
- Microscope (optional)
- Desk lamp
- Lab notebook
- Colored pencils

Object: To observe bird embryology by studying **The Chick Embryo**. Remember, however, that the changes and growth you see here are very similar to that which occurs in mammals. Thus, in some ways, this is also a lesson in mammal embryology.

Procedure:

A. Pull the micro slide of **The Chick Embryo** out of its cardboard booklet. Hold the slide in your right hand so that the copyright section is closest to your hand and so that you can read the words in the copyright section. Now, hold it up so that the light from the desk lamp shines through the images on the slide. With this illumination, you should be able to see the image in this section fairly well with your naked eye. To get a closer look, use your magnifying glass.

B. There are eight frames on the slide. Observe each section and read the information in the cardboard booklet about that section. Observe with both your naked eye and the magnifying glass. Be careful not to stare at the image for too long, as your eyes will tire of the light from the lamp. Draw and label the image in each frame. Remember, each frame represents a certain stage in the embryo's development. The total time it takes for all of this development to take place is only 96 hours!

C. Important things to note:
   1. The brain is forming in the 18th hour.

2. By the 21st hour, the mouth and digestive system have started.
3. In the 28th hour the circulatory system is beginning.
4. By the 38th hour, the heart is forming and beating. The brain now has 5 regions.
5. By the 56th hour, the 4 chambers of the heart are forming along with the ears and eyes. Plus, the organs (lungs, kidneys, etc.) are developing.
6. By the 96th hour, the chick has every part including the beginning of the wings.

D. (Optional) Observe each frame under the lowest power of your microscope. This slide was not really meant for a microscope, so you will have to hold it in your hands and move it up and down to find a good focus. Even though this method is a bit of a pain, it will allow you to see some really nice detail.

E. Clean up and put all equipment away leaving it in good condition.

ON YOUR OWN

16.12 Fresh samples of blood are taken from a bird and a reptile. If the samples were taken during a cool night, which creature's blood will be the warmest?

16.13 A biologist shows you a sample of blood taken from a creature's ventricle. If this blood is a mixture of oxygenated and de-oxygenated blood, did it come from an amphibian or a bird?

16.14 You find an egg that has a rough-feeling, flexible shell. Is it a bird egg or a reptile egg?

A Bird's Ability to Fly

As we mentioned before, humans have always been fascinated by a bird's ability to fly. Thus, we have studied birds a great deal, in order to unlock the secrets of flight. As a result, we have some pretty good ideas on what characteristics of birds make it possible for them to fly. The three most important of these are feathers, wings, and skeletal structure. We will first look at feathers, then we will discuss wings, and then we will look at skeletal structure.

Feathers

Regardless of whether or not a bird can fly, it must have feathers. Indeed, the presence of feathers is one of the unique characteristics that we listed for class Aves. The feathers of a bird, which are both very strong and very light, grow from small structures called **papillae** (puh pil' ee), which reside on the skin. The basic structure of a feather is shown in Figure 16.5.

FIGURE 16.5
The Basic Structure of a Feather

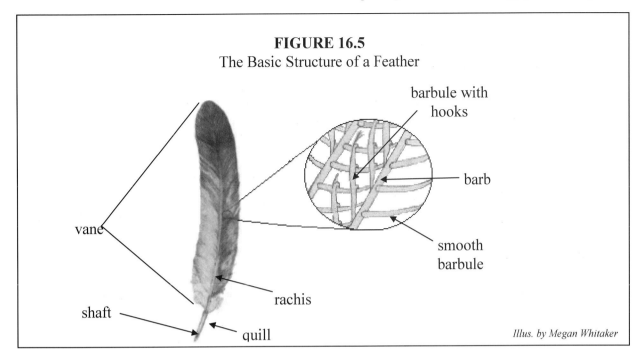

Illus. by Megan Whitaker

Notice from the figure that a feather is comprised of two basic parts: a **shaft** and a **vane**. The shaft is generally divided into two sections. The bare portion of the shaft that connects to the papilla (singular of papillae) is called the **quill**, while the portion of the shaft that holds the vane is called the **rachis** (ray' kis). The vane is comprised of parallel **barbs** that originate on the rachis and extend outward. These barbs have two types of **barbules** (barb' yoolz) that extend away from the barb. One type of barbule is smooth, and the other type has hooks. These barbules are arranged so that the hooks grab onto and slide along the smooth barbules.

Now think about this design for a moment. Since the hooked barbules attach to the smooth barbules, the feather is supported by a network of interlocking strands. This gives the feather a lot of strength. At the same time, however, the feather is flexible. After all, the hooked barbules can slide up and down the smooth barbules, allowing the feather to flex a great deal. If the feather flexes so much that the hooked barbules actually slip off the smooth barbules, that's no problem. The bird can re-attach the hooks to the smooth barbules by stroking the edge of the vane with its beak. In the end, then, the feather possesses the *ideal* compromise between weight, strength, and flexibility.

In order to retain ultimate flexibility in the wing, the hooked barbules must slide along the smooth barbules very easily. Thus, there needs to be some sort of lubricant covering the barbules, allowing them to slide against each other freely. Well, it turns out that a bird produces oil in a gland near the base of the tail. When a bird wants to "oil" its feathers, it dips its bill into that gland and then runs the bill over its feathers. This is called **preening**, and by doing this, the bird can spread oil all over its wings. This allows the hooked barbules to slide freely along the smooth barbules and, as a side benefit, the oil that the bird spreads on its feathers is waterproof. This keeps the bird's wings from getting wet, which in turn keeps the bird warm. It also keeps the bird from getting so waterlogged that it becomes too heavy to fly.

Now it turns out that there are really two types of feathers: **down feathers** and **contour feathers**.

<u>Down feathers</u> - Feathers with smooth barbules but no hooked barbules

<u>Contour feathers</u> - Feathers with hooked and smooth barbules, allowing the barbules to interlock

Obviously, then, the feather pictured in Figure 16.5 is a contour feather. Down feathers, since they have no hooked barbules, do not really have vanes. After all, the barbules on a down feather cannot interlock like they do on a contour feather. Thus, down feathers are fluffy and soft rather than strong and flexible. They are found underneath the layer of contour feathers on a bird, and they mostly serve as insulation for the bird. Humans, however, use down feathers as pillow or cushion stuffing because they are amazingly soft. Contour feathers are the feathers you are used to seeing on a bird. The contour feathers on the bird's body give it shape and color, while the contour feathers on its wings are essential for flight.

Like scales, feathers are made of non-living tissue. Thus, they cannot grow as the bird grows. As a result, birds must molt their feathers like reptiles molt their scales. Unlike reptiles, however, birds cannot molt their feathers all at once. After all, if a bird loses all of its feathers at once, it will not be able to fly. In fact, flight is such a difficult skill that a bird cannot even lose more feathers on one side of its body than on the other. If that were to happen, the bird would become unbalanced and could not stay in the air for any length of time! To avoid such problems, then, the molting of a bird is a very precise, delicately-balanced process. Basically, a bird molts its feathers in pairs. Each feather on one side of the bird's body will molt *at exactly the same time as the corresponding feather on the other side of its body*! This process occurs one pair at a time, so that the bird is never really missing more than one pair of feathers at any given time.

Wings

Feathers are, indeed, marvelously engineered structures. They provide strength and flexibility at a fraction of the weight that even the most advanced structures made by human science cannot come close to matching. Nevertheless, feathers do not allow for flight until they are attached to wings. Scientists recognized rather early that a bird's wings were the key to its flight ability, and they had the mistaken notion that wings were the sole key to a bird's ability to fly. As a result, they made all manner of contraptions with wings attached, hoping that these wings would give their contraptions the ability to fly as well. Of course, it took centuries of painstaking study to determine *how* a bird's wings give it the ability to fly. After we learned this secret from studying birds, we were able copy the design and apply it to airplanes that actually could fly.

Now it turns out that the principle behind how wings allow birds to fly is just a bit too complicated to cover in this course. If you take physics in college, you will learn about something called Bernoulli's principle which describes how a bird's wing or an airplane's wing makes flight possible. Even though a discussion of this principle is beyond the scope of this course, there is one very important aspect of a bird's wing that we want to discuss here.

As we said before, scientists and engineers studied bird wings and bird flight in order to determine how to make airplanes fly. Through the genius of the Wright brothers at Kitty Hawk, man was finally able to leave the confines of the ground and fly through the air , albeit relatively slowly and only for a few minutes. Eventually, however, airplane technology began to improve, and airplanes began flying faster and for longer periods of time. As planes got to traveling faster and faster, however, engineers suddenly discovered a problem.

You see, airplanes experience **turbulence**, which is essentially "rough air." This turbulence shakes the plane as the plane passes through it. If you have ever ridden on a plane, you have probably experienced episodes of turbulence. Well, it turns out that turbulence is particularly bad when you are near the surface of the earth. Thus, when a plane takes off and lands, it experiences a lot of turbulence. So much so that when early airplanes began traveling at relatively high speeds, the turbulence at takeoff or landing was so severe that the planes' wings would shake violently and eventually fall off. Clearly, this was a problem! Engineers fretted and worried over this problem and could not figure out how to fix it, until someone had the bright idea to go back and study birds again. Clearly, birds didn't have the problem that man-made airplanes had. How, then, did birds get around the problem? Figure 16.6 shows you how.

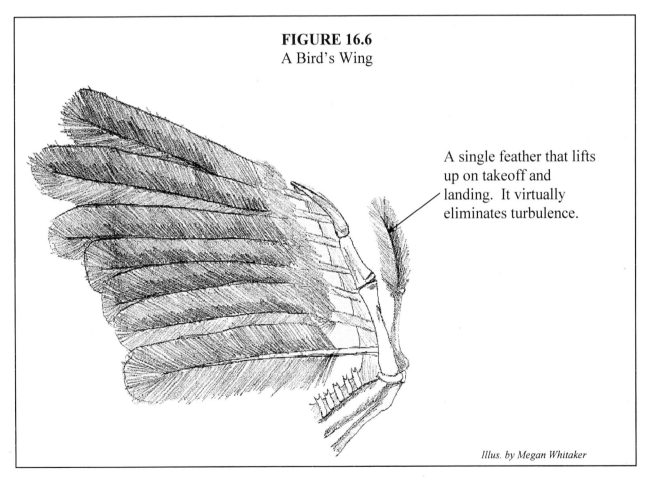

FIGURE 16.6
A Bird's Wing

A single feather that lifts up on takeoff and landing. It virtually eliminates turbulence.

Illus. by Megan Whitaker

A bird's wing is equipped with a single feather that lies parallel to the bird's wing arm. When the bird takes off and lands, the feather is lifted so that a gap exists between the wing and the feather. In flight, this same feather lies flush against the wing. Engineers had *no idea* how this got rid of the turbulence problem, but since the action of the feather was related to takeoffs and landings, they decided that it *must* do something! Thus, they created a similar mechanism on an airplane's wing and guess what? The turbulence problem was solved! Now remember, even though the mechanism that engineers copied from the bird's wing had fixed the problem, they still *had no idea why it worked.* They only knew that it did work. Not until 1989, using the power of a supercomputer, did physicists finally explain *how* this mechanism reduced air turbulence. The explanation is far too complex to discuss here. The point should be clear, however. The bird possesses mechanisms for flight that we could not understand without years of effort and mammoth computing power!

Skeletal Structure

One of the characteristics of birds is that they have a skeleton which is made up of porous, light bones. This is not only a way of identifying a creature's membership in class Aves, but it is an essential part of a bird's ability to fly. You see, one of the really tricky parts about flight is the fact that any flying object must be very lightweight but at the same time very strong. We've already seen that feathers meet that condition very nicely, but if a creature's skeleton is too heavy, all of the feathers in the world will not make a difference. It turns out that a skeleton made of vertebrate bones like that which we studied in Module #13 would simply be too heavy for flight. Thus, the bones of birds must be unique in subphylum Vertebrata.

Unlike other vertebrates, birds have skeletons which are comprised of bones that are full of air-filled cavities. These cavities make the bones very light, so light that the skeleton does not hinder a bird's ability to fly! Now you might wonder if a bone that is full of air-filled cavities can be strong. Well, the answer is no, it cannot. However, a bird's bones are full of thin strands of bone tissue which traverse the air cavities. These strands serve as struts which strengthen the bones. It turns out that in order to build viable airplanes, this is another design feature that engineers had to copy from birds. Engineers realized early on that they needed to make airplanes out of lightweight materials, so they used hollow tubes, etc. to build the frames of airplanes. As airplanes got bigger and faster, however, these tubes would collapse, because they simply were not strong enough. When engineers studied bird bones, however, they discovered these "struts" in the air-filled cavities, and they constructed similar struts, which they called "Warren's trusses," in their hollow tubes. These trusses kept the tubes from collapsing during flight.

Now think about all of these facts for a moment. The bird has elegant feathers that have a better combination of strength, flexibility, and weight than anything that modern human science can build. In addition, human scientists and engineers could not design a working plane until they copied the wing design, the turbulence-reducer design, and the skeletal design from birds. Even with all of that help, human science cannot create a plane that is anywhere near as efficient as a bird! Shouldn't that tell you something about the bird? In the end, all that human scientists did was copy the design of the Creator in order to make airplanes. Unfortunately, human science cannot copy the design as well as God can create, so the most sophisticated product of human science still cannot compare to what God's Creation has always had!

ON YOUR OWN

16.15 A bird's feathers begin to lose their flexibility. What should the bird do in order to fix this problem?

16.16 Air turbulence is affected by speed. The faster you go, the worse turbulence gets. If a bird's wing does not have the special feather pointed out in Figure 16.6, what can you conclude about the speed at which it takes off and lands?

16.17 Two bones look identical. The person showing you the bones says that one comes from a reptile and the other from a bird. How could you tell which came from the bird without damaging the bones in any way?

Classification in Class Aves

There are many, many orders of birds. Table 16.1 lists the major orders and examples of the birds which are found in them.

TABLE 16.1
The Major Orders of Birds

Order	Examples of Birds In This Order
Struthioniformes (struh thee' on uf for meez')	ostriches
Sphenisciformes (fen uh suh' for meez')	penguins
Procellariiformes (pro sel' ar ih for meez')	pelicans, albatrosses, petrels
Ciconiiformes (sih' con ih for meez')	herons, flamingoes
Anseriformes (an' ser uh for meez')	swans, geese, ducks
Falconiformes (fal' cuhn uh for meez')	eagles, hawks, falcons, vultures
Galliformes (gal uh for meez')	turkeys, quails, pheasants
Columbiformes (ko luhm' buh for meez')	pigeons, doves
Strigiformes (strih guh for meez')	owls
Apodiformes (uh pod' uh for meez')	hummingbirds
Passeriformes (pass er uh for meez')	robins, finches, wrens, sparrows, nightingales
Psittaciformes (sit' uh suh for meez')	parrots, parakeets

Now don't get wrapped up in trying to memorize these orders and their members. We present them just so you get an idea of bird classification. Examples of birds in some of these orders are shown in Figure 16.7.

FIGURE 16.7
Members of Class Aves

Ostrich
order *Struthioniformes*

Photo by Kathleen J. Wile

Penguin
order *Sphenisciformes*

Photo by Dr. Jay L. Wile

Bald Eagle
order *Falconiformes*

Photo from the MasterClips collection

Flamingo
order *Ciconiiformes*

Photo from the MasterClips collection

Owl
order *Strigiformes*

Photo from the MasterClips collection

Hummingbird
order *Apodiformes*

Photo by Kathleen J. Wile

Quail
order *Galliformes*

Photo by Kathleen J. Wile

Parrot
order *Psittaciformes*

Photo from the MasterClips collection

Pelican
order *Procellariiformes*

Photo from the MasterClips collection

The majority of flightless birds are found in two orders: Struthioniformes and Sphenisciformes. These birds, such as **penguins** and **ostriches**, have all of the characteristics of birds but simply do not fly. Instead, penguins are graceful and powerful swimmers, and ostriches are powerful runners. Unlike the popular myth, ostriches do not bury their head in the sand when threatened. They are, in fact, fierce fighters, able to cut off a person's fingers with a quick bite from their sharp bills. The myth of ostriches hiding their head in the sand actually comes from the fact that when an ostrich is surprised, it will often lower its head to the ground, trying to get a different perspective on the situation.

The members of orders Falconiformes and Strigiformes are often called **birds of prey**. These are the hunting birds. They have keen eyesight which they use to spot their prey from a tall perch or while in flight. When they have their prey in sight, they swoop down on it in a high-speed (up to 70 mph) dive and grasp the unfortunate creature in their sharp talons. Usually, the fast impact of the birds' talons on the prey instantly stuns or kills the victim, and the birds then take their feast somewhere that they can eat in peace. There are exceptions to this general rule, however. Vultures do not employ any of these means to catch prey. Instead, they feast on dead creatures. Thus, there is no need to employ the high-speed tactics of the other birds in these orders.

Although we generally think of birds of prey capturing mice and other land-dwelling creatures, a large number of these birds actually hunt fish. They fly over bodies of water looking for a fish swimming near the surface of the water. They then swoop down and grab the fish before it can swim away. The **bald eagle**, symbol of the United States of America, typically feeds on fish. The name "bald eagle" is a bit misleading. The bald eagle is not bald. Instead it has distinctive white feathers on both its head and tail. The term "bald" comes from an obsolete use of the word "piebald," which can mean "blotched with white." It turns out that the population of bald eagles is dwindling, mostly because of overhunting. Although it is illegal to kill a bald eagle, hunters typically shoot them by accident, because a bald eagle does not develop its distinctive white head and tail feathers until it is 3 or 4 years old. As a result, many young bald eagles are shot down by hunters who honestly do not recognize them as bald eagles.

Members of orders Galliformes and Columbiformes often are called **game birds**. They have tender white meat that is, in fact, composed of their flying muscles. Found predominantly in the breast, these tissues have fewer blood vessels running through them than typical bird muscles. As a result, these muscles tire quickly, and game birds are only capable of quick bursts of flight, after which they must rest their muscles. This makes the meat derived from these tissues more tender than typical bird meat. Since their meat is so tender, game birds are the most common birds to find on a dinner table. Typically, members of order Galliformes, such as **quail**, spend the majority of their time on the ground. Members of order Columbiformes, such as **doves**, fly a little more often and spend the majority of the rest of their time on a perch off of the ground.

Three orders of birds inhabit water. The first order, Anseriformes, contains **ducks**, **geese**, and the like. Members of this order are typically called **swimming birds**, because they are at home swimming in the water. The oil they produce for preening is especially waterproof, so that they dry off quickly after leaving the water. Their feet are webbed for efficient swimming. The next order, Procellariiformes, contains **pelicans**, **albatrosses**, and so on. These birds, often called **diving birds**, are specially designed to dive into the water and catch fish. They do not grab onto their prey with talons; instead, they dive headlong into the water and catch their prey with their bills. Finally, order Ciconiiformes contains birds such as the **heron** and the **flamingo**. These birds, called **wading birds**, wade in the water, looking for small fish and other aquatic organisms to eat. Although some biologists consider penguins (order Sphenisciformes) to be water birds because they are such excellent swimmers, we believe that they are better placed with the flightless birds, because that is probably their most distinguishing characteristic.

The order Passeriformes contains what many people call the **song birds**. All birds have the ability to make a chirping sound because of a special organ called the **syrinx** (sihr' inks). Members of this order, such as sparrows and nightingales, tend to have well-developed syrinxes which are capable of making beautiful sounds.

Order Apodiformes contains **hummingbirds**. The term "hummingbird" is actually a generic name for more than 300 species of small (as little as 3 inches long) birds that feed on the nectar found in certain flowers. These birds are known for the rapid rate at which they beat their wings. This fast motion causes a humming sound from which they derive their common name. They feed by hovering near a flower and probing it for nectar and small insects. It turns out that members of this order are the only creatures in Creation that have the ability to fly backwards. This is a necessary skill for hummingbirds as they must back away from a flower after feeding on it.

To get an idea of what birds inhabit your region of the world, perform the following experiment.

EXPERIMENT 16.2
Bird Identification

<u>Supplies:</u>

- Bird field guides (available at your local library)
- Binoculars if available
- Bird seed
- Lab notebook

<u>Object:</u> To become familiar with birds that are common to your locality

<u>Procedure:</u>

A. This exercise is two fold. The first part should be done at your own home. Buy a small bag of bird seed. Spread some of the seed on a large cleared area on the ground. Place some in a higher place like a bird feeder or a window sill.

B. Observe for several days. Note the birds that come to feed on the ground and the ones at the higher place. You should observe that some birds tend to be ground feeders and others will eat at the feeder or window sill.

C. Use your field guides to identify these birds. Note the birds that you see and their classification in your lab notebook.

D. Plan a trip to a state or city park where they have an outside bird sanctuary. Plan to spend at least an hour observing the various types of birds and their habits. Note the information in your lab notebook.

ON YOUR OWN

16.18 The kingfisher is a bird that dives headlong into the water for its prey. In which order would you place it?

16.19 The red jungle fowl has tender white meat when cooked. It also spends most of its time on the ground. In which order would you place it?

Class Mammalia

We conclude our biology course with a discussion of the class which contains humans, class Mammalia. It turns out that class Mammalia is one of the smaller classes in subphylum Vertebrata, but it nevertheless is probably the class that most people associate with the word "animal." This is probably because mammals occupy virtually every ecosystem with which people are familiar. As a result, "animal" has come to mean "mammal" in many people's minds. It is important that you do not make such a mistake, however. The term "animal" also includes such creatures as sponges, hydra, fish, amphibians, reptiles, and birds. The following characteristics separate mammals from the other vertebrates.

- Mammals have hair covering their skin.

- Mammals reproduce with internal fertilization and are usually viviparous.

- Mammals nourish their young with milk secreted from specialized glands.

- Mammals have a four-chambered heart.

- Mammals are endothermic.

Notice that the last two characteristics of mammals are the same as those of birds. The first three characteristics, however, set mammals apart from any other creatures in phylum Chordata.

The one characteristic of mammals that most students learn in grade school is that mammals have **hair**. In most mammals, the hair is really obvious. In some mammals, however, it is much harder to see. Elephants have hair, but unless you examine them closely, you might miss the ragged tufts that make a patchwork of hair across the skin. Whales are mammals, and it is even harder to find their hair. It is there, nevertheless.

Hair, like the scales on reptiles and the feathers on birds, is made of non-living cells. These cells are produced by **hair follicles**, which are tiny structures deep in the skin. Most mammals have two types of hair, **underhair** and **guardhair**. As its name implies, underhair is found under the guardhair. It is a soft, insulating layer of fur next to the animal's skin. The hair that we see when we look at mammals is typically the guard hair. This hair is coarser than the underhair and is typically longer as well. This is the hair that gives most mammals their colors and distinctive markings.

Hair has many functions. Underhair insulates the mammal and keeps it warm. In a beaver or sea otter, the underhair is so dense and well-insulating that these mammals can swim in ice-cold water without decreasing their body temperature! Hair can also provide camouflage. Young deer, for example, tend to blend in with their surroundings because of the coloring and markings of their hair. Hair can also be used as a defense mechanism, as is the case with the porcupine, whose guard hair is stiff and pointed. Hair can also be used to aid the senses. The whiskers on a dog and cat are very sensitive to touch. They are used to detect objects in the dark or when the objects are outside the mammal's field of vision. A cat, for example, will never go through a passage (such as a hole in a fence) unless he can move through it without his whiskers being touched. If the cat's whiskers can pass through the hole, then the cat knows that its entire body can pass through the hole. When dogs and cats are threatened, hairs on their neck or tail (or both) will stand up. When this happens, we usually say that the hair "bristles." Bristled hair is incredibly sensitive, and if that hair gets touched, the mammal's immediate instinct is to jump away then turn and attack in the direction of the touch.

Mammals reproduce with internal fertilization. In the vast majority of mammals, the embryo's development is viviparous. There are exceptions (of course) to the viviparous rule, however. One order of mammals (which we will discuss in a moment) is oviparous, while another (which we will discuss in a moment) has a pouch in which the embryo develops. These two orders of mammals are called **non-placental** (pluh sent' uhl) **mammals**. The other thirteen orders all contain **placental mammals**, whose young develop viviparously.

In placental mammals, a **placenta** (pluh sent' uh) forms when the embryo begins development inside the mother.

Placenta - A structure that allows nutrients and gases to pass between the mother and the embryo

The placenta is rich in blood vessels from both the developing embryo and the mother. These blood vessels get very close to one another, close enough to exchange gases and nutrients, but the blood of the mother and the blood of the developing offspring never actually mix. The mother's blood brings oxygen and nutrients to the placenta and there allow the offspring's blood vessels to absorb them. At the same time, the offspring's blood vessels give waste products to the mother's blood vessels so that they can exit via the mother's excretory system. The blood of the infant travels back up from the placenta through an **umbilical** (um bil' ih kul) **cord** which connects the developing offspring to the placenta.

As the offspring develops inside the mother, we say that it is in **gestation**.

Gestation - The period of time during which an embryo develops before being born

The gestation time for mammals varies greatly. Mice have a gestation period of about 3 weeks while horses have a gestation period of almost one full year. The practical result of long gestation periods is that the offspring is much more developed when it is born. Mice, which have a short gestation period, are born hairless, blind, and weak. They grow hair, develop sight, and gain strength after birth. Horses, however, are born with a full coat of hair and, within hours of their birth, they can stand, walk, and even run.

Once the offspring is born, all mammals (placental or not) care for their young. The female provides milk to nourish the young. This milk is produced in **mammary** (mam' ur ee) **glands** which are specially designed to produce milk.

Mammary glands - Specialized organs in mammals that produce milk to nourish the young

The mammary glands of female mammals are always present, but they tend to be active only shortly before and for some length of time after birth. When a female mammal's mammary glands are active, we say that the female is **lactating** (lak' tate ing). Typically, the young get their mother's milk by sucking on nipples which attach to the mother's mammary glands. Most mammals have some sort of "family" structure in which the parents teach the young survival skills before the young leave the protection of the parents.

As is the case with birds, mammals are endothermic and have four-chambered hearts. Thus, they are able to regulate their own internal temperature and keep it constant. Typically, a mammal's underhair aids in the process, insulating the mammal so that only a small amount of the internally-generated warmth escapes in cold climates and so that only a tiny fraction of outside heat enters the body in hot climates. Because mammals have a four-chambered heart, de-oxygenated blood never mixes with oxygenated blood in the blood vessels.

ON YOUR OWN

16.20 Which mammal would have thicker underhair, a polar bear (which lives in a very cold climate) or a prairie dog (which lives in a warm climate)?

16.21 A cat with its whiskers cut off tends to run into objects a lot more frequently than the same breed of cat whose whiskers are not cut off. Why?

16.22 Two different species of mammals have different gestation periods. The first has a long gestation period and the second has a short gestation period. Which mammal will have offspring that are more developed at birth?

Classification in Class Mammalia

The members of class Mammalia are divided into 15 orders. They are summarized in the table below:

TABLE 16.2
The Major Orders of Mammals

Order	Examples of Mammals In This Order
Monotremata (mon' uh truh mah' ta)	duck-billed platypuses, spiny anteaters
Marsupialia (mar soop ee ah' lee uh)	kangaroos, wallabies, koalas, opossums
Chiroptera (ki rop' tur uh)	bats
Carnivora (kar' nuh vor uh)	lions, tigers, cats, dogs, bears, pandas
Sirenia (sir' ee nee uh)	manatees and other sea cows
Cetacea (suh tah' see uh)	whales, dolphins, porpoises, seals
Proboscidea (pro boh seed' ee uh)	elephants, mammoths (extinct)
Perissodactyla (pehr' uh suh dak til uh)	horses, zebras, rhinos, tapirs
Artiodactyla (art' ee oh dak til uh)	sheep, deer, bison, goats, pigs, cattle
Rodentia (roh' den she uh)	squirrels, rats, mice, porcupines, guinea pigs
Lagomorpha (lag' uh more fah)	rabbits, hares, pikas
Edentata (ed' in tah tuh)	anteaters, tree sloths, armadillos
Tubulidentata (tuh boo' luh den tah' tuh)	African aardvark
Insectivora (in sek' tuh vor uh)	shrews, moles, hedgehogs
Primates (pri' mates)	monkeys, lemurs, gibbons, baboons, humans

Once again, do not worry about memorizing these orders; they are shown for the sake of completeness. Examples of creatures from some of these orders are shown in Figure 16.8.

FIGURE 16.8
Members of Class Mammalia

Seal
order *Cetacea*

Photo by Kathleen J. Wile

Deer
order *Artiodactyla*

Photo from the MasterClips collection

Koala
order *Marsupialia*

Photo from the MasterClips collection

Chipmunk
order *Rodentia*

Photo from the MasterClips collection

Elephants
order *Proboscidea*

Photo from the MasterClips collection

Rhinocerous
order *Perissodactyla*

Photo from the MasterClips collection

Rabbit
order *Lagomorpha*

Photo from the MasterClips collection

Tiger
order *Carnivora*

Photo from the MasterClips collection

Lemur
order *Primates*

Photo from the MasterClips collection

Members of order Monotremata are often called the **egg-laying mammals**. These mammals lay eggs and attend them like birds do. Once the offspring hatch, they are nourished by their mother's milk. Interestingly enough, however, members of this order do not have nipples. Instead, their mammary glands are evenly distributed along the female's underside. These glands dump milk directly onto the skin, and the young lap the milk from the mother's fur.

The best known member of order Monotremata is the **duck-billed platypus**. When Europeans originally made it to Australia, where the duck-billed platypus lives, they reported seeing this odd creature which is covered with fur yet has a bill like a duck, lays eggs, and has webbed feet. Scientists in Europe did not believe the travelers, because they could not imagine a mammal like that. Even when a dead one was brought to Europe, scientists still thought it was a fake. They would not believe that such a strange creature existed until live specimens were brought back to Europe. The duck-billed platypus uses its bill to scoop invertebrates out of the mud to eat. Sought for its novel fur, it was nearly hunted to extinction. Now that the Australian government protects the duck-billed platypus, their numbers are on the rise.

Order Marsupialia contains the other group of non-placental mammals. In these creatures, such as **kangaroos**, **koalas**, and **opossums**, the fertilized egg develops within the mother for only a few days. Because this phase of development is short, no placenta is formed. Instead, in just a few days, the tiny, immature offspring crawls out of the mother and makes its way to a pouch, which contains nipples. The offspring must attach its mouth to a nipples and then continue their development in the pouch, drawing nourishment from the mammary gland to which the nipple is attached.

Most members of this order, such as **kangaroos**, **wallabies**, and **wombats**, live in or near Australia. In the United States, we do have one marsupial, called the opossum. This mammal, which resembles a large rat, is best known for its tendency to play dead. When approached by a potential predator, opossums tend to flop over on their sides, appearing dead. Their eyes take on a glassy look and their faces contort into a grimace. This tends to make the predator think that the animal has been dead for some time, and most predators instinctively avoid such easy prey, because long-dead animals are often carriers of disease. Although opossums play dead, they are also fierce fighters if forced into a fight. They typically prey on small mammals, birds, or insects.

Members of order Chiroptera are often called the **flying mammals**, and the best known example is the **bat**. Bats look like mice with wings. At one time, they were actually classified in the same order as mice, but today, they are considered unique enough to deserve their own order. Bats tend to live in dark environments like caves, barns, or dark forests. They live in these environments because of their unique ability to "see" in the dark, which gives them a distinct advantage over the other animals.

Bats "see" in the dark with an elegantly designed sonar system. They emit sound waves from their mouth. If the sound waves encounter an obstruction or another creature, these sound waves bounce back to the bat. The bat receives these reflected sound waves and interprets them in order to determine the size and shape of what they bounced off of. Somehow, the bat knows which sounds are its own and which ones come from other sources, even other bats. Laboratory experiments indicate that even when their reflected sound waves are 2,000 times weaker than other sounds reaching the bat's ears, the bat can still recognize them and use them to "see." This sonar system is so efficient that a bat can pinpoint the precise location of a fruit fly (one of the tiniest flies in Creation) from as far as 100 feet away! Bats can use this sonar so effectively that they can eat up to 5 fruit flies each minute! In the words of Michael Pitman, author of *Adam and Evolution*, "Ounce for ounce, watt for watt, it is millions of times more efficient and sensitive than the radars and sonars contrived by man." (Michael Pitman, *Adam and Evolution*, Rider & Company, p. 220). The bat stands as another constant testimony that life was *designed*.

Members of order Carnivora include **cats**, **dogs**, and other mammals that eat meat. These mammals have distinctive teeth. They have **canine fangs** which are designed to tear meat, and sharp molars that are designed to chew flesh. In addition, their feet typically have sharp claws used to capture prey. Cats such as **lions**, **tigers**, and even domesticated cats use their claws to pounce on prey and hold them. **Bears**, on the other hand, often use their claws to catch fish right out of streams.

Each family in order Carnivora has its own, distinctive behaviors that allow its members to catch prey. Cats, for example, rely on stealth and cunning to sneak up on their prey. Dogs, on the other hand, require teamwork. Dogs like **wolves**, for example, hunt in packs and share the food that the pack catches. Bears are actually omnivorous. They eat tree bark, honey, plants, and animals as well.

Aquatic mammals belong to orders Sirenia and Cetacea. All of these mammals breathe with lungs; thus they must come to the surface in order to breathe. **Whales** and **dolphins**, for example, have blow holes on the top of their heads which they use to breathe air. Order Cetacea contains the largest known animal in Creation, the **blue whale**, which reaches weights of 170 tons and lengths of 110 feet. Even though the blue whale is large, it is a gentle creature which feeds on plankton in the water. **Killer whales**, on the other hand, are fierce predators and have been known to break through ice in order to attack seals and people!

Perhaps the most interesting aquatic mammal is the **porpoise**. Porpoises look like dolphins, but there are distinguishing characteristics that separate the two. Porpoises are usually smaller than dolphins, and they have rounded heads. Dolphins, on the other hand, have pointier heads that end in a pointed mouth resembling a beak. Also, porpoises have triangular dorsal fins whereas dolphin dorsal fins are hooked. The porpoise has two unique abilities.

First, like the bat, porpoises navigate with sonar. They must do this because they swim very quickly, much too quickly to navigate by sight. This is because light travels poorly in water, so the range of sight is short under water. Sound, however travels quickly and efficiently in water, therefore the porpoise can use its sonar as effectively as the bat does.

Second, the porpoise can swim faster than any other creature in the sea. This is partially due to the fact that it uses sonar to navigate. However, even if another sea creature wanted to swim as fast as a porpoise and not worry about navigation, it simply couldn't. This is because water tends to drag against any body moving in it. As a result, it slows the motion of the body down. Now it turns out that the faster a body moves in water, the harder water drags against it. In the end, there comes a point at which no matter how much more energy is expended, speed cannot be increased, because all of the extra energy just goes into overcoming the drag of the water. Thus, for most water creatures, there is an inherent limit to how fast they can swim. Porpoises get around this problem because a spongy material within their loose, finely-laced, layered skin beats rhythmically with the motion of the water to *strongly reduce water drag*. This allows the porpoise to travel quickly, outswimming all other creatures in the sea. As with most of the elegantly designed structures in Creation, human science cannot develop anything similar to this material. If the navy could make a similar substance, ships and submarines could run up to 40 times faster!

Order Proboscidea contains the **elephants**. These creatures are distinguished by their long trunks. This boneless, muscular feature of elephants is actually a greatly elongated upper lip and nose. It is used to pick up grasses, leaves, and water and bring these substances to the elephant's mouth. An extremely versatile organ, the trunk is also used to trumpet calls, pull down trees, rip off foliage, and draw up water for bathing. It is also a highly sensitive organ, which the animals raise into the air to detect wind-borne scents. By means of finger-like lobes on the end of the trunk and by the sucking action of the two nostrils, elephants can pick up and examine small objects. Very hairy versions of these creatures, the mammoths and mastodons, are extinct today.

Orders Perissodactyla and Artiodactyla contain the **hoofed mammals**. The two orders are distinguished from one another by the type of hooves that their members possess. The mammals in order Perissodactyla have either a single hoof that encompasses the entire foot, or three enlarged toes that make up the hoof. As a result, these mammals, such as the **horse**, **zebra**, and **rhino**, are sometimes called the **odd-toed hoofed mammals**. Members of order Artiodactyla, such as **sheep**, **deer**, **bison**, **goats**, **pigs**, and **cattle**, have two or four toes which make up their hooves. Thus, they are often called the **even-toed hoofed mammals**.

Order Rodentia, which contains squirrels, rats, mice, porcupines, and the like, has the largest number of species. These creatures are distinguished by large, sharp incisor teeth in both the upper and lower jaws. These teeth grow continually, and members of order Rodentia use them to gnaw and bite. In fact, rodents must continually gnaw with their incisors, or those incisors will grow so long that they will interfere with the rodent's ability to close its own mouth! Thus, rodents have an instinct to gnaw and gnaw and gnaw. As a result, members of this order are often called the **gnawing mammals**.

Most rodents have essentially no defense mechanisms against predators. Consequently, they are constantly being eaten by birds and carnivorous mammals. How, then, do they keep from dying out? They reproduce in huge numbers! **Field mice**, for example, reproduce several times throughout the course of their life. The young are completely independent from their mother in as little as fours weeks after birth. At the end of six weeks of life, most field mice have produced their first litter of offspring! In laboratory experiments, some species of mice have produced as many as *80 offspring per year!*

Rabbits and **hares**, which belong to order Lagomorpha, are often called the **rodent-like mammals**. They look and behave much like the mammals of order Rodentia, but they are not similar enough to belong to the same order. What's the difference between a rabbit and a hare? Actually, the distinguishing characteristic is related to how they are born. The offspring of rabbits are born with no hair and with their eyes still closed. Hares, on the other hand, are born covered with fur (many believe that's where the name originated) and with their eyes open. In adults, hares can usually be distinguished from rabbits because a hare has longer ears and is usually larger than the average rabbit. Like rodents, they survive simply because they reproduce prolifically. The typical rabbit can begin producing litters about six months after birth. They produce 3-8 offspring in each litter, with a gestation period of a little more than a month. Since the average rabbit or hare lives about 10 years, you can see how many offspring just one individual can produce!

Order Edentata contains such creatures as the **anteater**. This interesting mammal has a long head with a long, tubular mouth and long tongue, but no teeth. It tends to live in forests and swampy areas and avoids highly populated regions. The animal actually walks on its knuckles, using its claws only for defense or for tearing apart anthills or termite mounds. Once it has torn the insects' home apart, its long tongue flicks rapidly in and out of its small mouth opening, scooping up termites or other insects on its sticky surface.

Members of order Tubulidentata, the **aardvarks**, are often mistakenly called anteaters. This is because they tend to eat ants and termites like the anteater and have a long mouth like an anteater. An aardvark's mouth, however, is not nearly as long as an anteater's. One of the main characteristics that separate aardvarks from anteaters, however, is the fact that aardvarks have teeth, while anteaters do not.

Members of order Insectivora include **shrews**, **moles**, and **hedgehogs**. These creatures have pointed snouts, designed for burrowing into the earth to find insects. Shrews and moles actually spend a great deal of their lives underground, burrowing tunnels and the like. Hedgehogs rarely burrow tunnels. As their name implies, they tend to live in hedges, hiding from predators during the day and emerging at night to hunt for food. While the defense mechanisms of moles and shrews are basically related to running underground to avoid predators, the hedgehog uses its guard hair which, like the porcupine's, is sharp and strong. When the hedgehog is threatened, it will actually roll itself up in a ball, and its spiny hairs will stick out in all directions, making it very difficult for a predator to attack without getting pricked.

The last order of mammals is the order Primates. This order includes **monkeys**, **apes**, **lemurs**, and **human beings**. These mammals are often called the **erect mammals**, because they all have the ability to walk on two legs. Humans are the only mammals whose natural position is to walk on hind legs. Most primates spend the majority of their time walking on all fours, but they have the ability to stand erect if they want to.

Although you might object to being classified with monkeys, apes, and the like, you actually have several characteristics in common with these mammals. All primates have vision that allows them to see depth, and they are the only mammals with such vision. While most mammals see the world in flat, two-dimensional terms (as if they were looking at a picture of what is before them), primates can judge the third dimension of depth. In addition, all primates have a sense of smell that is poor when compared to the other mammals. Most primates are omnivorous, although the vast majority of primates concentrate on eating vegetation and only occasionally eat meat. All primates have nails on their fingers and toes, and most have opposable thumbs.

Now, even though we have many characteristics in common with apes and monkeys, there is no reason to think that we are *related* to them. That is the mistake that evolutionists make. They see similarities between animals and immediately think that these similarities come from common genes in a common ancestor. As we learned in Module #9, however, all genetic information that we have been able to acquire indicates that this simply isn't the case. The similarities between animals are the result of a common Designer, not a common ancestor.

ON YOUR OWN

16.23 A mammal has a long snout, eats ants and termites, and has teeth. To which order does it belong?

16.24 A mammal lays eggs. To which order does it belong?

16.25 A mammal has depth perception. To which order does it belong?

Summing It All Up

Well, believe it or not, you have reached the end of your first high school biology course. We hope that you have learned a great deal about the world around you, but most of all, we hope that you have developed a deep appreciation for the Creation that God has given you. If you think back for a moment on the things that you have learned, you will hopefully get a glimpse of how much creative power and energy the Almighty used to create this awesome world. From the inner workings of the smallest microorganism to the life processes of the mammals, you should see the work of an magnificent Designer. Indeed, if biology teaches us anything, it is that the world around us is simply too complex to have come about by chance. Consider the words of Dr. Robert Gange, a physicist of impeccable academic credentials:

> Everything we know tells us that machines are structures intelligence designs, and that accidents destroy. Therefore, accidents do not design machines. Intellect does. And the myriad of biological wonders that sprinkle our world testify to the design ingenuity of a Supreme Intellect. (Robert Gange, *Origins and Destiny*, Word Publishing, p.100)

Indeed, Creation cries out the wonders of God, and it is up to the scientist to recognize these cries and seek out the Creator!

ANSWERS TO THE ON YOUR OWN PROBLEMS

16.1 <u>The first is a snake and the second is a fish</u>. Reptiles have dry, leathery scales, while fish scales are slimy due to a mucus that covers them.

16.2 <u>The embryo will not live. It will die of suffocation</u>. The allantois allows for the respiration of the embryo, because it has blood vessels that are near the shell, where air is seeping in.

16.3 <u>The allantois is the one on the right</u>. As time goes on, the yolk sac grows smaller because the yolk is being used up. The allantois does not get used up, so it either stays the same size or grows with the embryo.

16.4 <u>The baby water snake can drown</u>. Tadpoles are the larva stage of an amphibian. Thus, they have gills and can breathe water. Reptiles, on the other hand, must breathe through lungs. Thus, the snake can drown.

16.5 <u>If the tuatara laid out in the sun too long, it could die of overheating</u>. Remember, reptiles are ectothermic. They must warm themselves in the sun. However, if they lay out in the sun too much, they will warm themselves too much.

16.6 <u>The reptile is a snake</u>. Lizards can hear and have eyelids.

16.7 <u>Although humans do not hunt tuatara, they brought rats to the tuatara's habitat, where rats did not exist. These rats are destroying the tuatara population because they eat tuatara eggs</u>.

16.8 <u>The snake is poisonous</u>. Only pit vipers have heat-sensing pits, and pit vipers are poisonous.

16.9 <u>They use their tongues and Jacobson's organs to enhance their sense of smell, so, in a way, they are smelling with their tongues</u>.

16.10 <u>It is a turtle</u>. If a member of order Testudines has flippers, it must live in the water. All members of Testudines that live in the water are called turtles.

16.11 <u>It is a crocodile</u>. Alligator teeth fit in an alligator's closed mouth; crocodile teeth do not fit into a crocodile's closed mouth.

16.12 <u>The bird's blood will be warmer</u>. Since birds are endothermic, their blood is always at a constant, warm temperature. Reptiles are ectothermic, so their blood temperature does vary with the outside temperature. The cooler the outside temperature, the cooler you expect the blood of a reptile to be.

16.13 <u>It came from an amphibian</u>. A bird has a four-chambered heart and the oxygenated blood does not mix with the de-oxygenated blood.

16.14 <u>It is a reptile egg</u>. Bird eggs have hard, lime-containing shells.

16.15 <u>The bird should start preening itself</u>. In order to make sure the hooked barbules slide smoothly along the smooth barbules, the feathers must be oiled. Birds do this by preening.

16.16 <u>The bird must be a slow flyer</u>. If a bird flies quickly, it must have this specialized feather, or its wings would be destroyed by turbulence.

16.17 <u>Weigh the bones. The lighter one comes from the bird</u>. Birds have air-filled cavities in their bones that make them lighter than similar bones in other vertebrates.

16.18 <u>It would go into order Procellariiformes</u>. These birds are called the diving birds, so the kingfisher seems to fit into that order. It actually turns out that there is some controversy on this. Many biologists give kingfishers their own order.

16.19 <u>It should go into order Galliformes</u>. The game birds have white meat, and they belong to order Galliformes and Columbiformes. The members of order Galliformes spend most of their time on the ground, while those in order Columbiformes spend a lot of time in the air or on perches.

16.20 <u>The polar bear would have thicker underhair</u>. The primary function of underhair is insulation. Clearly a polar bear needs that more than a prairie dog.

16.21 <u>A cat uses its whiskers to sense things outside its field of vision. Without the ability to do that, it will run into things that are outside its normal vision range</u>.

16.22 <u>The first will have offspring that are more developed</u>. The longer the gestation time, the more developed the offspring is at birth.

16.23 <u>It belongs to order Tubulidentata</u>. Since it behaves and looks like an anteater, you might think that it belongs with the anteater. However, it has teeth, so it belongs with the aardvark.

16.24 All egg-laying mammals are in order <u>Monotremata</u>.

16.25 Only mammals in order <u>Primates</u> have depth perception.

STUDY GUIDE FOR MODULE #16

1. Define the following terms:

a. Amniotic egg
b. Neurotoxin
c. Hemotoxin
d. Endothermic
e. Down feathers

f. Contour feathers
g. Placenta
h. Gestation
i. Mammary glands

2. State the five characteristics that set reptiles apart from other vertebrates.

3. In this module, we studied reptiles, birds, and mammals. For each class, indicate whether they are ectothermic or endothermic.

4. Identify the parts of the amniotic egg:

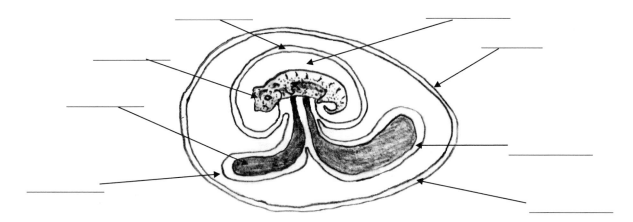

5. State the function of the yolk, the allantois, and the egg white.

6. What do arthropods and reptiles have in common?

7. What are the two most important functions of reptile scales?

8. These are the reptile orders that contain currently living reptiles:

Rhynchocephalia, Squamata, Crocodilia, Testudines

Place the following types of reptiles into their appropriate order:

a. snakes   b. tuataras   c. lizards   d. tortoises   e. alligators   f. turtles

9. State the six characteristics that set birds apart from other vertebrates.

10. Do all birds fly? no

11. A blood sample comes from the ventricle of an animal that is either an amphibian or a bird. How can you tell which?

12. Which is harder, the egg of a reptile or the egg of a bird? bird

13. You see some barbs from a feather. You have no idea whether they came from a down feather or a contour feather. Looking at the barbs under the microscope, however, you see that there are no hooked barbules. What kind of feather did the barbs come from? down

14. What type of feather (down or contour) is used for flight? contour What kind are used for insulation? down

15. What is a bird actually doing when it is preening?

16. What is unique about a bird's method of molting?

17. What three things (at least) did flight engineers have to learn from birds to make practical flight possible?

18. Which is heavier, a bird's bone or the same size bone from an amphibian?

19. State the five characteristics that set mammals apart from other vertebrates.

20. What is the principal function of underhair?

21. What do we usually see when we look at a mammal, underhair or guardhair?

22. Name a non-placental mammal.

23. What is the main difference between offspring born after a long gestation period and offspring born after a short gestation period?

GLOSSARY

The numbers in parentheses refer to the page number on which the definition is presented

Abdomen - The body region posterior to the thorax (374)

Abiogenesis (a' bye oh jen uh sis) - The theory that, long ago, very simple life forms spontaneously appeared through random chemical reactions (16)

Absorption - The transport of dissolved substances into cells (167)

Activation energy - Energy necessary to get a chemical reaction going (188)

Active transport - Movement of molecules through the plasma membrane (typically opposite the dictates of osmosis or diffusion) aided by a chemical process (184)

Aerial hypha - A hypha that is not imbedded in the material upon which the fungus grows (102)

Aerobic (ehr oh' bik) respiration - Respiration that requires oxygen (43)

Allele (uh leel') - One of a pair of genes that occupies the same position on homologous chromosomes (242)

Alternation of generations - A lifecycle in which sexual reproduction gives rise to asexual reproduction, which in turn gives rise to sexual reproduction (467)

Amebocytes (uh mee' buh sites) - Cells in a sponge that perform digestion and transport functions (343)

Amniotic egg - An egg in which the embryo is protected by a membrane called an amnion. In addition, the egg is covered in a hard or leathery covering. (510)

Anadromous - A lifecycle in which creatures are hatched in fresh water, migrate to salt water as adults, and then go back to fresh water in order to reproduce (417)

Anaerobic (an uh ro' bik) respiration - Respiration that does not require oxygen (43)

Annual plants - Plants that live for only one year (445)

Anterior end - The end of an animal that contains its head (353)

Antibiotic (an tie bye ah' tik) - A chemical secreted by a living organism that kills or reduces the reproduction rates of other organisms (119)

Antibodies - Specialized proteins that aid in destroying infectious agents (228)

Anticodon - A three-nucleotide sequence on tRNA (195)

Appendicular (ah pen dihk' you luhr) skeleton - The portion of the skeleton that attaches to the axial skeleton and has the limbs attached to it (413)

Arteries - Blood vessels that carry blood away from the heart (413)

Asexual reproduction - Reproduction accomplished by a single organism (7)

Atrium (ay' tree uhm) - A heart chamber that receives blood (428)

Autosomal inheritance - Inheritance of a genetic trait not on a sex chromosome (260)

Autosomes - Chromosomes that do not determine the sex of an individual (257)

Autotrophs (aw' toh trohfs) - Organisms that are able to make their own food (6)

Axial skeleton - The portion of the skeleton that supports and protects the head, neck, and trunk (413)

Biennial plants - Plants that live for a year, lay dormant for a year, and then live and grow for one more year (445)

Bilateral (bye lat' uh ruhl) symmetry - An organism possesses bilateral symmetry if it can only be cut into two identical halves by a single longitudinal cut along its center which divides it into right and left halves. (340)

Bile - A mixture of salts and phospholipids that aids in the breakdown of fat (426)

Binomial nomenclature (bye no' mee ul   no' mun klay chur) - Naming an organism with its genus and species name (29)

Biomass - A measure of the mass of organisms within a region divided by the area of that region (314)

Biosynthesis (bye oh sin' the sis) - The process by which living organisms produce molecules (146)

Bivalve - An organism with two shells (366)

Bone marrow - A soft tissue inside the bone that produces blood cells (412)

Botany - The study of plants (445)

Capillaries - Tiny, thin-walled blood vessels that allow the exchange of gases and nutrients between the blood and cells (413)

Carnivores (kar nih' vors) - Organisms that eat only organisms other than plants (3)

Catalyst - A substance that alters the speed of a chemical reaction but does not get used up in the process (145)

Cell wall - A rigid substance on the outside of certain cells, usually plant and bacteria cells (171)

Cellulose - A substance made of sugars. It is common in the cell walls of many organisms. (86)

Central vacuole - A large vacuole that rests at the center of most plant cells and is filled with a solution which contains a high concentration of solutes (176)

Centromere (sen troh' mer) - Constricted region of a chromosome and the point at which duplicate DNA strands attach themselves (212)

Cephalothorax (sef uh loh thor' aks) - A body region comprised of a head and a thorax together (374)

Cerebellum - The lobe that controls involuntary actions and refines muscle movement (415)

Cerebrum (suh ree' bruhm) - The lobes of the brain that integrate sensory information and coordinate the creature's response to that information (414)

Change in chromosome number - A situation in which abnormal cellular events in meiosis lead to either none of a particular chromosome in the gamete or more than one chromosome in the gamete (262)

Change in chromosome structure - A situation in which the chromosome loses or gains genes during meiosis (262)

Chemical change - A change that alters the makeup of the elements or molecules of a substance (136)

Chitin (ky' tin) - A chemical that provides both toughness and flexibility (103)

Chlorophyll (klor' oh fill) - A pigment necessary for photosynthesis (76)

Chloroplast (klor' oh plast) - An organelle containing chlorophyll for photosynthesis (76)

Chromatin (kroh' muh tun) - Clusters of DNA and proteins in the nucleus (178)

Chromoplasts (kroh' muh plasts) - Organelles that contain pigments used in photosynthesis (175)

Chromosome (krohm' uh zohm) - A strand of DNA coiled around and supported by proteins, found in the nucleus of the cell (210)

Cilia (sil' ee uh) - Numerous short extensions of the plasma membrane used for locomotion (78)

Circulatory system - A system designed to transport food and other necessary substances throughout a creature's body (355)

Closed circulatory system - A circulatory system in which the oxygen-carrying blood cells never leave the blood vessels (413)

Cohesion (coh he' shun) - The phenomenon that occurs when individual molecules are so strongly attracted to each other that they tend to stay together, even when exposed to tension (481)

Collar cells - Flagellated cells that pump water into a sponge (343)

Complete metamorphosis - Insect development consisting of 4 stages: egg, larva, pupa, and adult (395)

Compound eye - An eye made of many lenses, each with a very limited scope (375)

Concentration - A measurement of how much substance exists within a certain volume (140)

Conjugation - A temporary union of two organisms for the purpose of DNA transfer (47)

Consumers - Organisms that eat living producers and/or other consumers for food (4)

Contour feathers - Feathers with hooked and smooth barbules, allowing the barbules to interlock (526)

Cotyledon - A "seed leaf" which develops as a part of the seed. It provides nutrients to the developing seedling and eventually becomes the first leaf of the plant. (471)

Cytology (sigh tahl' uh jee) - The study of cells (169)

Cytolysis (sigh tahl' us sis) - The rupturing of a cell due to excess internal pressure (185)

Cytoplasm (sy' tuh plaz uhm) - A jelly-like fluid inside the cell in which the organelles are suspended (172)

Cytoplasmic streaming - The motion of the cytoplasm which results in a coordinated movement of the cell's organelles (173)

Deciduous (duh sid' you us) plant - A plant that loses its leaves before winter (457)

Decomposers - Organisms that break down the dead remains of other organisms (4)

Dehydration (dee hye dray' shun) reaction - A chemical reaction in which molecules combine by ejecting water (149)

Diffusion - The random motion of molecules from an area of high concentration to an area of low concentration (140)

Digestion - The breakdown of absorbed substances (167)

Dihybrid cross - A cross between two individuals concentrating on two definable traits (251)

Diploid cell - A cell whose chromosomes come in homologous pairs (219)

Diploid chromosome number (2n) - The total number of chromosomes in a diploid cell (219)

Disaccharides (dye sak' uh rides) - Carbohydrates that are made up of two monosaccharides (149)

Dominant allele - An allele that will determine phenotype if even one is present in the genotype (243)

Dominant generation - In alternation of generations, the generation that occupies the largest portion of the lifecycle (468)

Double fertilization - A fertilization process that requires two sperm to fuse with two eggs (499)

Down feathers - Feathers with smooth barbules but no hooked barbules (526)

Ecological pyramid - A diagram that shows the biomass of organisms at each trophic level (314)

Ecology - The study of relationships among organisms in ecosystems (310)

Ecosystem - An association of living organisms and their physical environment (309)

Ectoplasm (ek' toh plas uhm) - The thin, watery cytoplasm near the plasma membrane of some cells (73)

Ectothermic (ek toh thur' mik) - Lacking an internal mechanism for regulating body heat (428)

Egestion - The removal of non-soluble waste materials (168)

Element - All atoms that contain the same number of protons (132)

Endoplasm (en' doh plas uhm) - The dense cytoplasm found in the interior of many cells (73)

Endoplasmic reticulum (en do plaz' mik rih tik' yuh lum) - An organelle composed of an extensive network of folded membranes which perform several tasks within a cell (174)

Endoskeleton (en doh skel' uh tuhn) - A skeleton on the inside of a creature's body, typically composed of bone or cartilage (411)

Endospore - The DNA of a bacterium that is coated with several hard layers (50)

Endothermic - A creature is endothermic if it has an internal mechanism by which it can regulate its own body temperature, keeping it constant. (522)

Environmental factors - Those "non-biological" factors that are involved in a person's surroundings such as the nature of the person's parents, the person's friends, and the person's behavioral choices (209)

Epidermis (ep uh dur' miss) - An outer layer of cells designed to provide protection (342)

Epithelium (ep ih thee' lee uhm) - Animal tissue consisting of one or more layers of cells that have only one free surface, because the other surface adheres to a membrane or other substance (347)

Eukaryotic cell (yoo kehr ee aht' ik sell) - A cell with distinct, membrane-bound organelles (19)

Excretion - The removal of soluble waste materials (168)

Exoskeleton - A body covering, typically made of chitin, that provides support and protection (373)

External fertilization - The process by which the female lays eggs and the male fertilizes them once they are outside of the female (416)

Extracellular digestion - Digestion that takes place outside of the cell (100)

Eyespot - A light-sensitive region in certain protozoa (76)

Fermentation - The anaerobic (without oxygen) breakdown of sugars into alcohol, carbon dioxide, and lactic acid. (112)

Flagellate (flah' gel ates) - A protozoan that propels itself with a flagellum (75)

Foot - A muscular organ that is used for locomotion and takes a variety of forms depending on the animal (365)

Fossils - Preserved remains of once-living organisms (282)

Fruit - A mature ovary which contains a seed or seeds (500)

Gametes - Haploid cells (n) produced by diploid cells (2n) for the purpose of reproduction (220)

Ganglia (gan' glee uh) - Masses of nerve cell bodies (356)

Gemmule (jem' yool) - A cluster of cells encased in a hard, spicule-reinforced shell (344)

Gene - A section of DNA that codes for the production of a protein or a portion of protein, thereby causing a trait (209)

Genetic disease carrier - A person who is heterozygous in a recessive genetic disorder (261)

Genetic factors - The general guideline of traits determined by a person's DNA (209)

Genetics - The science that studies how characteristics get passed from parent to offspring (207)

Genotype (jee' nuh tipe) - Two-letter set that represents the alleles an organism possesses for a certain trait (242)

Gestation - The period of time during which an embryo develops before being born (535)

Girdling - The process of cutting away a ring of inner and outer bark all the way around a tree trunk (464)

Golgi (gole' jee) bodies - The organelles where proteins and lipids are stored and then modified to suit the needs of the cell (175)

Gonad - A general term for the organ that produces gametes (382)

Gravotropism (grav' uh trohp iz uhm) - A growth response to gravity (484)

Greenhouse effect - The process by which certain gases (principally water, carbon dioxide, and methane) trap heat that would otherwise escape the earth and radiate into space (328)

Haploid cells - Cells that have only one of each chromosome (219)

Haploid chromosome number (n) - The number of homologous pairs in a diploid cell (219)

Haustorium - A hypha of a parasitic fungus which enters the host's cells, absorbing nutrition directly from the cytoplasm (102)

Hemotoxin - A poison that attacks the red blood cells and blood vessels, destroying circulation (515)

Herbivores (hur bih' vors) - Organisms that eat plants exclusively (3)

Hermaphroditic - Possessing both the male and the female reproductive organs (356)

Heterotrophs (het' er uh trohfs) - Organisms that depend on other organisms for their food (6)

Heterozygous (het uh roh zy' gus) genotype - A genotype with two different alleles (243)

Hibernation - A state of extremely low metabolism (435)

Holdfast - A special structure used by an organism to anchor itself (89)

Homeostasis (ho mee oh stay' sis) - Maintaining the status quo in a cell (168)

Homozygous (ho muh zy' gus) genotype - A genotype in which both alleles are identical (243)

Hormones - Chemicals that affect the rate of cellular reproduction and the development of cells (483)

Hydrogen bond - A strong attraction between hydrogen atoms and certain other atoms (usually oxygen or nitrogen) in specific molecules (160)

Hydrolysis - Breaking down complex molecules by the chemical addition of water (150)

Hydrophobic - Lacking any affinity to water (153)

Hypertonic (hi pur tahn' ik) solution - A solution in which the concentration of solutes is greater than that of the cell which resides in the solution (185)

Hypha (hi' fuh) - Filament of fungal cells (100)

Hypothesis - An educated guess that attempts to explain an observation or answer a question (9)

Hypotonic (hi puh tahn' ik) solution - A solution in which the concentration of solutes is less than that of the cell which resides in the solution (185)

Imperfect flowers - Flowers with either stamens or carpels, but not both (491)

Incomplete metamorphosis - Insect development consisting of 3 stages: egg, nymph, and adult (395)

Inheritance - The process by which physical and biological characteristics are transmitted from the parent (or parents) to the offspring (7)

Internal fertilization - The process by which the male places sperm inside the female's body, where the eggs are fertilized (416)

Interphase - The time interval between cellular reproduction (212)

Invertebrates (in vur' tuh brates) - Animals that lack a backbone (339)

Ions - Substances in which at least one atom has an imbalance of protons and electrons (172)

Isomers - Two different molecules that have the same chemical formula (148)

Isotonic (eye suh tahn' ik) solution - A solution in which the concentration of solutes is essentially equal to that of the cell which resides in the solution (184)

Karyotype (kehr' ee uh tipe) - The figure produced when the chromosomes of a species during metaphase are arranged according to size (218)

Leaf margin - The characteristics of the leaf edge (448)

Leaf mosaic - The arrangement of leaves on the stem of a plant (448)

Leucoplasts (loo' kuh plasts) - Organelles that store starches or oils (175)

Loam - A mixture of gravel, sand, silt, and clay (480)

Lysosome (lye' soh soam) - The organelle in animal cells responsible for hydrolysis reactions which break down proteins, polysaccharides, disaccharides, and some lipids (173)

Macroevolution - The hypothesis that the same processes which work in microevolution can, over eons of time, transform an organism into a completely different kind of organism (280)

Mammary glands (mam' ur ee) - Specialized organs in mammals that produce milk to nourish the young (535)

Mantle - A sheath of tissue that encloses the vital organs of a mollusk, makes the mollusk's shell, and performs respiration (365)

Matter - Anything that has mass and takes up space (129)

Medulla oblongata (muh dul' uh ahb lawn gah' tuh) - The lobes that coordinate vital functions, such as those of the circulatory and respiratory systems, and transport signals from the brain to the spinal cord (415)

Medusa - A free-swimming cnidarian with a bell-shaped body and tentacles   (345)

Meiosis - The process by which a diploid (2n) cell forms four gametes (n) (220)

Membrane - A thin covering of tissue (107)

Mesenchyme (mes' uhn kime) - The jelly-like substance that separates the epidermis from the inner cells in a sponge (342)

Mesoglea (mez uh glee' uh) - The jelly-like substance that separates the epithelial cells in a cnidarian (347)

Messenger RNA - The RNA that performs transcription (195)

Metabolism (muh tab' uh liz um) - The process by which a living organism takes energy from its surroundings and uses it to sustain itself, develop, and grow (2)

Microevolution - The theory that natural selection can, over time, take an organism and transform it into a more specialized species of that organism (280)

Microorganisms - Living creatures that are too small to see with the naked eye (14)

Microtubules - Spiral strands of protein molecules that form a rope-like structure (177)

Middle lamella (luh mel' uh) - The thin film between the cell walls of adjacent plant cells (171)

Mitochondria (my tuh kahn' dree uh) - The organelles in which nutrients are converted to energy (173)

Mitosis (mye toh' sis)- The duplication of a cell's chromosomes to allow daughter cells to receive the exact genetic makeup of the parent cell (212)

Model - An explanation or representation of something that cannot be seen (130)

Molecules - Chemicals that result from atoms linking together (134)

Molt - To shed an old exoskeleton so that it can be replaced with a new one (374)

Monohybrid cross - A cross between two individuals concentrating on only one definable trait (251)

Monosaccharides (mahn uh sak' uh rides) - Simple carbohydrates that contain three to ten carbon atoms (149)

Mother cell - A cell ready to begin reproduction, containing duplicate DNA and centriole (213)

Mutation - A radical chemical change in one or more alleles (261)

Mutation - An abrupt and marked difference between offspring and parent (8)

Mycelium (my sell' ee uhm) - The part of the fungus responsible for extracellular digestion and absorption of the digested food (100)

Nastic (nas' tik) movement - Movement in a plant caused by changes in turgor pressure (478)

Nematocysts (nih mat' uh sists) - Small capsules that contain a toxin which is injected into prey or predators (347)

Nervous system - A system of sensitive cells that respond to stimuli such as sound, touch, and taste (356)

Neurotoxin - A poison that attacks the nervous system, causing blindness, paralysis, or suffocation (515)

Notochord (no' tuh kord) - A rod of tough, flexible material that runs the length of a creature's body, providing the majority of its support (407)

Nuclear membrane - A highly-porous membrane that separates the nucleus from the cytoplasm (178)

Nucleus (new' clee us) - The region of a eukaryotic cell which contains the DNA (72)

Olfactory lobes - The regions of the brain that receive signals from the receptors in the nose (414)

Omnivores (ahm nih' vors) - Organisms that eat both plants and other organisms (3)

Open circulatory system - A circulatory system that allows the blood to flow out of the blood vessels and into various body cavities so that the cells are in direct contact with the blood (376)

Optic lobes - The regions of the brain that receive signals from the receptors in the eyes (415)

Organic Molecule - A molecule that contains carbon and at least one of the following: hydrogen, oxygen, nitrogen, sulfur, and/or phosphorous (146)

Osmosis - The tendency of a solvent to travel across a semipermeable membrane into areas of higher solute concentration (141)

Ovaries (oh'vuh reez) - The organ that produces eggs (349)

Oviparous (oh vip' ur us) development - development that occurs in an egg which is hatched outside the female's body (416)

Ovoviviparous (oh vo vye vip' ur us) development - development that occurs in an egg which is hatched inside the female's body (416)

Parasite (pehr' uh syte) - An organism that feeds on a living host (42)

Passive transport - Movement of molecules through the plasma membrane according to the dictates of osmosis or diffusion (184)

Pathogen (path' uh jen) - An organism that causes disease (37)

Pedigree - A diagram that follows a particular species' phenotype through several generations (248)

Pellicle (pel' ick ul) - A firm, flexible coating outside the plasma membrane (76)

Peptide bond - A bond that links amino acids together in a protein (154)

Perennial plants - Plants that grow year after year (445)

Perfect flowers - Flowers with both stamens and carpels (491)

Phagocytic(fag uh' sih tik) vacuole - A vacuole that holds the matter which a cell engulfs (176)

Phagocytosis (fag uh' sigh toh' sis) - The process by which a cell engulfs foreign substances or other cells (176)

Phenotype (fee' nuh tipe) - The observable expression of an organism's genes (243)

Phloem - Vascular tissue that carries substances downward in a plant (446)

Phospholipid - A lipid in which one of the fatty acid molecules has been replaced by a molecule which contains a phosphate group (181)

Photosynthesis (foh' toh sin thuh' sis) - The process by which a plant uses the energy of sunlight and certain chemicals to produce its own food. Oxygen is often a by-product of photosynthesis. (3)

Phototropism (foe' toe trohp iz uhm) - A growth response to light (484)

Physical change - A change that affects the appearance but not the chemical makeup of a substance (136)

Physiology (fiz ee awl' uh gee) - The study of life processes that occur in the daily life of an organism (477)

Phytoplankton (fye toe plank' ton) - Tiny floating photosynthetic organisms, primarily algae (85)

Pinocytic (pin uh sih tik') vesicle - Vesicle formed at the plasma membrane to allow the absorption of large molecules (177)

Placenta (pluh sent' uh) - A structure that allows nutrients and gases to pass between the mother and the embryo (534)

Plankton (plank' ton) - Tiny organisms that float in the water (85)

Plasma membrane - The semipermeable membrane between the cell contents and either the cell wall or the cell's surroundings (172)

Plasmolysis (plaz mahl' uh sis) - A collapse of the cell's cytoplasm due to lack of water (185)

Pollen - A fine dust that contains the sperm of seed-producing plants (470)

Pollination - The transfer of pollen grains from the anther to the carpel in flowering plants (496)

Polyp - A sessile, tubular cnidarian with a mouth and tentacles at one end and a basal disk at the other (345)

Polysaccharides (pahl ee sak' uh rides) - Carbohydrates that are made up of more than two monosaccharides (149)

Pore spaces - Spaces in the soil which determine how much water and air the soil contains (479)

Posterior end - The end of an animal that contains the tail (353)

Primary consumer - An organism that eats producers (311)

Producers - Organisms that produce their own food (4)

Prokaryotic cell (pro kehr ee aht' ik   sell) - A cell that has no distinct, membrane-bound organelles (19)

Pseudopod (soo' doe pod) - A temporary, foot-like extension of a cell, used for locomotion or engulfing food (72)

Radial symmetry - An organism possesses radial symmetry if it can be cut into two identical halves by any longitudinal cut through its center. (340)

Radula - A organ covered with teeth that mollusks use to scrape food into their mouths (365)

Receptors - Special structures or chemicals that allow living organisms to sense the conditions of their surroundings (6)

Recessive allele - An allele that will not determine the phenotype unless the genotype is homozygous with that allele (243)

Regeneration - The ability to re-grow a missing part of the body (362)

Reproduction - Producing more cells (168)

Reproductive plant organs - The flowers, fruits, and seeds of a plant (446)

Respiration - The breakdown of food molecules with a release of energy (167)

Respiration - The process by which food is converted into useable energy for life functions (42)

Rhizoid hypha - A hypha that is imbedded in the material on which the fungus grows (101)

Ribosomes - Non-membrane-bound organelles responsible for protein synthesis (174)

Rough endoplasmic reticulum - ER that is dotted with ribosomes (174)

Saprophyte (sap' roh fyte) - An organism that feeds on dead matter (41)

Saturated fat - A lipid made from fatty acids which have no double bonds between carbon atoms (154)

Scientific law - A theory that has been tested by and is consistent with generations of data (10)

Secondary consumer - An organism that eats primary consumers (311)

Secretion - The release of biosynthesized substances for use by other cells (168)

<u>Secretion vesicle</u> - Vesicle that holds secretion products so that they can be transported to the plasma membrane and released (177)

<u>Seed</u> - An ovule with a protective coating, encasing a mature plant embryo and a nutrient source (500)

<u>Semipermeable membrane</u> - A membrane that allows some molecules to pass through but does not allow other molecules to pass through (141)

<u>Sessile Colony</u> - A colony that uses holdfasts to anchor itself to an object (89)

<u>Sex chromosomes</u> - Chromosomes that determine the sex of an individual (257)

<u>Sex-linked inheritance</u> - Inheritance of a genetic trait located on the sex chromosomes (261)

<u>Sexual reproduction</u> - Reproduction that requires two organisms, a male and a female (7)

<u>Shell</u> - A tough, multilayered structure secreted by the mantle. It is usually used for protection, but sometimes for body support (365)

<u>Simple eye</u> - An eye with only one lens (375)

<u>Smooth endoplasmic reticulum</u> - ER that has no ribosomes (174)

<u>Species</u> - A unit of one or more populations of individuals that can reproduce under normal conditions, produce fertile offspring, and are reproductively isolated from other such units (22)

<u>Spherical symmetry</u> - An organism possesses spherical symmetry if it can be cut into two identical halves by any cut that runs through the organism's center. (340)

<u>Spiritual factors</u> - The quality of a person's relationship with God (209)

<u>Spore</u> - A reproductive cell with a hard, protective coating (80)

<u>Sporophore</u> (spor' uh for) - Specialized aerial hypha that produces spores (102)

<u>Statocyst</u> (stat' uh sist) - The organ of balance in a crustacean (382)

<u>Steady state</u> - A state in which members of a population die as quickly as new members are born (46)

<u>Stolon</u> (sto' lun) - An aerial hypha that asexually reproduces to make more filaments (102)

<u>Strains</u> - Organisms from the same species that have markedly different traits (57)

<u>Symbiosis</u> (sim by oh' sis) - Two or more organisms of different species living together so that each benefits from the other (78)

Taxonomy (tak sahn' uh mee) - The science of classifying organisms (29)

Tertiary (ter' she air ee) consumer - An organism that eats secondary consumers (312)

Testes (test' ez) - The organ that produces sperm (349)

Thallus - The body of a plant-like organism that is not divided into leaves, roots, or stems (86)

The immutability of species - The idea that each individual species on the planet was specially created by God and could never fundamentally change (279)

Theory - A hypothesis that has been tested with a significant amount of data (9)

Thigmotropism (thig' muh trohp iz uhm) - A growth response to touch (484)

Thorax (thor' aks) - The body region between the head and the abdomen (374)

Transcription - The process in which mRNA produces a negative of a strand of DNA (195)

Transformation - The transfer of a "naked" DNA segment from a nonfunctional donor cell to that of a functional recipient cell (49)

Translation - The process by which proteins are formed in the ribosome according to the negative in mRNA (197)

Translocation - The process by which organic substances move down the phloem of a plant (482)

Transpiration (tran spuh ray' shun) - Evaporation of water from the leaves of a plant (323)

True breeding - If an organism has a certain characteristic that is always passed on to all of its offspring, we say that this organism bred true with respect to that characteristic. (236)

Undifferentiated cells - Cells that have not specialized in any particular function (446)

Univalve - An organism with a single shell (365)

Unsaturated fat - A lipid made from fatty acids that have at least one double bond between carbon atoms (154)

Vaccine - A weakened or inactive version of a virus that stimulates the body's production of antibodies which can destroy the virus (228)

Vacuole (vac' you ol) - A membrane bound "sac" within a cell (72)

Vegetative organs - The stems, roots, and leaves of a plant (446)

Veins - Blood vessels that carry blood back to the heart (413)

Ventricle (ven' trih kul) - A heart chamber from which blood is pumped out (428)

Vertebrae (ver' tuh bray) - Segments of bone or some other hard substance that are arranged int a backbone (407)

Vertebrates (vur' tuh brates) - Animals that possess a backbone (339)

Virus - A non-cellular infectious agent that has two characteristics: (1) It has genetic material inside a protective protein coat (2) It cannot reproduce itself (225)

Visceral (vis' ur ul) hump - A hump that contains a mollusk's heart, digestive, and excretory organs (365)

Viviparous (vye vip' ur us) development - development that occurs inside the female, allowing the offspring to gain nutrients and vital substances from the mother through a placenta (pluh sent' uh) (416)

Waste vacuoles - Vacuoles that contain the waste products of digestion (176)

Watershed - An ecosystem where all water runoff drains into a single river or stream (323)

Xylem - A vascular tissue that carries substances upward in a plant (446)

Zooplankton (zoo plank' ton) - Tiny floating organisms that are either small animals or protozoa (85)

Zygospore (zie go' spor) - A zygote surrounded by a hard, protective covering (115)

Zygote (zie' goht) - The result of sexual reproduction when each parent contributes half of the DNA necessary for the offspring (115)

APPENDIX

Criteria for life:

1. All life forms contain deoxyribonucleic acid (DNA).

2. All life forms have a method by which they extract energy from the surroundings and convert it into energy that sustains them.

3. All life forms can sense changes in their surroundings and respond to those changes.

4. All life forms reproduce.

A Simple Biological Key:

A Simple Biological Key	
1. Microscopic..	**2**
Macroscopic (visible with the naked eye).................................	**3**
2. Eukaryotic cell..	*kingdom Protista*
Prokaryotic cell..	*kingdom Monera*
3. Autotrophic...........................*kingdom Plantae*.....	**4**
Heterotrophic..	**5**
4. Leaves with parallel veins ....*phylum Anthophyta*......	*class Monocotyledoneae*
Leaves with netted veins ....*phylum Anthophyta*.....	*class Dicotyledoneae*
5. Decomposer..	*kingdom Fungi*
Consumer.......................................*kingdom Animalia*....	**6**
6. No Backbone...	**7**
Backbone..*phylum Chordata*....	**22**
7. Organism can be externally divided into equal halves (like a pie), but it has no distinguishable right and left sides.........................	**8**
Organism either can be divided into right and left sides that are mirror images or cannot be divided into two equal halves...........	**9**
8. Soft, transparent body with tentacles	*phylum Cnidaria*
Firm body with internal support; covered with scales or spiny plates; tiny, hollow tube feet used for movement........................	*phylum Echinodermata*
9. External plates that support and protect.......*phylum Arthropoda*..	**14**
External shell or soft, shell-less body...............................	**10**

10. External Shell...*phylum Mollusca....* **11**
    No external shell.. **12**
11. Coiled shell..*class Gastropoda*
    Shell made of two similar parts.....................................*class Pelecypoda*
12. Worm-like body without tentacled receptors on head...................*phylum Annelida*
    Non-worm-like body or tentacled receptors
    on head...*phylum Mollusca.......* **13**
13. Worm-like body with tentacled receptors on head.......................*class Gastropoda*
    Non-worm-like body but 8 or more tentacles used for grasping.... *class Cephalopoda*
14. More than 3 pairs of legs... **15**
    3 pairs of walking legs.............................*class Insecta......* **16**
15. 4 pairs of walking legs, body in two divisions............................*class Arachnida*
    More than 4 pairs of walking legs...*class Malacostraca*
16 Wings.. **17**
    No wings... **21**
17. All wings transparent... **18**
    Non-transparent wings.. **19**
18. Capable of stinging from back of body.. *order Hymenoptera*
    Cannot sting (may be able to bite).. *order Diptera*

19. Large, sometimes colorful wings.. *order Lepidoptera*
    Thick, hard, leathery wings... **20**
20. Pair of hard wings covering a pair of folded, transparent wings.... *order Coleoptera*
    Pair of leathery wings covering a pair of transparent wings.......... *order Orthoptera*
21. Piercing, sucking mouthparts for obtaining blood........................ *order Siphonaptera*
    Mouthparts for chewing.. *order Hymenoptera*
22. Jaws or beak... **23**
    No jaw or beak.. *class Agnatha*
23. Skin covered with scales... **24**
    No scales on skin... **26**
24. Fins and gills... **25**
    No fins, breathes with lungs... *class Reptilia*
25. Mouth on lower part of body... *class Chondrichthyes*
    Mouth on front part of body.. *class Osteichthyes*
26. No scales, no hair, no feathers; skin is slimy......*class Amphibia...* **27**
    Feathers or hair.. **28**
27. Tail... *order Caudata*
    No tail... *order Anura*
28. Feathers on body... *class Aves*
    Hair on body.......................................*class Mammalia...* **29**
29. Hooves... **30**
    No hooves.. **31**
30. Odd number of toes... *order Perissodactyla*
    Even number of toes.. *order Artiodactyla*

31. Carnivore.. **32**
     Herbivore.. **33**
32. Teeth.. *order Carnivora*
     No teeth, eats insects... *order Insectivora*
33. Enlarged front teeth for gnawing...................................... **34**
     No enlarged front teeth for gnawing................................. **35**
34. Legs for crawling... *order Rodentia*
     Hind legs for jumping... *order Lagomorpha*
35. Enlarged trunk, used for breathing and grasping................ *order Proboscidea*
     Tendency to stand erect on two hind limbs...................... *order Primates*

TABLE 5.1
Biologically Important elements

Element Name	Abbreviation
carbon	C
hydrogen	H
oxygen	O
nitrogen	N
phosphorus	P
sulfur	S

Phase Changes

$$\text{SOLID} \underset{\substack{\text{TAKE ENERGY}\\\text{AWAY}}}{\overset{\text{ADD ENERGY}}{\rightleftarrows}} \text{LIQUID} \underset{\substack{\text{TAKE ENERGY}\\\text{AWAY}}}{\overset{\text{ADD ENERGY}}{\rightleftarrows}} \text{GAS}$$

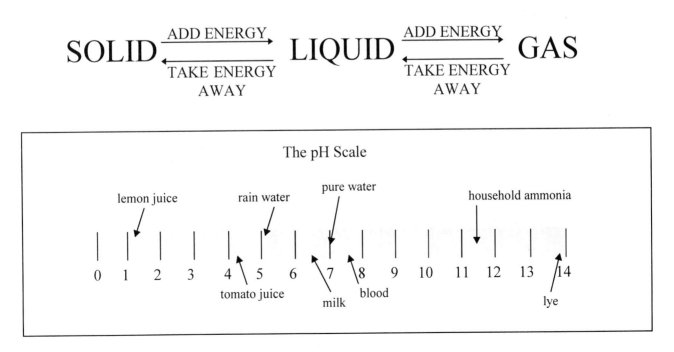

The pH Scale

Schematic Representations of an Animal Cell (left) and a Plant Cell (right)

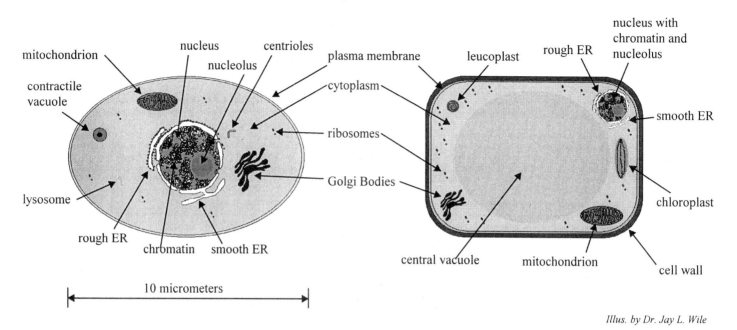

Illus. by Dr. Jay L. Wile

A Simplified Schematic of Cellular Respiration

STAGE 1: GLYCOLYSIS
Occurs in Cytoplasm - Needs no oxygen - Requires 2 ATPs to start, but produces 4 ATPs in the end, for a net gain of 2 ATPs. It also produces 4 hydrogen atoms, and 2 pyruvic acid molecules

STAGE 2: KREBS CYCLE
Occurs in Mitochondria - Requires 3 oxygen molecules and 2 pyruvic acid molecules from stage 1- Produces 2 ATPs, 8 hydrogen atoms, and 6 carbon dioxide molecules

$C_6H_{12}O_6 \rightarrow 4H$ (to electron transport system)

$+ 2C_3H_4O_3$ (to Krebs cycle)

(The released energy can make 4 ATPs from ADP and phosphate in cytoplasm. Since the process needs 2 ATPs to start, the net gain is 2 ATPs.)

$2C_3H_4O_3 + 3O_2 \rightarrow 8H$ (to electron transport system)

$+ 6CO_2$

(released energy can make 2 ATPs from ADP and phosphate in mitochondria)

STAGE 3: ELECTRON TRANSPORT SYSTEM
Occurs in Mitochondria - Requires 3 oxygen molecules and 12 hydrogen atoms (4 from stage 1 and 8 from stage 2) - Produces 32 ATPs and 6 water molecules

$12H$ (4 from stage 1 and 8 from stage 2) $+ 3O_2 \rightarrow 6H_2O$

(released energy can make 32 ATPs from ADP and phosphate in mitochondria)

MENDEL'S PRINCIPLES OF GENETICS

1. **The traits of an organism are determined by its genes.**

2. **Each organism has two alleles that make up the genotype of a given trait.**

3. **In sexual reproduction, each parent contributes ONLY ONE of its alleles to the offspring.**

4. **In each genotype, there is a dominant allele. If it exists in an organism, the phenotype is determined by that allele.**

The Major Data That Relate To Macroevolution

Data Set	Summary
The geological column	This data is inconclusive as far as macroevolution is concerned. If you believe that the geological column was formed according to the speculations of Lyell, then it is evidence for macroevolution because it shows that life forms early in earth's history were simple and gradually got more complex. If you believe that the geological column was formed by natural catastrophe, then it is evidence against macroevolution. Since geologists have seen rock strata formed each way, it is impossible to tell which belief is scientifically correct.
The fossil record	This data is strong evidence against macroevolution. There are no clear intermediate links in the fossil record. The very few that macroevolutionists can produce are so similar to one of the two species they supposedly link, it is more scientifically sound to consider them a part of that species.
Structural homology	This data is strong evidence against macroevolution. The similar structures are not a result of inheritance from a common ancestor, because the similar structures are determined by quite different genes.
Molecular biology	This data is strong evidence against macroevolution. There are no evolutionary patterns in the sequences of amino acids of common proteins.

Data That Relate to Global Warming

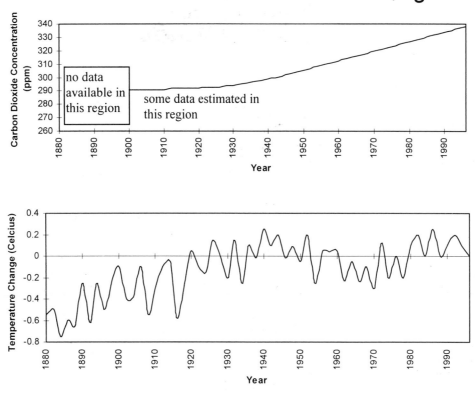

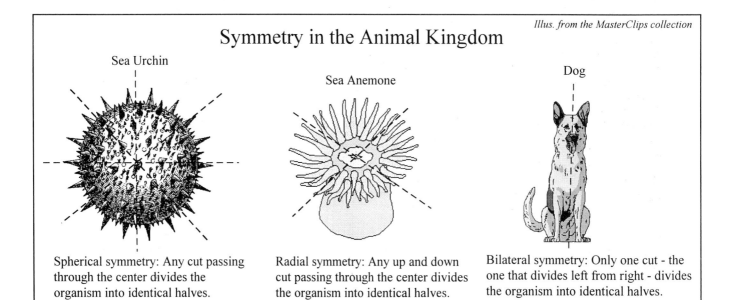

Symmetry in the Animal Kingdom

Illus. from the MasterClips collection

Sea Urchin

Spherical symmetry: Any cut passing through the center divides the organism into identical halves.

Sea Anemone

Radial symmetry: Any up and down cut passing through the center divides the organism into identical halves.

Dog

Bilateral symmetry: Only one cut - the one that divides left from right - divides the organism into identical halves.

Index

—A—

abdomen, 374
abiogenesis, 16, 17, 169
abscisic acid, 485
abscission layer, 457
absorption, 167, 169
acetic acid, 136
achene, 501
Achnanthes, 70
acid, 151
acid group, 151
activation energy, 188
active transport, 184, 185
active transport site, 183
adenine, 160, 161, 194
adenosine, 189
adenosine diphosphate, 189
adenosine triphosphate, 189
ADP, 189, 190
aeciospores, 109
aerial hypha, 101, 127
aerobic respiration, 43
afferent brachial arteries, 428
Africa, 12
African sleeping sickness, 77
agar, 92
aggregate fruit, 500
Agnatha, 417
AIDS, 227
air bladder, 92, 427
alba, 29
albatrosses, 532
alcohol, 112, 113
alcoholism, 208
algae, 56, 68, 71, 79, 85, 86, 88,
    123, 175, 343
algae bloom, 86
algae, blue-green, 20
algae, green, 86
algin, 91
alginic acid, 70, 91
allantois, 511
Allegemeine Krakenhaus, 10
allele, 242, 252, 256, 257, 259, 260,
    281
alligator, 518
alternation of generation, 469
alternation of generations, 467
Amanita, 108

Amastigomycota, 104, 105, 115,
    117, 118, 127
amebocyte, 343, 345, 351
American chestnut, 114
amine group, 152
amino acid, 154, 155, 156, 173,
    196, 316, 350
amino acid sequence, 295
ammocoetes, 418
amnion, 511
amniotic egg, 522
amniotic eggs, 510
amoeba, 20, 69, 71, 72, 73, 74, 176
Amoeba proteus, 71, 72, 74
Amphibia, 434
amphibian, 434
amphioxus, 409
anaconda, 515
anadromous, 417
anaerobic, 192
anaerobic bacteria, 43
anaerobic respiration, 43, 192
anal pore, 79
Analipus, 70
Anamalia, 21
anaphase, 213, 214, 215, 220, 223
anaphase I, 221, 223
anaphase II, 221
anatomy, 425
anemone, 316, 345
animal cell, 170, 171, 172, 173,
    176, 178
Animalia, 339
Annelida, 352, 363
annual, 445
anole, 517
Anoxyphotobacteria, 55
ant, 103, 396
antenna, 375, 377
antennule, 377
anterior, 353
anterior cardial vein, 428
anterior dorsal fin, 421, 425
anther, 491, 494
anthocyanin, 455
Anthophyta, 471
antibiotic, 119, 127
antibody, 228
Anura, 436, 437
anus, 353, 382, 410, 426
aortic arches, 355, 436
Apostle Paul, 489

appendicular skeleton, 413
aquatic mammals, 539
Arachnida, 387
Archaebacteria, 55, 56, 58
Archaeopteryx, 291
Aristotle, 13
arthropod, 374
Arthropoda, 373
Artiodactyla, 540
asci, 111
Ascomycetes, 105, 111, 112, 114
ascospore, 111, 114
asexual reproduction, 7
aster, 213, 221
Atlantic salmon, 430
atom, 129, 130, 131, 132, 134, 135,
    319
atoms, 133
ATP, 189, 190, 191
atrial cavity, 410
atriopore, 410
atrium, 428, 436
Aurelia, 351
Austin, Thomas, 309
Australia, 309, 310
Australopithecus, 289
Australopithecus afarensis, 289,
    290, 291
autosomal inheritance, 260
autosome, 257
autotroph, 6, 21
autotrophic, 25
auxin, 483
Aves, 521
axial skeleton, 413

—B—

Babel, 275
bacillus, 39, 50
backbone, 25, 407, 410, 413
bacteria, 4, 5, 10, 20, 37, 39, 79, 80,
    112, 119, 150, 171, 192, 343
bacteria, anaerobic, 43, 56
bacteria, autotrophic, 42
bacteria, chemosynthetic, 56
bacteria, heterotrophic, 42
bacteria, non-photosynthetic, 54
bacteria, parasitic, 42
bacteria, pathogenic, 54, 58
bacteria, photosynthetic, 55, 56
bacterial colony, 50, 51

bacterial growth, conditions for, 58
bacterial photosynthesis, 42
bacteriophage, 226, 227
bacterium, 38, 41, 67, 176, 227
baker's yeast, 112
Balantidium coli, 80
bald eagle, 531
barb, 525
barberry, 109
barbule, 525
barely, 109
bark, 463
barnacle, 384
basal body, 40
base, 151
basidia, 106, 108, 109
Basidiomycetes, 105, 106, 109, 111
basidiospore, 108, 109
basidiospores, 106
bat, 538
Baumgartner, Andreas, 235
beak, 25
Bears, 539
beaver, 534
beer, 112
berry, 500
betta, 430
Bible, 15, 16, 88, 273, 275, 276
biennial, 445
bilateral symmetry, 340
bile, 426
binomial nomenclature, 29
Biochemstry, 1
biological key, 22, 28
biomass, 314
Biosphere 2, 44
biosynthesis, 146, 168, 169, 173,
    175
bird, 21, 522
birds of prey, 531
bison, 540
bivalve, 365
black widow, 387
black-necked cobra, 516
blade, 447
blind shrimp, 317
bloom, 86
blue phycobilin, 145
blue whale, 539
blue-green algae, 55, 56
blue-streak wrasse, 318, 319
blue-streak wrasses, 319
boa constrictor, 515
Bohr model, 131
bombardier beetle, 398
bone, 411
bone marrow, 411
bony fish, 424
book lungs, 387, 391
Borrelia burgdorferi, 54

botulism, 57
Brachiosaurus, 521
brain, 414
brain lobe, 414
Brazil, 324
bread, 99
bread mold, 21, 115, 116
brown algae, 91
brown recluse, 387
budding, 112
butter, 38
butterfly fish, 431
buttermilk, 38
button, 107, 124
button stage, 107

—C—

caladium plant, 454
calcium carbonate, 342
calculus, 131
Cambridge, 274
Camembert, 119
Canada, 463
cancer, 227
canine fangs, 539
cap, 101, 106, 108
caps, 108
capsule, 39, 40, 58, 501
carapace, 377
carbohydrate, 146, 149, 150, 153,
    157, 172
carbon, 133, 134, 135, 142, 154
carbon cycle, 326, 332
carbon dioxide, 112, 134, 135, 142,
    188, 328
carbon monoxide, 135
caritenoids, 86
Carnivora, 539
carnivore, 3, 4, 5, 17, 25
carotenoid, 455
carotid arch, 436
carp, 298, 430
carpel, 490, 493, 498
carpels, 491
carples, 292
cartilage, 411, 412
cartilaginous fish, 420
cat, 21, 210
cat's-cry disease, 262
catalase, 398
catalyst, 145, 156, 158, 174
caterpillar, 395
cats, 539
cattle, 540
caudal fin, 421, 425
Caudata, 436
cell, 19, 167, 168, 169, 170, 171,
    172, 173, 174, 176, 177, 193,
    197

cell wall, 38, 40, 45, 48, 54, 56,
    171, 172, 176, 185, 216
cell, eukaryotic, 19, 20, 21
cell, prokaryotic, 19, 20, 37
cellular division, 177, 212
cellular fermentation, 192
cellular respiration, 188, 191, 192,
    199
Cellulose, 70, 77, 86, 150, 171, 315
central vacuole, 176, 185
centriole, 213, 221, 223
centrioles, 177
centromere, 212, 217, 218, 221
Cephalochordata, 407, 409
cephalothorax, 374, 377
Ceratium, 70, 89
cerebellum, 415
cerebrum, 414
Cetacea, 539
chain structure, 148
chameleon, 517
charge, 131, 173
cheddar cheese, 38
cheese, 38
cheliped, 377
chemical, 2
chemical bond, 147
chemical change, 137, 143
chemical changes, 136
chemical equation, 143
chemical equations, 144
chemical formulae, 134, 135
chemical reaction, 143
chestnut blight, 114
Chicago Field Museum of Natural
    History, 286
chicken pox, 227
chitin, 103, 373
Chlorella, 87
chlorine, 85
chlorophyll, 86, 145, 454
Chlorophyta, 68, 70, 86
chloroplast, 175
cholorplast, 87
cholorplasts, 76, 86
Chondrichthyes, 420
Chondrus, 92
Chordata, 25, 339, 407
chorion, 511
Christ's College, 274
Christian, 236, 273, 274, 275
chromatin, 178
chromatin material, 214
chromatophore, 430
chromoplast, 175
chromosome, 210, 212, 213, 217,
    218, 221, 223, 242, 244, 257,
    262
Chrysophyta, 68, 70, 88, 89
Chytridiomycetes, 117
chytrids, 117

Cilia, 69, 78, 79, 80, 177
ciliate, 79, 80
ciliates, 78
Ciliophora, 68, 69
cilium, 177
circulatory system, 355
clam, 90, 364
class, 18, 22
classification, 3, 5, 6, 17, 19, 28
Claviceps purpurea, 114
clay, 479
clitellum, 353, 359
closed circulatory system, 413
Clostridium botulinum, 57
clownfish, 316, 350
club fungi, 106
Cnidaria, 345, 347, 350
cobra, 516
coccus, 39, 50
cockroach, 83
cocoon, 356
codon, 198
coelacanth, 12
cohesion, 481
cohesion-tension theory, 480, 482
Coleoptera, 398
collagen, 411
collar cell, 343
collenchyma, 453
colony, 77
colony, of bacteria, 50, 51
common cold, 227
compact bone tissue, 411
complete metamorphosis, 395
composite flower, 492
compound eye, 375
compound fruit, 500
concentration, 140
conidiophore, 102, 103
conjugation, 47, 49, 79
conjugation tube, 48
consumer, 4, 5, 17, 25, 311
contour feathers, 526
contracile vacuole, 79
contractile vacuole, 176
contractile vacuoles, 72, 75, 185
conus arteriosus, 436
copperhead, 516
coral, 319, 351
coral reef, 351
coral snake, 516
cork cambium, 463
cork cell, 463
corn, 109
cortex, 460, 462
cosmetics, 91
cottage cheese, 38
cotyledon, 471
cough syrup, 91
cow, 5, 280
crayfish, 376

creation, 4, 5, 6, 7, 41, 43, 44, 78,
124, 129, 133, 134, 136, 145,
216, 236, 279, 285, 313, 315,
319, 320, 321, 325, 333, 339,
349, 354, 381, 388, 389, 397,
477, 497, 509, 515, 521, 539
Creator, 15, 78, 121, 129, 273, 281,
320, 325, 350
cri-du-chat, 262
criteria for life, 1
crocodile, 518
Crocodylia, 513
crop, 354
Crustacea, 374, 376
crustacean, 103
Cryphonectria parasticia, 115
Cup fungi, 114
curds, 38
cuticle, 355, 452
cyanobacteria, 20, 55, 56
cyst, 74, 80
cysteine, 199
cytochrome C, 294, 295, 297, 298
cytokinin, 485
cytology, 169
cytolysis, 185
cytoplasm, 39, 40, 73, 76, 77, 81,
101, 172, 173, 174, 178, 188,
190
cytoplasmic streaming, 173, 177
cytosine, 160, 161, 194

—D—

Darwin's Finches, 276
Darwin, Charles R., 273, 274, 275,
276
Darwin, Henrietta, 275
daughter cell, 214
deciduous plant, 457
decomposer, 4, 5, 17, 21, 25, 41,
99, 315
deer, 540
deforestation, 324
dehydrated food, 59
dehydration, 149, 150
dehydration reaction, 149, 150, 154
dental disease, 55
Denton, Dr. Michael, 293
deoxyribose, 160, 193
design, 41, 303, 319
designer, 320, 453
Desmid, 87
diatom, 88
diatomous earth, 88
dicot, 450, 461, 471
Dicotyledonae, 450, 471
diffusion, 140, 141, 142
digestion, 73, 167, 168, 169
digestion, extracellular, 100, 127
dihybrid cross, 251, 253, 256

dinoflagellate, 70, 89, 90
dinosaur, 519
diploid, 219, 220, 223, 244
diploid cell, 219, 220
diploid chromosome number, 219
Diptera, 399
disaccharide, 149, 150, 173
Discovery Bay, 410
diving birds, 532
division, 18
DNA, 1, 8, 9, 39, 40, 45, 48, 49, 52,
107, 160, 177, 178, 193, 194,
195, 197, 208, 209, 210, 212,
215, 220, 221, 226, 293, 295
dog, 279, 280
dogma, 274
dogs, 539
dogwood tree, 454
dolphins, 539
dominant, 252
dominant factor, 239
dominant generation, 468
Doppler effect, 235
Doppler. J.C., 235
dorsal aorta, 428
dorsal nerve chord, 407
double bond, 147
double fertilization, 499
dove, 531
Down's syndrome, 262
down feathers, 526
Drosophila, 258
drupe, 500
dry fruit, 500
duck-billed platypus, 538
ducks, 531
dust, 37
Dutch elm disease, 114
Dynobryon, 89

—E—

eagle, 278, 280
earthworm, 353
Echinodermata, 366
ecological pyramid, 313, 315
ecologist, 324
ecology, 310, 333
ecosystem, 309, 310, 311, 313, 315,
318, 320, 321, 324, 326, 328,
330, 333
ectoplasm, 73
ectothermic, 428, 435, 509
Edinburgh, University of, 274
efferent brachial arteries, 428
egestion, 168, 169
egg, 223, 244
egg cell, 224
egg white, 511
egg-laying mammals, 538
Eldredge, Dr. Niles, 289

electron, 129, 130, 131, 132, 133, 135
element, 132, 134, 135
elephant, 24, 25
elephants, 540
elongation region, 460
embryo sac, 491, 493, 495, 498
endodermis, 460
endoplasm, 73
endoplasmic reticulum, 174
endoskeleton, 411
endosperm, 499
endospore, 50, 58
endothermic, 522
energy, 1, 2, 3, 4, 5, 19, 138, 167, 173, 188, 310, 313
England, 274, 309
English, 29
Entamoabe histolytica, 74
Entamoeba, 74
Entamoeba coli, 74
Entamoeba gingivalis, 74
environmental factors, 209
enzyme, 156, 157, 158, 174, 189, 192
Eohippus, 288
epidermis, 342, 452, 460, 462
epiphytic orchid, 458
epithelium, 346
equatorial plane, 213, 221
Equus, 288
ergot of rye, 114
Escherichia Coli, 57, 74
esophagus, 426
ethylene, 485
etiolation, 484
Euclid, 274
Euglena, 69, 71, 75, 76, 86, 145, 172, 177
eukaryotic, 74, 90
eukaryotic cell, 167, 174, 177
eukaryotic, 67
even-toed hoofed mammals, 540
evolution, 273, 276, 278, 280
excretion, 168, 169
exoskeleton, 373
extensor muscles, 382
external fertilization, 416
extracellular digestion, 100, 127, 347
eyespot, 76

—F—

factor, 239
fairy ring, 108
family, 18, 22
fang, 390
Faraday, Michael, 275
fat, 152, 153, 172
fatty acid, 152, 154, 173, 181

feather, 524
feather worm, 358
feathers, 25
feces, 83
feeding frenzy, 422
female, 223
fermentation, 112
fern, 469
fertile, 22
fertilization, 244, 245, 262
fibrous root system, 458
fibrovascular bundle, 462
field study, 438
filament, 40, 87
finch, 276, 277, 278, 279, 280
Firmibacteria, 55, 57
Firmicutes, 55, 57
flagella, 103, 104, 117, 177
flagellate, 74, 79
flagellum, 39, 69, 74, 76, 177, 223, 343
flamingo, 532
flatworms, 361
Fleming, Alexander, 119
fleshy fruit, 500
flexor muscles, 382
flood, 88
floor polish, 91
Florida, 314
florigen, 485
flower, 21
fluke, 363
flying mammals, 538
food chain, 312
food poisoning, 57
food vacuole, 72, 73, 79, 176
food web, 312
foot, 365, 370
fossil, 282
fossil record, 286, 287
fox, 309
frog, 437
fructose, 148, 149
fruit, 471, 500
fruit fly, 258
fruiting body, 103, 106, 107, 108, 114, 122
Fucus, 92
Fungi, 18, 21, 99, 100, 101, 103, 104, 105, 106, 109, 114, 116, 118, 119, 121, 123, 124, 150
fungus, 4, 5, 100, 101, 104, 105, 108, 114, 116, 117, 118, 123, 124
fungus root, 123

—G—

galactose, 150
Galactosemia, 260
Galapagos Archipelago, 276, 282

Galapagos Islands, 518
gall bladder, 427
game birds, 531
gamete, 220, 223, 244, 245, 252, 256, 257
gametophyte, 468
ganglia, 356, 382
garter snake, 515
gastric ceca, 394
gecko, 517
geese, 531
gemmule, 344
gene, 209, 219
Genesis, 8
Genesis, book of, 273
genetic code, 281, 300
genetic disease, 260
genetic disease carrier, 261
genetic disorder, 260
genetic factors, 209
genetic tendency, 208
genetics, 207, 281, 293, 294, 299, 300
genetics, father of, 236
genotype, 242, 243, 244, 245, 252
genus, 18, 22, 29
geological column, 283, 284, 285
geology, 274, 285
germ, 10
germination, 503
gestation, 535
giant tortoise, 518
gibberellin, 485
Gila monster, 517
gills, 101, 108, 378
girdling, 464
gizzard, 354, 394
global warming, 330, 332
glucose, 135, 144, 145, 147, 148, 149, 150, 188
glycerin, 181
glycerol, 152
glycine, 295
glycogen, 150, 153
glycolysis, 189, 190, 192
goats, 540
goby, 317
God, 4, 5, 7, 8, 9, 11, 41, 44, 78, 119, 124, 129, 133, 134, 136, 145, 150, 192, 216, 229, 236, 273, 275, 276, 281, 309, 311, 315, 319, 321, 333, 339, 349, 355, 375, 379, 381, 390, 397, 477, 493, 496, 497
goldfish, 117
Golgi bodies, 175
gonad, 382
Gracilicutes, 54, 55, 57
grafting, 488
grain, 501
Gram stain, 53, 54

Gram, Hans Christian, 53
Gram-negative, 54
Gram-positive, 54
grass, 21
grasshopper, 21, 392, 393, 399
gravel, 479
gravotropism, 484
Great Lakes, 419
great potato famine, 117
Great Salt Lake, 55
green algae., 86
green gland, 380
greenhouse effect, 328, 330, 331, 332
guanine, 160, 161, 194
guard cells, 452
guardhair, 534
guinea pig, 249
gullet, 79, 80
gum disease, 55
gut, 410
Gymnodinium, 90
Gymnodinium brevis, 90

—H—

hair, 25, 136, 533
hair follicles, 534
haploid, 219, 220, 221, 223, 244
haploid cell, 220, 221
haploid cells, 219
haploid chromosome number, 219
haustoria, 123
haustorium, 102
hawk, 309
head, 374
heat-sensing pits, 516
helium, 131, 132, 133, 135
hemoglobin, 294, 413
hemotoxin, 515
herbivore, 3, 4, 5, 17, 25, 311
hermaphroditic, 356
heterotroph, 6, 21
heterotrophic, 25
heterozygous, 243
hibernation, 435
HIV, 226, 227
HMS Beagle, 274, 275, 276, 279
holdfast, 89
homeostasis, 168, 169
homologous chromosome, 242
homologous chromosome pair, 217
homologous chromosomes, 244, 257
homologous pair, 219, 221, 223
homologue, 217, 218, 221
homologues, 221
homology, 293
homozygous, 243
hoofed mammals, 540
hook, 40

hooves, 25
Hope, Sir James, 275
hormone, 483
horned toad, 517
horny wings, 392
horse, 210, 280, 297, 298, 540
horse, fake evolution of, 288
housefly, 83
Hoyle, Sir Frederick, 320
humerus, 292
hummingbird, 532
Huntington's disorder, 261
Hutchinson-Gilford progeria syndrome, 261
hydra, 345, 347, 349, 350
hydrogen, 133, 135, 142, 147, 154
hydrogen bonding, 160
hydrogen peroxide, 398
hydrogen transport system, 189, 190
hydrolysis, 150, 156, 477, 478
hydrophilic, 183
hydrophobic, 153, 183
hydroquinone, 398
hydrothermal vents, 55
Hymenoptera, 396
hypertonic solution, 184
hypha, 100, 101, 103, 127
hypha, aerial, 101, 127
hypha, rhizoid, 101
hypha, septate, 101
hyphae, 100, 101, 103, 107
hyphae, nonseptate, 101
hypothesis, 9, 10, 12, 13
hypotonic solution, 185

—I—

ice, 137
ice cream, 86, 91
immutability of species, 279
imperfect flowers, 491
Imperfect Fungi, 104, 118, 119, 124, 127
incomplete metamorphosis, 395
India, 324
Indian Ocean, 12
Indiana Ocean, 12
influenza, 227
information, 1
infusion, 14
inheritance, 7
inner bark, 463
insect, 83
Insecta, 392
insectivorous plant, 486
intermediate link, 286, 289, 299
internal fertilization, 416
interphase, 212, 214
intestine, 57, 354
invertebrate, 339, 341
iodine, 86

ion, 173
Ireland, 117
Irish moss, 86
irritability, 168, 169
isomer, 148
isotonic solution, 184
Israel, 489

—J—

Jacobson's organs, 514
Jamaica, 410
jaw, 25
jawless fish, 417
jelly beans, 91
jellyfish, 345, 351, 352
Jew, 273

—K—

kangaroos, 538
karyotype, 218
kelp, 91
Kentucky, 274
kidney, 428
killer whales, 539
kingdom, 18, 19, 21, 22
koalas, 538
Krebs cycle, 189, 190, 192

—L—

lactating, 535
Lactobacilus, 38
lactose, 150, 157
lactose intolerance, 260
lactose intolerant, 157
lamprey, 298
lamprey eel, 418
lancelet, 409
larva, 395
late blight of potato, 117
lateral line, 422, 425
Latin, 29
law, scientific, 10, 11
leaf, 449, 451
leaf mosaic, 448
leafy shoot, 467
leatherback sea turtle, 518
leather-like wings, 392
leeche, 357
Lepidoptera, 396
leucine, 295
leucoplast, 175
lichen, 123, 124
life, 1
life, criteria for, 1
lime, 342, 343
limitations (of science), 9

limitations of the scientific method, 12
Lincoln, Abraham, 274
lionfish, 430
lions, 539
lipid, 152, 153, 173, 181
liver, 82, 426
loam, 479
lobster, 103
locomotion, 71, 73, 74, 80, 103
locomotion, of bacteria, 39
locomtion, 76
Lucy, 290
Lumbricus, 353
lungs, 434
Lyell, Sir Charles, 274, 278, 284
Lyme disease, 54
lymph node, 228
lysosome, 173, 174
lytic pathway, 226, 227

—M—

Macrocytis, 91
macroevolution, 280, 281, 282, 284, 285, 286, 287, 288, 289, 293, 297, 298, 299, 300, 302, 319
macronucleus, 79
macroscopic, 25
maggot, 13, 395
maggots, 13
malaria, 81, 399
male, 223
Malpighian tubules, 394
Malthus, Robert, 274, 278
Mammalia, 533
mammary glands, 535
mandible, 394
mantle, 365, 370
margin, 448
Marine, 70, 75, 85, 92
marsh gas, 56
Marsupialia, 538
Massachusetts, 114
Mastigomycota, 104, 117, 118, 127
Mastigophora, 68, 69, 74, 77
maturation region, 460
maxillae, 379
McMillen, S. I., 11
measles, 227
medulla oblongata, 415
medusa, 345, 351
megaspore, 495
meiosis, 220, 223, 244, 252, 261, 300
meiosis I, 220, 257
meiosis II, 220, 221, 224
membrane, 19, 107, 109
membranous wings, 392
Mendel's experiments, 237, 238, 239, 240, 241

Mendel, Gregor, 207, 208, 235
Mendel, Johann, 235
Mendelian genetics, 236, 293, 294, 299
Mendosicutes, 54, 55
meristematic region, 460
meristematic tissue, 446
Merychippus, 288
mesenchyme, 342
Mesohippus, 288
messenger RNA, 195, 197
metabolism, 2, 54, 56, 72, 75, 79, 294
metabolism, of bacteria, 39, 40, 41
metacarples, 292
metamorphosis, 395
metaphase, 213, 214, 215, 217, 220
metaphase I, 221
metaphase II, 221
methane, 56, 135, 142
metric system, 40
Mexico, 324
Microbiology, 1
microevolution, 280, 281
micrometer, 40
micronucleus, 79
microorganism, 14, 16, 37, 67
microorganisms, 20
microscope, 10, 29
microspore, 494
microtubule, 213, 221
microtubules, 177
middle lamella, 171, 216
mildew, 117
milk, 157
milt, 427
minerals, 86, 124
missing link, 286, 287, 349
mistletoe, 458
mitochondria, 173, 174, 177, 189, 215
mitochondrion, 188, 189, 190
mitosis, 212, 213, 214, 215, 216, 217, 220, 221, 223, 300
mnemonic (for classification), 18
model, 130
modified berry, 500
mold, 99
molecular biology, 294
molecule, 134, 135, 137, 147, 156
molecules, 2
Mollicutes, 55
Mollusca, 364
molt, 374
monastery, 235
Monera, 18, 20, 29, 37, 38, 41, 47, 53, 56, 169
monocot, 450, 461, 471
Monocotyledonae, 450, 471
monohybrid cross, 251
monosaccharide, 149, 150, 173, 188

Monotremata, 538
Morel, 114
Morris, Dr. Henry, 285
Morse code, 178
Mosasarus, 521
mosquito, 82, 310, 394, 399
moss, 467
mother cell, 213, 214
motile, 117
motile spore, 104
motorcycle, 8
Mount St. Helens, 285
mouth pore, 79
movement, 168, 169
mRNA, 195, 196, 197
mucus, 425
mullosk, 90
multicellular, 21
multicellular organism, 90
multiple fruits, 500
mumps, 227
mushroom, 21, 99, 100, 101, 105, 106, 107, 108, 124
mutatation, 261
mutation, 8, 300
mycelia, 103, 106, 107
mycelium, 101, 102, 103, 107, 117, 124, 127
mycorrhizae, 123, 124
Myxococcus xanthus, 51
Myxomycota, 104, 119, 127

—N—

nastic movement, 478
natural gas, 142
natural selection, 275, 276, 277, 278, 279, 280, 293
naturalist, 274, 275
Needham, John, 14, 17
nematocysts, 347, 350
neo-Darwinism, 299, 300
nephfridiopores, 354, 359
nephridia, 354
nerve chord, 382
nervous system, 356
netted venation, 449
neurotoxin, 515
neutron, 129, 130, 131, 132, 133, 134, 135
New Testament, 489
Newton, Sir Isaac, 275
nitrogen, 86, 133, 135, 154
nitrogen dioxide, 136
nitrogen monoxide, 136
Noah's Flood, 285
Nobel Prize, 119
non-motile spore, 104
non-placental mammals, 534
nonseptate hyphae, 101
notochord, 407, 409

nuclear membrane, 178
nuclear reactor, 37, 56
nucleolus, 178
nucleotide, 178, 193, 197
nucleus, 72, 75, 79, 177, 193, 210, 212
nucleus, atomic, 130, 131
Numbers, 11
nut, 501
nymph, 395

—O—

oak, 29
odd-toed hoofed mammals, 540
Old Testament, 11, 275
olfactory lobes, 414
omnivore, 3, 4, 5, 17
Oomycetes, 117
opercula, 425
Ophiostoma ulmi, 115
opossums, 538
optic lobes, 415
oral cirri, 410
oral groove, 79
orb web, 389
order, 18, 22
organelle, 19, 112, 215, 224
organic molecule, 146
oriental sweetlips, 318, 319
Orthoptera, 399
osmosis, 141, 142, 184
Osteichthyes, 424
ostrich, 531
outer bark, 463
ovaries, 349
ovary, 382, 471, 491, 498
oviducts, 356
oviparous, 416
ovipositor, 397
ovoviviparous, 416
ovule, 491, 498
oxygen, 20, 85, 133, 134, 135, 142, 147, 154, 188
oxygen cycle, 324, 325, 326
Oxyphotobacteria, 55
oyster, 90, 364
ozone, 112, 135

—P—

Pacific salmon, 430
paleontologist, 12
paleontology, 289
palisade mesophyll, 453
palmate venation, 449
paper, 91
papillae, 524
parallel venation, 449
paramecia, 20, 177
Paramecium, 69, 79

parasite, 21, 42
parenchyma tissue, 453
parietal eye, 513
passive transport, 184
Pasteur, Louis, 14, 16
pathogen, 37, 228
pathogenic, 37, 39
pea, 235
pea plant, 236, 238, 245
pectin, 171
pectoral fin, 421, 425
pectoral girdle, 413
pedicel, 490
pedigree, 248
pelicans, 532
pellicle, 76, 79, 172
pelvic fin, 421, 425
pelvic girdle, 413
penguin, 531
penicillin, 119
Penicillium, 119
peppered moth, 282
peptide bond, 154
perennial, 445
perfect flowers, 491
pericardial sinus, 380
pericycle, 461
Perissodactyla, 540
peroxidase, 398
petal, 491
Peter the Great, 114
petiole, 447, 457
pH, 152
pH scale, 152
Phaeophyta, 68, 70, 90, 91
phagocytic cell, 228, 349
phagocytic vacuole, 176
phagocytosis, 176
pharynx, 354, 410, 426
phase, 137
phenotype, 243, 245, 258, 259
phlanges, 292
phloem, 446, 453, 462, 480, 482, 483
phosphate, 189, 190
phospholipid, 181, 183
phosphorous, 154
photosynthesis, 2, 3, 54, 85, 123, 135, 144, 145, 326, 328, 477, 482
photosynthesis, of bacteria, 42
phototropism, 484
phycobilin, 145
phylum, 18, 21
physical change, 136, 137, 138
physical changes, 136
physiology, 477
Phytophthora infestans, 117
phytoplankton, 85, 326
pigeon, 279, 297, 298
pigs, 540

pill bug, 376
pilus, 39, 40, 48
pinnate venation, 449
pinocytic vesicle, 176
pinocytosis, 177
pit viper, 516
pith, 462
placenta, 534
placental mammals, 534
plankton, 85
plant cell, 170, 171, 172, 173, 175, 176, 178, 185
Plantae, 18, 21, 87, 445
planula, 352
plasma membrane, 39, 40, 50, 71, 172, 175, 176, 177, 181, 183, 184, 221
plasmid, 48, 49
Plasmodium, 69, 81, 82, 83, 122, 399
plasmolysis, 184
plastid, 175, 455
Platyhelminthes, 361
pneumonia, 55
pod, 501
poison gland, 390
polar body, 224
pollen, 470
pollen grain, 494
pollen grains, 491, 493
pollen sacs, 494
pollen tube, 498
pollination, 496
pollution, 310
Polychaeta, 357
polyp, 345, 351
polysaccharide, 149, 150, 173
pome, 500
Population Biology, 1
population growth, 46
porcupine fish, 431
pore spaces, 479
Porifera, 341
Porphyra, 70
porpoise, 539
posterior, 353
posterior cardial vein, 428
posterior dorsal fin, 421, 425
potash, 86
potato, 117
praying mantis, 392, 399
precipitation, 323
predator, 309, 313, 315, 316
preening, 525
present, key to past, 275, 278
primary cell wall, 171
primary consumer, 311, 312, 313, 314, 321
principles of genetics, 239, 243
Proboscidea, 540

producer, 4, 5, 17, 311, 313, 314, 321
product, 143
Prokaryota, 37
prokaryotic, 38, 67, 74
prokaryotic cell, 174
propeller, 40
prophase, 213, 214, 215, 220
prophase I, 221
prophase II, 221
protein, 39, 154, 155, 156, 172, 173, 174, 175, 177, 178, 181, 183, 192, 193, 196, 197, 208, 209, 210, 212, 213, 294, 295
protein synthesis, 175
prothalus, 469
Protista, 18, 20, 29, 67, 71, 90, 91, 121, 122, 145, 169
proton, 129, 130, 131, 132, 133, 135
Protozoa, 68, 71, 80, 150, 175, 185
protozoan, 68
Pseudopod, 69, 73
psuedopod, 71
PTC, 256
Pterophyta, 469
pudding, 86, 91
puffball, 109
pulmocutaneous arteries, 436
pulmonary vein, 436
punctuated equilibrium, 299, 300
Punnett square, 245, 252, 253, 255, 256, 257, 258, 260
pupa, 395
pyloric ceca, 426
Pyrophyta, 68, 70, 89
pyruvic acid, 189, 192
python, 515

—Q—

quadrate bone, 516
quail, 531
quantum mechanical model, 131
Quercus, 29
quill, 525

—R—

rabbit, 309, 310
rachis, 525
radial symmetry, 340
radius, 292
radula, 365
rainbow, 275
rattlesnake, 516
Raup, Dr. David, 286
ray, 423
reactant, 143
receptacle, 491
receptor, 6

recessive, 243, 252
recessive allele, 258
red algae, 92
red blood cell, 20
red blood cells, 82
red oak, 29
red phycobilin, 145
red tide, 90
Redi, Francesco, 13
remora, 423
reproduce, 7, 22
reproduction, 7, 8, 169, 177, 207, 213, 215, 218
reproduction, asexual, 7, 44, 45, 215
reproduction, of bacteria, 44
reproduction, sexual, 7, 47
reproduction, sexual, in bacteria, 48
reproduction, speed, 46
reproductive organ, 446
Reptilia, 509
respiration, 42, 56, 134, 135, 168, 169
respiration, aerobic, 43
respiration, anaerobic, 43
rhino, 540
rhizoid, 467
rhizoid hypha, 101
Rhizopus, 116
Rhodophyta, 68, 70, 92, 145
Rhodospirillum rubrum, 296
Rhynchocephalia, 513
ribonuclease, 156
ribonucleic acid, 178, 193
ribose, 193
ribosome, 39, 174, 193, 195
ring structure, 148
RNA, 178, 193, 194, 195, 226
rockweed, 92
root, 101, 123, 458
root cap, 460
Roquefort, 119
rubra, 29
rust, 109
rye, 109

—S—

sac fungi, 111
Saccharomyces cerevisiae, 112
salad dressing, 91
salamander, 434, 436
Salem witch trials, 114
saliva, 82
salivary gland, 82
salmon, 430
Salmonella, 57
Salmonella enteriditis, 57
Salmonella typhimurium, 57
samara, 501
San Francisco State University, 303

sand, 479
saprophyte, 41
Sarcodina, 68, 69, 71, 72, 79
saturated fat, 154
sauerkraut, 38
scaled wings, 392
scientific law, 10, 11, 12, 13, 15
scientific method, 9, 10, 11, 12
scientific method, limitations, 12
scion, 488, 489
Scotobacteria, 54, 57
scripture, 273, 274
scutes, 514
sea anemone, 316, 345, 350
sea anemones, 316
sea otter, 534
sea snake, 516
sea squirt, 408
sea urchin, 366
secondary cell wall, 171
secondary consumer, 311, 312, 313, 314, 321
secret ingredient (of life), 8
secretion, 168, 169
secretion vesicle, 176
Sedgewick, Adam, 274
seed, 500
seed cone, 470
segmented worm, 352
self-fertilization, 470
self-pollination, 238
self-reassembly, 184
seminal vesicles, 356
semipermeable, 172
semipermeable membrane., 141
Semmelweis, Ignaz, 10
sepal, 490, 491
septate hypha, 101
sessile colony, 89
setae, 354
sex, 257
sex chromosomes, 257
sex-linked genetic trait, 257
sex-linked genetic traits, 257, 259
sex-linked inheritance, 261
sexual reproduction, 7, 220
shaft, 525
shark, 421, 422, 423
sheep, 540
sheet web, 389
shelf fungi, 109
shell, 365, 370
Shrewsbury, 274
silica, 342, 343
Silicon Dioxide, 70, 88
silk glands, 390
silkworm moth, 297
silt, 479
Silver Springs, 314
simple eyes, 375
simple fruit, 500

simple sugar, 149
sinus venus, 436
Sirenia, 539
skate, 423
skin, 7
slime mold, 99, 104, 121, 122, 124
slime tube, 356
smooth ER, 174
smut, 109
snail, 364
snake, 514
social insects, 397
solute, 138
solution, 138
solvent, 138
song birds, 532
soredia, 123
sorri, 469
Spallanzani, Lazzaro, 14
spawn, 430
spawning, 427
species, 18, 22, 29
species (definition), 22
sperm, 223, 244
sperm cell, 224
spherical symmetry, 340
spicule, 342, 343
spider, 103, 387
spinal chord, 415
spindle, 213, 221
spinnerets, 390
spiny puffer, 430
spiracles, 393
spirillum, 39, 50
spiritual factors, 209
Spirogyra, 70, 87
sponge, 341, 343, 344
spongin, 343, 344
spongy bone tissue, 411
spongy mesophyll, 453
spontaneous generation, 13, 14, 15,
    16, 17, 41
sporangia, 116, 122
sporangiophore, 102, 103, 105, 116
spore, 80, 81, 83, 100, 103, 104,
    106, 107, 109, 111, 112, 118,
    122, 124
spore, motile, 104
spore, non-motile, 104
sporophore, 102, 127
sporophyte, 468, 469
Sporozoa, 68, 69
Squamata, 513
squid, 364
stain, 53
stalk, 101, 106
stamen, 490, 491, 493, 494
staphylobacillus, 51
starch, 150, 478
starfish, 366
statocyst, 382

steady state, 46
Stentor, 80
sternal sinus, 380
stigma, 491
stingray, 423
stipe, 106, 108
stipules, 447
stock, 488, 489
stolon, 102, 103
stomach, 426
stomata, 452
strain, 57
strata, 282, 285
stroma, 175
structural formula, 147
structural homology, 293
structural tissue, 446
style, 491
sucrose, 149
suicide sac, 174
sulfur, 133, 154
sun, 2, 3, 4, 5, 310, 313
sunlight, 3, 144
swimmerets, 377
swimming birds, 531
symbiosis, 78, 123, 315, 318, 320
symmetry, 339, 340
syrinx, 532

—T—

tadpole, 434
tangle web, 389
tannic acid, 457
tapeworm, 363
taproot system, 458
tarantula, 388
taste buds, 426
taxonomy, 118
teeth, 25
teliospore, 109
telophase, 213, 214, 215, 220
telophase I, 221, 223
telophase II, 221, 223, 224
telson, 377
Tenericutes, 54, 55
tension, 481
termite, 77, 315
tertiary consumer, 311, 313, 314
testes, 349
testis, 382
Testudines, 513, 517
Thallobacteria, 55
thallus, 86
the amino acid, 196
The Origin of Species, 273, 275
the plasma membrane, 45
theology, 274
theory, 9, 10, 12, 13, 15
thigmotropism, 484, 485
thorax, 374

thymine, 160, 161
tiger, 5
tigers, 539
toad, 437
tomato, 117
tongue, 426
tooth decay, 39
toothpaste, 91
tortoise, 517
Toxoplasma, 83
toxoplasmosis, 83
tracheae, 393
transcription, 195, 197
transfer RNA, 195
transformation, 49
translation, 196, 197
translocation, 482, 483
transpiration, 323
transport, 477
trap door spider, 389
tree, 21
Trichonympha, 77, 315
triple bonds, 147
tRNA, 195, 196, 197
trophic level, 311, 312, 313, 314
trumpet, 80
truncus arteriosus, 436
trunk, 25
Trypanosoma, 77
tsetse fly, 77
tuatara, 513
tube nucleus, 494
tuber, 464
tuna, 297
tunicate, 408
turgor pressure, 176, 477, 478
turtle, 298, 517
Tyrannosaurus, 521

—U—

ulna, 292
umbilical cord, 535
underhair, 534
undifferentiated cell, 446
United States, 463
univalve, 365
unsaturated fat, 154
unsaturated fats, 154
uracil, 193, 194
uredospore, 109
Urochordata, 407, 408
uropod, 377

—V—

vaccine, 228, 229
vacuole, 72, 112, 176
vacuole, contractile, 72
vacuole, food, 72
van Helmont, Jean Baptist, 13, 17

van Leeuwenhoek, Anton, 14, 37
vascular cambium, 461, 463
vascular chamber, 460
vascular tissue, 446, 453
vegetative organ, 446
vegetative reproduction, 487
venae cavae, 436
venation, 448
ventral aorta, 428
ventral nerve cord, 356
ventricle, 428, 436
vertebrae, 407, 410
vertebral column, 413
Vertebrata, 407, 410
vertebrate, 339, 407, 410
vertebrates, 407
vesicle, 176, 216
Vienna, 10
vinegar, 38, 136
viper, 516
virus, 225, 226, 227, 228, 229, 310
visceral hump, 365, 370
Vitamin K, 57
vitamins, 86
viviparous, 416
Volvox, 77

—W—

wading birds, 532
wallabies, 538
warts, 227
waste vacuole, 176
water, 137, 321, 322, 323
water cycle, 322, 324, 326
water moccasin, 516
water mold, 117
water strider, 392
watershed, 323, 324
western diamondback, 516
Westminster Abbey, 275
whales, 539
wheat, 109
wheat rust, 109
whey, 38
whiskey, 113
white birch tree, 463
white blood cell, 172, 228
white oak, 29
wine, 112
wolves, 539
wombats, 538
Word of God, 15
worm, segmented, 352
wrasse, 318, 319
wrasses, 319

—X—

X chromosome, 257, 258, 259
xylem, 446, 453, 462, 480, 482, 483

—Y—

Y chromosome, 257, 258, 259
yeast, 99, 111, 112, 297
yeast, baker's, 112
yolk, 511
yolk sac, 511

—Z—

zebra, 540
zooplankton, 85
Zygomycetes, 105
zygospore, 115, 116
zygote, 115, 220, 225, 244